THE WISDOM OF
SOLOMON

The Genius and Legacy of Solomon Golomb

THE WISDOM OF SOLOMON

The Genius and Legacy of Solomon Golomb

Editors

Beatrice A Golomb
University of California San Diego, USA

Guang Gong
University of Waterloo, Canada

Alfred W Hales
IDA Center for Communications Research, la Jolla, USA

NEW JERSEY · LONDON · SINGAPORE · BEIJING · SHANGHAI · HONG KONG · TAIPEI · CHENNAI · TOKYO

Published by

World Scientific Publishing Co. Pte. Ltd.

5 Toh Tuck Link, Singapore 596224

USA office: 27 Warren Street, Suite 401-402, Hackensack, NJ 07601

UK office: 57 Shelton Street, Covent Garden, London WC2H 9HE

Library of Congress Cataloging-in-Publication Data

Names: Golomb, Beatrice Alexandra, editor. | Gong, Guang, 1956– editor. |
   Hales, Alfred W., editor.
Title: The wisdom of Solomon : the genius and legacy of Solomon Golomb /
   edited by Beatrice A Golomb, University of California San Diego, USA,
   Guang Gong, University of Waterloo, Canada,
   Alfred W Hales, IDA Center for Communications Research, la Jolla, USA.
Description: Hackensack, New Jersey : World Scientific, [2023] |
   Includes bibliographical references.
Identifiers: LCCN 2022026288 | ISBN 9789811234361 (hardcover) |
   ISBN 9789811234378 (ebook for institutions) | ISBN 9789811234385 (ebook for individuals)
Subjects: LCSH: Golomb, Solomon W. (Solomon Wolf) | Mathematicians--
   United States--Biography. | Electrical engineers--United States--Biography. |
   Mathematics. | Electrical engineering.
Classification: LCC QA29.G597 W57 2023 | DDC 510.92 [B]--dc23/eng20220829
LC record available at https://lccn.loc.gov/2022026288

British Library Cataloguing-in-Publication Data
A catalogue record for this book is available from the British Library.

Copyright © 2023 by World Scientific Publishing Co. Pte. Ltd.

*All rights reserved. This book, or parts thereof, may not be reproduced in any form or by any means,
electronic or mechanical, including photocopying, recording or any information storage and retrieval
system now known or to be invented, without written permission from the publisher.*

For photocopying of material in this volume, please pay a copying fee through the Copyright Clearance
Center, Inc., 222 Rosewood Drive, Danvers, MA 01923, USA. In this case permission to photocopy
is not required from the publisher.

For any available supplementary material, please visit
https://www.worldscientific.com/worldscibooks/10.1142/12211#t=suppl

Desk Editors: Soundararajan Raghuraman/Steven Patt

Typeset by Stallion Press
Email: enquiries@stallionpress.com

© 2023 World Scientific Publishing Company
https://doi.org/10.1142/9789811234378_fmatter

Preface

Solomon Wolf Golomb passed away on May 1, 2016, at the age of 83. He left an unparalleled legacy of accomplishment and recognition for his work in the applications of mathematics to coding and communication theory. His legacy also includes a host of individuals whose lives and careers were immeasurably enriched by their interactions with him. We three co-editors, members of this host, have attempted in this volume to collect many articles/essays, of a not-too-technical sort, by a representative collection of such individuals. Among our motivations is the wish to convince the reader of the incredibly prescient choice of his given name.

The first section consists of three articles which have already appeared elsewhere, by noted experts in the fields where the majority of Sol's contributions lie — mathematics, computer science and electrical engineering.

In the second section we have placed our own three articles in memory of and tribute to Sol. As will be clear, we each owe a tremendous debt of gratitude to him, for a wide variety of reasons.

The third section contains 30–35 articles by Sol's family, friends, students, colleagues, etc. They cover a great range of topics and vary widely in length, technicality and other aspects.

Finally, in the last section, we have contributions by Sol himself, covering various aspects of his career and interests, as well as other related material.

We thank all the contributors and other individuals who have assisted us in this project. In addition, we want to express regret for the contributions which might have been included were it not for those who have passed on or otherwise were unable to participate.

About the Editors

Beatrice A. Golomb MD, PhD, had the inestimable privilege of being the daughter of Solomon (and his wife Bo) — growing up with Sol's depth and breadth of knowledge, lightning intellect, rigor of thought, originality, and truly, the wisdom of Solomon. Almost equally remarkable were his kindness, devotion to family and friends, playful humor, and seemingly limitless generosity with his time and help to all. Beatrice is currently a Professor of Medicine at UC San Diego.

Guang Gong is a Professor in the Department of Electrical and Computer Engineering, University Research Chair at the University of Waterloo, and an IEEE Fellow. She worked with Sol from 1996–1998 as a Post-doc (Research Associate), since then she has co-authored with Sol many research papers including the book *Signal Design for Good Correlation*. She served as a co-guest editor for the *IEEE Transactions on Information Theory* (IEEE-IT) Special Issue in memory of Solomon Golomb, and Associate Editor for *Sequences* for IEEE-IT (2005–2008, 2020–).

Alfred W. Hales is Professor Emeritus of Mathematics at UCLA and a former Director of IDA's Center for Communications Research in La Jolla. He worked under Sol Golomb at JPL early in his career and was a junior coauthor of part of Golomb's *Shift Register Sequences*. He is best known for his work in Ramsey Theory for which he shared SIAM's Polya Prize in Combinatorics. He is an AAAS Fellow and an AMS Inaugural Fellow.

© 2023 World Scientific Publishing Company
https://doi.org/10.1142/9789811234378_fmatter

List of Contributors

Daniel Arovas is a Professor of Physics at UC San Diego. He is interested in strongly correlated quantum systems, especially in low dimensions where quantum fluctuations can lead to interesting and exotic new states of matter. He is a longtime family friend via Sol's daughter.

Elwyn Berlekamp was a Professor of Mathematics and EECS at the University of California, Berkeley. He was noted for his work in computer science, coding theory and combinatorial game theory. He invented an algorithm for factoring polynomials and was a co-inventor of the Berlekamp-Welch algorithm and Berlekamp-Massey algorithm for error correction. He was a consultant at JPL, a business partner with Golomb, and a Shannon Award recipient.

Richard Blahut is the Henry Magnuski Professor Emeritus of Electrical Engineering at the University of Illinois Urbana-Champaign and the author of ten books on coding and signal processing. These include the 1983 *Theory and Practice of Error Control Codes*. He was a recipient of the IEEE Shannon Award and the IEEE Alexander Graham Bell Medal. He is a member of the National Academy of Engineering.

Gary S. Bloom was a Professor of Computer Science at CUNY (New York). He received his PhD from USC in 1976 under advisor Sol Golomb. His major research interests were graph theory and applications. He co-ran the 1992 60th birthday event for Sol Golomb. Coincidentally his name G. S. Bloom is an anagram of S. Golomb.

Tor Bu has been Professor in Mathematics at the University of Bergen and at the Stavanger Regional College. He visited Sol at USC for one sabbatical year in the end of the 1970's. He later worked some years at Norsk Hydro. He has also been the IT-Director of the University of Bergen.

Leo Buxbaum who fled Nazi Germany as a child was raised in Upstate New York. He majored in chemistry at Cornell and received his medical degree at University of Rochester in 1956, followed by residency and a hematology fellowship in the UCLA system. He launched the GI department at Whittier Hospital Medical Center and practiced until age 80. He continues to teach introduction to clinical medicine at USC.

John Dillon is a leading mathematician at the National Security Agency. In addition to his extensive classified work, he has authored many papers in combinatorics, especially ones on difference sets, Hadamard matrices and sequences with ideal 2-level autocorrelation.

Tuvi Etzion is a Professor at the Henry and Marilyn Taub Faculty of Computer Science at the Technion-Israel Institute of Technology. He works on information theory and coding theory, and visited Sol for 2 years during 1985–1987. He was the organizer of the memorial session for Sol at the Information Theory (ITA) Workshop in 2017 in San Diego.

Jeremiah Farrell, professor emeritus of mathematics at Butler University, is well known to word aficionados as a regular contributor of crossword puzzles to the New York Times and as a designer of Will Shortz's favorite crossword puzzle "the 1996 Election Day" crossword. He and wife Karen helped organize the biannual "Gathering for Gardner," an invitation-only event in honor of Martin Gardner, commencing in 1993, which Sol avidly attended. In 2006, he and wife Karen Farrell took over as editors and publishers of the quarterly publication *Word Ways: the Journal of Recreational Linguistics* (est. 1968), to which Sol subscribed and contributed. (It was issued in print during Sol's life, but is now on-line only.)

Aviezri S. Fraenkel is Professor of Computer Science and Applied Mathematics at the Weizmann Institute in Israel. He is particularly noted for his work in combinatorial game theory, for which he received the Euler Medal. He also received the IEEE WEIZAC Medal, for his role in designing and constructing the WEIZAC computer.

Harold Fredricksen is an emeritus Professor of Applied Mathematics at the Naval Postgraduate School in Monterey, CA. His research interests are cryptography, coding and combinatorial mathematics and he has extensive publications on de Bruijn graphs and sequences. He was Sol's doctoral student at USC and contributed to Sol's *Shift Register Sequences* book.

Jonathan I. Hall is a Professor Emeritus of Mathematics at Michigan State University. He is an AMS Fellow and works on group theory, combinatorics, incidence geometry and algebraic coding theory. The first edition of Sol's book *Shift Register Sequences* was dedicated to Jonathan's father Marshall Hall, Jr. Tingyao Xiong (coauthor of Hall's contribution) was his PhD student, now an Associate Professor at Radford University.

Jon Hamkins is the Chief Technologist of the Communications, Tracking and Radar Division at the Jet Propulsion Laboratory (JPL). For 15 years he was Technical Supervisor of the Information Processing Group, which Sol Golomb had led earlier. Though he did not overlap with Sol at JPL, they met to compare their experiences.

Robert Hanisee (MA Economics, Berkeley) began his career on Wall Street as a securities analyst with a focus on high tech, defense, and aerospace —

for JP Morgan, Merrill Lynch, and Crowell Weedon, then his own "institutional research boutique" firm, Seidler AMDEC Securities. Following a stint in venture capital (high tech firms), he became managing director and head of research at Trust Company of the West (TCW). He served on boards of multiple companies chairing the audit committees for Titan, EDO, and Orbital Sciences. He was a member of the Commercialization Council at JPL from 1999–2001, and served on the NASA Advisory Council, chairing the Audit and Finance Committee under two heads of NASA. He met Sol through shared membership on company boards.

Larry Harper is Professor Emeritus of Mathematics at UC Riverside and worked at JPL under Sol Golomb while a graduate student. He has written many papers in combinatorics and is perhaps best known for his work on discrete isoperimetric problems.

Tor Helleseth is a Professor Emeritus at the Selmer Center, a research center in reliable and secure communication at the Department of Informatics, the University of Bergen. He has led the Selmer Center from the very beginning, is an IEEE Life Fellow, and a member of Det Norske Videnskaps-Akademi.

Jonathan Jedwab is a Professor in the Department of Mathematics at Simon Fraser University. He works on combining combinatorial, algebraic and analytical techniques to solve such classical and emerging problems of digital communications as Golay complementary sequences for peak- to-average power reduction.

Scott Kim is a puzzle designer, mathematical artist, and recreational mathematician. His background includes studies in mathematics, computer science, music and graphic design at Stanford University. He is best known for his book *Inversions*, which was the first published collection of symmetrical lettering designs called ambigrams. As a puzzle designer he has worked on many games and puzzle products, including Tetris and Rush Hour.

Rochelle Kronzek is Executive Editor at World Scientific Publishing. She was instrumental in the publishing of the third edition of Golomb's *Shift Register Sequences* book.

Abraham Lempel was Professor Emeritus and the Andrew and Erna Finci Viterbi Chair of Information Systems in the Faculty of Computer Science at the Technion-Israel Institute of Technology. He was the co-inventor of the LZ77 algorithm for data compression and a recipient of IEEE's Golden Jubilee Award and its Hamming Medal. He also received the ACM Kanellakis Award and the 2010 Rothschild Prize.

Nadav Levanon is a Professor in the Department of Electrical Engineering — Systems at Tel Aviv University. He was a recipient of the 2016 IEEE Dennis J. Picard Medal for Radar Technology and Applications, and is a Life Fellow of the IEEE.

C. L. Max Nikias was the 11th University of Southern California president (August 2010–August 2018). He holds the Malcolm R. Currie Chair in Technology and the Humanities and is president emeritus of the university.

Robert Rosenstein: Robert "Bob" Rosenstein met Sol when both were studying in Norway in 1955–1956. He was with Sol in Denmark when Sol first met Bo. (Sol traveled to Copenhagen because Norway did not have a mathematics library.) A polymath, and by profession an x-ray crystallographer at University of Pittsburgh for many years, he was a lifelong friend to Sol and Bo. Bob "retired" to work at The Scripps Research Institute in La Jolla, (with regular trips to UC Berkley), where he continued to have an office until his death at age 94. Born in 1922, he died in February of 2017. An obituary can be found at https://history.amercrystalassn.org/robert-rosenstein. This contribution had been written by Bob for Sol and Bo's 40th anniversary.

Edna Rebecca Sharoni (née Golomb) was the sole sibling of Sol, five years his elder (born in 1927). She grew up in Baltimore, graduated from Goucher College at age 20, and moved to what is now Israel when Sol was 15. She participated in code-breaking efforts in the nascent days of that state. After raising four children, she taught English literature at Bar-Illan University for over 35 years. Sol visited her annually, and they remained close throughout his life.

Stephen Schloss (born 1931) received a master's degree in mathematics from Harvard. He worked in Aerospace prior to receiving his PhD in engineering with Sol at the University of Southern California. Perceiving that a starting academic salary was inadequate to meet the needs of his family blessed with five children, he shifted careers, and, at Merrill Lynch, became Sol's stockbroker, in which capacity he had "near monthly" meetings with Sol, and remained a friend until Sol's death. He characterizes his PhD with Sol as the "best time of my life."

Robert Scholtz is Professor Emeritus of Electrical and Computer Engineering at USC. He was Chair of the Electrical Engineering Systems Department, co-author of the book; *Basic Concepts in Engineering and Coding* with Sol Golomb and R. Peile, a recipient of the IEEE Donald G. Fink Prize Paper Award, and a co-recipient of the IEEE Eric E. Sumner Award.

Terrence J. Sejnowski (PhD physics, Princeton) is the Francis Crick Professor at the Salk Institute, and is a distinguished Professor at UC San Diego in the Departments of Biology, Computer Science, Bioengineering, Psychology, and Neuroscience (and previously, physics). He was a pioneer in neural networks and computational neuroscience. He is a member of the National Academies of Sciences, of Engineering, and of Medicine. He came to know Sol as Sol's son-in-law.

Hong-Yeop Song is a Professor of Electrical and Electronic Engineering at Yonsei University in South Korea. He was a PhD student of Sol at USC. He served as a former chair of the School of Electrical and Electronic Engineering. His research focuses on PN sequences and error-correcting codes for digital communications.

Robert C. Tausworthe, now retired, was a Senior Research Engineer and Chief Technologist of Information Systems at JPL, initially working alongside, and then under, Sol Golomb there. Sol was also his Caltech PhD thesis advisor. They coauthored a number of reports at JPL, now still classified.

List of Contributors

Andrew Viterbi is a co-founder and retired Vice Chairman and Chief Technical Officer of Qualcomm Incorporated, a Presidential Chair Visiting Professor at the University of Southern California and a distinguished Visiting Professor at the Technion-Israel Institute of Technology. He was the recipient of IEEE Medal of Honor, the U.S. National Medal of Science, the Franklin Medal and the Charles Stark Draper Prize of the National Academy of Engineering.

Lloyd Welch is Professor Emeritus of Electrical Engineering at USC, and hence was a colleague of Golomb at USC and (earlier) was a colleague of his at JPL. He is best known for the Baum-Welch algorithm (for which he received the IEEE Shannon Award) and also for the Berlekamp-Welch error correction algorithm.

Alan E. Willner, the Andrew and Erna Viterbi Professorial Chair and Distinguished Professor of Electrical and Computer Engineering at USC, is known for his work on optical communications. He received the IEEE Eric E. Sumner Technical Field Award, is a member of the US National Academy of Engineering, and is an International Fellow of the UK Royal Academy of Engineering.

Stephen Wolfram (PhD 1979) is the creator of Mathematica, a software computing system launched in 1988, Wolfram|Alpha, and the Wolfram Language, the founder and CEO of Wolfram Research and the originator of the Wolfram Physics Project. He is the author of *A New Kind of Science* and *Metamathematic: Foundations & Physicalization*, among other books — including *Idea Makers*, from which his article about Solomon Golomb is taken.

Yannis C. Yortsos is the Dean of the USC Viterbi School of Engineering and the Zohrab Kaprielian Chair in Engineering, a position he holds since 2005. He is a member of the National Academy of Engineering since 2008 and Associate Member of the Academy of Athens since 2013, and co-winner of the Gordon Prize of the NAE in 2022.

Neal Zierler received his doctorate in mathematics from Harvard University and has made fundamental contributions to the theory of codes and sequences in communications, including work on shift register sequences. He was a colleague of Sol Golomb at JPL, and later did critically important classified work at Princeton's Center for Communications Research, an NSA funded branch of IDA.

© 2023 World Scientific Publishing Company
https://doi.org/10.1142/9789811234378_fmatter

Contents

Preface		v
About the Editors		vii
About the Contributors		ix

Part I		1
1.	Introduction to Solomon W. Golomb *E. Berlekamp*	3
2	Solomon Wolf Golomb 1932–2016 *Andrew J. Viterbi*	7
3.	Solomon Golomb (1932–2016) *Stephen Wolfram*	13

Part II		41
4.	A Career in Engineering *Solomon W. Golomb and Beatrice A. Golomb*	43
5.	Golomb's *Shift Register Sequences* — Work with a Great Mind *Guang Gong*	75
6.	Memories of Sol *Alfred W. Hales*	105

Part III		119
7.	The Most Interesting Man in the World *Daniel Arovas*	121

xvi *The Wisdom of Solomon*

8. The Issue of *Shift Register Sequences*   123
   Richard E. Blahut

9. Solomon Golomb — Some Personal Reflections Plus Random
   Biographical and Bibliographical Notes   131
   Gary S. Bloom

10. Golomb's Norske Forbindelser   137
    Tor Bu

11. Sol Golomb, My Friend   141
    Leo Buxbaum

12. Ode to Sol   143
    John Dillon

13. Puzzles and Tilings — A Fascinating Kingdom of Solomon Golomb   145
    Tuvi Etzion

14. Solomon Wolf Golomb May 30, 1932–May 1, 2016   161
    Jeremiah Farrell and Karen Farrell

15. Solomon W. Golomb's Enlightening Games   165
    Aviezri S. Fraenkel

16. My Life with Sol   169
    Harold Fredricksen

17. Focus and Contrast   173
    Jonathan I. Hall and Tingyao Xiong

18. Taped Conversation re Sol and JPL   191
    Jon Hamkins

19. Board Game to Board Room   197
    Robert Hanisee

20. The Wisdom of Solomon   215
    Lawrence Hueston Harper

21. The Norwegian Connections of Solomon W. Golomb   227
    Tor Helleseth

22. Sol Golomb and a Twice-in-a-Lifetime Celestial Event   237
    Jonathan Jedwab

23.	Solomon Golomb in Pentominoes	241
	Scott Kim	
24.	Solomon Wolf Golomb: A Man of Humility, Integrity and Compassion	243
	Rochelle Kronzek	
25.	Sol Golomb — My Hero and My Idol	247
	Abraham Lempel	
26.	Golomb and Radar Waveforms	251
	Nadav Levanon	
27.	Funeral Oration — Sol; and A Tribute to Dr. Solomon Golomb	263
	C. L. Max Nikias	
28.	Memories on 40th Anniversary of Bo and Sol: May 18, 1996	271
	Robert Rosenstein	
29.	From Professor Solomon Golomb's Math PhD Advisee to His Stock Market Guru; & Orbits and Calendrics	273
	Stephen Schloss	
30.	Bob Scholtz's Memories of Sol	289
	Robert Scholtz	
31.	Solomon Golomb: Wise Man, Biblical Scholar and Father-in-law	291
	Terrence J. Sejnowski	
32.	Professor S. W. Golomb was My Advisor and Mentor	293
	Hong-Yeop Song	
33.	Solomon Golomb in the Faye Zlok Days at JPL	305
	Robert C. Tausworthe	
34.	Early Days with Sol Golomb	329
	Lloyd Welch	
35.	The Consummate Scholar and Gentleman: Professor Solomon Wolf Golomb	331
	Alan Eli Willner	
36.	A Tribute to Sol Golomb, the Genius who Transformed USC	347
	Yannis Yortsos	
37.	Meeting Young Sol	353
	Neal Zierler	

xviii — *The Wisdom of Solomon*

Part IV		355
38.	Elhanan Golomb — Introduction by Michal Reuven	357
39.	Sol's Historical Artifact: National Honor Society Letter	363
40.	Early Sol and Family	367
41.	JPL Section 331 — The Space Age Begins	377
42.	*Life* Magazine — Explorer One	379
43.	Sol's Trip Report re Marshall Hall Visit	381
44.	Comma Free Codes and Sol's Tribute to Max Delbrück	385
45.	Radar Measurements of the Planet Venus	415
	L. R. Malling and S. W. Golomb	
46.	Hadamard Matrix Discovery	423
47.	Eight JPL *Lab Oratory* Issues	431
48.	Shift Register Coders — The Golomb Algorithm	443
49.	The PapaVerse	451
50.	Palindrominologist Gets Him Coming, Going	457
	Solomon W. Golomb	
51.	Puzzles from *LA Times*	459
52.	Several Pieces from Golomb's Gambits	469
53.	Marketing in USA	481
54.	*Word Ways* Selections	483
55.	Mathematics After Forty Years of the Space Age	509
	Solomon W. Golomb	
56.	Golomb's Reminiscences	517
	Solomon W. Golomb	
57.	USC Engineer From JPL to USC and Beyond	529
58.	Periodic Binary Sequences: Solved and Unsolved Problems	535
59.	Sol — Digital Pioneer, Sol Golomb Celebrates 50 Years at USC	543
60.	USC Trojan "Busy Signals": National Medal of Science	547
61.	The "Golomb-Dickman Constant" — E-mail re Lagarias Article	549
62.	Letter to Granville; Erdos Number "Minus One"	551
	Solomon W. Golomb	
63.	Sol — Too Famous for a Stamp	555

Contents

64. Shizef Raphaeli Obituary Translation   557
   Edna Sharoni

65. In Memoriam: Solomon W. Golomb   559

66. Puzzles in Memory of Solomon Golomb   563
   Joe Buhler, Paul Cuff, Al Hales, and Richard Stong

67. Final Oral Exam   569
   Solomon W. Golomb

Part I

© 2023 World Scientific Publishing Company
https://doi.org/10.1142/9789811234378_0001

INTRODUCTION TO
SOLOMON W. GOLOMB*

E. Berlekamp
University of California, Berkeley

I feel greatly honored to write this Introduction to the Proceedings of the Golomb70-fest.[1] Prof. Golomb is well known as the recipient of the highest honors possible at University of Southern California, the Information Theory Society, and the US engineering profession.[2]

Solomon W. Golomb was born on May 30, 1932.[3] His father was a rabbi and a linguist. Sol soon developed a precocious appetite for mathematics as well as for languages and a wide range of classical literary works. Sol completed his undergraduate studies at Johns Hopkins in two years, then obtained a PhD in mathematics from Harvard, and went to Norway on a Fulbright fellowship, where he met his future bride.[4] Along the way, he spent a summer working for Martin Aircraft Company, where he also became interested in a variety of engineering problems in aerospace electronics. In pursuit of this interest, he moved to Southern California to begin his first full-time job with the Jet Propulsion Laboratory. This was in the early heyday of NASA, immediately following sputnik. Sol soon emerged as the leader of JPL's Space Communications efforts.

My first encounter with Golomb's name was as the author of a paper published in the proceedings of a 1962 symposium, entitled, "Mathematical Theory of Discrete Classification." I thought this was a brilliant paper, and it had a big impact on me. I was surprised and flattered when I was invited to meet Sol at JPL in January 1965, and only thereafter did I come to realize the wide breadth of his work, which included many other papers and books that were generally regarded as even more significant than the one that had so strongly impressed me. By January 1965, Sol had already moved from JPL to his professorship in both mathematics and electrical engineering at the University of Southern California, but he remained an influential eminence grise at JPL. Largely on his recommendation, JPL hired me as a consultant. I visited there weekly for the next year and a half, and I was able to work personally with an

vii

---

*This chapter was published (actually republished) in 2017 in the book *Shift Register Sequences*, 3rd Revised Edition, by Solomon W. Golomb, World Scientific Publishing. See Note 1.

viii   Introduction

extraordinary cast of characters, many of whose names appear in this
volume. The two who worked most closely with me, Gus Solomon and
Ed Posner, are now both deceased. There was also a bright array of luminary consultants, including Lloyd Welch, Andy Viterbi, Irwin Jacobs,
Marshall Hall, and some Caltech students, including Bob McEliece and
Richard Stanley. Sol was heavily involved in the recruitment of most
of these people, as well as many others. On the occasion of Sol's 60th
birthdayfest in 1992, Gus Solomon proposed a memorable toast to "the
man who brought modern combinatorial mathematics to Southern California." At the time, I was shocked by the boldness of the claim. After some further reflection, I was also shocked by the surprisingly large
amount of truth it contained.

Sol has written landmark books on a wide variety of topics. "Shift
Register Sequences" are used in radar, space communications, cryptography and now cell phone communications. This book has long been
a standard reading requirement for new recruits in many organizations,
including the National Security Agency and a variety of companies that
design anti-jam military communication systems. "Polyominoes" defined that subject and established Sol as a leader in the broad field of
recreational mathematics. Both "shift register sequences" and "polyominoes" have become subject headings in the classification of mathematics
used by Mathematical Reviews. "Digital Communications with Space
Applications," written with Baumert, Easterling, Stiffler, and Viterbi,
was among the earliest and most influential books on that subject. "Information Theory and Coding," written with Peile and Scholtz, is a
novel text for graduate course which has become very popular at USC
and elsewhere.

Sol maintains a strong interest in elementary number theory. Many
of his research papers deal with questions concerning prime numbers.
Sol also has a great interest in teaching, both advanced and elementary
courses, and in promoting popular interest in mathematics. He is an
avid collector, solver, and composer of problems. He has authored the
Problems Section (sometimes known as Golomb's corner) in periodicals
including the Newsletter of the IEEE Information Theory Society, the
Alumni Magazine of Johns Hopkins University, and the Los Angeles
Times. Following a change of publishers some years ago, Sol's column
was discontinued. But, he noted, this was totally consistent with the
new editorial policy then being adopted, which also discontinued the
entire Science News section and increased the coverage of astrology.

Sol's publications also include some provocative commentaries on the
philosophy and history of mathematics. His 1998 critique of G. H.
Hardy's famous 1940 "Mathematician's Apology" ("Mathematics Forty

Introduction ix

Years After Sputnik") is a well-documented and articulate exposition of the practical values of "pure" mathematics, including even those topics discovered by purists like Hardy. In some respects, Sol's 1982 obituary of Max Delbrook may be a more objective view of the rise of modern molecular biology than could have been written by any of the major pioneers in that field. Golomb's interest in biology predates even his own paper, "On the plausibility of the RNA code," published in the 1962 issue of Nature, long before the idea of a lengthy digital basis of inheritance had appeared on the mental radar screens of most biologists.

Sol has also been a senior academic statesman. He served as President of USC's Faculty Senate in 1976-1977, as Vice Provost for Research in 1986-1989, and as Director of Technology at USC's Annenberg Center for Communications in 1995-1998. He founded the "National Academies Group" at USC, and restored the influence of the faculty on the governance of that university on certain occasions when the administrations of the day had been veering off course.

I have personally had the great fortune to be placed in positions which allowed me to see certain other aspects of Sol's multi-faceted intellect. Annoyed by the commercial misappropriation of one of his polyominoe games, in the late 1960s,Sol incorporated one of his hobbies into a company called "Recreational Technology, Inc." I accepted his invitation to become a founding director of this venture. Although this venture never had any outside investors, nor any significant sales or earnings, it became the dry run for another venture called Cyclotomics, which, at Sol's urging, I founded in December 1973. Sol was my founding outside director and primary business mentor and confidante for the next 15 years.

References

[1] S. W. Golomb, "Mathematical Theory of Discrete Classification", in *Information Theory*, Proceedings of the Fourth (1960) London Symposium, Colin Cherry, Editor, Butterworths, London, 1961.

[2] S. W. Golomb, "Mathematics Forty Years After Sputnik", *American Scholar*, vol. 67, no. 2, Spring, 1998.

[3] S. W. Golomb, "Mathematics After Forty Years of the Space Age", Mathematical Intelligencer, vol. 21, no. 4, Fall, 1999.

[4] S. W. Golomb, "Max Delbruck - An Appreciation", American Scholar, vol. 51, no. 3, Summer, 1982.

x   Introduction

Editors' Notes

1. This Introduction from 2003 is reprinted here from *Mathematical Properties of Sequences and Other Combinatorial Structures*, edited by Jong-Seon No, Hong-Yeop Song, Tor Helleseth, P. Vijay Kumar, in The Springer International Series in Engineering and Computer Science, Volume 726 (2003) with permission of Springer.
2. These honors include, among others, USC's Presidential Medallion, memberships in the National Academy of Science and of Engineering, the IEEE Hamming Medal and Shannon Award, the Benjamin Franklin Medal, and the National Medal of Science — the country's highest distinction for contributions to scientific research.
3. May 31, 1932 was his actual birth date.
4. Sol met his wife Bo in her native Denmark, not Norway (though it was during Sol's Norway Fulbright term). As Sol has elsewhere detailed, he visited Copenhagen for their mathematics library. Friend Robert Rosenstein, also studying in Norway, was with Sol when he met Bo (and remained a friend to both for the remainder of their lives). JPL colleague Richard Arnold Epstein, who died shortly after Sol, was the sole person to attend Sol and Bo's wedding in Denmark on May 18, 1956, the day after Sol's May 17 submission of his Ph.D. thesis to Harvard.

© 2023 World Scientific Publishing Company
https://doi.org/10.1142/9789811234378_0002

Solomon Wolf Golomb 1932–2016*

Andrew J. Viterbi

Solomon W. Golomb was born into an intellectually gifted family in Baltimore, MD. His father, a polyglot, was a professor at Baltimore Hebrew College and Sol grew up speaking several European and Middle Eastern languages. In adulthood he would add at least two East Asian languages to the point of being able to lecture in both. With this background he was also expert in the etymology of numerous words and expressions in a variety of language families and he never missed an opportunity to so enlighten conversation with friends and colleagues.

Beyond this facility supported by his remarkable memory which served him to the end of life, Sol's greatest intellectual passion was for mathematics, particularly of the discrete variety: combinatorics, number theory, abstract algebra. All of these he applied throughout his career to practical problems of communication and information theory, radar, and cryptography along with other disciplines critical to space exploration. Since the 1960's, Sol was one of the first "pure" mathematicians to counter the orthodoxy of previous generations of mathematicians who blindly preached the inapplicability of mathematics, ignorant of rich application areas such as those which Sol pioneered.

Sol began his college studies at Johns Hopkins University in 1949 and prior to his nineteenth birthday in 1951 he had graduated with a BA in mathematics with honors, as he said "by taking twice as many courses as everyone else." His graduate student experience began with a year at Hopkins followed by three years at Harvard. He then won a Fulbright Fellowship for pursuing his doctoral research on Twin Primes at the University of Oslo. During his graduate school years Sol achieved two breakthroughs which were to bring him the most renown throughout his entire prolific career. The first came as a result of having spent several summers as a graduate student Intern at the Martin Company in a Baltimore suburb. During his last period of summer employment, he was introduced to the topic of linear feedback shift registers as generators of sequences with interesting properties, perhaps the most valuable of which were their "pseudo-randomness" endowing them with the principal characteristics of coin-tossing binary sequences, but with the ability to be constructed and reproduced in a deterministic manner. Shift register sequences played important roles in electronic security and cryptography before becoming the current "workhorse" of the majority of digital mobile phones. Applying the abstract algebra fundamentals which Sol had mastered as a doctoral student, he developed practically the entire theory underlying the important communication properties of this class of sequences. This early work led to a multitude of applications throughout the fields of digital communication and radar, which he was to pioneer and pursue throughout his career. Also during his graduate education, Golomb, a "puzzle-solver" all his life, posed a problem involving

---

*This chapter was originally published in *IEEE Transactions in Information Theory*, April 2018, Vol. 64, No. 4, Part II, pp. 2837–2838.

8   *The Wisdom of Solomon*

contiguous shapes fully covering a checkerboard in all but its four corners. From this came the theory and the game of polyominoes. Originally published in the *American Math Monthly* in 1954 and as a book in 1965, it gained a large lay audience through its appearance in Martin Gardner's "Mathematical Games" column in the May 1967 issue of *Scientific American*. Sol's love of puzzles later also extended to his service to the Information Theory Society for whose newsletter he served as mathematical puzzle editor, providing intellectually challenging and satisfying exercise for longer than anyone can remember.

As already noted, to proceed with his doctoral research Sol received a Fulbright Fellowship to spend the 1955-56 year at the University of Oslo where he collaborated with Viggo Brun, the distinguished Norwegian mathematician whose earlier work on the Twin Prime problem influenced the young American. Sol thus completed his doctoral dissertation on the subject, receiving his Harvard Ph.D. early the following year. The Norwegian sojourn was rendered doubly successful by his meeting Bo, a Danish student. After a short courtship, this led to a long, fruitful marriage and the birth in the next few years of two lovely, bright and accomplished daughters, Astrid and Beatrice.

Returning to the States with his new Danish bride in the summer of 1956, Sol accepted an offer to join the Communication Research Section of Caltech's Jet Propulsion Laboratory (JPL). In October 1957, the Soviets' launch of Sputnik into orbit caused the shift of the lab's focus from secure communication with guided missiles to communication and control of satellites and space probes. Three months later this effort contributed to the launch of the first U.S. satellite, Explorer I. Beginning with the verification of the successful launch and the determination of the orbit parameters of Explorer, at JPL Sol rapidly became the undisputed leader of digital communication and information theory research. Over the next half century this was to impact government and commercial satellites for communication and navigation uses ranging from military and NASA applications to consumer products. The latter include terrestrial navigation (GPS), digital satellite television broadcasting and cellular phones. Among early successes was the project to obtain a radar echo off the planet Venus, which led to a major correction to the previously believed value of the Astronomical Unit, critical to the success of planetary exploration.

In early 1963, Golomb joined the faculty of the Electrical Engineering Department of the University of Southern California (USC), where he remained for well over half a century playing a leading university-wide role in elevating the institution from its early regional standing to international repute and ranking among the world's top universities. On the administrative side, he was an advisor and mentor to five Deans of Engineering from Zohrab Kaprelian to Yannis Yortsos and an equal number of University Presidents from Norman Topping to Max Nikias. On the academic side, Sol taught a multitude of students including scores of doctoral candidates,

Solomon Wolf Golomb 1932–2016

some of whom have achieved renown, academically at major research universities or in high level positions in the corporate world. Sol's greatest passion always remained for his research and teaching which continued in all the areas of applied mathematics which impact digital communication and signal processing. His early contributions to interference suppressing codes, based on number theory and abstract algebra, blossomed into the much wider applications of multiple-access wireless, cryptography and even the genetic code. Among the now common expressions and terms which he created are the concept of pseudo-randomness based on the Golomb "randomness properties", as well as the Golomb Constant and the Golomb Ruler. But Sol's interests and knowledge extended well beyond science and technology: particularly as noted, his mastery of languages and linguistics enabling him to lecture in over half a dozen languages, including three Scandinavian languages, Hebrew, Japanese and Chinese. The two Asian languages he learned by immersing himself during successive summers in preparation to lecture in the corresponding countries. His mastery of Hebrew was in evidence annually by his expert chanting of Torah at the Hillel High Holiday services. His command of Swedish was also put to good use, as I will reveal momentarily. For both science and the humanities, Sol's sharp mind and prodigious memory remained undiminished into his ninth decade.

Solomon Golomb received many honors in his illustrious career. These include membership in three U.S. based academies, the National Academy of Sciences, the National Academy of Engineering and the American Academy of Arts and Sciences. He received the two most prestigious awards for Information Theory, the IEEE Claude Shannon Award and the IEEE Richard Hamming Award. He was also honored by two Russian Academies and was awarded honorary doctorates by universities worldwide. More recently he received the Franklin Medal for Electrical Engineering and the National Medal of Science from the President of the United States.

I have been asked to add a few personal anecdotes about my friendship with Sol which lasted nearly six decades, during which time Sol grew from the brightest kid in the room to the wisest man on campus. I first met Sol on June 14, 1957, the day I arrived in southern California to begin my engineering career at JPL in Pasadena. I called in to my Section's phone number and it was my good fortune that the phone was answered by Sol, alone in the office. I explained that I'd be reporting in a couple of days right after I had found an adequate residence for myself and my elderly parents who had accompanied me west. To which Sol replied "Nonsense, come in and get on the payroll and I'll help you find a place." So began our friendship, which included Bo, who shortly thereafter became a dear friend of my soon-to-be bride, Erna. And their offspring, Astrid and Beatrice, also befriended our first-born, Audrey, coming into the world just months after Beatrice.

After a couple of years, Sol was promoted to Deputy Section Chief and I reported to him for the rest of my JPL years. More importantly, he exposed me to the value, and indeed the beauty, of discrete mathematics. My MIT Master's degree had prepared me with the applied mathematical tools needed for communication, control and signal processing, but in those benighted early days of the computer era, such niceties of discrete math as group theory and number theory were left to the pure mathematicians who trained only their own kind. As was previously noted, Sol was a pioneer in breaking down the barrier and inspired me to take courses taught by notable math faculty members at USC. But perhaps the most memorable event in both our JPL careers was the evening in late January 1958 when we stood by a primitive data terminal breathlessly awaiting the appearance at our site of the tiny satellite Explorer I. Armed with a large sheaf of graphs of pre-computed range and velocity values, we validated its success and determined the parameters of the orbital ephemeris of our new nearest astronomical neighbor. A few days later a *Life magazine* photographer came to the Lab to get us to re-enact the action of that historical evening, producing a photo which appeared in the magazine in February, 1958. (Young Sol was fully visible in the photo at the back of the room, but as for me, only the back of my head appeared.)

I could recount many more of my interactions with Sol over the years, after we left JPL for academia at the two ends of the city of Los Angeles, and even after I moved south to co-found two startup companies. I'll end my reminiscences with an important event at the last one, Qualcomm. Sometime in the middle 1990's the eminent and powerful Swedish telecommunication company, L.M. Ericsson, filed an infringement suit against little Qualcomm based on irrelevant prior art. Sol was retained by us as an Expert Witness. The evening before his court appearance he plowed through the mountain of documents which we had received from Ericsson under subpoena. Many of those were in Swedish, which our adversaries assumed would be unread. With his linguistic prowess, Sol fooled them and quickly he discovered written proof of the invalidity of Ericsson's claims. Further the next day, under cross-examination he overwhelmed Ericsson's attorney so decisively that when our counsel was offered the standard opportunity to re-examine the witness, he responded that it was not needed since Dr. Golomb had already done far better in exposing the truth than he ever could. Ultimately Ericsson settled in a manner most beneficial to Qualcomm, which has since prospered and grown into its current leading position in semiconductors for mobile devices.

Let me end here by just stating what we all feel: that Sol's warmth, intellect and wisdom is sorely missed by all who knew and admired him.

Andrew Viterbi with Sol – Sol's 80th birthday, 2012

© 2023 World Scientific Publishing Company
https://doi.org/10.1142/9789811234378_0003

11/5/2020 Solomon Golomb (1932–2016)—Stephen Wolfram Writings

≡ STEPHEN WOLFRAM Writings

RECENT | CATEGORIES | Q

Solomon Golomb (1932–2016)*

May 25, 2016

The Most-Used Mathematical Algorithm Idea in History

An octillion. A billion billion billion. That's a fairly conservative estimate of the number of times a cellphone or other device somewhere in the world has generated a bit using a maximum-length linear-feedback shift register sequence. It's probably the single most-used mathematical algorithm idea in history. And the main originator of this idea was Solomon Golomb, who died on May 1—and whom I knew for 35 years.

Solomon Golomb's classic book *Shift Register Sequences*, published in 1967—based on his work in the 1950s—went out of print long ago. But its content lives on in pretty much every modern communications system. Read the specifications for 3G, LTE, Wi-Fi, Bluetooth, or for that matter GPS, and you'll find mentions of polynomials that determine the shift register sequences these systems use to encode the data they send. Solomon Golomb is the person who figured out how to construct all these polynomials.

He also was in charge when radar was first used to find the distance to Venus, and of working out how to encode images to be sent from Mars. He introduced the world to what he called polyominoes, which later inspired Tetris ("tetromino tennis"). He created and solved countless math and wordplay puzzles. And—as I learned about 20 years ago—he came very close to discovering my all-time-favorite rule 30 cellular automaton all the way back in 1959, the year I was born.

How I Met Sol Golomb

Most of the scientists and mathematicians I know I met first through professional connections. But not Sol Golomb. It was 1981, and I was at Caltech, a 21-year-old physicist who'd just received some media attention from being the youngest in the first batch of

https://writings.stephenwolfram.com/2016/05/solomon-golomb-19322016/ 1/28

---

*The following essay was reprinted with permission from *Idea Makers: Personal Perspectives on the Lives & Ideas of Some Notable People* and www.wolfram.com.

13

14   *The Wisdom of Solomon*

MacArthur award recipients. I get a knock at my office door—and a young woman is there. Already this was unusual, because in those days there were hopelessly few women to be found around a theoretical high-energy physics group. I was a sheltered Englishman who'd been in California a couple of years, but hadn't really ventured outside the university—and was ill prepared for the burst of Southern Californian energy that dropped in to see me that day. She introduced herself as Astrid, and said that she'd been visiting Oxford and knew someone I'd been at kindergarten with. She explained that she had a personal mandate to collect interesting acquaintances around the Pasadena area. I think she considered me a difficult case, but persisted nevertheless. And one day when I tried to explain something about the work I was doing she said, "You should meet my father. He's a bit old, but he's still as sharp as a tack." And so it was that Astrid Golomb, oldest daughter of Sol Golomb, introduced me to Sol Golomb.

The Golombs lived in a house perched in the hills near Pasadena. I learned that they had two daughters—Astrid, a little older than me, an aspiring Hollywood person, and Beatrice, about my age, a high-powered science type. The Golomb sisters often had parties, usually at their family's house. There were themes, like the flamingoes & hedgehogs croquet garden party ("recognition will be given to the person who appears most appropriately attired"), or the Stonehenge party with instructions written using runes. The parties had an interesting cross-section of young and not-so-young people, including various local luminaries. And always there, hanging back a little, was Sol Golomb, a small man with a large beard and a certain elf-like quality to him, typically wearing a dark suit coat.

I gradually learned a little about Sol Golomb. That he was involved in "information theory". That he worked at USC (the University of Southern California). That he had various unspecified but apparently high-level government and other connections. I'd heard of shift registers, but didn't really know anything much about them.

Then in the fall of 1982, I visited Bell Labs in New Jersey and gave a talk about my latest results on cellular automata. One topic I discussed was what I called "additive" or "linear" cellular automata—and their behavior with limited numbers of cells. Whenever a cellular automaton has a limited number of cells, it's inevitable that its behavior will eventually repeat. But as the size increases, the maximum repetition period—say for the rule 90 additive cellular automaton—bounces around seemingly quite randomly: 1, 1, 3, 2, 7, 1, 7, 6, 31, 4, 63, ... A few days before my talk, however, I'd noticed that these periods actually seemed to follow a formula that depended on things like the prime factorization of the

Solomon Golomb (1932–2016)

15

number of cells. But when I mentioned this during the talk, someone at the back put up their hand and asked, "Do you know if it works for the case $n=37$?" My experiments hadn't gotten as far as the size-37 case yet, so I didn't know. But why would someone ask that?

The person who asked turned out to be a certain Andrew Odlyzko, a number theorist at Bell Labs. I asked him, "What on earth makes you think there might be something special about $n=37$?" "Well," he said, "I think what you're doing is related to the theory of linear-feedback shift registers," and he suggested that I look at Sol Golomb's book ("Oh yes," I said, "I know his daughters..."). Andrew was indeed correct: there is a very elegant theory of additive cellular automata based on polynomials that is similar to the theory Sol developed for linear-feedback shift registers. Andrew and I ended up writing a now-rather-well-cited paper about it (it's interesting because it's a rare case where traditional mathematical methods let one say things about nontrivial cellular automaton behavior). And for me, a side effect was that I learned something about what the somewhat mysterious Sol Golomb actually did. (Remember, this was before the web, so one couldn't just instantly look everything up.)

The Story of Sol Golomb

Solomon Golomb was born in Baltimore, Maryland in 1932. His family came from Lithuania. His grandfather had been a rabbi; his father moved to the US when he was young, and got a master's degree in math before switching to medieval Jewish philosophy and also becoming a rabbi. Sol's mother came from a prominent Russian family that had made boots for the Tsar's army and then ran a bank. Sol did well in school, notably being a force in the local debating scene. Encouraged by his father, he developed an interest in mathematics, publishing a problem he invented about primes when he was 17. After high school, Sol enrolled at Johns Hopkins University to study math, narrowly avoiding a quota on Jewish students by promising he wouldn't switch to medicine—and took twice the usual course load, graduating in 1951 after half the usual time.

From there he would go to Harvard for graduate school in math. But first he took a summer job at the Glenn L. Martin Company, an aerospace firm founded in 1912 that had moved to Baltimore from Los Angeles in the 1920s and mostly become a defense contractor—and that would eventually merge into Lockheed Martin. At Harvard, Sol specialized in number theory, and in particular in questions about characterizations of sets of prime numbers. But

16   *The Wisdom of Solomon*

every summer he would return to the Martin Company. As he later described it, he found that at Harvard "the question of whether anything that was taught or studied in the mathematics department had any practical applications could not even be asked, let alone discussed". But at the Martin Company, he discovered that the pure mathematics he knew— even about primes and things—did indeed have practical applications, and very interesting ones, especially to shift registers.

The first summer he was at the Martin Company, Sol was assigned to a control theory group. But by his second summer, he'd been put in a group studying communications. And in June 1954 it so happened that his supervisor had just gone to a conference where he'd heard about strange behavior observed in linear-feedback shift registers (he called them "tapped delay lines with feedback")—and he asked Sol if he could investigate. It didn't take Sol long to realize that what was going on could be very elegantly studied using the pure mathematics he knew about polynomials over finite fields. Over the year that followed, he split his time between graduate school at Harvard and consulting for the Martin Company, and in June 1955 he wrote his final report, "Sequences with Randomness Properties"— which would basically become the foundational document of the theory of shift register sequences.

Sol liked math puzzles, and in the process of thinking about a puzzle involving arranging dominoes on a checkerboard, he ended up inventing what he called "polyominoes". He gave a talk about them in November 1953 at the Harvard Mathematics Club, published a paper about them (his first research publication), won a Harvard math prize for his work on them, and, as he later said, then "found [himself] irrevocably committed to their care and feeding" for the rest of his life.

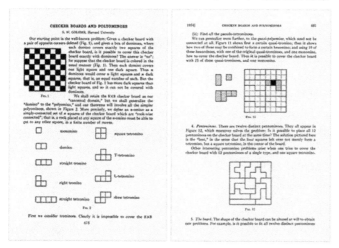

In June 1955, Sol went to spend a year at the University of Oslo on a Fulbright Fellowship—partly so he could work with some distinguished number theorists there, and partly so he could add Norwegian, Swedish and Danish (and some runic scripts) to his collection of language skills. While he was there, he finished a long paper on prime numbers, but also spent time traveling around Scandinavia, and in Denmark met a young woman named Bo (Bodil Rygaard)—who came from a large family in a rural area mostly known for its peat moss, but had managed to get into university and was studying philosophy. Sol and Bo apparently hit it off, and within months, they were married.

When they returned to the US in July 1956, Sol interviewed in a few places, then accepted a job at JPL—the Jet Propulsion Lab that had spun off from Caltech, initially to do military work. Sol was assigned to the Communications Research Group, as a Senior Research Engineer. It was a time when the people at JPL were eager to try launching a satellite. At first, the government wouldn't let them do it, fearing it would be viewed as a military act. But that all changed in October 1957 when the Soviet Union launched Sputnik, ostensibly as part of the International Geophysical Year. Amazingly, it took only 3 months for the US to launch Explorer 1. JPL built much of it, and Sol's lab (where he had technicians building electronic implementations of shift registers) was diverted into doing things like making radiation detectors (including, as it happens, the ones that discovered the Van Allen radiation belts)—while Sol himself worked on using radar to determine the orbit of the

18   *The Wisdom of Solomon*

satellite when it was launched, taking a little time out to go back to Harvard for his final PhD exam.

It was a time of great energy around JPL and the space program. In May 1958 a new Information Processing Group was formed, and Sol was put in charge—and in the same month, Sol's first child, the aforementioned Astrid, was born. Sol continued his research on shift register sequences—particularly as applied to jamming-resistant radio control of missiles. In May 1959, Sol's second child arrived—and was named Beatrice, forming a nice A, B sequence. In the fall of 1959, Sol took a sabbatical at MIT, where he got to know Claude Shannon and a number of other MIT luminaries, and got involved in information theory and the theory of algebraic codes.

As it happens, he'd already done some work on coding theory—in the area of biology. The digital nature of DNA had been discovered by Jim Watson and Francis Crick in 1953, but it wasn't yet clear just how sequences of the four possible base pairs encoded the 20 amino acids. In 1956, Max Delbrück—Jim Watson's former postdoc advisor at Caltech—asked around at JPL if anyone could figure it out. Sol and two colleagues analyzed an idea of Francis Crick's and came up with "comma-free codes" in which overlapping triples of base pairs could encode amino acids. The analysis showed that exactly 20 amino acids could be encoded this way. It seemed like an amazing explanation of what was seen—but unfortunately it isn't how biology actually works (biology uses a more straightforward encoding, where some of the 64 possible triples just don't represent anything).

In addition to biology, Sol was also pulled into physics. His shift register sequences were useful for doing range finding with radar (much as they're used now in GPS), and at Sol's suggestion, he was put in charge of trying to use them to find the distance to Venus. And so it was that in early 1961—when the Sun, Venus, and Earth were in alignment—Sol's team used the 85-foot Goldstone radio dish in the Mojave Desert to bounce a radar signal off Venus, and dramatically improve our knowledge of the Earth-Venus and Earth-Sun distances.

With his interest in languages, coding and space, it was inevitable that Sol would get involved in the question of communications with extraterrestrials. In 1961 he wrote a paper for the Air Force entitled "A Short Primer for Extraterrestrial Linguistics", and over the next several years wrote several papers on the subject for broader audiences. He said that "There are two questions involved in communication with Extraterrestrials. One is the mechanical

Solomon Golomb (1932–2016)

issue of discovering a mutually acceptable channel. The other is the more philosophical problem (semantic, ethic, and metaphysical) of the proper subject matter for discourse. In simpler terms, we first require a common language, and then we must think of something clever to say." He continued, with a touch of his characteristic humor: "Naturally, we must not risk telling too much until we know whether the Extraterrestrials' intentions toward us are honorable. The Government will undoubtedly set up a Cosmic Intelligence Agency (CIA) to monitor Extraterrestrial Intelligence. Extreme security precautions will be strictly observed. As H. G. Wells once pointed out [or was it an episode of *The Twilight Zone*?], even if the Aliens tell us in all truthfulness that their only intention is 'to serve mankind,' we must endeavor to ascertain whether they wish to serve us baked or fried."

While at JPL, Sol had also been teaching some classes at the nearby universities: Caltech, USC and UCLA. In the fall of 1962, following some changes at JPL—and perhaps because he wanted to spend more time with his young children—he decided to become a full-time professor. He got offers from all three schools. He wanted to go somewhere where he could "make a difference". He was told that at Caltech "no one has any influence if they don't at least have a Nobel Prize", while at UCLA "the UC bureaucracy is such that no one ever has any ability to affect anything". The result was that—despite its much-inferior reputation at the time—Sol chose USC. He went there in the spring of 1963 as a Professor of Electrical Engineering—and ended up staying for 53 years.

Shift Registers

Before going on with the story of Sol's life, I should explain what a linear-feedback shift register (LFSR) actually is. The basic idea is simple. Imagine a row of squares, each containing either 1 or 0 (say, black or white). In a pure shift register all that happens is that at each step all values shift one position to the left. The leftmost value is lost, and a new value is "shifted in" from the right. The idea of a feedback shift register is that the value that's shifted in is determined (or "fed back") from values at other positions in the shift register. In a linear-feedback shift register, the values from "taps" at particular positions in the register are combined by being added mod 2 (so that $1\oplus1=0$ instead of 2), or equivalently XOR'ed ("exclusive or", true if either is true, but not both).

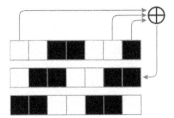

If one runs this for a while, here's what happens:

Obviously the shift register is always shifting bits to the left. And it has a very simple rule for how bits should be added at the right. But if one looks at the sequence of these bits, it seems rather random—though, as the picture shows, it does eventually repeat. What Sol Golomb did was to find an elegant mathematical way to analyze such sequences, and how they repeat.

If a shift register has size n, then it has 2^n possible states altogether (corresponding to all possible sequences of 0s and 1s of length n). Since the rules for the shift register are deterministic, any given state must always go to the same next state. And that means the maximum possible number of steps the shift register could conceivably go through before it repeats is 2^n (actually, it's 2^n-1, because the state with all 0s can't evolve into anything else).

In the example above, the shift register is of size 7, and it turns out to repeat after exactly $2^7-1=127$ steps. But which shift registers—with which particular arrangements of taps—will produce sequences with maximal lengths? This is the first question Sol Golomb set out to investigate in the summer of 1954. His answer was simple and elegant.

Solomon Golomb (1932–2016)

The shift register above has taps at positions 7, 6 and 1. Sol represented this algebraically, using the polynomial $x^7 + x^6 + 1$. Then what he showed was that the sequence that would be generated would be of maximal length if this polynomial is "irreducible modulo 2", so that it can't be factored, making it sort of the analog of a prime among polynomials—as well as having some other properties that make it a so-called "primitive polynomial". Nowadays, with Mathematica and the Wolfram Language, it's easy to test things like this:

In[1]:= **IrreduciblePolynomialQ$[x^7 + x^6 + 1,$ Modulus → 2$]$**

Out[1]= **True**

Back in 1954, Sol had to do all this by hand, but came up with a fairly long table of primitive polynomials corresponding to shift registers that give maximal length sequences:

The Prehistory of Shift Registers

The idea of maintaining short-term memory by having "delay lines" that circulate digital pulses (say in an actual column of mercury) goes back to the earliest days of electronic computers. By the late 1940s such delay lines were routinely being implemented purely digitally, using sequences of vacuum tubes, and were being called "shift registers". It's not clear when the first feedback shift registers were built. Perhaps it was at the end of the

22   *The Wisdom of Solomon*

1940s. But it's still shrouded in mystery—because the first place they seem to have been used was in military cryptography.

The basic idea of cryptography is to take meaningful messages, and then randomize them so they can't be recognized, but in such a way that the randomization can always be reversed if you know the key that was used to create it. So-called stream ciphers work by generating long sequences of seemingly random bits, then combining these with some representation of the message—then decoding by having the receiver independently generate the same sequence of seemingly random bits, and "backing this out" of the encoded message received.

Linear-feedback shift registers seem at first to have been prized for cryptography because of their long repetition periods. As it turns out, the mathematical analysis Sol used to find things like these periods also makes clear that such shift registers aren't good for secure cryptography. But in the early days, they seemed pretty good—particularly compared to, say, successive rotor positions in an Enigma machine—and there's been a persistent rumor that, for example, Soviet military cryptosystems were long based on them.

Back in 2001, when I was working on history notes for my book *A New Kind of Science*, I had a long phone conversation with Sol about shift registers. Sol told me that when he started out, he didn't know anything about cryptographic work on shift registers. He said that people at Bell Labs, Lincoln Labs and JPL had also started working on shift registers around the same time he did—though perhaps through knowing more pure mathematics, he managed to get further than they did, and in the end his 1955 report basically defined the field.

Over the years that followed, Sol gradually heard about various precursors of his work in the pure mathematical literature. Way back in the year 1202 Fibonacci was already talking about what are now called Fibonacci numbers—and which are generated by a recurrence relation that can be thought of as an analog of a linear-feedback shift register, but working with arbitrary integers rather than 0s and 1s. There was a little work on recurrences with 0s and 1s done in the early 1900s, but the first large-scale study seems to have been by Øystein Ore, who, curiously, came from the University of Oslo, though was by then at Yale. Ore had a student named Marshall Hall—who Sol told me he knew had consulted for the predecessor of the National Security Agency in the late 1940s—possibly about shift registers. But whatever he may have done was kept secret, and so it fell to Sol to discover and publish the

Solomon Golomb (1932–2016) 23

story of linear-feedback shift registers—even though Sol did dedicate his 1967 book on shift registers to Marshall Hall.

What Are Shift Register Sequences Good For?

Over the years I've noticed the principle that systems defined by sufficiently simple rules always eventually end up having lots of applications. Shift registers follow this principle in spades. And for example modern hardware (and software) systems are bristling with shift registers: a typical cellphone probably has a dozen or two, implemented usually in hardware but sometimes in software. (When I say "shift register" here, I mean linear-feedback shift register, or LFSR.)

Most of the time, the shift registers that are used are ones that give maximum-length sequences (otherwise known as "m-sequences"). And the reasons they're used are typically related to some very special properties that Sol discovered about them. One basic property they always have is that they contain the same total number of 0s and 1s (actually, there's always exactly one extra 1). Sol then showed that they also have the same number of 00s, 01s, 10s and 11s—and the same holds for larger blocks too. This "balance" property is on its own already very useful, for example if one's trying to efficiently test all possible bit patterns as input to a circuit.

But Sol discovered another, even more important property. Replace each 0 in a sequence by −1, then imagine multiplying each element in a shifted version of the sequence by the corresponding element in the original. What Sol showed is that if one adds up these products, they'll always sum to zero, except when there's no shift at all. Said more technically, he showed that the sequence has no correlation with shifted versions of itself.

Both this and the balance property will be approximately true for any sufficiently long random sequence of 0s and 1s. But the surprising thing about maximum-length shift register sequences is that these properties are always exactly true. The sequences in a sense have some of the signatures of randomness—but in a very perfect way, made possible by the fact that they're not random at all, but instead have a very definite, organized structure.

It's this structure that makes linear-feedback shift registers ultimately not suitable for strong cryptography. But they're great for basic "scrambling" and "cheap cryptography"— and they're used all over the place for these purposes. A very common objective is just to "whiten" (as in "white noise") a signal. It's pretty common to want to transmit data that's got long sequences of 0s in it. But the electronics that pick these up can get confused if they see what amounts to "silence" for too long. One can avoid the problem by scrambling the original data by combining it with a shift register sequence, so there's always some kind of "chattering" going on. And that's indeed what's done in Wi-Fi, Bluetooth, USB, digital TV, Ethernet and lots of other places.

It's often a nice side effect that the shift register scrambling makes the signal harder to decode—and this is sometimes used to provide at least some level of security. (DVDs use a combination of a size-16 and a size-24 shift register to attempt to encode their data; many GSM phones use a combination of three shift registers to encode all their signals, in a way that was at first secret.)

GPS makes crucial use of shift register sequences too. Each GPS satellite continuously transmits a shift register sequence (from a size-10 shift register, as it happens). A receiver can tell at exactly what time a signal it's just received was transmitted from a particular satellite by seeing what part of the sequence it got. And by comparing delay times from different satellites, the receiver can triangulate its position. (There's also a precision mode of GPS, that uses a size-1024 shift register.)

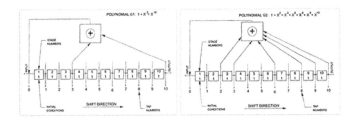

A quite different use of shift registers is for error detection. Say one's transmitting a block of bits, but each one has a small probability of error. A simple way to let one check for a single error is to include a "parity bit" that says whether there should be an odd or even number of 1s in the block of bits. There are generalizations of this called CRCs (cyclic redundancy checks) that can check for a larger number of errors—and that are computed essentially by feeding one's data into none other than a linear-feedback shift register. (There are also error-correcting codes that let one not only detect but also correct a certain number of errors, and some of these, too, can be computed with shift register sequences—and in fact

Solomon Golomb (1932–2016)

Sol Golomb used a version of these called Reed–Solomon codes to design the video encoding for Mars spacecraft.)

The list of uses for shift register sequences goes on and on. A fairly exotic example—more popular in the past than now—was to use shift register sequences to jitter the clock in a computer to spread out the frequency at which the CPU would potentially generate radio interference ("select Enable Spread Spectrum in the BIOS").

One of the single most prominent uses of shift register sequences is in cellphones, for what's called CDMA (code division multiple access). Cellphones got their name because they operate in "cells", with all phones in a given cell being connected to a particular tower. But how do different cellphones in a cell not interfere with each other? In the first systems, each phone just negotiated with the tower to use a slightly different frequency. Later, they used different time slices (TDMA, or time division multiple access). But CDMA uses maximum-length shift register sequences to provide a clever alternative.

The idea is to have all phones essentially operate on the same frequency, but to have each phone encode its signal using (in the simplest case) a differently shifted version of a shift register sequence. And because of Sol's mathematical results, these differently shifted versions have no correlation—so the cellphone signals don't interfere. And this is how, for example, most 3G cellphone networks operate.

Sol created the mathematics for this, but he also brought some of the key people together. Back in 1959, he'd gotten to know a certain Irwin Jacobs, who'd recently gotten a PhD at MIT. Meanwhile, he knew Andy Viterbi, who worked at JPL. Sol introduced the two of them —and by 1968 they'd formed a company called Linkabit which did work on coding systems, mostly for the military.

Linkabit had many spinoffs and descendents, and in 1985 Jacobs and Viterbi started a new company called Qualcomm. It didn't immediately do especially well, but by the early 1990s it began a meteoric rise when it started making the components to deploy CDMA in cellphones—and in 1999 Sol became the "Viterbi Professor of Communications" at USC.

Where Are There Shift Registers?

It's sort of amazing that—although most people have never heard of them—shift register sequences are actually used in one way or another almost whenever bits are moved around

The Wisdom of Solomon

in modern communication systems, computers and elsewhere. It's quite confusing sometimes, because there are lots of things with different names and acronyms that all turn out to be linear-feedback shift register sequences (PN, pseudonoise, M-, FSR, LFSR sequences, spread spectrum communications, MLS, SRS, PRBS, ...).

If one looks at cellphones, shift register sequence usage has gone up and down over the years. 2G networks are based on TDMA, so don't use shift register sequences to encode their data—but still often use CRCs to validate blocks of data. 3G networks are big users of CDMA —so there are shift register sequences involved in pretty much every bit that's transmitted. 4G networks typically use a combination of time and frequency slots which don't directly involve shift register sequences—though there are still CRCs used, for example to deal with data integrity when frequency windows overlap. 5G is designed to be more elaborate—with large arrays of antennas dynamically adapting to use optimal time and frequency slots. But half their channels are typically allocated to "pilot signals" that are used to infer the local radio environment—and work by transmitting none other than shift register sequences.

Throughout most kinds of electronics it's common to want to use the highest data rates and the lowest powers that still get bits transmitted correctly above the "noise floor". And typically the way one pushes to the edge is to do automatic error detection—using CRCs and therefore shift register sequences. And in fact pretty much every kind of bus (PCIe, SATA, etc.) inside a computer does this: whether it's connecting parts of CPUs, getting data off devices, or connecting to a display with HDMI. And on disks and in memory, for example, CRCs and other shift-register-sequence-based codes are pretty much universally used to operate at the highest possible rates and densities.

Shift registers are so ubiquitous, it's a little difficult to estimate just how many of them are in use, and how many bits are being generated by them. There are perhaps 10 billion computers, slightly fewer cellphones, and an increasing number of billions of embedded and IoT ("Internet of Things") devices. (Even many of the billion cars in the world, for example, have at least 10 microprocessors in them.)

At what rate are the shift registers running? Here, again, things are complicated. In communications systems, for example, there's a basic carrier frequency—usually in the GHz range—and then there's what's called a "chipping rate" (or, confusingly, "chip rate") that says how fast something like CDMA is done, and this is usually in the MHz range. On the

Solomon Golomb (1932–2016)   27

other hand, in buses inside computers, or in connections to a display, all the data is going through shift registers, at the full data rate, which is well into the GHz range.

So it seems safe to estimate that there are at least 10 billion communications links, running for at least 1/10 billion seconds (which is 3 years), that use at least 1 billion bits from a shift register every second—meaning that to date Sol's algorithm has been used at least an octillion times.

Is it really the most-used mathematical algorithm idea in history? I think so. I suspect the main potential competition would be from arithmetic operations. These days processors are doing perhaps a trillion arithmetic operations per second—and such operations are needed for pretty much every bit that's generated by a computer. But how is arithmetic done? At some level it's just a digital electronics implementation of the way people have done arithmetic forever.

But there are some wrinkles—some "algorithmic ideas"—though they're quite obscure, except to microprocessor designers. Just as when Babbage was making his Difference Engine, carries are a big nuisance in doing arithmetic. (One can actually think of a linear-feedback shift register as being a system that does something like arithmetic, but doesn't do carries.) There are "carry propagation trees" that optimize carrying. There are also little tricks ("Booth encoding", "Wallace trees", etc.) that reduce the number of bit operations needed to do the innards of arithmetic. But unlike with LFSRs, there doesn't seem to be one algorithmic idea that's universally used—and so I think it's still likely that Sol's maximum-length LFSR sequence idea is the winner for most used.

Cellular Automata and Nonlinear Shift Registers

Even though it's not obvious at first, it turns out there's a very close relationship between feedback shift registers and something I've spent many years studying: cellular automata. The basic setup for a feedback shift register involves computing one bit at a time. In a cellular automaton, one has a line of cells, and at each step all the cells are updated in parallel, based on a rule that depends, say, on the values of their nearest neighbors.

To see how these are related, think about running a feedback shift register of size n, but displaying its state only every n steps—in other words, letting all the bits be rewritten before one displays again. If one displays every step of a linear-feedback shift register (here with two taps next to each other), as in the first two panels below, nothing much happens at each step, except that things shift to the left. But if one makes a compressed picture, showing only every *n* steps, suddenly a pattern emerges.

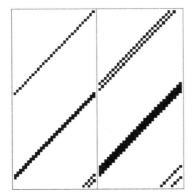

It's a nested pattern, and it's very close to being the exact same pattern that one gets with a cellular automaton that takes a cell and its neighbor, and adds them mod 2 (or XORs them). Here's what happens with that cellular automaton, if one arranges its cells so they're in a circle of the same size as the shift register above:

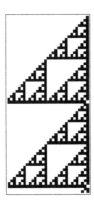

At the beginning, the cellular automaton and shift register patterns are exactly the same—though when they "hit the edge" they become slightly different because the edges are handled differently. But looking at these pictures it becomes less surprising that the math of shift registers should be relevant to cellular automata. And seeing the regularity of the nested patterns makes it clearer why there might be an elegant mathematical theory of shift registers in the first place.

Typical shift registers used in practice don't tend to make such obviously regular patterns, though. Here are a few examples of shift registers that yield maximum-length sequences. When one's doing math, like Sol did, it's very much the same story as for the case of obvious nesting. But here the fact that the taps are far apart makes things get mixed up, leaving no obvious visual trace of nesting.

So how broad is the correspondence between shift registers and cellular automata? In cellular automata the rules for generating new values of cells can be anything one wants. In linear-feedback shift registers, however, they always have to be based on adding mod 2 (or XOR'ing). But that's what the "linear" part of "linear-feedback shift register" means. And it's also in principle possible to have nonlinear-feedback shift registers (NFSRs) that use whatever rule one wants for combining values.

And in fact, once Sol had worked out his theory for linear-feedback shift registers, he started in on the nonlinear case. When he arrived at JPL in 1956 he got an actual lab, complete with racks of little electronic modules. Sol told me each module was about the size of a cigarette pack—and was built from a Bell Labs design to perform a particular logic operation (AND, OR, NOT, …). The modules could be strung together to implement whatever nonlinear-feedback shift register one wanted, and they ran pretty fast—producing about a million bits per second. (Sol told me that someone tried doing the same thing with a general-purpose computer—and what took 1 second with the custom hardware modules took 6 weeks on the general-purpose computer.)

When Sol had looked at linear-feedback shift registers, the first big thing he'd managed to understand was their repetition periods. And with nonlinear ones he put most of his effort into trying to understand the same thing. He collected all sorts of experimental data. He told me he even tested sequences of length 2^{45}—which must have taken a year. He made summaries, like the one below (notice the visualizations of sequences, shown as oscilloscope-like traces). But he never managed to come up with any kind of general theory as he had with linear-feedback shift registers.

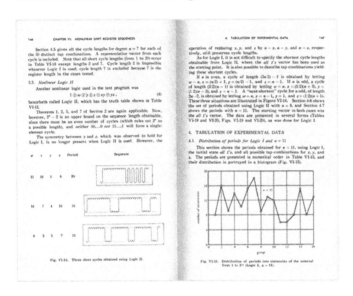

It's not surprising he couldn't do it. Because when one looks at nonlinear-feedback shift registers, one's effectively sampling the whole richness of the computational universe of possible simple programs. Back in the 1950s there were already theoretical results—mostly based on Turing's ideas of universal computation—about what programs could in principle do. But I don't think Sol or anyone else ever thought they would apply to the very simple—if nonlinear—functions in NFSRs.

And in the end it basically took until my work around 1981 for it to become clear just how complicated the behavior of even very simple programs could be. My all-time favorite example is rule 30—a cellular automaton in which the values of neighboring cells are combined using a function that can be represented as $p+q+r+qr$ mod 2 (or p XOR (q OR

r)). And, amazingly, Sol looked at nonlinear-feedback shift registers that were based on incredibly similar functions—like, in 1959, $p+r+s+qr+qs+rs$ mod 2. Here's what Sol's function (which can be thought of as "rule 29070"), rule 30, and a couple of other similar rules look like in a shift register:

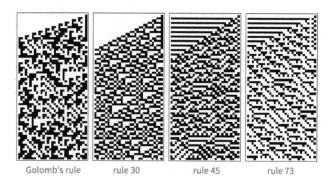

Golomb's rule   rule 30   rule 45   rule 73

And here's what they look like as cellular automata, without being constrained to a fixed-size register:

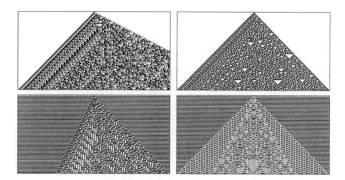

Of course, Sol never made pictures like this (and it would, realistically, have been almost impossible to do so in the 1950s). Instead, he concentrated on a kind of aggregate feature: the overall repetition period.

Sol wondered whether nonlinear-feedback shift registers might make good sources of randomness. From what we now know about cellular automata, it's clear they can. And for example the rule 30 cellular automaton is what we used to generate randomness for

32 The Wisdom of Solomon

Mathematica for 25 years (though we recently retired it in favor of a more efficient rule that we found by searching trillions of possibilities).

Sol didn't talk about cryptography much—though I suspect he did quite a bit of government work on it. He did tell me though that in 1959 he'd found a "multi-dimensional correlation attack on nonlinear sequences", though he said that at the time he "carefully avoided stating that the application was to cryptanalysis". The fact is that cellular automata like rule 30 (and presumably also nonlinear-feedback shift registers) do seem to be good cryptosystems —though partly because of confusions about whether they're somehow equivalent to linear-feedback shift registers (they're not), they've never been used as much as they should.

Being a history enthusiast, I've tried over the past few decades to identify all precursors to my work on 1D cellular automata. 2D cellular automata had been studied a bit, but there was only quite theoretical work on the 1D case, together with a few specific investigations in the cryptography community (that I've never fully found out about). And in the end, of all the things I've seen, I think Sol Golomb's nonlinear-feedback shift registers were in a sense closest to what I actually ended up doing a quarter century later.

Polyominoes

Mention the name "Golomb" and some people will think of shift registers. But many more will think of polyominoes. Sol didn't invent polyominoes—though he did invent the name. But what he did was to make systematic what had appeared only in isolated puzzles before.

The main question Sol was interested in was how and when collections of polyominoes can be arranged to tile particular (finite or infinite) regions. Sometimes it's fairly obvious, but often it's very tricky to figure out. Sol published his first paper on polyominoes in 1954, but what really launched polyominoes into the public consciousness was Martin Gardner's 1957 Mathematical Games column on them in *Scientific American*. As Sol explained in the introduction to his 1964 book, the effect was that he acquired "a steady stream of correspondents from around the world and from every stratum of society—board chairmen of leading universities, residents of obscure monasteries, inmates of prominent penitentiaries..."

Game companies took notice too, and within months, for example, the "New Sensational Jinx Jigsaw Puzzle" had appeared—followed over the course of decades by a long sequence of other polyomino-based puzzles and games (no, the sinister bald guy doesn't look anything like Sol):

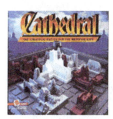

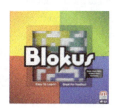

Sol was still publishing papers about polyominoes 50 years after he first discussed them. In 1961 he introduced general subdividable "rep-tiles", which it later became clear can make

34 *The Wisdom of Solomon*

nested, fractal ("infin-tile"), patterns. But almost everything Sol did with polyominoes involved solving specific tiling problems with them.

For me, polyominoes are most interesting not for their specifics but for the examples they provide of more-general phenomena. One might have thought that given a few simple shapes it would be easy to decide whether they can tile the whole plane. But the example of polyominoes—with all the games and puzzles they support—makes it clear that it's not necessarily so easy. And in fact it was proved in the 1960s that in general it's a theoretically undecidable problem.

If one's only interested in a finite region, then in principle one can just enumerate all conceivable arrangements of the original shapes, and see whether any of them correspond to successful tilings. But if one's interested in the whole, infinite plane then one can't do this. Maybe one will find a tiling of size one million, but there's no guarantee how far the tiling can be extended.

It turns out it can be like running a Turing machine—or a cellular automaton. You start from a line of tiles. Then the question of whether there's an infinite tiling is equivalent to the question of whether there's a setup for some Turing machine that makes it never halt. And the point then is that if the Turing machine is universal (so that it can in effect be programmed to do any possible computation) then the halting problem for it can be undecidable, which means that the tiling problem is also undecidable.

Of course, whether a tiling problem is undecidable depends on the original set of shapes. And for me an important question is how complicated the shapes have to be so that they can encode universal computation, and yield an undecidable tiling problem. Sol Golomb knew the literature on this kind of question, but wasn't especially interested in it. But I start thinking about materials formed from polyominoes whose pattern of "crystallization" can in effect do an arbitrary computation, or occur at a "melting point" that seems "random" because its value is undecidable.

Complicated, carefully crafted sets of polyominoes are known that in effect support universal computation. But what's the simplest set—and is it simple enough that one might run across by accident? My guess is that—just like with other kinds of systems I've studied in the computational universe—the simplest set is in fact simple. But finding it is very difficult.

A considerably easier problem is to find polyominoes that successfully tile the plane, but can't do so periodically. Roger Penrose (of Penrose tiles fame) found an example in 1994. My book *A New Kind of Science* gave a slightly simpler example with 3 polyominoes:

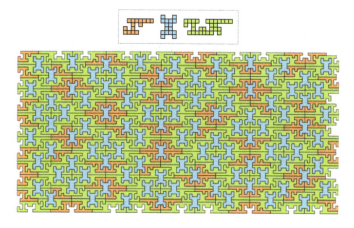

The Rest of the Story

By the time Sol was in his early thirties, he'd established his two most notable pursuits—shift registers and polyominoes—and he'd settled into life as a university professor. He was constantly active, though. He wrote what ended up being a couple of hundred papers, some extending his earlier work, some stimulated by questions people would ask him, and some written, it seems, for the sheer pleasure of figuring out interesting things about numbers, sequences, cryptosystems, or whatever.

Shift registers and polyominoes are both big subjects (they even each have their own category in the AMS classification of mathematical publication topics). Both have had a certain injection of energy in the past decade or two as modern computer experiments started to be done on them—and Sol collaborated with people doing these. But both fields still have many unanswered questions. Even for linear-feedback shift registers there are bigger Hadamard matrices to be found. And very little is known even now about nonlinear-feedback shift registers. Not to mention all the issues about nonperiodic and otherwise exotic polyomino tilings.

36 *The Wisdom of Solomon*

Sol was always interested in puzzles, both with math and with words. For a while he wrote a puzzle column for the *Los Angeles Times*—and for 32 years he wrote "Golomb's Gambits" for the Johns Hopkins alumni magazine. He participated in MegaIQ tests—earning himself a trip to the White House when he and its chief of staff happened to both score in the top five in the country.

He poured immense effort into his work at the university, not only teaching undergraduate courses and mentoring graduate students but also ascending the ranks of university administration (president of the faculty senate, vice provost for research, etc.)—and occasionally opining more generally about university governance (for example writing a paper entitled "Faculty Consulting: Should It Be Curtailed?"; answer: no, it's good for the university!). At USC, he was particularly involved in recruiting—and over his time at USC he helped it ascend from a school essentially unknown in electrical engineering to one that makes it onto lists of top programs.

And then there was consulting. He was meticulous at not disclosing what he did for government agencies, though at one point he did lament that some newly published work had been anticipated by a classified paper he had written 40 years earlier. In the late 1960s —frustrated that everyone but him seemed to be selling polyomino games—Sol started a company called Recreational Technology, Inc. It didn't go particularly well, but one side effect was that he got involved in business with Elwyn Berlekamp—a Berkeley professor and fellow enthusiast of coding theory and puzzles—whom he persuaded to start a company called Cyclotomics (in honor of cyclotomic polynomials of the form x^n-1) which was eventually sold to Kodak for a respectable sum. (Berlekamp also created an algorithmic trading system that he sold to Jim Simons and that became a starting point for Renaissance Technologies, now one of the world's largest hedge funds.)

More than 10,000 patents refer to Sol's work, but Sol himself got only one patent: on a cryptosystem based on quasigroups—and I don't think he ever did much to directly commercialize his work.

Sol was for many years involved with the Technion (Israel Institute of Technology) and quite devoted to Israel. He characterized himself as an "non-observant orthodox Jew"—but occasionally did things like teach a freshman seminar on the Book of Genesis, as well as working on decoding parts of the Dead Sea Scrolls.

Solomon Golomb (1932–2016)   37

Sol and his wife traveled extensively, but the center of Sol's world was definitely Los Angeles —his office at USC, and the house in which he and his wife lived for nearly 60 years. He had a circle of friends and students who relied on him for many things. And he had his family. His daughter Astrid remained a local personality, even being portrayed in fiction a few times—as a student in a play about Richard Feynman (she sat as a drawing model for him many times), and as a character in a novel by a friend of mine. Beatrice became an MD/PhD who's spent her career applying an almost mathematical level of precision to various kinds of medical reasoning and diagnosis (Gulf War illness, statin effects, hiccups, etc.)—even as she often quotes "Beatrice's Law", that "everything in biology is more complicated than you think, even taking into account Beatrice's Law". (I'm happy to have made at least one contribution to Beatrice's life: introducing her to her husband, now of 26 years, Terry Sejnowski, one of the founders of modern computational neuroscience.)

In the years I knew Sol, there was always a quiet energy to him. He seemed to be involved in lots of things, even if he often wasn't particularly forthcoming about the details. Occasionally I would talk to him about actual science and mathematics; usually he was more interested in telling stories (often very engaging ones) about personalities and organizations ("Can you believe that [in 1985] after not going to conferences for years, Claude Shannon just showed up unannounced at the bar at the annual information theory conference?", "Do you know how much they had to pay the president of Caltech to get him to move to Saudi Arabia?", etc.)

In retrospect, I wish I'd done more to get Sol interested in some of the math questions brought up by my own work. I don't think I properly internalized the extent to which he liked cracking problems suggested by other people. And then there was the matter of computers. Despite all his contributions to the infrastructure of the computational world, Sol himself basically never seriously used computers. He took particular pride in his own mental calculation capabilities. And he didn't really use email until he was in his seventies, and never used a computer at home—though, yes, he did have a cellphone. (A typical email from him was short. I had mentioned last year that I was researching Ada Lovelace; he responded: "The story of Ada Lovelace as Babbage's programmer is so widespread that everyone seems to accept it as factual, but I've never seen original source material on this.")

Sol's daughters organized a party for his 80th birthday a few years ago, creating an invitation with characteristic mathematical features:

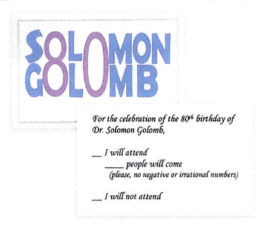

Sol had a few medical problems, though they didn't seem to be slowing him down much. His wife's health, though, was failing, and a few weeks ago her condition suddenly worsened. Sol still went to his office as usual on Friday, but on Saturday night, in his sleep, he died. His wife Bo died just two weeks later, two days before what would have been their 60th wedding anniversary.

Though Sol himself is gone, the work he did lives on—responsible for an octillion bits (and counting) across the digital world. Farewell, Sol. And on behalf of all of us, thanks for all those cleverly created bits.

Posted in: Historical Perspectives, Mathematics

Join the discussion

+ 9 comments

Related Writings

The Empirical Metamathematics of Euclid and Beyond
September 28, 2020

How We Got Here: The Backstory of the Wolfram Physics Project
April 14, 2020

Announcing the Rule 30 Prizes
October 1, 2019

A Book from Alan Turing… and a Mysterious Piece of Paper
August 27, 2019

Popular Categories

Artificial Intelligence

Big Picture

Companies and Business

Computational Science

Computational Thinking

Data Science

Education

Future Perspectives

Historical Perspectives

Language and Communication

Life and Times

Life Science

Mathematica

Mathematics

New Kind of Science

New Technology

Personal Analytics

Physics

Software Design

Wolfram|Alpha

Wolfram|One

Wolfram Language

Other

Writings by Year

2020 | 2019 | 2018 | 2017 | 2016 | 2015 | 2014 | 2013 | 2012 | 2011 | 2010 | 2009 | 2008 | 2007 | 2006 | 2004 | 2003 | All

© Stephen Wolfram, LLC | Terms | RSS

Part II

A Career in Engineering*

Solomon W. Golomb and Beatrice A. Golomb

(Invited Paper)

Abstract—During the summers (1951–1954) that I was a graduate student in "pure mathematics," I worked in the Systems Engineering Section of the Glenn L. Martin Company. I began to notice the applicability of supposedly pure topics like prime number theory and finite field theory to problems in communications. My first major applied effort involved developing the theory of "Shift Register Sequences." (My book with this title will soon see its third edition, with a third publisher.) Much of my work has been in response to practical questions I was asked, for which I had the necessary mathematical tools. These topics have included comma-free codes, Costas arrays, Tuscan squares, Golomb rulers, zero-sidelobe radar, etc. My shift register work has had the broadest impact: to cell phone signals, the GPS system, error-correcting codes, radar, cryptography, etc. I am fortunate to have lived long enough to get some recognition for my work.

Index Terms—Shift register sequence, comma-free codes, Costas arrays, Tuscan squares, Golomb rulers, zero-sidelobe radar, and cryptography.

I. Introduction

WHAT will follow this "prologue" – really, reminiscences and a bit of commentary – is an article based on a talk that my father, Solomon Golomb, gave at USC (Nov. 7, 2014) and the University of Waterloo (Aug. 12, 2015).

Professor Guang Gong (U. of Waterloo), whom Sol affectionately denoted his "mathematical daughter," kindly entered Sol's address.

Guang Gong on the right, from GolombFest 70, USC, May 2002, with Heiko Harborth and Richard Guy. (Richard Guy coauthored, with Elwyn Berlekamp and John Conway, the classic book *Winning Ways*.) Photo from Hong-Yeop Song.

Guang asked that I – Sol's other kind of daughter – write a prologue. (Sol's elder daughter, Astrid, also merits mention among his progeny. And I have heard Solomon referred to as the "father of modern digital communications," so perhaps there is a further sibling in the mix.)

I shall strive (ineptly) to convey a tiny taste of the wonder that was Solomon Golomb – of his unrivalled intellectual élan, his deeply kind nature, and his ever ready sense of humor. Those who knew him knew that he possessed a rare further gift – wisdom – that led his counsel to be sought by so many.

Understand, since I myself have not Sol's gifts, I can not do justice in striving to convey his.

Sol received his PhD in mathematics from Harvard, after completing his undergraduate studies at Hopkins while still 18, a graduating sophomore. But tales of Sol's intellectual prowess predate that time. His high school biology teacher, seeking to astonish the class with the number of potential genetic combinations that two parents could produce, stated that there were two options from each of 23 chromosomes from each of two parents, so 2 to the 24th power "and you all know what that is." Sol promptly replied "sixteen million seven hundred and seventy seven thousand two hundred and sixteen." She said "Hah hah. The real number is –"; looking at her paper, she was forced to pronounce, "16,777,216." I have spoken to some of Sol's classmates, but Sol affirmed the tale to me. By his reckoning it wasn't such a feat: He simply recognized that 2 to the 24th was equivalent to 8 to the 8th; and he had, naturally, memorized N to the N up to some extravagant N.**

Sol's facility in complex calculations was the stuff of legend – and a probable reason he long abjured calculators and computers, which would have been cheating. (He'd mentioned to me that he had encountered but one contender to his prowess in mental calculations – Persi Diaconis, who, as a magician cum mathematician and erstwhile carnival worker, had made a living astounding audiences with his feats.)

As a youngster, Sol was in the audience for a quiz show, and the audience was asked the following question:
 "One hundred and one by fifty divide
 And now if a cipher be rightly applied
 And then if the answer you duly divine
 You'll find the result equals one out of nine."
It was Sol who, as a youth, in real time, gave the answer: "CLIO."

One hundred and one was "divided" by fifty via the intercession of Roman numerals (CI split by L). One meaning of "cipher" is zero ("0") – applied on the right. CLIO was one out of nine – muses, in Greek mythology. This event illustrates that Sol's precocity in matters, um, numerical, was not so narrowly circumscribed.

---

*This chapter was originally published in *IEEE Transactions in Information Theory*, April 2018, Vol. 64, No. 4, Part II, pp. 2805–2836.

**BG corrigendum to 2^{24} story: In recalling the 2^{24} tale from decades past, I could not reconstruct how the 2^{24} value came to be with only 23 human chromosomes. I could only presume that the biology professor made an error, though then Sol would presumably have noticed this. Only later I recalled that Sol had included, as an aside, that at the time there were thought to be 24 human chromosomes. I relayed this to Al Hales, who found in two science books inherited from his parents (from the 1930s and 1940s) that this was the case. So in the corrected tale the professor related that each parent provided two options for each of 24 chromosomes, providing 2^{24} possible combinations, "and you all know what that is."

(Did I mention Sol's broad knowledge base? His vocabulary?)

Sol's fund of knowledge was ever prodigious. He used to joke (not really joking) that he could give a half hour lecture on any subject whatsoever. (This was long before internet and Wikipedia made it easy to find information.) And it wasn't *just* that he seemed somehow to know everything; his brain reformulated, organized and understood it – with his massive knowledge base and powers of reasoning and synthesis to draw on – in a way not elsewhere to be found. I recall, as one exemplar, when an engineering colleague lamented to Sol that he had been roped into serving on a panel on jurisprudence, a subject about which he knew nothing. Sol proceeded (in illustration of his claim) to sit the colleague down, and deliver a half hour disquisition on jurisprudence. After the friend participated in the panel, he triumphantly told Sol that – thanks purely to Sol's tuition – he had known more about jurisprudence than any on the panel, including judges.

That benison has been bestowed on many of us with the fortune to have known Sol. Sol, having already inexplicably acquired and reformulated the understanding in some random sphere about which we besought information, shared his revelations in easily assimilated form, and we as the recipient, by simply later repeating a shred of what he had shared, were deemed brilliant for our knowledge and insight.

But, back to the quiz show. Whether Sol's éclat on that occasion inspired or more likely followed from it, Sol had a lifelong love of puzzles, words, logic and numbers – and information. (Not just information *theory* – but its application!) For decades he wrote a puzzle column – "Golomb's Gambits" – in the Johns Hopkins alumni magazine, another puzzle column in the IEEE Information Theory Society newsletter, and for a spell, a third puzzle column for the *LA Times*. (Sol's dear friend Elwyn Berlekamp – like Sol a recipient of the Hamming Award and who knows what more – had a "rival" puzzle column in the Mathematical Sciences Research Institute newsletter, *Emissary* – "M S R I".) Since Sol's Hopkins column ran for decades, few readers will recall early columns if ever they saw them; and since a number of alumni reported that Sol's was the first column they turned to, Hopkins' magazine is now reprinting the older columns (albeit in small size).

all mathematics was useful. Give him any beautiful finding, he contended, and he could find an application for it.

Sol on chessboard 1948 (age 15-16). Courtesy Edna Sharoni.

Several allusions in Sol's address call to my mind instances of the enchantment of words, numbers, and logic – entwined with humor – by or involving Sol. (Persons referenced may be best placed in context *after* reading Sol's address.)

In his address Sol mentions Golomb's constant, noted to be 0.6243296.... This evoked the occasion when Sol told me would be nice to have a more *commonly* used constant named for him. Like, he suggested: zero.

His allusion to Marshall Hall reminds me of the time when, telling me of Marshall Hall, Sol mentioned (as an aside) an auditorium or atrium named for that man – Marshall Hall Hall – providing for three consecutive "Hall"s.

He mentions Gus Solomon. (G. Solomon and Solomon G. in the same group at JPL must have spawned confusion.) This conjures the time, in Sol's early days at JPL when he suggested to Gus (also an eminent cryptographer) that they collaborate on a code. Gus demurred, on grounds it would be surely be called the Golomb Solomon code. Sol assured him: "Oh, no. I'd be happy to give you first billing."

{Gus recognized a good theme and riffed on it. According to Jack Stiffler, Gus, later coauthor of the fabled Reed Solomon code, threatened to change his first name to Reed.}

Elwyn Berlekamp with Sol

Sol's love of puzzles and his love of mathematics were inextricably intertwined. When once asked about his affinity for recreational mathematics, Sol replied that to him, *all* mathematics was recreational. Conversely, he claimed that

Left: Sol's wife Bo dances with the wrong Solomon
Right: Irving Reed and Gus Solomon
Both March 1990, Cal Tech Athenaeum

1963: Al Ingersoll (dean), Solomon Golomb, Irving Reed, Nasser Nahi, Zohrab Kaprielian. Courtesy George Bekey.

Paul Erdös is not mentioned in the address, but I will note as a matter of historical interest that Erdös dined at Chez Golomb on Nov. 22, 1963, the night President John F. Kennedy was killed. Readership will doubtless be familiar with the concept of Erdös numbers: A person is accorded the Erdös number of one if they published directly with prolific mathematician Paul Erdös, two if they published with someone who published with Erdös, etc. Mathematicians vie for the lowest Erdös number. Sol observed that in the usual course of affairs, mathematicians received an Erdös number of (+)1 when Erdös solved a problem they had posed; then the paper was written up jointly. Therefore since Sol had published the solution to a problem Erdös had posed, he claimed his Erdös number was *minus* one.

(It is well known that Erdös called small children Epsilons – including myself at dinner that fated night. So my Erdös number, Epsilon, is especially authentic as it was bestowed by Erdös himself. As I see it, this means I am arbitrarily close to *being* Paul Erdös.)

Sol mentions his friend Nobel laureate Max Delbrück. Max often played chess with Sol at our home, with Sol's two minutes on the chess clock pitted against Max's hour.

Postcard from Francis Crick to Sol, England 1960

I recall the time when Max, again vanquished despite his time advantage, accused Sol of cheating: "You've been thinking on my time."

In Sol's address he mentions the early days of the genetic code, and his involvements with Francis Crick and Sydney Brenner.

Pictured is a postcard from Francis to Sol, both in England in 1960, referencing Leslie (Orgell) and Sydney (Brenner), the latter also to receive a Nobel Prize. All three – Francis, Leslie and Sydney – later settled in La Jolla, and I came to know them all (originally through Francis, but later, especially Sydney, independently). At this writing, only the indomitable Sydney remains alive and very much kicking, in Singapore, though he has finally given up travel. (Terry Sejnowski – Sol's begotten son-in-law, also in the NAE – recently attended Sydney's 90th there.)

Francis invited Cambridge mathematician Graeme Mitchison to join him at the Salk Institute in La Jolla for a summer, during a time I worked with Francis (reviving Francis' dormant relationship with Sol and Bo).

When, after Sol's death, I wrote to beseech people's memories of Sol (please share yours, if you have not!), the request made its way to Graeme. Graeme writes beautifully – and his reflections sparkle through their specificity.

"I first met Sol and Bo in the early 1980s. With characteristic generosity, Sol had taken Beatrice and several of her friends out to lunch in La Jolla, and I had been greatly impressed by his omniscience. He was the obvious person to write to when Richard Durbin and I needed some advice on a combinatorial puzzle – how does one write the numbers from 1 to N^2 in an N by N square in such a way as to minimise the sum of absolute values of differences between nearest neighbours? It turned out that Sol had formulated the puzzle in the 1960s, and his reply introduced us, amongst other things, to 'boustrophedon numbering', which harks back to ancient Greek civilizations and the oxen ploughing their fields.

"He subsequently invited me to their house in Los Angeles, and in my memory this event still seems to shine with his brilliance. Amongst other things, he told me about the connection between Rubik's cube and the eightfold way in particle physics, and about the existence of small sets of integers that have the same pattern of differences (alas I have lost the piece of paper with the numbers on it). As we were walking through his garden I pointed out a vine and said it was called a Bignonia. 'Ah yes', said Sol, 'it is related to a very rare plant called a Littleknownia'. The vine has never looked the same since.

"My next meeting with Sol was when he and Bo came to lunch at my house in Cambridge. The photo shows him in my garden, expounding the rules for interpreting Chinese characters to Richard Durbin while Bo relaxes in the summer heat. Alas, that was the last time we met, but his influence in my life is recorded in ironwork in my garden (see the second photo), with the gold-painted bars depicting the solution to his combinatorial puzzle."

Photos referred to by Graeme Mitchison

Sol's humor was not littleknownia, but rather ever at the ready, playful with words or logic.

During the epoch when Monica Lewinsky was much in the news, for a time I was told by the head of Health Defense at the "thinktank" RAND (where I consulted) that I was to brief then-president Bill Clinton on a "best-selling" report I had written. When I told Sol, he quipped: "As long as it's a briefing and not a debriefing."

(I did come to sit in Monica's chair in the Pentagon.)

Sol with Zsa Zsa Gabor in press attention to the Cube on its US debut. Sol and (especially) his colleague and former PhD student Herb Taylor trained Minh Thai, a talented local high school student (recently hailing by boat from Vietnam), on the algorithms they developed. Minh Thai became the (televised) national champion, and then world champion Rubik's Cube solver. (Minh went on to attend USC.) Sol published on the shared group theoretic properties of Rubik's Cube and Quarks.
Photo Nick Ut.

Sol could enliven ordinary discourse with scientific concepts. Before my wedding, to still-/first-husband Terrence Sejnowski (at which Graeme Mitchison and Stephen Wolfram, below, were groomsman and Francis bestowed a blessing), I informed my (mortified) sire of the convention of the father-daughter dance. Terpsichorean proficiencies were not

Graeme's mention of Rubik's cube is an excellent pretext for sharing photos of Sol assaulting the cube. Before the Cube's US debut, Astrid, then at Oxford, brought Sol a Cube from England, whereupon Sol promptly unriddled its mysteries. Sol took pleasure in glancing briefly at a scrambled cube to observe its configuration, then solving it behind his back. Photos Nick Ut; Graham Martin.

Sol implements simple harmonic motion
Right: Bo secures the right Solomon Cal Tech Athenaeum, March 24, 1990

among Sol's manifold gifts. I illustrated that he could simply step from left foot to right and back. He brightened and said: "Oh, it's just simple harmonic motion."

In 2011, Sol was to receive an award from the Technion in Haifa (Sol served for many years on the Technion Board of Governors), and I timed some talks in Israel, to attend the honor. My luggage went astray, and – two afternoons later with the event imminent – it had yet to arrive. I set out by public transport to find one of each article of apparel – shirt, pants, jacket, etc – and a pair of shoes to replace the Crocs (*horresco referens*) I had worn on the plane. Returning to the hotel, I found the last shuttle – with Sol in it – preparing to depart for the event. I pleaded "give me *two* minutes," dashed in, emerged momentarily in fresh raiment, and sat by Sol as the conveyance departed. I looked down to discover – with dismay – that I had successfully changed *one* of my shoes – for this event honoring Sol. With shame I told Sol. He said with a twinkle: "If anyone notices, tell them you have another pair just like it at home."

I should like to propose an effort to rectify this in one instance, and popularize "Golombinoes" (with or without the "b" which in any case is silent, "Golomb" rhyming with "aplomb"), as a synonym for polyominoes – or specifically pentominoes.

Golominoes also make excellent cookies (from Sol's 80th)

Sol receives an Honorary Doctorate from the Technion, 2011. Photo courtesy of Technion/ President Peretz Lavie

Sol used his facility with words in service of his work, dubbing tilings of the plane that repeated in some fashion "reptiles;" and ones in which each tile comprised infinite repetitions of the same shape on successively smaller scale (imagine squares of successive half size, spiraling in, for instance) "infintiles." He did observe the downside of creating good names: the thing was not then named for you.

The 12 distinct pentomino shapes are each used once; reflections are not deemed distinct. (c) Scott Kim, scottkim.com.

Sol in Pentominoes, by Ken Knowlton (with permission)

There are many more tales of Sol's humor – and prowess. His extraordinary memory abetted his linguistic virtuosity, his adroitness in legion languages and alphabets, modern and ancient; and his Biblical scholarship, with which linguistic knowledge is entwined. (A couple vignettes are included with photos at the end.)

I here shift theme completely, to address another facet of Sol.

In his address (the one to follow), Sol states "I could give many more examples where someone asked me about a problem, and I happened to have the right tools for a solution." So Graeme's example illustrates.

My friend Stephen Wolfram (he of Mathematica and Wolfram Alpha), who wrote an obituary for Sol – in which he observed Sol responsible for the most used algorithm in history, with a conservative estimate of an octillion times deployed (http://blog.stephenwolfram.com/2016/05/solomon-golomb-19322016/) – remarked to me that Sol appeared to have been most productive when other people were giving him problems they needed solved. I believe this a just insight. In my mind, this does not diminish Sol, to the contrary. He was deeply motivated by the desire to do for others when asked, and it is part and parcel of what made him not just a great mathematician and scientist, but a great man.

Sol with Stephen Wolfram (in red shirt), ~1990

I cannot count the number of people who approached me after Sol's death, to say, in heartfelt timbre, how grave was their debt to Sol, how greatly and effectually Sol had helped them through some crisis at a critical juncture. These people, now giants in their fields – having quite manifestly overcome whatever the cataclysm – were startlingly unforthcoming about just what the problem had been, or just how Sol had exerted himself on their behalf – and Sol would have well approved this reticence. (But, I would love to know the details.) Each of us just knew how he helped *us*. Many after Sol's death, hearing *others'* gratitude, remarked that given the devotion of time and effort he sacrificed to their aid, they assumed they were the only one.

I do know specifics of one instance of aid, because I was the conduit through whom Sol learned of the impending calamity, to a friend whom Sol knew (and had played chess with). Sol instantly grasped the gravity and urgency of the situation (I had not comprehended the sheer immediacy of action required), understood as others would not what might be done to avert catastrophe, acted swiftly, decisively, and effectively (with the beset party's full knowledge and blessing, need I say), and forestalled a desperate outcome, forging in its place – through dint of long care – a spectacular success. Because I know specifics of that case (plus snippets of still more – through aided parties, not through Sol), although no fault redounded to the individual then in peril, I can see that broadcasting that period of darkness might be felt to cast a shadow on the individual's ultimate success, and I too must remain mum on the particulars.

It is only after Sol's death, as I spoke to Sol's beloved sister Edna Sharoni (five years Sol's senior, remarkable at 90, mentally sharp, very smart, wonderfully kind, living in Israel – where she just this last year gave an address to that nation's president), that I came to appreciate that Sol's deep drive to fix or solve problems of and for others who came to him for help – in life as in mathematics – seems drawn from his forebears, by whatever dint of genes and environment. Sol's father and paternal grandfather, both rabbis, were each, I have learned, not merely renowned for their scholarship, but much sought for their wisdom – and each, as Edna emphasized, for their striking effectiveness – in helping others. Sol's grandfather was called "the wise one" by locals. Sol's own father, was similarly revered. Sol himself died sometime in the night of April 30/May 1 2016. This was, I learned from Edna, the night of the 60th anniversary of his own father's death.

Sol had many interwoven careers – mathematics, communications, business, linguistics, puzzles, and polymathy. Through all this, helping others who wished it, in ways small and large, exercised, I think, his most core motive force. Yet it could not be equally his legacy, since a part of the gift he gave – and its effectiveness – lay in his own role remaining unheralded.

Yet some hint of this, inevitably, became known. Following Sol's funeral, former USC Engineering doctoral student Terry Lewis shared a moving parable. After a storm, thousands of crabs had washed ashore, and a man was seen casting first one crab, then another into the sea to save them. Asked why he bothered, given the vast number, the impossibility of rescuing them all, the man, beholding the crab he held, said "But I can save this one." Per Lewis, in this parable, Sol was that man. And he was one among those saved.

A half year after Sol's death (Nov.18, 2016), I happened into a used bookstore in Pasadena, and the delightful proprietress introduced me to a regular patron, announcing that he – a Tom Tomlinson – was a former Professor of History at USC. I mentioned that my father had also been on the USC faculty and Tomlinson asked his name. When told, he replied: "Oh, Sol, of course I knew him. Everyone on campus knew if they had any problem, he was the one to go to."

Sol loved his colleagues, and loved USC. He was gratified to have played a role in USCs rise as a University. Regarding that role, perhaps this is indiscreet to say, but several of Sol's colleagues have told me, *sotto voce*, that – reflected in Sol's influence in choosing USC's last two presidents and the weight carried by his opinion – he was, at USC, esteemed a "kingmaker."

2013: USC President C. L. Max Nikias honors National Medal of Science recipient Solomon Golomb. Photo, courtesy USC.

Photo courtesy USC.

Sol did "Fight on!" for his beloved USC. (He quipped that USC should retire his number.) Photos courtesy of USC. Note from Sol: 80 is "Π" in Greek (which like Hebrew, assigns numerical values to letters)

Andrew Viterbi with Sol

At USC, Sol was proud to have been the Viterbi Professor in the Viterbi School of Engineering, both named for his longtime friends since JPL days, Andrew and Erna Viterbi.

Erna and Andrew Viterbi

Joan and (the back of) Irwin Jacobs, Sol and Phil Sotel

Sol had been delighted by the success of two good friends whom he had introduced to one another – and the company they founded, Qualcomm. {A few tales are shared later, among Sol's courtroom feats of memory, mathematics and linguistics – in that instance enlisting Sol's proficiency in Swedish, Norwegian, and Japanese – at a parlous juncture for *that* company, during a lawsuit with Ericsson.} Indeed, the success of these friends is enshrined in the names of the Engineering Schools at both Sol's University (the Viterbi School at USC) and my own (the Jacobs School at UCSD), named for these respective founders.

In addition to its Viterbi School, and the Viterbi Chair, USC has an annual Viterbi lecture, and the fourth to deliver it, Toby Berger (among the keepers of Sol's courtroom tales), captured the affection and awe that Sol stirred in so many who knew him.

In an email to colleagues after Sol's death, passed to me by a third party, Toby said: "I 'LOVED' Sol and worshiped his brilliance... He is irreplaceable. A genius among our geniuses, he was a man who somehow knew almost EVERYTHING in an era in which that seems to be impossible, but it was true…Perhaps never before has so much knowledge left the earth at the same instant. Sol is utterly and incontrovertibly irreplaceable!"

USC President C.L. Max Nikias said of Sol, "He was one of the most incredible intellects in the history of the American academy. He was one of the great geniuses of his generation…"

From the standpoint of sheer intellect, I think, of any generation.

A Career in Engineering

Sol receiving the (Benjamin) Franklin Award, April 2016.

President Barack Obama presents Sol with the National Medal of Science. White House, Feb. 1, 2013.
Photo courtesy Ryan K. Morris, National Science & Technology Medals Foundation.

A Career in Engineering

Sol received many national and international awards. But Sol valued those from USC no less for this. Here, I believe he receives the Exceptional Academic Achievement Award, USC 2012

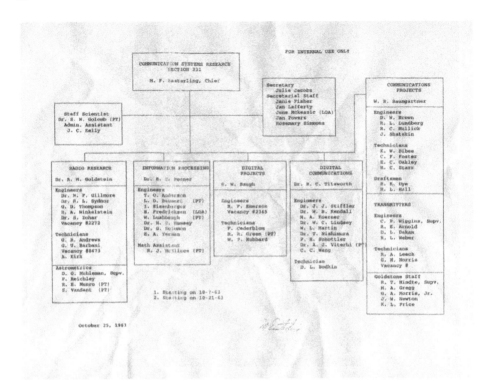

JPL Section 331 Org Chart (point in time). Sol, "Staff Scientist", is the lone scientist in tier 2, with oversight over the researchers. Many went on to distinguished careers, NAE memberships and other honors. Charmingly, the Secretarial Staff are positioned above the scientists in the Org Chart. Shared by Robert Tausworthe (here as Titsworth, his former name).

A Career in Engineering

At Sol's 75th birthday celebration at USC in 2007, when a number of the JPL Section 331 members were present. From left: Robert Tausworthe, William Lindsey, Solomon Golomb, Herbert Taylor, Andrew Viterbi, Harold (Hal) Fredricksen, Elwyn Berlekamp, William Hurd, Robert McEliece, and Laif Swanson. From Bob Tausworthe.

USC colleagues Irving Reed, Andrew Viterbi, William Lindsey, Lloyd Welch, Robert Scholtz, and Solomon Golomb.

Sol with Claude Shannon (center), 1985, on occasion of Sol's delivery of the Shannon lecture – the only Shannon lecture Shannon attended, other than his own. Pictured, from left: Robert and Lolly Scholtz, Betty and Claude Shannon, and Solomon. Person on far right is unidentified. IEEE Symposium on Information Theory, the Metropole Hotel, Brighton, England, June 23-28

Biblical Scholarship

Sol (standing, 2nd from right) at summer camp.
From Edna Sharoni

When Douglas Hofstadter's book *Gödel Escher Bach: the Eternal Golden Braid* – later to receive a Pulitzer – was first published (1979), a friend gave me a copy. I showed it to Sol who methodically turned to a ∼frontispiece, bearing what professed to be the beginning of Genesis in ancient Hebrew. Sol said: "First of all, it's upside down." (He inverted the book to right the text.) "Second, the script is not ancient Hebrew, but ancient Samaritan. Third, this is not the first lines of Genesis, but only the first several words of each of those lines." He proceeded to read it in the original, then give real-time translation. (I believe later editions corrected or removed this.)

As a child Sol had attended a Jewish summer camp. The older of two age-stratified groups of children were asked if any could cite the passage in the (Hebrew) Bible where three successive words commence with the Hebrew letter "mem;" none could do so. After the approved answer was purveyed, Sol – from the younger group (see photo above) – piped up to proclaim that there was another such instance, which he recited – to the astonishment of the preceptors. (Sol's consecutive words crossed a sentence boundary.)

A similar talent enabled Sol, as an adult, to contribute to the Dead Sea Scroll effort (on which an academic colleague was working). Viewing a fragment that had not yet yielded its secrets, Sol – recognizing the alphabet used, and noting the repetition of a sound in proximity to (beneath) itself on the fragment, relayed that the relative positioning suggested a specific locus in Genesis. Other fragments from the site were found to confirm Sol's thesis, and I understand Sol received a credit on the Scrolls.

Languages

The count of languages Sol "knew" is imperfectly defined. His memory remained remarkable relative to mere mortals in older age; but relative to his own baseline, age and health conditions (including a small stroke) had their effect. The count he gave, if asked, declined correspondingly (to ∼12, from a peak of ∼30). His linguistic gifts were phenomenal. When I was young, my mother Bo, a native Dane (abiding in Denmark till after her marriage to Sol), would ask *Sol* for words in Danish – *and he would provide them*. Bo's sister Hanna, who remained in Denmark, underscored to me that as a young adult, "Sol's Danish was *perfect*; I mean, *absolutely perfect*."

Sol viewed his Norwegian as stronger still (encompassing old Norwegian as well as "Nynorsk"). I shared with Sol some works of Norwegian crime novelist (and economist) Jo Nesbø, whose ouvre Sol enjoyed, in part for insights into modern Norway. Dissatisfied with aspects of the translation, Sol read several of Nesbø's books in the original Norwegian, including Nesbø's then-latest, in Sol's last year of life.

Innumerable languages that Sol professed *not* to know, he nonetheless knew a great deal about. Exemplifying this, in the latter 1990s I'd mentioned a forthcoming trip to Finland for a research collaboration, and Sol, appealing solely to his Brain for reference materials, promptly penned a multipage suite of rules about Finnish (adding key vocabulary). I showed it to my academic host in Finland: She said in amazement that all his rules were absolutely correct – but she had never before been aware of them. And Finnish was a language Sol did *not* know. (Indeed, unlike other Scandinavian tongues, Finnish is not even of Proto-Indo-European descent, but is from the quite distinct Finno-Ugric language family.)

Sol's friend Robert Hanisee – who served with Sol on several Boards of Directors – retains the extensive multipage language primers – in Japanese and in Hebrew – that Sol composed for *him*, prior to trips *he* took, to Japan and Israel.

The Al-gorithm

In Sol's address (to follow), he references an algorithm named for Xiao and Massey, which his own work preceded (by a quarter century). Sol had laid out the basis for what might rather be called the Golomb algorithm, in a book of which both Xiao and Massey had copies. Sol notes that he had omitted *direct* reference to the cryptanalytic implications in the chapter, because at the time he *could* not mention this.

What Sol did not state in his talk, but had shared with me, is that Sol could not mention this because he had elucidated it (in the 1950s) in a classified document written for a government agency that no longer exists (complicating the process of requesting declassification). I have asked Sol's friend and colleague Al Hales (who had worked at JPL – then UCLA Mathematics, now IDA/CCR-La Jolla) if there was a process to request declassification in such a case.

Al Hales, bedight in mathematical tie and pentomino pin (bestowed on those at Sol's 80th)

Al generously undertook the challenge and has made notable progress: A copy of a 1959 classified report by Sol, that there is cause to believe purveys that algorithm and (explicitly) its relation to cryptography/ cryptanalysis, has been located at an NSA contractor's facility. Hales has sent a request to the person at NSA who oversees requests for the declassification of mathematical items. The assessment for declassification must evidently await digitization of the archive now on microfiche, to occur perhaps this year (2018).

Harvard "Memories"

Sol with Howard Laitin at Sol's 80th

Per Howard Laitin*, Sol was easygoing, funny, very sociable – the "anti-stereotype" of a Harvard genius, in their mutual time sojourning at Harvard for graduate school; they met in 1952. He underscores that these tales just cannot capture how impressive the feats were. He emphasizes that Sol did not put himself forward, but was kindly as people sought to challenge his memory.

Students in the dorms all ate in the Commons dining room – science and social science, law students etc. A "game" evolved in which a student would bring their language dictionary – Greek, Russian, French, German, Spanish and maybe Latin were represented – and select any page. Sol would methodically peruse the left then right leaf, and return the tome. His student interrogator would ask "what is the fifth word in the 2nd column on the left page" (or whatsoever they chose) and Sol would correctly recite the word and its definition. "Sol never wavered." Remarkably, a question could be posed from a *previously* viewed page/dictionary, seen perhaps 3 months earlier - with intervening dictionary games. "Without hesitation Sol would perform perfectly. This made Sol's feat even more amazing."

Laitin states Sol's memory was photographic, though to my knowledge Sol did not claim so. (Per Sol's sister Edna, by age 2 his parents enjoyed showing him off to guests as Sol recited books verbatim, pretending to read them, and turning each page at the correct word. This may be a nonrare stage in learning to read, but not by age 2. Also, until old age Sol could recite the year, date, day of the week, and other events of the day, for meetings, visits and events he attended decades previously. Clearly other memory gifts, beyond visual, were also singular.)

As one Harvard pleasure, Sol and Laitin would traipse from room to room in the graduate dorms, eliciting from each student their special topic of study. As discussion progressed (Laitin states), irrespective of their field, it was soon apparent that Sol knew more about each student's specialty than they did.

Per Laitin, Sol's aid to others was also evident, assisting PhD efforts in fields as diverse as Middle English, history, genetics, and physics. For travails referred through Laitin: when the student hit what they viewed as an insurmountable analytical and/or directional obstacle in their PhD research process, Laitin would question them sufficiently to "sort of understand their problem," then seek out Sol "who always came up with brilliant solutions." Laitin would "sort of explain" the solution to the party in peril. "The results were always much more outstandingly favorable than the questioner had ever imagined."

*PhD Econ; was Chief Scientist at Hughes Aircraft & at Raytheon, and consulted for RAND, HHS, DOD, GAO, OMB, etc.

Courtroom Panache

In a major lawsuit in which Qualcomm battled Swedish company Ericsson over intellectual property, a lawsuit that could have changed the course of history had it gone less favorably for Qualcomm, Sol served as an expert on the Qualcomm side. Sol, who relished games like chess – battles of wits – enjoyed himself immensely. He shared droll courtroom events contemporaneous with their occurrence. (Toby Berger, also on Qualcomm's team, tells his own overlapping tales of Sol's courtroom virtuosity.) These help illumine the sorts of ways Sol's gifts regularly had real-world application. (Perhaps Sol's technical contributions were more important still – but for these, I would not have been the right audience.)

A stout stack of technical papers in Swedish required translation. (Google Translate and the like were years into the future.) A technical translation company *had* a Swedish translator, but assigned to another job. Sol said "I don't really consider Swedish among the languages I know, but maybe I can help." He hefted the stack home Friday, returning Monday with the articles arrayed in four piles. The largest bore those which, on review, Sol deemed irrelevant to the lawsuit. These Sol had not translated. The next comprised those that were somewhat relevant, for which Sol had translated the abstracts. The third were more germane, and these Sol translated in necessary depth. Of the smallest pile, Sol said: "These are in Norwegian, not Swedish; but they were not relevant."

Later, a new fat stack – this time in Japanese – required translation. The technical translation company didn't even have a Japanese translator. Sol said: "I don't really consider Japanese among the languages I know well, but maybe I can help." He heaved the stack home Friday, and returned Monday with the papers again in four groupings: not relevant, a bit relevant (abstract translated), more relevant (more in depth translation), and the smallest, a mere couple papers, of which Sol said: "these are in Korean, not Japanese; but they were not relevant."

Through discovery, Qualcomm had access to Ericsson emails – but in Swedish. Sol scrutinized these and identified internal statements at odds with courtroom claims by Ericsson – manifestly helpful to Qualcomm's case. Swedish is little known outside Sweden, and Sol did not broadcast his understanding of the language. Ericsson lawyers thus spoke freely in his presence, enabling Sol to share with Qualcomm counsel key Ericsson statements and strategies. (On Sol's final day in court, Sol, with a twinkle, addressed the Ericsson lawyers in Swedish – to their manifest alarm.)

With Sol on the stand, Ericsson claimed the first page of their patent bore a mistranslation: "conventional" should have been "convolutional." Sol, having read the patent application, responded that the word "conventional" was also used on page 6 (or whichever – mine is not Sol's memory); was that instance too meant to be "convolutional"? This led to visible confusion among the Ericsson lawyers, necessitating a break as they settled their response.

The Ericsson lawyer asked Sol a question about time-varying nonlinear convolutional codes. Sol asked in which sense the lawyers meant this. The lawyers expressed confusion with the question. Sol stated "It's like 'pretty little girls school': Which term does "pretty" modify? Is it pretty little, pretty girls, or a pretty school? Are they little girls, or is it a little school?" This precipitated more commotion on the Ericsson side, and I believe, another time out.

Sol had averred that some quantity (call it B) was one fourth more than another (A). On cross-examination, Ericsson parried that per their mathematician expert, Sol's assertion was incorrect: The actual ratio of A to B was 12 to 15. Sol responded "last I heard, 5/4 is the same as 15/12." The Ericsson lawyer (and their mathematician expert) were acutely embarrassed.

The judge appeared hugely amused.

Toby Berger and Sol each shared this event: The Ericsson lawyer, bested repeatedly in his efforts to rout Sol, finally decreed "no further questions." The judge was heard to audibly intone: "that's the smartest thing you've said."

Sol's memory came into play, in a stage of the trial in which Sol was not himself testifying. A mathematician expert for Ericsson made an assertion adverse to Qualcomm's position. Sol had, years previously, perused a book by that expert, and recalled a specific relevant statement. He conveyed the book thither. Qualcomm counsel displayed an assertion to the court in direct opposition to the expert's testimony and asked if the expert agreed with it. "Absolutely not" he said – whereupon its source – it was verbatim from the man's own book – was displayed.

Sol, always kindly (though enjoying intellectual digladiation), felt genuinely sad that the other mathematician viewed him less benevolently after these events.

Solomon Golomb and Robert Scholtz, Powell Hall, USC Aug 20, 1973. Courtesy of USC.

Sol with USC EE colleague Lloyd Welch

A Career in Engineering

Sol, I believe in his digs in Boston, during the 1959-60 time Sol notes the Jacobs' were close friends

Unidentified JPL object

Sol at home with JPL Object (Mariner 2 per Wolfram) Jan. 1963

Sol + Object (+ Epsilons), June 1963

II. A Career in Engineering

By my senior year in high school, I had decided that I wanted to get a Ph.D. in mathematics. My special interest was prime number theory. In 1949, when I was 17, I submitted my first to-be-published result as a Problem, to the *American Mathematical Monthly*:

"Prove that there are infinitely many twin primes if and only if there are infinitely many positive integers n not of the form $6ab \pm a \pm b$, where a and b are positive integers, and all combinations of the $\pm$ signs are allowed."

This problem appeared in the *Monthly* in May, 1951, and had only one solver besides the Proposer. This is a very elementary result that merely restates the problem, but it has been subsequently rediscovered, and published, usually by people with the delusion that it will help them prove the still-unsolved "twin prime conjecture" that for infinitely many prime numbers p, the number $p + 2$ is also prime.

I was an undergraduate at the Johns Hopkins University from September, 1949, to May, 1951, a period of 21 months, taking my last final exam just before my 19th birthday, graduating with an A average, Phi Beta Kappa, with "general and departmental honors", with a BA in mathematics. (I met the full "units" requirement for graduation, by taking twice as many courses as everyone else, never taking notes but instead paying close attention in class, and maintaining a strict, efficient schedule.)

The summer of 1951 was the first of four summers that I worked in the "Systems Engineering" Section of the Glenn L. Martin Co., at its original location in Middle River, Maryland, just east of my home town of Baltimore. That first summer, I worked in a Controls group, and was exposed to Nyquist stability criteria, Laplace transforms, and poles in the left half plane. The subsequent three summers I worked in a Communications group at Martin, a subject I found much more interesting.

Having been a "graduating sophomore", I hadn't applied to graduate schools, so I spent academic 1951-52 as a math grad student at Hopkins. That year was more valuable than I realized, since the emphasis at Hopkins was more specific and less abstract than what I encountered the next three years at Harvard, where I found it a great advantage (for my way of learning) to have specific examples of abstract concepts. At Harvard, most math grad students learned "Abstract Harmonic Analysis" without knowing what a Fourier series was!

I continued to study "prime number theory" on my own. My most important discovery was a formula involving the "von Mangoldt function",

$$\Lambda(n) = \begin{cases} \log_e p, & \text{if } n = p^k, p \text{ prime, } k \text{ a positive integer} \\ 0, & \text{otherwise} \end{cases}$$

which plays a key role in all known proofs of the fundamental "Prime Number Theorem", that $\pi(x)$, the number of prime numbers $\leq x$, satisfies

$$\pi(x) \sim x/\log_e x \text{ as } x \to \infty$$

Since

$$\log_e n = \sum_{d|n} \Lambda(d)$$

where the summation is over all positive integer divisors d of n,

by a classic process called "Möbius Inversion",

$$\Lambda(n) = \sum_{d|n} \mu(d) \log_e d,$$

where the "Möbius function" is defined by

$$\mu(n) = \begin{cases} 1 & \text{if n} = 1 \\ (-1)^k & \text{if n is a product of k } \textit{distinct} \text{ prime} \\ & \text{numbers} \\ 0 & \text{if n is divisible by the square of any prime} \\ & \text{number} \end{cases}$$

My new result, now called "Golomb's Lemma", states

$$\prod_{i=1}^{k} \Lambda(a_i) = \frac{(-1)^k}{k!} \sum_{d|A} \mu(d) \log_e d, \quad \text{where } A = \prod_{i=1}^{k} a_i$$

when all $a_i > 1$ and when g.c.d. $(a_i, a_j) = 1$ for all $i \neq j$.

In recent years, "Golomb's Lemma" has been used by others to get new results on the sometimes-closeness of prime numbers; first, about ten years ago, in a theorem of Goldston, Yildirim and Pintz; and more dramatically, in 2013, by Yitang Zhang, who proved that infinitely often, there are two prime numbers within a bounded distance B. Zhang's original result had $B = 70{,}000{,}000$; but a world-wide effort, and some new ideas, has reduced B to 246, much smaller than 70 million, but still much larger than 2.

PROBLEMS IN THE DISTRIBUTION OF THE PRIME NUMBERS

A thesis presented by

Solomon W. Golomb

to

The Department of Mathematics

in partial fulfillment of the requirements for the degree of Doctor of Philosophy in the subject of Mathematics

Harvard University
Cambridge, Massachusetts

May, 1956

Figure 1. The title page of my Harvard Ph.D. thesis.

At both Hopkins and Harvard, I had absorbed the then-prevalent notion, expounded in G.H. Hardy's influential 1940 book *A Mathematician's Apology*, that "good" mathematics has to be "pure", and that "applied mathematics" is "uninteresting" mathematics. Yet, at the start of my fourth summer at the Martin Co., in June, 1954, the leader of the Communications Group, Thomas E. Wedge, introduced me to two things that changed my life. He had attended a short course at MIT on a recent new subject called "Information Theory". I read an early book on the subject by Woodward, titled *Probability, Information Theory, and Applications to Radar*. I already knew everything he wrote about probability, and discovered "Information Theory" to be a natural extension of probability theory; and I was surprised to discover that "radar" had some mathematical structure.

Tom Wedge also told me about a problem that was discussed at a meeting he attended. He described it as involving a "tapped delay line with feedback"; but I would call it a binary linear shift register with feedback. For some combinations of taps along the delay line, which were then summed modulo 2 and fed back into the delay line, very long binary sequences would be produced, but for other combinations of taps, much shorter output sequences resulted. Was this a problem I could help with? I said I thought so, and that was how I got involved with shift register sequences. My approach was to look at power series generating functions over finite fields, which yields more information than other linear algebra (matrix theory) approaches. This was the first of many examples where the "pure mathematics" I had been studying was directly applicable to an important applied problem. My career has been largely defined by being told of important practical problems that I had the right tools from "pure mathematics" to solve. Most of these problems have been related to communications (or radar) and as both computation and communication have become digital, my knowledge of "discrete" mathematics has been a key.

In addition to prime numbers and shift registers, my years at Harvard (1952-1955) involved one more essential component to my career. I generalized a puzzle problem about putting dominoes on a checkerboard from which a pair of opposite corners had been removed, and created the subject of *Polyominoes*. My talk, in November, 1953, titled "Checkerboards and Polyominoes", was awarded the Rogers Prize in Mathematics, and became my first published article, in the *American Math Monthly*, December, 1954. My subsequent (1965) book *Polyominoes* (Charles Scribner Sons) appeared in a British edition (George Allen and Unwin), a Russian translation, a second (Princeton University Press) hardcover (1994) and paperback (1996) edition, and a Japanese translation (2014); and a soon-to-exist third (Princeton) edition. {BG Comment: Because of his death, this did not attain publication.}

Martin Gardner featured my Polyominoes in his May, 1967, "Mathematical Games" column in *Scientific American* magazine (and in numerous later articles), giving them a worldwide audience. A young Russian, A. Pazhitnov, reading the Russian translation of *Polyominoes*, was inspired, using my

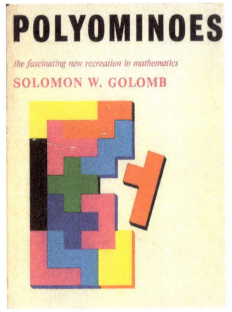

Figure 2: Polyominoes. Original 1965 Scribner's Edition   Figure 3: British "Allen and Unwin" Edition.

A Career in Engineering

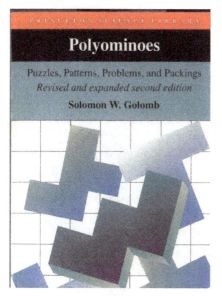

Figure 4: Second Edition, 1994, Princeton University Press, Hardcover.

Figure 6: Russian translation (from First Edition).

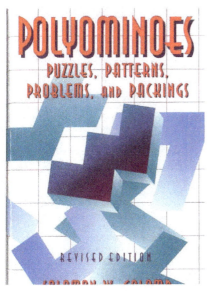

Figure 5: Second (1996) Edition in Paperback.

Figure 7: Japanese (2014) translation from Second Edition.

The Wisdom of Solomon

"tetromino" shapes, to invent the computer game *tetris*. More people have heard of me from polyominoes than from anything else I have done.

I had a consulting contract from the Martin Co. to continue my research on shift registers during my final year in residence at Harvard (1954-55), and in June, 1955, my opus *Sequences with Randomness Properties* appeared as a widely circulated Martin Company report. It subsequently became *Chapter 3* of my book *Shift Register Sequences* (Holden-Day, Inc., 1967; second edition Aegean Park Press 1982; soon to appear third edition, World Scientific Publishing Co.).

{BG update: The third edition was released in 2017. Of note, the Glenn L. Martin Company, for which Sol had worked, is now part of Lockheed Martin.}

Also in June, 1955, I left for a year at the University of Oslo, Norway, on a Fulbright Fellowship. The Math Department there had two very distinguished senior mathematicians, Viggo Brun (who had obtained the first interesting results on twin primes in the 19-teens, using Brun's "sieve method") and Thoralf Skolem (responsible for many deep results in algebraic number theory and about diophantine equations); and a younger distinguished mathematician, Ernst S. Selmer, who got interested in linear shift register sequences from me. We subsequently exchanged post-docs, when he became head of the Math. Dept. at the University of Bergen (Norway), which now houses the *Selmer Institutt av Informatikk* in his memory. When Andrew Wiles was trying to repair a serious flaw in his first attempt to prove Fermat's Last Theorem, the salvation came in the form of the "Selmer Groups".

When I returned from Norway, in July, 1956, with my Scandinavian bride, I interviewed at three places, and selected the Caltech Jet Propulsion Laboratory as a place to work. At JPL, the Telecommunications Section (Section 8) was headed by Dr. Eberhardt Rechtin. (I had met the previous head, Frank Lehan, when Lehan had visited the Martin Co. in the summer of 1954. Lehan (who remained a life-long friend and mentor) had left JPL to join a new company headed by Simon Ramo and Dean Wooldridge (later named TRW), but Rechtin had heard of me, and even visited me at Harvard in January, 1955.) I had been all over Europe before journeying further west than Frederick, Maryland, for my summer, 1956, interview at JPL.

I worked at JPL from August 1956, until January, 1963, when I left JPL for USC. My first two years, I was in the Communications Research Group, as a Senior Research Engineer. Lloyd Welch was in that group while working on a Ph.D. in Math at Caltech. In June, 1957, Andy Viterbi, with a fresh MS from MIT in Communication Engineering, joined the group, and I hired a Pasadena City College student, Harold M. Fredricksen, as a part time assistant.

The Russians launched Sputnik 1 on October 4, 1957, and the US Army was then given permission, denied by the Eisenhower Administration some 15 months earlier, to launch a US satellite. During a period of only 80 days from when we were told to start, with a booster rocket provided by Werner von Braun's group in Huntsville, Alabama, JPL provided the second, third and fourth stages, and the payload, for what became known as Explorer I, the first non-Soviet earth orbiting satellite. It was launched from Cape Canaveral, Florida, on January 31, 1958, and even though the configuration had never been tested, it succeeded on the first launch attempt! The payload included radiation detectors, designed and built in my lab, using my technicians (Yes, I had my own lab and technicians, where I experimented on finding the periods of nonlinear shift registers sequences) by a graduate student of James van Allen, and Explorer I discovered the "van Allen radiation belts".

In the special organization chart at JPL for Explorer I, our group was assigned to the "early orbit determination" of the satellite. That was an exciting adventure, too long to go into here. In February, 1957, I went back to Harvard to pass the Final Oral Exam for my Ph.D. in Mathematics. The "Committee" consisted of the entire Math Department, and the subject I was to be examined on was "all of mathematics". I had spent a month of intensive study of "All of Mathematics" and passed with little difficulty. (The details could fill another essay.)

In May, 1958, I was made head of a new group, called the Information Processing Group. JPL had been reorganized, and what had been Section 8 was now Division 33, headed by Rechtin, with two sections, 331 for Communication Systems, headed by Walter (Walt) Victor, and 332 for Antennas and Propagation, headed by Robertson (Bob) Stevens. I was in Victor's section, and in May, 1960, he promoted me to Deputy Section Chief, in charge of all the research, around 25 people.

During fall, 1959, I spent over 2 months on leave from JPL, visiting the Communications Group at MIT. I sat in on Claude Shannon's course on Information Theory, and had lunch with him three times a week. I even solved a math problem he asked me about. On several occasions I had lunch with Norbert Wiener; and I had frequent conversation with Peter Elias, Robert Fano, David Huffman, William Siebert and other members of the Communications Group headed by Fano.

I was in Cambridge, MA, with my wife Bo, and our two infant daughters, Astrid and Beatrice. Our closest friends socially were Irwin and Joan Jacobs. (Irwin had obtained his Ph.D. at MIT the previous June, and had stayed on for a year as an Instructor.) I invited Irwin to visit our group at JPL and when he did, the next summer, I introduced him to Andy Viterbi.

During fall, 1959, at MIT, there was a weekly seminar on the newly developing field of algebraic coding. Active participants included Neal Zierler, my friend from my frequent visits to MIT's Lincoln Labs during my last year at Harvard before Norway, and J. Wesley Peterson, who published the first book on *Error-Correcting Codes*. (This included several errors due to his unfamiliarity with finite-field arithmetic. In subsequent editions, Peterson was joined by Ned Weldon as co-author, and the book became more reliable.)

In fall, 1956, a few months after I arrived at JPL, Lloyd Welch told me about a problem he was asked to look into by a biology professor at Caltech named Max Delbrück (who later received a Nobel prize as the principal founder of Molecular Genetics, which revolutionized the entire field of biology). Francis Crick (not yet famous to the general public for the Watson-Crick model of DNA; Watson was a post-doc of Delbrück the year before he joined Crick in Cambridge) had an idea about the nature of the genetic code. An earlier model, proposed by the cosmologist George Gamow, had many attractive features, but was shown by an experiment (by Sidney Brenner, a later Nobelist), to not possibly be correct.

Crick's model became known as a "comma-free code", which said that if ABC and DEF were both "codewords" for amino acids, then the overlaps of two codewords, if ABC and DEF occurred consecutively, namely BCD and CDE, could not be codewords. Like Gamow's discredited model, Cricks model also had the "advantage" that using 3-letter "codewords" from the four symbols (A,C,G,T) of DNA, the 64 "triplets" could code for exactly 20 amino acids, believed (correctly) to be the number of amino acids to be needed for all living organisms.

In addition to Lloyd Welch, another Math Ph.D. student at Caltech, Basil Gordon (who was a fellow undergraduate of mine at Hopkins; and we first met on opposite sides of a chess board representing our rival Baltimore high schools) was also enlisted by Delbrück to look at "comma-free codes".

Instead of confining ourselves to the biological case of word-length $k = 3$ and alphabet size $n = 4$, we considered general (n, k) comma-free codes, and found bounds on the maximum size of comma-free dictionaries. Our first paper "Comma-Free Codes", by Golomb, Gordon and Welch, was published in 1958 in the *Canadian Journal of Mathematics*. But Delbrück urged me to look specifically at the case $k = 3$; $n = 4$. One of my results showed that in a maximum-sized 20 word comma-free "dictionary", in a sequence of dictionary words, the same symbol could never occur four times in a row.

The limited available experimental data showed that in actual DNA sequences, the same symbol *did* occur four times in a row, and that killed the comma-free hypothesis for molecular genetics. The second paper, "Construction and Properties of Comma-Free Codes", by Golomb, Welch, and Delbrück (Gordon, having completed his Ph.D., was by then serving as a Private in the U.S. Army) appeared also in 1958, in the *Biologiske Meddelelser af det Kongelige Danske Videnskabernes Selskab* (the "Biological Proceedings of the Royal Danish Scientists Society"); and USC Professor Robert Scholtz (the first speaker in this PIONEER series) {BG: the USC series at which Sol first gave this address}, discovered a construction for comma-free codes that achieves the upper bound for all *odd* word lengths $k \geq 1$, and all alphabet sizes $n \geq 2$. (The case of even word-length is much more complicated, and the "basic bound" on dictionary size, *achieved* for odd word-length, cannot be achieved for even word length and "large" alphabet size. How large is "large" for this purpose has been successively improved, from 3^j to $2^j + j$, to a polynomial in j (where $j = k/2$), but the numerical evidence suggests that "$n = 4$" is always large enough – a problem awaiting a solution worth a Ph.D. thesis.)

During fall, 1954, while at MIT, I revisited two friends at Lincoln Labs, Paul Green and Bob Price, who were the pioneers in "planetary radar". A couple of years earlier they reported detecting a radar signal bounced off Venus (using radar modulated by a linear shift register sequence of period $2^{13} - 1 = 8191$) and reported in *Science* magazine that the probability that the detection was spurious was less than 10^{-7}. However at the next (most recent) inferior conjunction of Venus (its nearest approach to Earth) they were unable to repeat the previous "success". I reported this when I got back to JPL.

Walt Victor was convinced that JPL had the technology for a successful radar detection of Venus at the next favorable inferior conjunction, in March, 1961. The problem was that NASA, formed months *after* Explorer I, and now overseeing JPL, had decreed that JPL was to do *engineering*, but not *science*. Eb Rechtin, ever resourceful, pointed out that JPL was preparing to launch a space probe to the vicinity of Venus, and for this purpose it would be useful to know where Venus actually was.

Another recent member of our Communications Group was Richard Goldstein. Like most of the researchers who did not yet have the Ph.D.s, he had enrolled in a Ph.D. program. In his case, it was in EE at Caltech. He proposed to his Advisor, Professor Hardy Martel, that he would like to make the radar detection of Venus his thesis topic. Martel was convinced that this would not succeed, and agreed that it *could* be used as a Ph.D. thesis, but with the caveat "no echo, no thesis".

Walt Victor appointed me "Project Leader" on the Venus Radar detection project. I learned that in this capacity my main task was to make certain that each engineer contributed a detailed write-up to the Final Report, which (realistically) listed on the cover front it was by Victor, Stevens, and Golomb. Yes, Venus was successfully detected, by JPL, using our 85-foot radio antenna at Goldstone, California, in the Mojave Desert, on March 10, 1961. (I wrote the Preface to the Final Report.)

The basic yardstick for the Solar System, known as the Astronomical Unit (the A.U.), defined as the mean value of the semi-major axis of the elliptical orbit of the Earth around the Sun, or "the distance from Earth to Sun" for short, had a value adopted by the Astronomical Community, computed on the basis of two occurrences of the "transit of Venus" (across the face of the Sun) in the mid-nineteenth century. (The next such transit occurred in June, 2012.) Our Venus experiment, which *directly* measured, by radar, the distance between Earth and Venus, showed that the "official" value of the A.U. was off by one part in one-thousand, a *huge* error, which we were able to improve to a value correct within less than one part in one million! (The previous "observation" by Lincoln Labs had "located" Venus tens of thousands of miles away from where Venus actually was. All the non-JPL radar attempts in 1961 reported values more-or-less consistent with the *wrong* value of the A.U. which they thought they were confirming. Further doppler radar observation of Venus showed that the text book assertion about the rate of rotation of Venus on its axis was also wrong.

In May, 1960, at a major meeting of the International Union of Radio Science (U.R.S.I) in Washington. D.C., I presented a major address, titled "The Role of Ranging Experiments in Space Exploration." (The original invitation had gone to JPL Director William Pickering, who passed it on to Eb Rechtin, who passed it on to me.) I had designed an interplanetary ranging system at JPL, based on binary phase modulation

The Wisdom of Solomon

of an RF carrier, that basically counted RF-cycles, and thus potentially provided extreme range accuracy. In my U.R.S.I. talk, I mentioned "planetary radar", on which we were already working at JPL. I also described a sensitive test of General Relativity that could be performed.

Suppose we had a Mars probe, with a transponder for our ranging signal. When the Mars probe reached the opposite side of the Sun from the Earth, there would be *two* relativistic effects. The strong gravitational field of the Sun would *bend* the path of the RF signal, increasing the path length, and therefore the count of the RF cycles. Also, the Sun's gravity would give more energy to the photons as they passed the Sun, and since these photons were already traveling at the speed of light, their velocity could not increase, but instead their *frequency* would increase, thus shortening their wavelength, and further increasing the number of RF cycles along the path. So these two relativistic effects would reinforce, giving a sensitive test of Einstein's Relativity. This experiment, as I proposed it in 1960, was carried out using the Mariner 9 Mars probe in 1969 and provided the most sensitive test (up to that time), verifying Einstein's predictions with $\pm 1\%$, and contradicting the alternative models of gravity proposed by several others.

There is also a priority dispute involving the relativistic calculations, by Duane Muhleman of JPL (and later of the Caltech Faculty), to whom I had explained this idea when it first occurred to me; and Irwin Shapiro of MIT's Lincoln Labs, who Muhleman is convinced got the idea from him when Muhleman went to study for his Ph.D. at Harvard. I will omit the gory details.

At JPL, I did extensive research on both linear and nonlinear shift register sequences. I combined a number of papers to form the chapters in my book *Shift Register Sequences*, which first appeared in 1967. Chapter 8 was the text of a talk I gave at an Information Theory meeting held at UCLA in September, 1958, and which appeared in two IEEE TRANSACTIONS (on Information Theory and on Circuit Theory). It was titled "On the Classification of Boolean Functions." What it really described was the vulnerability of sequences from nonlinear shift registers to cryptanalysis based on their Boolean feedback functions, but in 1958 I couldn't mention that application explicitly.

Some 25 years later, my analysis was "rediscovered", in a paper by Guo-Zhen Xiao and James Massey, both of whom owned copies of my *Shift Register Sequences* book. While I believe there was no conscious plagiarism, there are numerous terms and themes in their paper that suggest that what became known as the Xiao-Massey Algorithm was simply a republication of my work.

In studying the sequences from non-linear shift registers, I used the behavior of "random permutations" as a model. One question I asked was "In a random permutation on n symbols, what fraction of the symbols are on the longest cycle?" I showed that this fraction has a limit λ as $n \to \infty$ and this $\lambda = 0.6243296 \ldots$ Following Donald Knuth, this λ is now known as "Golomb's Constant." It also occurs in several other statistical settings.

While at JPL, I taught part-time, first in the evening program at UCLA, where one of my students in a course I taught on

Combinatorial Analysis was Aviezri Fraenkel. After taking my course, he switched his major from Electrical Engineering to Mathematics (my course was not the main reason, he tells me), and his speciality became the mathematics of full-information games. He became a professor in the Applied Mathematics Department in Israel's Weizmann Institute of Science, where I have visited him on numerous occasions. (In June, 2014, presenting a paper at a session at Bar Ilan University on Mathematical Games to which he had invited me, I learned that he was the grandfather of Naftali Fraenkel, one of the three kidnapped Israeli teenagers (subsequently found murdered).) {BG: these events had been much in the news when Sol gave his USC address.}

The Caltech Electrical Engineering Department had no regular faculty in Systems Engineering; so I volunteered to teach their first ever course in Information Theory. This was deemed to be part of my JPL job, so I got no extra pay for this, although my workload at JPL did not decrease.

Two of my friends at JPL, Harry Lass and Carl Solloway, told me they were teaching in the evening program at USC, and suggested I contact someone at USC named Zohrab Kaprielian about this. Zohrab invited me to give a seminar talk to the USC E.E. Department, which I did in April 1961; and in fall, 1961, I started teaching at USC.

In Robert Vivian's History of the School of Engineering, he lists me as a "new faculty member" in EE, as of fall, 1961, which predates anyone still in EE/Systems in 1970.

In early 1962, while still at JPL, I chaired a special symposium at JPL on "Digital Communications." With the other speakers (Leonard Baumert, Mahlon Easterling, Jack Stiffler, and Andrew Viterbi) I co-wrote and edited the book *Digital Communications With Space Applications*, Prentice-Hall, 1964, the first book to have either *Digital Communication* OR *Space Communications* in its title!

At Caltech, I supervised two Ph.D. theses in EE, by Robert Tausworthe and Jack Stiffler. *De facto*, I largely supervised Leonard Baumert's thesis, but I told him it would look better on his résumé to have Marshall Hall, Jr., as his advisor of record. Also, Andy Viterbi's thesis work was closely linked to what he was doing for me at JPL, but I was not yet connected to USC, and his advisor of record was Greg Young.

At least a dozen members who were in the Communications Group at JPL between 1954 and 1965 were ultimately elected to the National Academy of Engineering, and several of us to the National Academy of Sciences as well. The newest list of recipients of the National Medal of Science (yet to be officially awarded) includes the third member from that JPL group, Thomas (Tom) Kailath, who joins Andy Viterbi and me as recipients.

By the fall of 1962, I decided it was time to leave JPL. (Eb Rechtin had left to take a position at the Pentagon; Andy Viterbi, with a fresh Ph.D. from USC, had taken a faculty position at UCLA; and others had also left. Gus Solomon had gone to TRW. Gus Solomon and Ed Posner would surely have been elected to the NAE had they lived longer.)

I got offers from the three schools where I had taught (UCLA, Caltech, and USC). I also had offers from Industry, at up to twice the academic salary offers. The question I

A Career in Engineering

asked myself was not "Which place could do the most for me?", but "Where do I have the best chance to make a difference?"

My long-term intention had always been to be at a University. Caltech and UCLA had far better reputations, and USC at that time was completely unranked. But I had a two-hour interview with USC President Norman Topping, and it was clear he had a vision of turning USC into a major research university. And my immediate boss would be Zohrab Kaprielian, ambitious for himself, for the EE Department, and for all of USC, who was a master of persuasion and a genius at getting things done. From my point of view, USC had other advantages. At Caltech, professor Charles Pappas (from whom Zohrab had gotten his Ph.D. at UC Berkeley) told me, "At Caltech, no one has any influence if they don't at least have a Nobel Prize." At UCLA, I was told, "The UC Bureaucracy is such that no one ever has any ability to affect anything."

I joined the USC faculty full-time, for the Spring Semester of 1963, and I have never looked back. With Zohrab's encouragement, I started recruiting right away. I visited Stanford, where I had a friend, Norman Abramson, on the faculty. (His book on *Information Theory* had made extensive use of the course notes I had sent him.) He strongly recommended an imminent Ph.D. student, Bob Scholtz, and Zohrab agreed to extend an offer. I had first met Irving Reed, briefly, when he was at Lincoln Labs, and after he came to Southern California, I had tried, unsuccessfully, to get him to come to JPL. But he *did* agree to join me at USC. Reed and Scholtz were both already on the faculty for fall, 1963. Lloyd Welch had left JPL to work at the Center for Communications Research in Princeton, a high-level mathematics research entity in support of the National Security Agency. Lloyd, originally from the Upper Peninsula of Michigan, didn't mind the New Jersey climate, but his wife Irene couldn't stand it. (She grew up in Glendale, CA.) I had visited Lloyd in Princeton on numerous occasions, and told him I thought I could get him a faculty position at USC. The biggest obstacle was his dearth of unclassified publications in the open literature. But I persuaded Zohrab, and Lloyd was hired.

I spent my first Sabbatical, a full academic year (1972-73) visiting the EE Department at Caltech (which shortened my commuting distance from my home in La Cañada). There, the two people I interacted with the most were John Pierce (who had just came to Caltech after retiring as Executive Director for Communications at the AT&T Bell Telephone Laboratories), and Joel Franklin in the Applied Math Department, who had been a consultant to my group when I was at JPL.

I was surprised to learn, in March, 1976, that I had been elected to the relatively new (then less than 20 year-old) National Academy of Engineering. I later found out that my nominator, Frank Lehan, had gotten Bill Pickering, John Pierce, and Eb Rechtin as "References", a list that could have elected almost anyone.

Within the next two or three years, I successfully nominated Lloyd Welch, Irving Reed, Andrew Viterbi, and Elwyn Berlekamp (who spent a month at USC between learning Bell Labs and starting on the faculty at UC Berkeley). (I later successfully nominated more than a dozen others from USC, including Bill Lindsey, Bob Scholtz, Len Silverman, Steven Sample, Max Nikias, and Yannis Yortos.)

I helped to recruit more people to USC. From JPL, I helped to bring Bill Lindsey, and Janos Laufer (who became chair of the Aeronautics Department). I also helped to recruit Physicist Sergio Porto, whom I had known at Hopkins, who later left USC to become President of the University of Campinas in his native Brazil. I also took Zohrab to meet Max Delbrück at Caltech, who had a list of suggestions of possible faculty to hire in "molecular genetics". Zohrab had looked for areas where Caltech and UCLA were weak, and had built EE at USC around Systems and Solid State Physics. He decided that Caltech (especially) and even UCLA were strong in molecular biology, and he would look for other areas in Biology. That was a mistake from which USC has never fully recovered. There *is no other area* in Biology; and in a field where dozens of Nobel Prizes have been awarded, and which dominates the *Proceedings of the NAS* (and *Science* and *Nature*), USC does not even have a single NAS member in experimental molecular biology.

(Zohrab made a similar dubious decision about Physics - observing Murray Gellmann and Richard Feynman at Caltech, and Julian Schwinger at UCLA, all Nobel laureates in "high energy physics", Zohrab decided to look for *other areas* of physics; but the bulk of physics Nobels go to high-energy physics.)

At USC, I had more time to concentrate on my research, and writing papers for publication. An early USC paper, "Run-Length Encoding," became a widely used "lossless" data compression technique, adopted, for example, for sending back scientific data from the Rover Mars landers. A pattern I devised, originally for coded pulse radar, was popularized in Martin Gardner's column as "Golomb Rulers"; and these have been applied in fields as far ranging as X-ray diffraction crystallography, and how to arrange the spacing of arrays of radio telescopes to maximize the information obtained from stellar radiation.

About 38 years ago, I received a letter (before the age of email) from a famous communications engineer at General Electric named John Costas. He had been trying to construct frequency hopping patterns for radar and sonar that would have optimum (inverted) "thumb-tack" ambiguity functions, but had only been able to construct examples up to size 12 x 12.

I named these patterns "Costas Arrays", and I discussed them with Lloyd Welch, my former post-doc Abraham Lempel, and my Ph.D. student Herbert Taylor. The two major infinite families of examples are now known as the "Welch constructions" and the "Golomb constructions". All the systematic families of constructions are based on "primitive roots" in "finite fields" – elementary concepts in the type of mathematics I was interested in since high school, but not part of the usual curriculum of engineers or even of applied mathematicians.

In early 1990, I spent part of a Sabbatical visiting the Electrical Engineering/Systems department at Tel Aviv University. There, Professor Nadav Levanon, son of a former mayor of Tel Aviv, and (both then and now) Israel's leading expert on all aspects of radar, told me of a problem that the world-wide radar community was wrestling with. I quickly recognized that I already knew the solution (in another context) and this led to a major article I published in the *IEEE Transactions on Aerospace and Electronic Systems*.

When I visited Tel Aviv University in June, 2013, Levanon showed me an ultra-sensitive radar they had implemented using these "Golomb codes"; and had another group demonstrate an application of these codes to optical communications, that achieved a 10dB improvement over the previous "standard method".

Also about 30 years ago, I invented a family of combinatorial designs that I named "Tuscan Squares". I was able to find a possible application to hold signal interference to a minimum in a "multi-user" environment, but this is a difficult application to describe.

In the summer of 2009, I was in Osaka, Japan, attending an annual meeting of puzzle enthusiasts, where I am known (exclusively) as the "inventor of polyominoes". (My invited talk was titled "PORIOMINO NO REKISHI", i.e. "The History of Polyominoes.") There, another puzzle enthusiast, Dr. Richard Hess (BS, Caltech; Ph.D. U.C. Berkeley, all in Physics) mentioned a puzzle-type problem he was having trouble with. There is a group of n people, and on consecutive evenings at dinner, you want each person seated next to every other person (at least once). How can this be achieved in a minimum number of evenings? The solution, I quickly realized, is a "natural" use of Tuscan squares, both for a King Arthur-type round table, and for a lunch-counter type of table with two ends. The article I wrote about this, listing Dr. Hess as a co-author, was accepted for publication by the Canadian journal *Ars Combinatoria*, but because of its heavy backlog, the article is not scheduled to appear until early 2015. {BG update: This article came out in 2016; volume 129, pp. 397-402.}

I could give many more examples where someone asked me about a problem, and I happened to have the right tools for a solution. The kind of "pure" mathematics I was most interested in, and which I learned extensively (often on my own) while a graduate student, turned out to be ideally suited for addressing many of the practical problems associated with the newly emerging fields of digital computation and digital communication. I never expected to be elected to the National Academy of Sciences, because what I worked on didn't neatly fit into any of its numerous sections. But around the year 2000, they formed a new section, number 34, for "Computer and Information Science". That was a good fit, and I was elected in 2003.

Many other awards, honorary degrees, etc., have followed. Probably the most impressive is the U.S. National Medal of Science, which was presented to me by President Obama in a colorful White House ceremony on February 1, 2013.

Acknowledgments: Guang Gong performed the initial transcription of Sol's talk from his pdf. Beatrice Golomb, with Leeann Bui and Hayley Koslik, proofread against the original. Photos with Astrid are omitted from the Prologue, at her preference. (She is the most photogenic.)

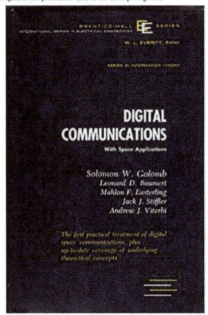

Figure 9: This Prentice-Hall book originally published in 1964 was the first book to feature either "Digital Communications" or "Space Communications" in its title

A Career in Engineering

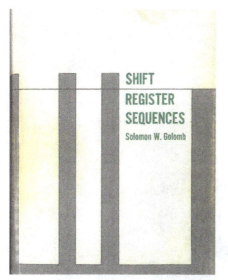

Figure 10: First edition (1967) of my Magnum Opus on Shift Register Sequences.

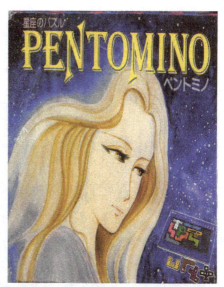

Figure 12: Box Cover of the Prior Picture. (There are some 50 different Pentomino sets that have been sold world-wide.).

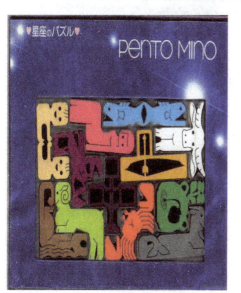

Figure 11: A Japanese edition of the 12 pentominoes (as signs of the zodiac).

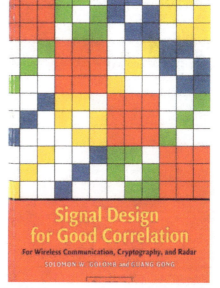

Figure 13: My most recent book, with former post-doc (now Professor) Guang Gong, Cambridge University Press, 2004.

Figure 14: With President Obama, for receipt of the National Medal of Science (Feb 1, 2013)

A Career in Engineering

Supplement. *The following is excerpted from a separate talk entitled "My Involvement with Shift Register Sequences," written by Solomon W. Golomb in March 2016, in relation to Sol's Franklin Medal Award (April 2016). It is for a different audience, so bears overlap with the above. But it provides further material of interest: It describes Sol's consideration of the National Security Agency or Lincoln Labs in lieu of JPL. It illustrates Sol's ready appeal to foreign language texts for his mathematical work (here citing yet a further foreign language), and provides another instance of Sol's then frequently-manifest real-time mathematical mental prowess (via his provision, on request, of the prime factorization of $2^{20} - 1$; he takes considerably longer demystifying his process than the moment or so it would have taken him to provide the answer). It discusses the launch of the first U.S. satellite (within 80 days of the OK to proceed on a satellite project), and its occurrence prior to the existence of NASA; the transfer of JPL from the U.S. Army Ballistic Missile Command to NASA, and the shift of responsibility for booster rockets into space from the Army's successful program to the Air Force. It characterizes the difference between radar signals used in the successful JPL radar detection of Venus, vs what proved to have been the false Lincoln Lab Venus detection.*

...in June 1951, I started my first of four summers working at the Glenn L. Martin Company (now part of Lockheed-Martin) at its original location, just east of my native Baltimore. That first summer I was in a Controls Group, where I learned about Nyquist diagrams and stability criteria (e.g. no poles in the right half plane).

My summers at the Martin Company coincided with my graduate work in pure mathematics at Harvard, where my professors proudly declared that nothing they were teaching had any possible practical applications. I've spent the past sixty-plus years proving them wrong. But this was the result of my summer jobs in the Systems Engineering Section of the Martin Company, at the dawn of the digital revolution.

Starting with my second summer at the Martin Company, I was in a Communications Group, which I discovered was much more to my liking. The group leader, Thomas E. Wedge, and another member, attended a short course at M.I.T. on a new subject, called Information Theory, introduced by Claude Shannon, at Bell Labs, in 1948. At Wedge's suggestion, I read a book titled *Probability and Information Theory, with Applications to Radar,* by an Englishman named Woodward. I already knew the probability part, and found that information theory was a natural extension of probability theory and I was surprised to learn that radar had a mathematical basis.

I think it was the summer of 1953 when Tom Wedge asked me, for no obvious reason, if I could tell him the factorization of $2^{20} - 1$. I said O.K., and mentally noted that $2^{20} - 1 = (2^{10} + 1)(2^{10} - 1) = 1025 \times 1023$. So I started with $5^2 \times 41$ (for 1025), and further mentally noted that $2^{10} - 1 = (2^5 + 1)(2^5 - 1) = 33 \times 31$, and then added that the remaining factors were 3, 11, and 31. He was sufficiently impressed to tell me what motivated his question.

He had recently attended a meeting where some people were talking about an interesting method for deterministically

generating seemingly random sequences of 1's and 0's. It involved having a tapped delay line, where 1's and 0's were running along the line, and at two integer points, the contents were sampled, or tapped, and added together with a half-adder (mathematically, a modulo-2 sum), and this sum was then fed back into the front end of the delay line. Depending on the positions of the taps the resulting sequence was always periodic, but for mysterious reasons, the period was sometimes fairly short, but sometimes as long as 2^n where n was the length of the delay line. It would be useful to be able to predict the period of the sequence, given the positions of the taps. Was this something I thought I could work on?

Of course, what he described was what we would now call a binary linear shift register with feedback. That was my first introduction to the subject! I said sure, that was a problem I was ideally suited to work on. I quickly realized that the model for this problem was the behavior of polynomials over a finite field, especially over the 2-element, or binary field. I found a 19th century book, in German, by Fricke, at the Hopkins Math library on this subject. I also found a section, a few pages long, in Volume II of B.L. van der Waerden's *Moderne Algebra* (also in German).

During academic year 1954-1955, while I was back at Harvard, I had a consulting contract from the Martin Company, to continue working on this subject, and to produce monthly reports. At the end of the academic year, I wrote a detailed Martin Company report titled *Sequences with Randomness Properties.* That report enjoyed a broad circulation, and ultimately became the basis for Chapter 3 of my 1967 book *Shift Register Sequences,* soon to appear in a third, updated edition with World Scientific Publications.

Soon after that report was finished, in June, 1955, I left for a year-long Fellowship in Oslo, Norway, where I continued to have a consulting agreement in this area with the Martin Company. In the previous (1954) summer at the Martin Co., there had been a meeting with someone from Lincoln Laboratories (Oliver Selfridge), and two people from the Jet Propulsion Laboratory (Frank Lehan and Henry Curtis) in far-off California, an organization I had not previously heard of.

During my subsequent year (1954-1955) at Harvard, I tried to visit Lincoln Labs once a week, where I interested Neal Zierler in shift-register sequences. He coined the term m-sequences for the binary linear shift register sequences having the maximum period of 2^n-1, and remained interested in this subject. Then, in January, 1995, I had a visit to my Harvard office from someone at JPL named Eberhardt Rechtin. He had heard of me and my shift-register work from Frank Lehan, whom he (Rechtin) had succeeded as head of Communications Research Section at JPL. He said he was interested in recruiting me for his section at JPL.

When I returned to the U.S. from Norway, in July, 1956, there were three places I visited for job interviews. One was Lincoln Labs, where I already knew many people. The second was the National Security Agency (NSA), still on Nebraska Avenue in Washington D.C., before they relocated to Fort Meade in Maryland, and where I had visited in June 1955, to give them a copy of my Martin Company report. And the third was JPL, in Pasadena, California, which at that time, before Sputnik and the creation of NASA, was supported by the U.S. Army Ballistic Missile Command.

All three places were interesting, but for me JPL was the obvious choice. I wanted to be able to publish what I did, and NSA would not allow that. Lincoln Labs was like a collection of very talented amateurs. I liked that working at JPL would be in support of a mission, but with very pleasant working conditions and broad freedom to decide how (and even when and where) to do your work. My original assignment at JPL was to work on secure radio-guidance for Army missiles. Linear shift register sequences were insufficiently secure against an enemy jammer, so my focus was on nonlinear shift register sequences.

I had been at JPL about 14 months when, on October 4, 1957, the Soviet Union launched Sputnik 1. Army General John B. Medaris, head of the Ballistic Mission Command, had proposed launching a US satellite (with a booster rocket from Werner von Braun's group at Huntsville, Alabama and a small payload from JPL) in September, 1956, more than a year ahead of the USSR, but had been forbidden to do so by the Eisenhower Administration. But a few weeks after the launch of Sputnik, Medaris was told he could begin work on a U.S. Army satellite. The new plan called for a booster rocket (a modified Redstone missile) from Huntsville, with second, third, and fourth stages (all solid-fueled) from JPL, and a payload from JPL.

For this project, a new temporary organization chart was created for all of JPL. Only 80 days after the authorization to begin was given, the new complicated satellite configuration, which had never even been tested, was successfully launched, on January 31, 1958, from Cape Canaveral, Florida. Miraculously it succeeded on the first attempt, and became known as Explorer I, the Free World's first artificial satellite. It had a radiation detector, assembled in my JPL lab, using my JPL technicians, by George Ludwig, a doctoral student of James A. Van Allen, and with it Explorer I discovered what became known as the Van Allen Radiation Belts.

Much later in 1958, the Eisenhower Administration decided to create a new organization, NASA (the National Aeronautics and Space Administration), and Vice President Richard Nixon was sent to JPL to tell us that JPL was being transferred from the U.S. Army to NASA. JPL Director William H. Pickering (himself a communications engineer) wisely decided to stake out unmanned exploration of the solar system as the future mission of JPL. Another decision, some months later, was to transfer responsibility for booster rocks into Space from the Army's successful Jupiter rockets program to the Air Force's usually failing Thor program. General Medaris was rewarded with another star, but his program was taken out from under him, and he retired.

I spent the fall of 1958 on leave from JPL, visiting MIT. When I visited Lincoln Labs, two of my friends there, Paul Green and Bob Price, were interested in radar detection of the planets. A couple of years earlier, using a radar signal modulated by an m-sequences of period $2^{13} - 1 = 8191$, they claimed to have detected a radar return from the planet Venus, and reported this in *Science* magazine.

But at the more recent inferior conjunction of Venus, they had been unable to repeat this success. I reported this to my immediate boss, Walter Victor, when I got back to JPL, and he said he was sure JPL had the technology to bounce a radar signal off Venus and detect it back on Earth. Victor's boss was now Eb Rechtin, head of the entire Telecommunications Division. NASA had decreed that JPL could do technology, but not science. {To circumvent this proscription,} Rechtin brilliantly argued that JPL was soon to launch a space probe toward Venus, and for that purpose, it would be useful to know where Venus actually was.

Walt Victor designated me as the project leader on the Venus Radar Project. It turned out what this meant was I had to make sure that everyone who worked on the project submitted their write-up for the Final Report on time. Using a continuous-wave (CW) radar signal, with a type of shift register modulation (rather than pulse radar, as Lincoln Labs had done), in April 1960, we detected a definite radar return from Venus, that enabled us to determine the distance between Earth and Venus, and from that, the Astronomical Unit (A.U.), a kind of average distance between from the Earth to the Sun. It turned out that the value of the A.U. accepted by the Astronomical Community was off by one part in a thousand, a huge error, which we were able to improve to an accuracy of better than one part in a million. This also showed that the earlier report of Venus radar detection by Lincoln labs was incorrect, because Venus wasn't close to where they said it was.

Using ideas I had developed from working with shift register sequences, I designed a high-precision interplanetary ranging system. I also realized that this ranging system could provide a sensitive test of General Relativity, as follows. My system basically counted the number of RF cycles in the signal as it made the round trip between earth and a transponder on the spacecraft. Now suppose a Mars-bound space probe was passing on the opposite side of the sun from the earth. There would be two relativistic effects. The gravity of the sun would bend the signal path, lengthening it, and thus increasing the number of RF cycles. But the photons of electromagnetic radiation would gain energy as they were pulled by the sun's gravitational field, and since already traveling at the speed of light they could not speed up, but instead would increase in frequency.

An increase in frequency means shorter wavelengths, thus further increasing the count of RF-cycles of the round-trip signal. I presented this proposed experiment in a talk I gave to a major Washington D.C. meeting of the International Union of Radio Science (U.R.S.I.) on May 2, 1960. The experiment as I proposed it was performed in 1969 using the Mars Mariner 9, and it verified the predictions of Einstein's General Relativity to within 1%, a far more accurate verification than any other test up to that time.

© 2023 World Scientific Publishing Company
https://doi.org/10.1142/9789811234378_0005

Golomb's Shift Register Sequences — Work with a Great Mind

Guang Gong

Department of Electrical and Computer Engineering, University of Waterloo
Waterloo, Ontario N2L 3G1, Canada
ggong@uwaterloo.ca

Sol Golomb was my supervisor, mentor, and long-term collaborator, and played an important role in my life. This article is written according to my talks at various occasions in memory of Sol for describing some experiences when I worked with Sol.

Keywords: Golomb, shift register sequences, randomness, 2-level autocorrelation, invariants, cryptography

1. Introduction

With unparalleled scholarly contributions and distinction to the field of engineering and mathematics, Sol's impact has been extraordinary, transformative and impossible to measure. His academic and scholarly work on the theory of communications built the pillars upon which our modern technological life rests.

— Yannis C. Yortsos
The Dean of the USC Viterbi School of Engineering

The content of this article is collectively written according to the following three talks. The first one, entitled "Golomb's Shift Register Sequences for Wireless Communication, Cryptography and Radar — Work with a Great Mind", which I gave at the *Memorial Session in Honor of Dr. Solomon W. Golomb* at the *International Conference on Sequences and Their Applications (SETA)* (organized by Tor Helleseth and myself), Chengdu, China, October 2016; the second talk "Golomb's Shift Register Sequences and Cryptography", at the *Golomb Symposium* at the *Event of Celebration of Life and Legacy of Sol Golomb* (organized by Alan Willner and myself), January 31, 2017, University of Southern California (USC), US; and the third talk "Golomb's Invariants and Modern Cryptology", at the event of *Remembering Sol Golomb* at *Information Theory and Application (ITA)*, February 2017, San Diego (organized by Tuvi Etzion).

2. Golomb's Discovery in 1954

How to generate random bits by a machinery method? This is the problem which was presented to Sol in June 1954 by Thomas Wedge, the leader of the Communications Group in the Glenn L. Martin Company (now Lockheed Martin) in Baltimore. Stephen Wolfram wrote the following appraisal at the opening of his blog "Solomon Golomb (1932–2016): The Most-Used Mathematical Algorithm Idea in History (May 25, 2016)" and later on collected in his book, titled as *Idea Makers*:

> An octillion. A billion billion billion. That's a fairly conservative estimate of the number of times a cellphone or other device somewhere in the world has generated a bit using a maximum-length linear-feedback shift register sequence. It's probably the single most-used mathematical algorithm idea in history.

The sequences generated by this machinery method are the so-called *linear feedback shift register sequences*. It works as follows. An n-stage linear feedback shift register (LFSR) consists of $n2$-state memory units (i.e., registers), numbered as $0, 1, \ldots, n-1$ with initially loaded n-bits. At each regular clock, it first computes the addition modulo 2, i.e., exclusive or (xor) of subsets of those n-bits (this operation is referred to as a *linear feedback function*), called *a feedback bit*, then it outputs the bit in register 0, and shifts the bit in the register 1 to register 0, register 2 to register 1, and so on. The bit in register $n-1$ is fed by the feedback bit. In this fashion, it outputs a bit stream. For example, a 3-stage LFSR is shown in Fig. 1.

For a 3-bit pattern, we have 8 choices. In Fig. 1, we load an initial state as

0	0	1

reg2, reg1, reg0 = (reg = register). At the first clock, it outputs 1, the content in reg0, and the feedback bit is 1, which is the xor of the contents in reg0 and reg1.

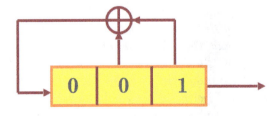

Fig. 1. A 3-stage LFSR.

The next state is updated as follows:

- the content in reg1 shifted to reg 0,
- the content in reg2 shifted to reg 1,
- and the content in reg2 is fed by the feedback bit.
- So, the next state is 100.

We continue this process to get a sequence, as shown below.

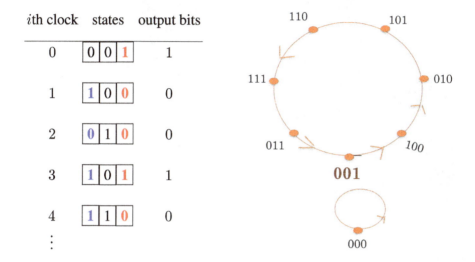

The red bit in the register 0 is the output bit, and the blue bit in register 2 is updated by a feedback bit. The output bits are 10010111001011.... This sequence is repeated after 7 bits. The right-hand side of the above figure is the state diagram, i.e., how a state is changed. For all 8 3-bit states, all nonzero 3-bit states are in one cycle and zero state is in another cycle.

If the feedback function is a nonlinear boolean function in n variables, the system is called a *nonlinear feedback shift register (NLFSR)*. For LFSRs, the question which Sol investigated is how to select a feedback function (i.e., the tap positions of the registers which contribute to the feedback bit) such that all the nonzero n-bit states will be in one cycle. In other words, it produces an output sequence with period $2^n - 1$. Such a sequence is called a *maximum-length LFSR register sequence*, shortened as an m-sequence. For example, the above LFSR generates an m-sequence of period 7.

> Result (Golomb, 1954) If a feedback function corresponds to a primitive polynomial, then it generates an m-sequence.

Those results are collected in Dr. Golomb's book, *Shift Register Sequences*, the first edition published in 1967, the second revised edition in 1982, and the third revised edition in 2017 thanks for Alfred Hales' efforts, for which the covers are shown in Fig. 2.

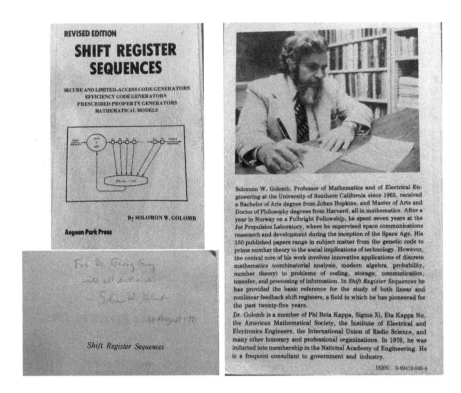

Fig. 2. The front cover and the back cover of Sol's book: *Shift Register Sequences*, the second revised edition, 1982, which are taken from the book presented to me by Sol on August 31, 1995, when we first met at USC about 3 months after I arrived US.

3. Golomb's Randomness Postulates

In the middle of 1950s, Sol also needed to answer the question how random are those sequences generated by LFSRs. Naturally, if one knows the current output bit is 0, the question asked would be: what is the probability of the next bit to be 1 or 0? If we wish those sequences to be random, then the probability of the next bit to be 1 or 0 should be $1/2$, i.e., it should be unpredictable. If the next bit being 1 occurs with probability $1/2$, then we further ask the question: what is the probability of k consecutive 0's or 1's for $k = 2, 3, \ldots$ (referred to as a *run* of 0's or 1's). How about if we compare the sequence of period N with its (non-zero) cyclic shift? What is the difference between the number of disagreements and the number of agreements (this leads to the definition of autocorrelation of a binary sequences)? Sol proposed three postulates for measuring the randomness of a sequence generated by a deterministic method (i.e., an algorithm), called *Golomb's three randomness postulates* since then. Intuitively, a sequence is qualified as a random sequence if the following Golomb's three randomness postulates are satisfied:

(1) Balance property (R1): the probability that each bit occurs is $1/2$ or equivalently, there are an almost equal numbers of 1's and 0's in the sequence;
(2) Run property (R2): k consecutive 1's or 0's occur equally likely;

(3) 2-level Autocorrelation (R3): the autocorrelation of the sequence at all (nonzero) cyclic shift is equal to a constant.

At this moment, we need a little bit of mathematical notation for autocorrelation. If we have a binary sequence $A = (a_0, a_1, \ldots, a_{N-1})$ with period N, then $(a_\tau, a_{\tau+1}, \ldots, a_{\tau+N-1})$ is a (cyclic) shift of τ from A where the indices are reduced modulo N. The autocorrelation of the binary sequence is defined as the sum (i.e., average) of a single bit $(-1)^{a_i}$ (transformed by mapping 0 to 1 and 1 to -1) multiplied the bit $(-1)^{a_{i+\tau}}$. In other words, the autocorrelation at shift τ is equal to the inner product of the mapped bit vector and the τ-shifted vector. For example, in the above example, $A = 1001011$ of period 7, so the autocorrelation of A at shift 1 is computed as follows.

$$
\begin{array}{ll}
A = \begin{array}{ccccccc} 1 & 0 & 0 & 1 & 0 & 1 & 1 \\ \downarrow & \downarrow & \downarrow & \downarrow & \downarrow & \downarrow & \downarrow \\ -1 & 1 & 1 & -1 & 1 & -1 & -1 \end{array} &
\text{1-shift of } A = \begin{array}{ccccccc} 0 & 0 & 1 & 0 & 1 & 1 & 1 \\ \downarrow & \downarrow & \downarrow & \downarrow & \downarrow & \downarrow & \downarrow \\ 1 & 1 & -1 & 1 & -1 & -1 & -1 \end{array}
\end{array}
$$

The autocorrelation value at shift 1 is given by

$(-1) \cdot 1 + 1 \cdot 1 + 1 \cdot (-1) + (-1) \cdot 1 + 1 \cdot (-1) + (-1) \cdot (-1) + (-1) \cdot (-1) = -1$

The other values of the autocorrelation function are all equal to -1 when τ is not a multiple of 7, as shown in the following figure.

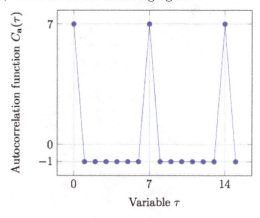

What are the pseudorandomness properties of binary m-sequences? Any m sequence generated by an n-stage LFSR posseses the randomness properties that one could hope for, i.e., all R1-R3, but it has the smallest linear span among all sequences of period $2^n - 1$, see Table 2.1. (*Note.* Linear span of a sequence is defined as *the length of the shortest LFSR* that generates the sequence.)

80   *The Wisdom of Solomon*

Table 2.1. Pseudorandomness properties of binary m-sequences of period $N = 2^n - 1$.

Golomb R1: Difference between number of 1's and 0's is 1.
Golomb R2: Each consecutive of 1's or 0's occurs equally likely except for the length $n-1$ and n.
Golomb R3: All autocorrelation values except at 0 are equal to -1.
Span n property (R4): Each nonzero n-tuple occurs exactly once.
Linear span: n, the number of stages in the LFSR.

The autocorrelation function of the m-sequence behaves like a Dirac delta function. This property was used in detection of signal sent back from Explorer I launched in 1958 as well as the distance calculation of the space probe that would arrive in the vicinity of Venus in 1961. Currently, there are many applications of autocorrelation of m-sequences such as spread-spectrum communication, code division multiple access (CDMA) cellular communication, location services provided by GPS, channel estimation in wireless communications, radar distance range, test vectors in computer hardware design, symmetric cryptography, random number generation, and so on, to just list a few.

The question Sol asked is whether there are other sequences having the same 2-level autocorrelation as m-sequences. The answer is a YES! All currently known 2-level autocorrelation sequences are collected in the book that I wrote with Sol. The book cover is designed by Sol, which is a Hadamard matrix of Williamson type with order 12.

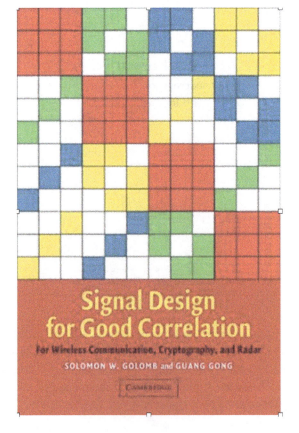

Golomb's Shift Register Sequences — Work with a Great Mind   81

viii   *Contents*

Appendix A: A Maple program for step 3 in Algorithm 3.1	72
Appendix B: Primitive polynomials	72
Appendix C: Minimal polynomials	77
Exercises for Chapter 3	79
4 Feedback Shift Register Sequences	**81**
4.1 Feedback shift registers	82
4.2 Definition of LFSR sequences in terms of polynomial rings	90
4.3 Minimal polynomials and periods	94
4.4 Decomposition of LFSR sequences	103
4.5 The matrix representation	106
4.6 Trace representation of LFSRs	108
4.7 Generating functions of LFSRs	112
Exercises for Chapter 4	114

Contents

Preface	*page* xi
Acknowledgments	xiii
Historical Introduction	xv
1 General Properties of Correlation	**1**
1.1 What is correlation?	1
1.2 Continuous correlation	2
1.3 Binary correlation	2
1.4 Complex correlation	3
1.5 Mutual orthogonality	3
1.6 The simplex bound on mutual negative correlation	4
1.7 Autocorrelation	6
1.8 Crosscorrelation	7
2 Applications of Correlation to the Communication of Information	**10**
2.1 The maximum likelihood detector	10
2.2 Coherent versus incoherent detection	12
2.3 Orthogonal, biorthogonal, and simplex codes	14
2.4 Hadamard matrices and code construction	15
2.5 Cyclic Hadamard matrices	18
3 Finite Fields	**22**
3.1 Algebraic structures	22
3.2 Construction of $GF(p^n)$	31
3.3 The basic theory of finite fields	34
3.4 Minimal polynomials	41
3.5 Trace functions	52
3.6 Powers of trace functions	58
3.7 The numbers of irreducible polynomials and coset leaders	69

5 Randomness Measurements and m-Sequences	**117**
5.1 Golomb's randomness postulates and randomness criteria	117
5.2 Randomness properties of m-sequences	127
5.3 Interleaved structure of m-sequences	135
5.4 Trinomial property	145
5.5 Constant-on-cosets property	148
5.6 Two-tuple balance property	152
5.7 Classification of binary sequences of period $2^n - 1$	155
Exercises for Chapter 5	159
6 Transforms of Sequences and Functions	**162**
6.1 The (discrete) Fourier transform	162
6.2 Trace representation	166
6.3 Linear spans and spectral sequences	174
6.4 One-to-one correspondence between sequences and functions	177
6.5 Hadamard transform and convolution transform	185
6.6 Correlation of functions	190
6.7 Laws of the Hadamard transform and convolution transform	193
6.8 The matrix representation of the DFT and the Hadamard transform	197
Exercises for Chapter 6	199
7 Cyclic Difference Sets and Binary Sequences with Two-Level Autocorrelation	**202**
7.1 Cyclic difference sets and their relationship to binary sequences with two-level autocorrelation	202
7.2 More results about C	207
7.3 Fourier spectral constraints	211
Exercises for Chapter 7	218

vii

Fig. 3. The content in our published book: Chapters 1–7.

We now begin our journey to look back of how this book is written in the next section!

4. An Exciting Journey

We started our plan for writing a book together in 1999. I had taught a graduate course titled as *Sequence Analysis* in the University of Waterloo in 1999 and Sol had accumulated a lot of materials for a potential largely extended new edition of his book on Shift Register Sequences. After we looked at those materials together, Sol suggested that we should write a new book together by combining those materials. Starting in the summer of 1999, our journey lasted for about 6 years until it was published in 2005 Cambridge Press.

4.1. *Narratives*

The original book outline, hand written by Sol is shown in Fig. 5, and the published one is presented in Figs. 3 and 4. Here I would like to recall some of the historic facts. I joined Sol as his post-doc in June 1996. Sol explained me the entire history of searching for 2-level autocorrelation sequences. As early as 1954, Sol asked the

Fig. 4. The content in our published book: Chapters 8–12.

question: What are all the balanced binary sequences with 2-level autocorrelation? Shortly after that, Sol realized that a construction of binary sequences with 2-level autocorrelation is equivalent to the construction of cyclic Hadamard difference sets in combinatorics. At that time, Sol knew only of the m-sequences and the quadratic residue (Legendre) sequences.

Before 1996, there were only four essentially different constructions for 2-level autocorrelation sequences of period $N = 2^n - 1$: 1) m-sequences, corresponding to the Singer constructions of cyclic Hadamard difference sets; 2) GMW (Gordon, Mill, Welch) sequences (1962) when n is composite; 3) the Quadratic Residue sequences (1932) when $2^n - 1$ is prime; 4) Hall's sextic residue sequences (1956) when $2^n - 1$ is prime and equal to $4a^2 + 27$. Exhaustive searches had been done for period $127 = 2^7 - 1$, $25 = 2^8 - 1$, and $511 = 2^9 - 1$ in 1971 by Baumert, 1983 by Cheng, and 1992 by Dreier, respectively. However, there was no explanation for several of the sequences found for these lengths which did not follow from the known constructions.

I computed the respective trace representations of those known 2-level auto correlation sequences for $n = 7, 8$, and 9. The trace representation describes that a binary sequence of period $2^n - 1$ can be represented by the sum of monomial trace terms where each monomial trace term corresponds to an n-stage LFSR. In this way, we can associate the sequence with a polynomial function mapping from the finite field of 2^n elements to the base field of two elements, i.e., $\{0, 1\}$. For example, an m-sequence is a 1-term sequence, since it is generated by one LFSR. So, we started our search along this direction. The first class I found was 3-term sequences for n odd and I verified its validity up to $n = 21$. Sol presented the conjecture of this family for an infinite number of n at *IEEE Information Theory Workshop 1997*.

At that time Lloyd Welch investigated the case for $n = 8$, i.e., the 2-level auto-correlation sequences of period $255 = 2^8 - 1$. There are only four such classes. They are m-sequences, GMW sequences, and two unknown classes which we may call the sequences A and B respectively. Lloyd found that A and B are related. Precisely, by complementing input to the polynomial function associated with A, we can obtain B. When I found the new 2-level autocorrelation sequence with 5 terms for $n = 10$, using Lloyd's method, I obtain a new 2-level autocorrelation sequence with 13 terms. Without Lloyd's work, it is impossible to get this new sequence, since the search complexity is beyond the computational power which even, we have now. However, for the case $n = 11$, i.e., period $2^{11} - 1 = 2,047$, this method did not work. Surprisedly, it worked when I also complement the output inside the trace function. Sol was so excited about this and named this method as *the Welch-Gong* (WG) *transformation*.

Till 1997, we got two more conjectures on the new families of binary 2-level auto-correlation sequences, namely, 5-term sequences (the sum of five LFSRs) and WG transformation sequences. Those appeared in *IEEE Transactions on Information Theory* in 1998.

In 1999, Hans Dobbertin and John Dillon proved a much more general construction for binary 2-level autocorrelation sequences, which also settled the validity of all three families of our conjectured sequences. Although Dobbertin and Dillon's paper was published a few year later, Hans Dobbertin had conveyed their writing to us. We have presented the constructions of all known 2-level autocorrelation sequences in depth in our book.

Unfortunately, in recent 20 years no new binary 2-level autocorrelation sequences have been found. The construction found by Dobbertin and Dillon remain the record. Optimistically, Sol amused to conjecture that all known binary 2-level auto-correlation sequences are known.

4.2. *Ties*

The following two figures, Figs. 6 and 7, are some discussions on our draft on Chapter 7 about the contents of the relationship between 2-level autocorrelation sequences and cyclic Hadamard difference sets. Further discussions on contents in Chapters 8 and 9 about 2-level autocorrelation sequences are shown in Figs. 8–10.

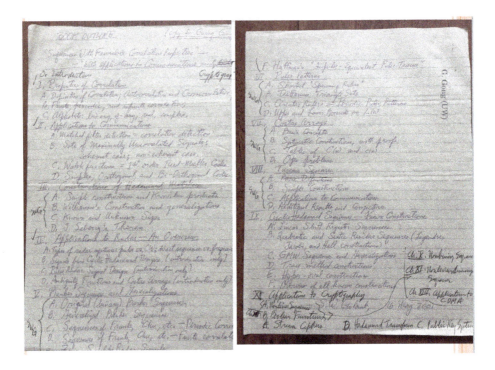

Fig. 5. Our book outline in 1999, hand written by Sol.

In fact, our writing process is also part of our research process. Although we did not solve the aforementioned mysterious fact in the caption of Fig. 9, we developed, for example, a new method to study and search for two-level auto correlation sequences for both binary and nonbinary cases. In seeking a better way to present Dobbertin and Dillon's proof in our book, we found that if we start with a single m-sequence, then we apply iteratively two operations: decimation and the Hadamard transform twice. In this way, we obtain all known 2-level autocorrelation sequences. We published those results, entitled as "The Decimation-Hadamard Transform of Two-Level Autocorrelation Sequences" in *IEEE Transactions on Information Theory* 2002.

Figures 11–13 show Sol's handwriting materials on the applications of shift register sequences on CDMA systems, the discussions on the selection of the materials, and some answers to the publisher's comments, respectively.

Along this journey, I have learned immensely a lot from Sol's thinking and writing style: from his enthusiasm and passion for research to his choices of ways for presenting the materials (e.g., the way that can inspire researchers to tackle unsolved problems), to his limitless energy for research, to numerous English usages that he corrected for me, and to a long list without end. Especially, Sol was always extremely carefully to treat each concept that we used and to look at the citation from the origin of the concept published (e.g., see Fig. 6).

He frequently left me voice messages on my office phone at his time in the midnight for the problems that we tackled. I would not venture that he could think that

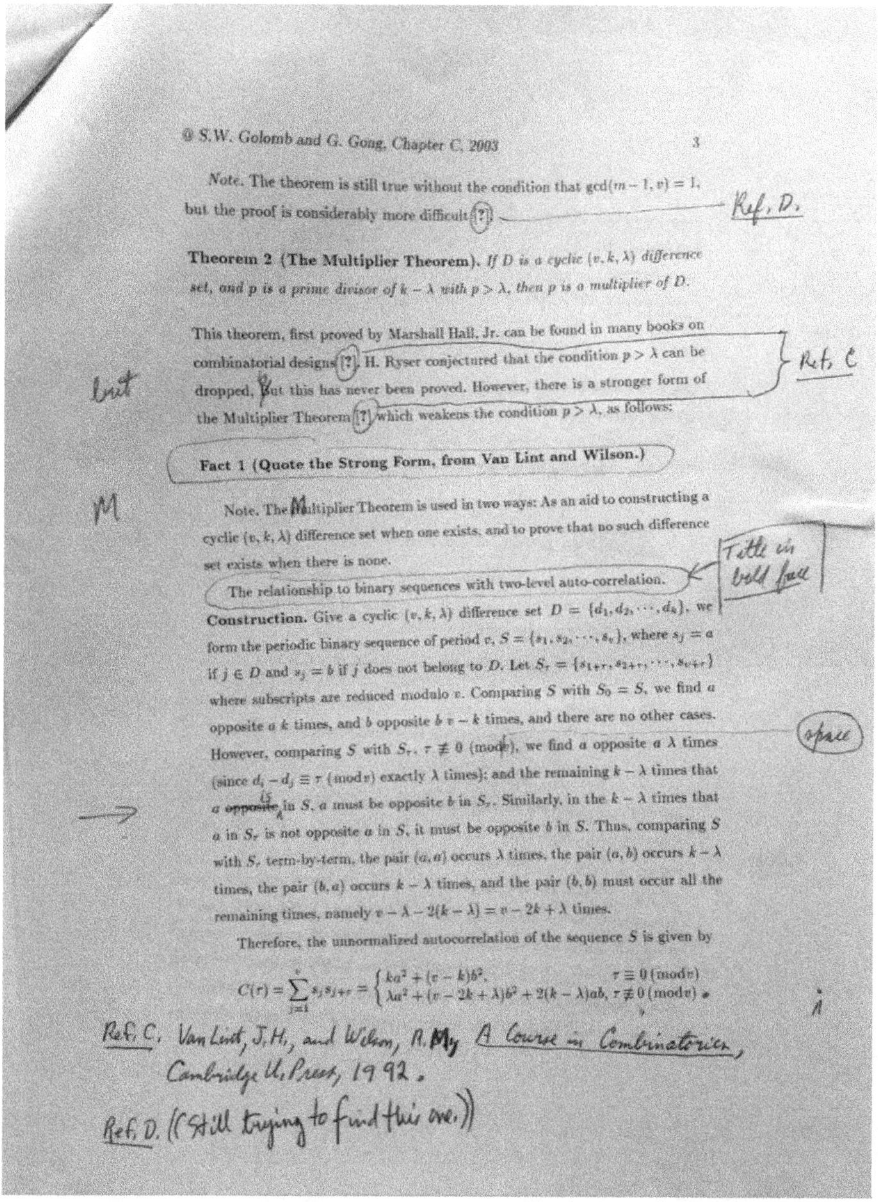

Fig. 6. Sol's comments and provision of the missing references on our draft.

I could be in office, since it was my time about 3:00 am. But when we worked on the publisher's proof-reading version with very tight time constrains, he wrote some notes for me at 4:00 am of his time! My heart was strongly touched by his devoted passion. Hereafter my soul is always echoed by the sound that Sol is a genius, but a hard worker as well, so I should have no excuses for being not hard working. I also shared my feeling to my students. Figures 14 and 15, show some of those notes sent by Sol with the time indicated in his notes.

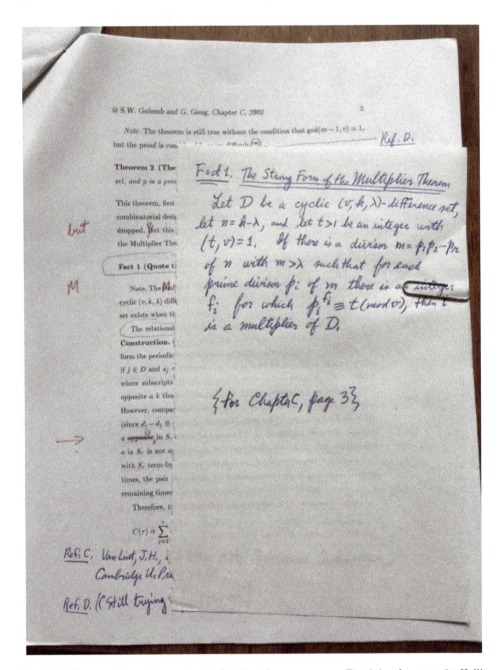

Fig. 7. Sol's suggestion on the content in Fact 1 at the same page as Fig. 6, i.e., how to write Hall's multiplier theorem in combinatorics which corresponds to the property of 2-level autocorrelation sequences under decimation two.

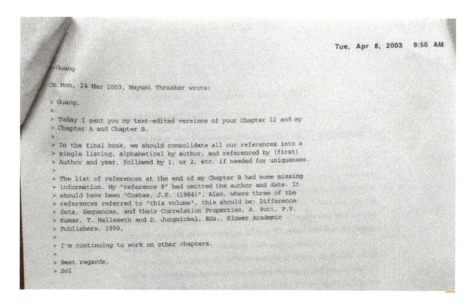

Fig. 8. Sol's email about the format to list the references.

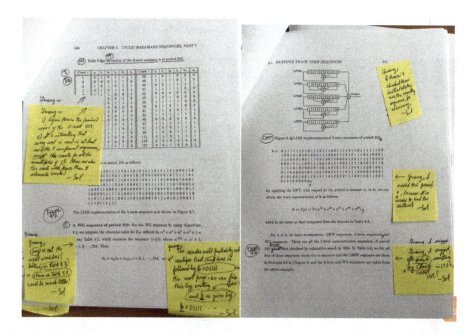

Fig. 9. Sol's comments on some contents in Chapter 9 on 2-level autocorrelation sequences. In the comments at the top of the left page, Sol observed that in the 5-term sequence all five bits being zeros occurs at the multiples of the index 17. This remains mystery till now. The right hand side page shows how this 5-term sequence is generated, i.e., it is equal to the sum of five LFSRs.

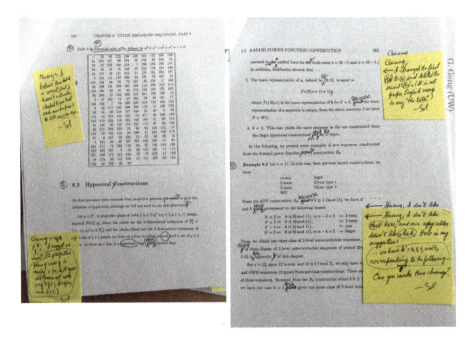

Fig. 10. Sol provided more editorial comments on Chapter 9.

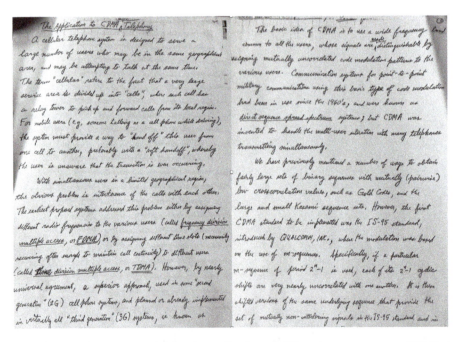

Fig. 11. Sol's handwriting materials on the applications of shift register sequences on CDMA, i.e., the contents in Chapter 12 of the book.

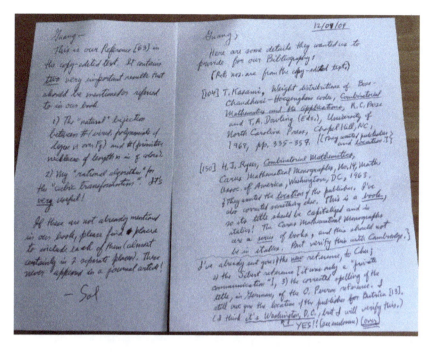

Fig. 12. Sol provided the materials requested by the publisher.

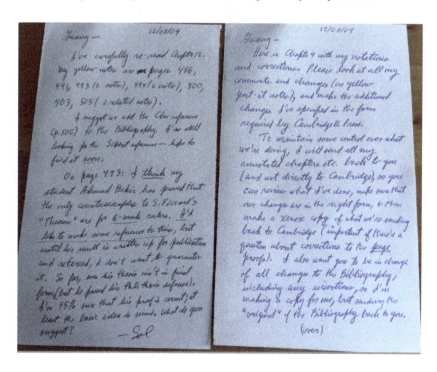

Fig. 13. Some comments on cyclic Hadamard difference sets.

90   *The Wisdom of Solomon*

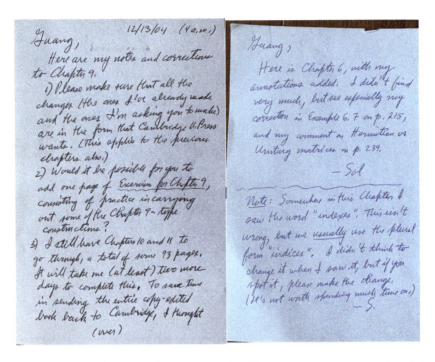

Fig. 14. The notes on the comments on the proof-reading version, which was written at 4 am.

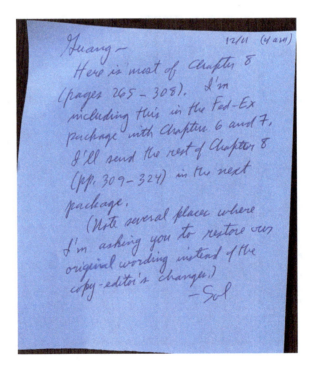

Fig. 15. The notes on the proof-reading version were written at 4 am.

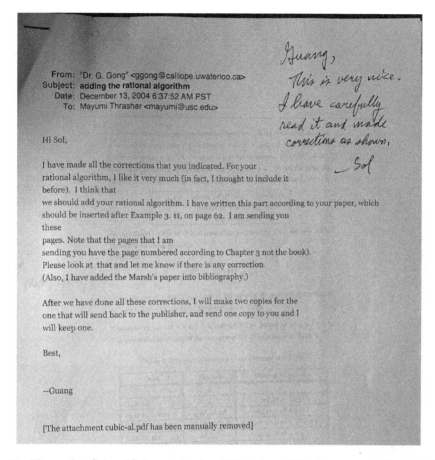

Fig. 16. My email to Sol and Sol's reply by fax. This is about Sol's elementary method (1969) to find all primitive polynomials starting with a given one (Algorithm 3.1's on page 48 in our book).

During our writing process, we have communicated through phones, emails, postmails, Faxes, and Fedex, namely, all the possible communication channels existed during that period of time, as revealed in Figs. 16–18.

5. Golomb's Invariance and Crypto

When I was in USC, Sol and I were very interested to apply shift register sequences to various applications in cryptography and secure communication. Sol supplied me many wonders for the cause of the work he published in 1959 about the concept and computation of the invariants of boolean functions. In that work, Sol said, what it really described was the vulnerability of sequences from nonlinear shift registers to cryptanalysis based on their Boolean feedback functions. However, in 1958, Sol could not mention that application explicitly.

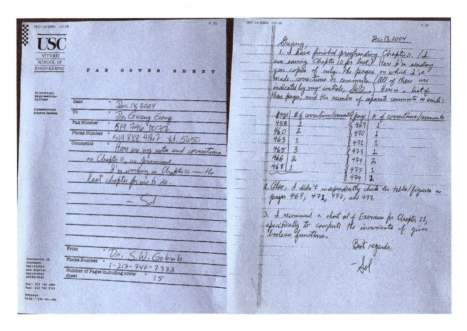

Fig. 17. The notes on the proof-reading version sending by Fax.

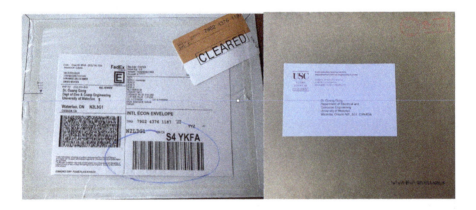

Fig. 18. The envelopes of Fedex and post-mail.

5.1. *One-time pad*

In secure communication, how to scramble message bits such that attacker cannot figure out the contents of the message (or plaintext) is referred to as *encryption*. One simple way to do so is to xor a message bit with a random bit, which results in a scrambled bit, called a *ciphertext bit*. This method is referred to as *stream cipher encryption* and those random bits, as *key stream bits*.

One-time-pad means that different messages are encrypted by different key streams in stream cipher model. Shannon (1948) proved that one-time-pad is un breakable. Thus, Sol's LFSRs are considered as the best choice for generation of key stream bits due to their randomness properties investigated by Sol. Evidently, about two decades of 1950–1960, a key stream generator was implemented by an LFSR where a pre-shared n bit key is loaded as an initial state of the LFSR (here after we mean the LFSRs which generate m-sequences). The LFSR only starts to output key stream bits after running certain times without output of any bits. This type of random bit generators dominated the encryption field until 1969.

Given any random bit stream, e.g., 001000101110..., does there exist an LFSR to generate the sequence? If so, how to find it? Berlekamp presented the following result in coding context in 1968 and Massey used it in LFSR synthesis in 1969: by knowing $2n$ consecutive bits of a sequence with linear span n, the rest of the bits of the sequence can be reconstructed. The algorithm to find the shortest LFSR is the one proposed by Berlekamp for decoding algebraic code in error correction code.

Applying this result to an m-sequence with period $2^{100} - 1$, the attacker only needs to know 200 consecutive bits, then the remaining bits can be reconstructed! This ends the monopoly life of LFSRs used as key stream generators (1954–1969). I would like to point it out that Sol mentioned to me that in fact this result is known for m-sequences much earlier than 1968. In practice, the data is limited to less than $2n$ bits for each encryption if the key stream bits are generated by an n-stage LFSR. However, by the end of 1960, computer networks had risen horizontally in both military applications and civilian applications. Thus, the data needing to be communicated is much larger than $2n$ by far, even $n > 200$ cannot fulfill such demands. Of course, it is not comparable with our current Internet with big data transmission and storage. So, a single LFSR for key stream generation is no longer in use.

5.2. *Two faces*

In the launch of Explorer 1, due to the hardware limitation in the late 1950s, an LFSR with the large number of stages could not be efficiently implemented. Sol started to test the randomness of the output sequences by using several small LFSRs combined by some functions, as shown in Fig. 19.

Sol explored this model for two applications: one for potential use in space communication, e.g., the distance calculation from the earth to Venus in 1961, in order to overcome the hardware limitation of implementing larger LFSRs, and the

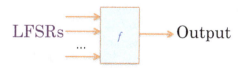

Fig. 19. Sol investigated this model before 1958. Currently, this configuration is referred to as a *combinatorial generator* if those LFSRs are different or a *filtering generator* if those LFSRs are same.

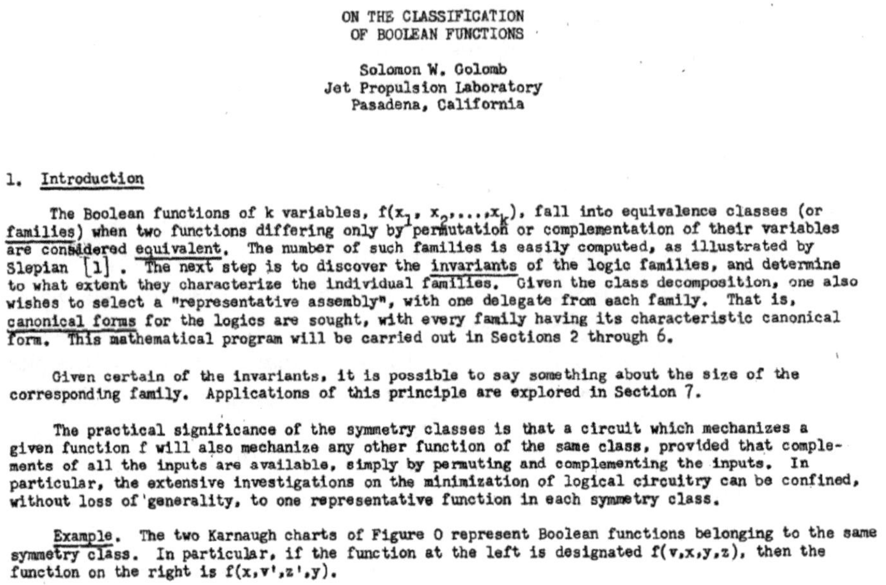

ON THE CLASSIFICATION
OF BOOLEAN FUNCTIONS

Solomon W. Golomb
Jet Propulsion Laboratory
Pasadena, California

1. Introduction

The Boolean functions of k variables, $f(x_1, x_2, \ldots, x_k)$, fall into equivalence classes (or families) when two functions differing only by permutation or complementation of their variables are considered equivalent. The number of such families is easily computed, as illustrated by Slepian [1]. The next step is to discover the invariants of the logic families, and determine to what extent they characterize the individual families. Given the class decomposition, one also wishes to select a "representative assembly", with one delegate from each family. That is, canonical forms for the logics are sought, with every family having its characteristic canonical form. This mathematical program will be carried out in Sections 2 through 6.

Given certain of the invariants, it is possible to say something about the size of the corresponding family. Applications of this principle are explored in Section 7.

The practical significance of the symmetry classes is that a circuit which mechanizes a given function f will also mechanize any other function of the same class, provided that complements of all the inputs are available, simply by permuting and complementing the inputs. In particular, the extensive investigations on the minimization of logical circuitry can be confined, without loss of generality, to one representative function in each symmetry class.

Example. The two Karnaugh charts of Figure 0 represent Boolean functions belonging to the same symmetry class. In particular, if the function at the left is designated f(v,x,y,z), then the function on the right is f(x,v',z',y).

Fig. 20. Golomb, On the Classification of Boolean Functions, *IRE Transactions on Information Theory*, May 1959.

other for secure communication. This model has been used for building blocks of modern symmetric key cryptography since 1970. Their cryptographic strength is dominated by boolean function f. In this regard, the question Sol asked is how to measure the crypto strength of f for applications in secure communication.

> Golomb (1959). Invariants of a boolean function measure the distances between the boolean function and linear combinations of its input variables and their complements.

Equivalently, the invariants measure the correlation between the output sequence and an input m-sequence with different initial states and their complements. When the k-order invariants are equal to 2^{n-1}, the output is uncorrelated with any k-subset of the input variables. Figure 20 shows the first page of Sol's paper, published in 1959, and Figs. 21 and 22 illustrate the definitions of the invariants in the paper.

In modern cryptography, it is termed as *k-order correlation immunity* of a boolean function if all the k-order invariants are equal to 2^{n-1}. The maximum value of all invariants is defined as the *nonlinearity* of the boolean function. They are the most important criteria to design cryptographic strong functions for symmetric-key cryptographic primitives including stream cipher, block cipher, hash functions, and message authentication code.

In distance ranging, one designs f to get a sequence with large periods from LFSRs with shorter periods. In order to facilitate the calculation of the distance between transmitter and receiver, f should be correlated with input variables!

Golomb's Shift Register Sequences — Work with a Great Mind

Note: The group G of allowable symmetries contains $k!$ permutations of the variables, 2^n complementations of the variables, and a two-fold choice of the constant b. Thus G consists of $2^{k+1} \cdot k!$ operators. In particular, the size of any family of equivalent Boolean functions is some factor of $2^{k+1} \cdot k!$ Since there are 2^{2^k} Boolean functions of k variables, the number of families is approximately $(2^{2^k})/(2^{k+1} \cdot k!)$.

Theorem 1: The quantity $T_o = \max (c_o, 2^k - c_o)$, where $c_o = \sum_{x_1,x_2,...,x_k} f(x_1,x_2,...,x_k)$, is

an **invariant** for the family containing $f(x_1,x_2,...,x_k)$. (The capital sigma denotes ordinary summation.)

Proof: The quantity c_o is the sum of all the entries in the truth table for $f(x_1,x_2,...,x_k)$. Permutation and complementation of the variables **rearranges** the entries in the truth table, but leaves their sum c_o unaltered. Complementing $f(x_1,x_2,...,x_k)$ replaces c_o by $2^k - c_o$, which means that half the operations of G leave c_o fixed, and the other half replace c_o by $2^k - c_o$. Thus $T_o = \max (c_o, 2^k - c_o)$ is invariant under **all** operations of G.

Corollary: Relative to the subgroup H of G which allows permutation and complementation of the variables but forbids complementation of the function (so that $[G/H] = 2$), the quantity c_o is itself an invariant of the equivalence classes.

Def. T_o is the **zero order invariant** of the logic family for the group G. (Likewise c_o is the **zero order invariant** for the group H.)

Theorem 2: For every i, $1 \leqq i \leqq k$, let $R_1^i = \max (c_1^i, 2^k - c_1^i)$, where

$$c_1^i = \sum_{x_1,x_2,...,x_k} (f(x_1,x_2,...,x_k) \oplus x_i).$$ Then the set of numbers $R_1^1, R_1^2,...,R_1^k$, when re-

arranged in descending order, forms a collection of k invariants $T_1^1, T_1^2,...,T_1^k$ for the family to which f belongs. (The symbol $\oplus$ denotes modulo 2 addition.)

Fig. 21. Golomb (1959)'s definition of zero order and 1-order invariants.

3. The Higher Order Invariants

Def. For every pair (i,j) with $1 \leqq i < j \leqq k$, define $c_2^{ij} = \sum_{x_1,x_2,...,x_k} (f(x_1,x_2,...,x_k) \oplus x_i \oplus x_j)$.

The complement of c_2^{ij} is $2^k - c_2^{ij}$. That ordering and complementing of variables and complementation of the function f, consistent with producing the $\{R_1^1\}$ in descending order, which makes the sequence of $\{c_2^{ij}\}$ numerically greatest, gives the second-order invariants $T_2^{1,2}, T_2^{1,3},...,T_2^{2,3},...,T_2^{k-1,k}$.

Fig. 22. Golomb (1959) defining high-order invariants.

In this way, the initials states of those LFSRs, which are correlated to f, can be recovered by computing correlation between the output sequence and the input sequence from each LFSR. Those initial states are then used in calculation of the distances. However, for applications in secure communication, if f is correlated with the inputs, then the initial states of those LFSRs, which are loaded as keys, can be recovered in the same way as for distance range. Thus, the system can be broken by computing correlation. This is referred to as *correlation attack* in modern cryptanalysis. Hence in crypto application, f should be uncorrelated with the input variables in order to prevent attackers from recovering keys!

Although Sol did not mention the application for cryptography in his paper on invariants, published in 1959, he received an award from the NSA in the middle of 90s for his contribution in cryptography.

5.3. *Hadamard transform*

Golomb's invariants or correlation immunity of boolean functions can be computed through the Hadamard transform, as introduced in his paper. From this point of view, the invariants measure the correlation or equivalently distance between the output sequence of f and an m-sequence with different initial states. When Sol and I discussed this problem again after I joined him in 1996, Sol murmured to me: which m-sequence, ..., which one? I suddenly got Sol's question. Since there are a number of distinct LFSRs which will generate shift distinct m sequences with period $2^n - 1$, it makes the sense that the crypto strength of a boolean function should be measured from all distinct LFSRs instead of just a single LFSR! This idea is sketched in Fig. 23.

In 1996, there is only one block cipher which is in public domain for civilian applications, i.e., Data Encryption Standard (DES), as shown in Fig. 24. Sol suggested that we should look at the functions used in DES first.

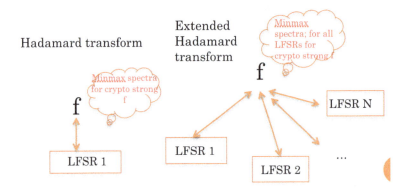

Fig. 23. Sol's suggestions to look at the distance of a boolean function to all distinct LFSRs.

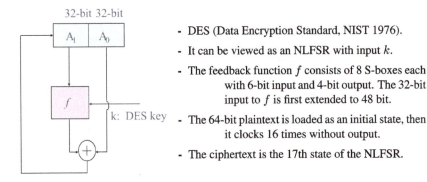

Fig. 24. DES is a two-stage NLFSR with input k where each register holds 32 bits.

In DES, f is implemented by 32 boolean functions in 6 variables. For degree 6, there are 6 shift-distinct m-sequences with period $2^6 - 1 = 63$. For each function, we computed their correlation with each m-sequence. Unexpectedly, each function has the same correlation spectra for all different m-sequences. We called this transform the *extended Hadamard transform* and published those findings for DES, entitled as "Transform Domain Analysis of DES", in *IEEE Transactions on Information Theory* in 1999. The functions used in DES are the currently only known class with this property in addition to the class of hyper bent functions, discovered in 2001.

5.4. *Applications*

The work on LFSR/NLFSRs investigated by Sol has led the way to design lightweight cryptographic schemes for securing Internet-of-Things (IoT) in which each device has limited computation power. Currently almost all known symmetric key schemes of lightweight cryptography employ them in one way or another! Those are crucial for securing IoT systems, since it is changing our ordinary life dramatically as shown in Fig. 25! An example of such a design is the lightweight WG stream cipher in Fig. 26.

6. My Two Sabbatical Leaves with Sol

In the fall of 2006, I had my first sabbatical leave from the University of Waterloo. I visited Sol for a half year at USC. In this precious time, after my 8 years service in the University of Waterloo, we co-instructed a graduate course at the Department of Electrical Engineering, namely, EE 599: Signal Design for Good Correlation, Spring 2007. We used our book as the text book. We also co-organized the International Workshop on Sequences, Subsequences, and Consequences (Los Angeles, May 31–June 2, 2007), and the proceedings of the work shop, co-edited with Tor Helleseth

Fig. 25. IoT has emerged as a new connected world and their security will affect each human being in this universe!

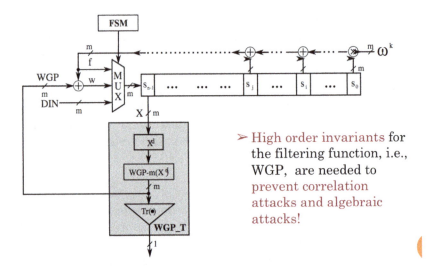

Fig. 26. A block diagram of WG stream cipher. The shaded block is served as a filtering function, which is implemented by the aforementioned WG (Welch-Gong) transformation and WGP represents the function inside the trace function, $Tr(\cdot)$, and the upper part is an LFSR over an extension field with m bits.

and Hongyeop Song, published by Springer in 2007 (see the cover of the proceedings in Fig. 26). Sol showed me how to write a proposal for getting grant to support a conference. This was my first time to learn how to apply for grants outside Canadian grant agencies. It greatly benefited me for later on my successfully obtaining research grants from US grant agencies to support my research on communication security.

In addition to those activities, we revisited some open problems on sequences and Costas arrays again. A Costas array of order n is known to be equivalent to a permutation $\{1, 2, \ldots, n\}$ for which the difference triangle contains no repeated elements in any row. It has only few known systematic constructions. We found that we can have some new constructions when we replace the integer computation by the finite field computation. This work was presented in our joint work, entitled as "The Status of Costas Arrays", published in *IEEE Transactions on Information Theory* in 2007 (Fig. 27).

I had my second sabbatical leave starting in the fall of 2013. I visited Sol in January–February 2014 for about 7 weeks. We still loved to look at our favorite, i.e., m-sequences. A sequence is said to have *perfect autocorrelation* if the autocorrelation function is equal to 0 for all nonzero shifts. If we allow the auto correlation function having nonzero value at one shift, then it is referred to as *almost perfect autocorrelation*. There are a very few known constructions for perfect autocorrelation sequences. Thus, a lot of attentions turn to the almost perfect autocorrelation. We started with playing around with ternary m-sequences to attempt construction of sequences with almost perfect autocorrelation (Fig. 28). We were not in vain!

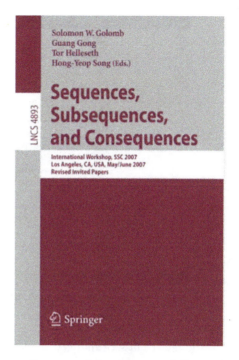

Fig. 27. The cover of International Workshop on Sequences, Subsequences, and Consequences, and the first page of our paper on Costas arrays.

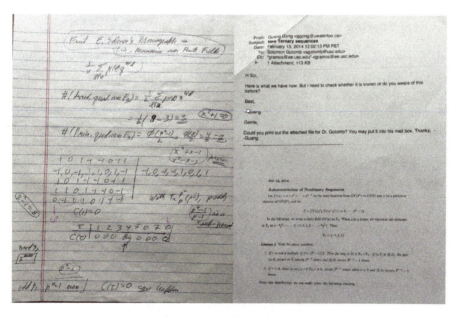

Fig. 28. Sol's sketch of some properties of ternary m-sequences, and my email to him to convey that we got the results where the ternary sequences are not new, but those quinary sequences are new.

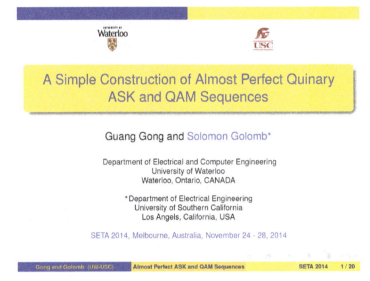

Fig. 29. This is our last joint work. Sol presented this work at SETA 2014

We have found a class of quinary sequences with almost perfect autocorrelation (Fig. 29).

7. Joyful Sharing

Sol shared many of his academic experiences with me. I have many "first time" from consulting activities with Sol. Examples include how to write an award winning research proposal (e.g., my first research proposal to NSERC Discovery Grant in 1999), how to write a research proposal for US grant agencies (as I mentioned before, the first practice was a small grant to support the SSC conference), how to sight new opportunities in new research areas, how to work with industrial partners from negotiating rate for contract grants to field visits to companies, how to inspire the graduate students to passionate the problems that they have being worked on, how to teach undergraduate courses with larger sizes, etc..

Sol was very excited whenever I had any achievements. When I was promoted to a full professor with tenure in 2004, Sol immediately called me for congratulating my promotion. He visited me in 1999, the second year I jointed Waterloo, and in 2015, the second to the last year of his life. I do not know whether there exists any connection between those two "seconds" scientifically, but I do have a strong feeling for that.

He also liked to share with me his happy moments. I received a USC magazine mailed me by Sol (Fig. 30). I was puzzled why he sent me this particular issue.

Fig. 30. A USC magazine mailed to me by Sol.

It was to celebrate his 50 years service in USC when I turned the pages (Figs. 31 and 32)!

Our last time together was in the occasion for Sol receiving the 2016 Benjamin Franklin Medal in Electrical Engineering on April 20, 2016, Philadelphia. I was honored to be invited to participate in this event and to give a talk "Sequences and Cryptography" at *the Workshop on Shift Register Sequences for Honoring*

Fig. 31. This issue has a special article: *Digital Pioneer Sol Golomb Celebrates 50 Years Service in USC.*

Fig. 32. Many different humorous postures of Sol are collected in this article.

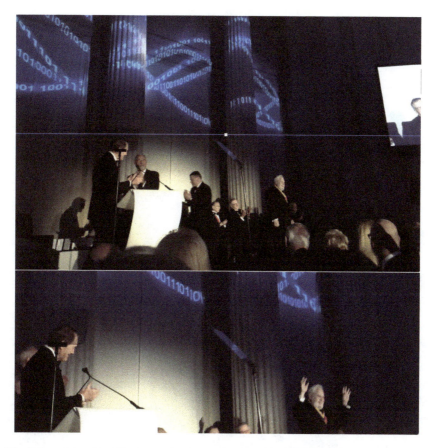

Fig. 33. Sol receiving 2016 Benjamin Franklin Medal in Electrical Engineering on April 20, 2016, Philadelphia.

Dr. Solomon Golomb to receive the award in Franklin Institute. I was so shocked when I received the sad news of his death since it was only less than 2 weeks after we were together. The above two photos, Fig. 33, were taken by me at the award ceremony.

8. From My Heart

In his latest 20 years (1996–2016), Sol Golomb was my supervisor, mentor, and long-term collaborator, and played an important role in my life. I read Dr. Golomb's book, *Shift Register Sequences*, when I studied towards my Master Degree. Since then, I have always wished that I could work with him. I arrived in the US in May 1995. While I was attending Crypto'95 in Santa Barbara in August 1995, I met one of his collaborators in the industry, who brought me to meet Sol in USC after the conference. Then I was lucky to get a chance to do research under his supervision in USC as his post-doc in June 1996–July 1998, and to be part of his life since then.

He is a hero in my heart!

© 2023 World Scientific Publishing Company
https://doi.org/10.1142/9789811234378_0006

Memories of Sol

Alfred W. Hales

I met Sol 62 years ago, in 1958, and he was (initially) my supervisor, and then mentor, colleague and friend until his untimely death in 2016. I have tried to organize my many memories of him chronologically, into three periods: first, while I worked part-time at JPL (1958–1970); next, while I was teaching at UCLA (1970–1992); and finally, while I was at IDA/CCR in La Jolla (1992 on).

Sol hired me in 1958, after my sophomore year at Caltech, to work under his supervision in the Information Processing group at JPL. My initial task, for the summer, was to factor trinomials $x^n + x^a + 1$, of small degree n, over the two-element field. For this I was given the very latest technology: a sharp pencil and squared paper! As it turns out, I was encouraged to make, verify and even prove conjectures (of my own and of others), so the task was quite fascinating — and led to my being included as a junior coauthor of a JPL report and of Sol's eventual classic "Shift Register Sequences".

One example of this: Sol had conjectured that $x^4 + x^3 + x^2 + x + 1$ never divides a trinomial modulo two. I found a proof of this. Much later on, following on many further results of this nature, a student of Sol's wrote a PhD thesis on this general topic.

I ended up working for Sol for three full summers while a Caltech student, plus another half summer: 1958, 1959, 1961, plus part of 1960. During this time, I learned a tremendous amount of mathematics from him, especially combinatorics, number theory and algebra. In the summer of 1959, he gave an informal series of lectures on combinatorics which were particularly valuable — I still have the notes from these.*

I also learned, by observing Sol, and without realizing it, how to run a research group. This stood me in very good stead 34 years later, when I took over as Director at IDA/CCR in La Jolla.

---

*In fact, based in part on these notes, I taught a course in combinatorics at Harvard University (Sol's PhD institution) in 1965. This was apparently the first combinatorics course ever taught in the Harvard Mathematics Department.

106   *The Wisdom of Solomon*

Through Sol, at JPL, I met quite a number of outstanding mathematicians and scientists, people he brought through as employees or visitors. Here are some of the names I remember: Len Baumert, Richard Epstein, Hal Fredricksen, Richard Goldstein, Ed Posner, Gus Solomon, Jack Stiffler, Bob Tausworthe, Herb Taylor, Andy Viterbi, Lloyd Welch and Neal Zierler. A number of these I had significant interactions with later on — see below.

Another individual, Basil ("Sandy") Gordon, I had already met at Caltech — he was teaching as a postdoc there when I was a freshman and was strikingly inspiring as a teacher. It turned out that he knew Sol well — they had met in high school playing "first board" for their respective school chess teams in Baltimore and then were undergrads together at Johns Hopkins. Sol and Basil and I, along with several others (including Len Baumert), collaborated on a paper about "comma-free codes" under the pseudonym of B. H. Jiggs. Later on, Basil and I became colleagues at UCLA.

One other coauthor on the above paper, Bob Jewett, I had met at Caltech already — we had math and volleyball in common. At JPL Bob and I coauthored another paper on "Regularity and Positional Games" — a topic in combinatorics (Ramsey Theory). Sol's precise role in this paper is unclear (due to our collective faulty memories), but he has always regarded himself as the paper's "godfather". More below about this Hales-Jewett theorem.

A bit later on, in 1968, Sol and I coauthored a paper on a group theory problem connected with the Burnside-Polya counting formula in combinatorics. This could be considered as stemming from his 1959 JPL lectures mentioned above.

At JPL, every week or so, Sol would send around a note with some new mathematical observation — formula or proof or application, or other item of interest. Two of these I particularly remember, dealing with information about mathematical luminaries who were visiting southern California at the time. One was about Marshall Hall, the famous group theorist then at Ohio State. A year or two later he moved to Caltech and I got to know him well, taking several courses from him. The other was Harvard's Andy Gleason, famous for solving one of Hilbert's problems, but also a noted combinatorial expert. Later I got to know him well while I was a postdoc at Harvard, and he also became my brother's thesis advisor. Even later we interacted through IDA/CCR.

While at JPL I (of course) learned about pentominoes from Sol. At the time he had a friend/neighbor make up ten or twenty really nice (inlaid) boards with matching dark wood pentomino pieces, and I bought one of the sets. This has withstood three generations of (sometimes rough) playing so far, and promises to last almost forever.

There are two very memorable stories from that era involving not Sol himself but members of his research group. One morning Bob Tausworthe was very late to work, and by the time he arrived we had already heard on the radio why he was late. Early that morning he had gone to the bathroom and discovered, in the toilet, the head of a boa constrictor! The rest of the boa went down, through a pipe, and up into the toilet in the bathroom of an adjacent apartment. Eventually a crew of

Memories of Sol 107

firemen (and maybe plumbers) had to dismantle both toilets to remove the snake. Several years later I happened to meet the "other end" of the story, the Caltech student from the adjacent apartment who owned the boa.

The other story involves golf. Richard Epstein and Dick Goldstein were arguing about what kind of score for 18 holes of golf one might expect from a "random" new golfer. One claimed it should be at most 120, and the other said it would surely be more than 120. Anyway, we all decided to test this — pick a "random" person and try it out. We picked "J.D.", and adjourned after work to the Arroyo Seco course. After NINE holes his score already passed the cutoff! Many years later, when J.D. retired and was asked by a reporter if he might take up golf, he replied "I tried golf once and it is not for me".

Finally, for this section, a couple of more personal items. Ginny and I were married in 1962, and shortly before this Sol and Bo (Sol's wife) invited us over to their house for a very nice dinner. At that occasion we met their two delightful daughters, Astrid and Beatrice, who were about 4 and 2, respectively, at the time. Shortly after that Sol and Bo came to our wedding, at St. Mark's in Altadena, and then to the reception at Ginny's home nearby. Both our families came from the Pasadena area, and so there were quite a few attendees. Sol asked me, at the occasion, "Al, do you really know all these people"?

When we returned to Southern California in 1966 and I took up a position at UCLA, I continued to consult for JPL until 1970. From then on, I was too busy with UCLA responsibilities. Also, Sol had moved to USC from JPL. Hence, for the next 20 or so years, our contacts were relatively infrequent — but there were some.

I had started classified consulting at IDA (an NSA contractor) in Princeton, and continued this while at UCLA. While doing this I had very substantial interactions with a number of former JPLers who also were involved in the classified IDA work: Len Baumert, Hal Fredricksen, Ed Posner, Lloyd Welch and Neal Zierler.

Among other things this led to a joint (unclassified) paper on Swan's theorem with Hal Fredricksen and Mel Sweet in 1986. I had first learned of Swan's theorem from Sol.

In 1972, Bob Jewett and I were fortunate enough to share in the SIAM "First Polya Prize" for our combinatorial paper mentioned above, the one which Sol "godfathered".

In 1982, the second edition of Sol's classic *Shift Register Sequences* book was published by Aegean Park Press. As mentioned above, I was listed as one of three junior coauthors.

Sol turned 60 in 1992, and I attended and spoke at his birthday conference that year in Oxnard.

I took early retirement from UCLA in 1992. to take up a new position as Director of the new IDA research center in La Jolla — a clone of the Princeton facility where I had consulted. Sol and Bo both attended my retirement dinner, where Sol's old friend and colleague Basil Gordon was also being honored on his retirement.

When I started as Director of IDA/CCR in La Jolla, I realized that I was taking on a role similar to Sol's at JPL 34 years earlier. Furthermore, the math involved, though mostly classified, was intimately connected in many cases to the JPL work. For this and other reasons my connections with/to Sol seemed to increase considerably. I learned at this point that Sol had received some special NSA award/recognition for work he did on a classified topic which he never could talk about.

I invited Sol down to give a talk to our summer SCAMP program in 1993, the first in our new building. At roughly the same time we hired his student Herb Taylor to our research staff — Herb worked here for 4 years and continued as a consultant.

In 1996 there was a symposium at IDA'S Princeton branch in honor of Neal Zierler's retirement — I attended this and Sol gave one of the invited talks. Then in 1997 there was a symposium at USC honoring Lloyd Welch's 70th birthday. Sol and I both spoke at this event.

Sol's *The American Scholar* article "Mathematics Forty Years After Sputnik" appeared in 1998, which I/we read with great pleasure and shared with our family and friends.

Sol and I collaborated on an article "Hypercube Tic-Tac-Toe" which appeared in an MSRI volume in 2000. This contained some older results along with new results, many related in a sense to the Hales-Jewett paper mentioned earlier. The new results involving drawing strategies required updating since we were using a faulty multiple precision routine. Joe Buhler, my successor as Director of IDA/CCR La Jolla, provided the new correct data. This is described in Sloane's OEIS #129935.

At this point, I can't resist mentioning the USC "recruiting brochure" (from some time in the 90's I think) featuring Sol on the cover fully attired in a USC football uniform!!!

There was a conference at USC in 2002 in honor of Sol's 70th birthday. I gave one of the invited talks, on a paper with NSA's Don Newhart, titled "Irreducibles of Tetranomial Type". This dealt with topics related to Swan's theorem (mentioned earlier), and was very much in the spirit of the JPL work I had done under Sol.

At about this time there was a new children's book *Chasing Vermeer* which we bought for a grandchild, and which featured pentominoes in it. Of course, I e-mailed Sol to tell him about this!

In 2005 I wrote a paper with Nora Hartsfeld (of Western Washington) on the directed genus of de Bruijn graphs. Not only was this a topic of great interest to Sol since JPL days, but also it turned out that the major tool we used in the paper was Mykkeltveit's 1972 solution to a conjecture made by Sol in 1967.

There was another conference at USC in 2007, in honor of Sol's 75th birthday. Again, I gave one of the invited talks, this time on a paper with Dan Goldstein (of IDA/CCR), on strongly primitive elements in finite fields. This paper, among other things, extended earlier work on a series of conjectures made by Sol in 1984. On this occasion I was able to help USC obtain some financial support from NSA (via IDA) for the conference.

In early 2012 Basil Gordon (Sol's old friend/colleague and my UCLA colleague) passed away. There was a memorial event for him at UCLA where Sol and I both

Memories of Sol

gave talks. At the time Sol made a comment to me about the number of mathematicians of his approximate age who had recently died.

There was a *New York Times* crossword puzzle sometime in 2012 where the black squares were in the shapes of the 12 pentominoes! Sol noted this but also received e-mails from a number of others about it.

Later in 2012 there was a conference in Waterloo, Canada, where Sol was honored in two respects. There was a special session celebrating his 80th birthday, and also, he was presented with Sigma Xi's Procter Prize for his distinguished achievements in science. I gave one of the talks in his session, on a paper with Tom Dorsey (of IDA/CCR La Jolla), on "Irreducible Coefficient Relations". This again was closely related to Sol's long-time interest in irreducible polynomials over finite fields. On the occasion Sol was also honored by Venus — the "transit of Venus" took place during the conference! Also, at this time Beatrice (Sol's daughter) made up a number of handsome "lapel pins" in Sol's honor, each displaying a small checkerboard tiled with colored pentominoes — I wear mine proudly on mathematical occasions.

There was an 85th birthday party for Lloyd Welch later in 2012. Sol was in Israel so could not attend, but his wife Bo was there, as were Ginny and I. This was the last time we saw Bo.

In 2013 Sol was a co-recipient of the National Medal of Science. This very prestigious honor was presented to him at the White House by President Barack Obama, with Sol's whole family looking on. The American Mathematical Society wrote this up in an article in their *Notices*, and asked me to contribute the mathematical content of the article. I was very pleased to do this, since it gave a chance to (partially) repay Sol for all he had done for me. Here is what I said:

"For the past 50+ years, Solomon Golomb has been a world leader in the development and application of mathematics for communications and coding theory — especially digital and space communications. In his remarkable career, first at the Jet Propulsion Laboratory and then at the University of Southern California (in Electrical Engineering and Mathematics), his research contributions have ranged over a wide spectrum of science and technology. Perhaps he is best known for his mathematical analysis of shift register sequences, in his classic book with that title and in numerous journal publications — such sequences are ubiquitous in radar, space communications, cryptography, cell phones, etc. Other noteworthy contributions in this direction include Golomb (entropy) coding, Golomb rulers, and the Golomb construction for Costas arrays. For all this work the IEEE has honored him with both its Hamming Medal and the Shannon Award in Information Theory. Golomb's Harvard PhD was in analytic number theory, and he has extensive (and seminal) publications in number theory, combinatorics, algebra and various other fields (including even molecular genetics). He is also a noted expert in mathematical game theory (polyominoes were his invention), and he continues to publish a number of puzzle columns in various journals. In addition to his many research contributions, Golomb has had a great influence on future generations through his generous and insightful mentoring of young people — his students, postdocs and many others."

The Golombs' 2014 "Holiday Card" featured a wonderful picture from this White House event! (See the following picture.)

There was a long article by Lagarias in a 2013 issue of the *AMS Bulletin*, titled "Euler's Constant". I called Sol's attention to this because one section of it focused on the so-called "Golomb-Dickman" constant, about which I had first learned from Sol's 1959 JPL lectures. This sparked quite a bit of back-and-forth communication between Sol and Lagarias, which Sol copied me on.

Also in 2013, Zhang announced his remarkable "bounded gap" result and Granville followed with his survey article "Bounded Gaps Between Primes". This led to a series of letters between Sol and Granville, dealing among other things with the result from Sol's thesis now called "Golomb's Lemma". (Sol's results played an important role in Zhang's and earlier work.) Sol copied me on these letters as well.

In January 2014, there was a special session on de Bruijn sequences at the AMS/MAA national meeting in Baltimore, and both Sol and I gave talks on this occasion.

Starting in 2015 I was asked by Sol and World Scientific Press first, to advise on the desirability of producing a third edition of Sol's classic *Shift Register Sequences*, and then to advise Sol on the appropriate changes and additions for Chapter IX, the "Selective Update". This led, finally, to a new version which was published posthumously in 2017.

Also, in 2015 I was asked to give one of the four invited talks on the occasion of Sol receiving yet another award, the Franklin Medal, in April 2016 in Philadelphia. I talked on "Coefficient Patterns for M-sequences", and the other speakers were Guang Gong, Tor Helleseth, and Andy Viterbi. This was a very nice occasion, but due to other commitments I had to leave after my lecture and hence missed the actual presentation and banquet the next day. This was the last time I actually saw Sol.

The week after I returned, we exchanged several e-mails, two about the third edition of his book and one about an upcoming conference on "50 Years of the Hales-Jewett Theorem". Here is that last e-mail:

> From sgolomb@usc.edu Wed Apr 27 15:54:50 2016
> Subject: Re: Several Things
> To: "Alfred W. Hales" <hales@ccr-lajolla.org.>
>
> Dear Al,
> Thanks for sending me the information about the conference to celebrate the Hales-Jewett Theorem. Unfortunately, pressing commitments here in Los Angeles will prevent me from attending. (I still consider myself the "godfather" of the Hales-Jewett Theorem.)
> All best wishes,
> Sol

Four days later he passed away.

To conclude this contribution, I enclose the slides from the last two talks I gave about Sol and his work. The first of these was my talk on the occasion of his Franklin Award in 2016. This talk focused on the theory underlying, and motivated by, shift register sequences — perhaps his most important and long-lasting scientific work. The reader should bear in mind that the "arithmetic" involved is binary, i.e., modulo two.

The second talk was at the USC memorial event in January 2017. This dealt with his first (mathematical) love, number theory. He was already fascinated by this in high school, contributing a problem in this area to a math journal. His PhD thesis at Harvard was on number theory and he continued to follow work in this area throughout his career. More recently, there were significant advances on an important problem which were based in part on his earlier work, which delighted him.

Coefficient Patterns for M-sequences*

Alfred W. Hales

April 2016

Binary LFSR's and M-sequences

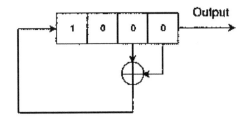

This shift register generates a sequence of bits $a_0, a_1, \ldots, a_t, \ldots$ satisfying the linear recurrence

$$a_{t+4} = a_t + a_{t+1}$$

More generally, we will have a sequence of bits satisfying

$$a_{t+n} = c_0 a_t + c_1 a_{t+1} + \cdots + c_{n-1} a_{t+n-1}.$$

The *driving polynomial* or *minimal polynomial* for this is

$$f(x) = x^n + c_{n-1} x^{n-1} + \cdots + c_1 x + c_0.$$

Theorem 1. *(Golomb) The maximal period of the sequence will be $N = 2^n - 1$ if and only if*

1) $f(x)$ *is irreducible, and*
2) $f(x)$ *does not divide any $x^d + 1$ where $d < N$ and d divides N.*

Such an f is called *primitive*, and the sequence generated an m-sequence (or PN-sequence).

Note that if $N = 2^n - 1$ is prime, then irreducible implies primitive.

---

*In honor of Sol Golomb (supervisor, mentor, colleague and friend) on the occasion of his Franklin Medal award in 2016.

Memories of Sol

Golomb's Randomness Postulates

(for bit sequences of length $2^n - 1$)

I. (Balance) In each period there are 2^{n-1} 1's and $2^{n-1} - 1$ 0's.

II. (Runs) In each period, for $1 \le k \le n - 2$, there are 2^{n-k-2} runs (of 0's or 1's) of length k. Also, there is one run of 0's of length $(n - 1)$ and one run of 1's of length n.

III. (Autocorrelation) The out-of-phase correlation under cyclic shift of the sequence is -1.

M-sequences satisfy all of these, and hence are called "pseudorandom." They are very useful in cryptography, random number generation, ranging, and many other areas as well.

Example: $000100110101111\ldots$ with period 15 and driving polynomial $x^4 + x + 1$

Finding Coefficients

Suppose $a_0, a_1, \ldots, a_t, \ldots$ does satisfy a linear recurrence. How can you determine the recurrence, i.e., find n and the coefficients $c_0, c_1, \ldots, c_{n-1}$?

If you know n, you can solve n equations in n unknowns. The approximate work is $O(n^3)$.

In general, you can try each n in turn. Then the work is $O(n^4)$.

This turns out to be a gross overestimate.

Berlekamp-Massey Algorithm

Berlekamp (1968): Algorithm for decoding BCH codes.

Massey (1969): Adapted above to coefficient finding.

Berlekamp-Massey Algorithm: A recursive algorithm for determining n, and the coefficients $c_0, \ldots, c_{n-1}$, using only $0(n^2)$ work.

It was realized later that the Berlekamp-Massey Algorithm is essentially the continued fraction algorithm applied to Laurent series.

And the continued fraction algorithm is essentially the Euclidean algorithm.

EUCLID (300 BC)!!!!

Trinomials and Swan's Theorem

An irreducible polynomial of degree 2 or more must have $c_0 = 1$, and an odd number of non-zero terms. Hence the "lightest" possible examples are trinomials $x^n + x^k + l$. These are the easiest to implement.

Swan's Theorem (1962): Suppose $1 < k < n$, and exactly one of k, n odd. Then the number of irreducible factors of $x^n + x^k + 1$ is even if and only if one of the following holds:

1. n is even, k odd, $n \ne 2k$, and $nk/2 = 0, 1$ mod 4.
2. n is odd, k even, $k \nmid 2n$, and $n = \pm 3$ mod 8.
3. n is odd, k even, $k \mid 2n$, and $n = \pm 1$ mod 8.

Hence if 8 divides n, then $x^n + x^k + 1$ cannot be irreducible!

Conjecture: There are infinitely many primitive trinomials.

The Great Trinomial Hunt

AMS Notices article, 2011.

Idea is to only consider "Mersenne exponents," n such that $N = 2^{n-1}$ is prime. Then irreducible implies primitive.

Some small examples: $x^2 + x + 1$

$$x^5 + x^2 + 1$$
$$x^{31} + x^3 + 1$$
$$x^{89} + x^{38} + 1$$

Status in 2002: $x^{6,972,593} + x^{3,037,958} + 1$.

Status in 2016: $x^{43,112,609} + x^{3,569,337} + 1$.

What about degree $n = 74,207,281$?

Heavy Polynomials and Tetranomials

Polynomials with *all* coefficients 1 are seldom irreducible, and never primitive (for degree > 2).

Polynomials $f(x)$ with all but one coefficient 1 can be both irreducible and primitive. But hard to implement?

Multiply by $(x+1)$. Then $(x+l)f(x)$ will have only four terms, a "tetranomial." This is easy to implement as a shift register, and the period will still be $N = 2^n - 1$ (when $f(x)$ is primitive of degree n).

So, search for primitives $f(x)$ with $(x+1)f(x)$ a tetranomial! Are they numerous?

A Swan-like Theorem

Theorem 2. *(H-Newhart, 2003/6) Let $x^n + x^a + x^b + 1$ be a square-free tetranomial, with r irreducible factors. Then r is even if and only if one of the following holds:*

(1) 8 divides n, 4 divides $(n-a)$, and $(n-b)$ is odd, Etc.

Too complicated to fully state here, but aids greatly in searching for irreducibles $f(x)$ of "tetranomial type".

Result: (1) For every degree $n \leq 5,000$ there is a tetranomial $(x+1)f(x)$ with $f(x)$ irreducible.

(2) For every degree $n \leq 500$ (except 12,96) there is a tetranomial $(x+1)f(x)$ with $f(x)$ primitive.

In contrast, irreducible/primitive trinomials in this range occur only about half the time.

Linear Relations

Might for instance $c_1 = c_2$ for all degree n irreducibles? Or $c_7 = 1$? What relations of this sort can occur?

$c_0 = 1$ always (since x does not divide f).

$c_0 + c_1 + \cdots + c_{n-1} = 0$ always (otherwise 1 would be a root, so $x + 1$ would divide f).

Question: For which n-tuples $(k_0, \ldots k_{n-1})$ is it true that $k_0 c_0 + \cdots + k_{n-1} c_{n-1}$ is constant for all irreducible f of degree n?

Theorem 3. *(Dorsey-H, 2012) Every linear relation*

$$k_0 c_0 + \cdots + k_{n-1} c_{n-1} = \text{``constant''}$$

holding for all irreducible f of degree n is a linear combination of $c_0 = 1$ and $c_0 + c_1 + \cdots + c_{n-1} = 0$.

This holds in more generality, say for irreducibles over finite fields F_q. Except the conclusion is more complicated:

"Every polynomial relation holding has degree at least $q - 1$, and if the degree is $q - 1$ then ..."

Approximate/Asymptotic Relations

What about "patterns" which appear most (though not all) of the time? For instance, might $c_1 = c_{n-1}$ a high percentage of the time for irreducible f?

Number Theory Digression:

A prime number (except 2, 5) must "end" in 1, 3, 7, or 9.

11, 13, 17, 19 prime

21, 33, 27, 39, not prime

Dirichlet (1837): "An infinite number of primes end in 1. Also 3. Also 7. Also 9."

"Furthermore, the number of each type is (approximately) the same!"

Artin (1924)

Hayes (1965)

Cohen (2005)

"Away from the center, the coefficients of irreducible polynomials are (approximately) random, i.e., all patterns are (about) equally likely."

$$1 + c_1 x + c_2 x^2 + \cdots + c_s x^s + \underline{\qquad} + c_{n-t} x^{n-t} + \cdots c_{n-1} x^{n-1} + x^n.$$

Fix s and t. Consider $c_1, \ldots c_s$ and $c_{n-1}, c_{n-2} \ldots, c_{n-t}$. For n large, all possible patterns for these $s + t$ coefficients occur (about) equally often for irreducible f of degree n.

Each pattern occurs (about) $1/2^{s+t}$ of the time.

1. Artin, Quadratische Korper im Gebiete der Hoheren Kongruenzen II, *Math. Z.* **19** (1924).
2. Berlekamp, *Algebraic Coding Theory*, McGraw-Hill, 1968.
3. Brent-Zimmerman, The great trinomial hunt, *AMS Notices* **58** (2011).
4. Cohen, Explicit theorems on generator polynomials, *Finite Fields Appl.* **31** (2005).

5. Dorsey-Hales, "Irreducible coefficient relations," in Sequences and Their Applications, *Springer LNCS* **7280** (2012)
6. Golomb *et al.*, *Shift Register Sequences*, Aegean Park Press, 1982 (Holden Day 1967).
7. Golomb-Gong, *Signal Design for Good Correlation*, Cambridge Univ. Press, 2005.
8. Hales-Newhart, Swan's theorem for binary tetranomials, *Finite Fields Their Appl.* **12** (2006).
9. Hayes, The distribution of irreducibles in GF$[q, x]$, *Trans. Amer. Math. Soc.* **114** (1965).
10. Massey, Shift-register synthesis and BCH decoding, *IEEE Trans. Inform. Theory* **15** (1969).
11. Mullen-Panario, *Handbook of Finite Fields*, CRC Press, 2013.
12. Swan, Factorization of polynomials over finite fields, *Pacific J. Math.* **12** (1962).

Number Theory — Sol's First Love

Alfred W. Hales,

*IDA Center for Communications Research**

QUOTE

"Since my PhD, my career has consisted mostly of applying all the topics my Harvard professors took great pride in proclaiming their inapplicability, to problems in signal design for digital communication (which has gotten me into NAE, the NAS, and most recently receiving the National Medal of Science), but my first love has remained prime number theory."

(Dec. 2013 letter from Sol to Andrew Granville.)

MATHSCINET
(database of reviews and abstracts)

Solomon W. Golomb 143 papers

58 papers

62 in information and communications

32 in combinatorics;

"The man who brought modern combinatorial mathematics to Southern California" (Gus Solomon toast at Sol's 1992 birthday fest.)

25 in number theory

and 24 in sixteen other areas.

NUMBER THEORY

Distribution of Primes:

1. Euclid ($\sim$300 BCE) There are infinitely many primes.
2. Gauss, Legendre ($\sim$1,800), Hadamard-de La Valle Poisson (1896) $\prod(n)$ is asymptotic to $n/\log n$.
3. *Conjecture*: There are infinitely many "twin primes," i.e., instances where p and $p+2$ are both primes.

That is, prime "gaps" of length 2 occur infinitely often.

---

*In memory of Sol Golomb, supervisor, mentor, colleague, and friend

The Wisdom of Solomon

Sol's Contributions:

1. (High school, 1949)

    "Prove that there are infinitely many twin primes if there are infinitely many positive integers n not expressible in the form $6ab \pm a \pm b$ (where a and b are positive integers, and the two $\pm$ signs are independent of each other)."

2. (Harvard Thesis, 1957): "Problems in the distribution of the prime numbers."
3. (*J. Number Theory*, 1970): "The lambda method in prime number theory."

$\Lambda_k(n) := \sum_{d|n} \mu(d)(\log n/d)^k$ is supported only when n has no more than k distinct prime factors.

GPY

Golston, Pintz, Yildirim (2005,2009), Annals of Math.

"Primes in tuples."

Theorem 1. *If p_n denotes the nth prime, then*

$$\lim \inf_{n \to \infty} \frac{p_{n+1} - p_n}{\log p_n} = 0$$

So, there are infinitely many cases where the gaps are arbitrarily small compared to the average spacing.

A major advance, *but* gaps could still grow and stay arbitrarily large.

ZHANG
ZHANG, etc.

Theorem 2. *(Zhang, 2014, Annals of Math.) There are infinitely many prime pairs p, q with difference at most* 70,000,000.

Uses Sol's work, and builds on GPY.

Shortly thereafter, "polymath" (Tao, etc.) lowered this to 4,680. Maynard lowered it to 600.

Further polymath work lowered it to 246.

Possible further lowering to 12 or even 6, using not-yet proved conjectures.

But lowering to 2 many take a year, or a decade or a century ...

"I feel privileged to have lived long enough to witness recent progress: GPY, primes in arbitrarily long arithmetic progress, and the existence of two (or more) primes infinitely often in bounded intervals. But the original twin prime conjecture still seems to remain just out of reach." (Sol's 2013 letter to Granville.)

Part III

© 2023 World Scientific Publishing Company
https://doi.org/10.1142/9789811234378_0007

The Most Interesting Man in the World

Daniel Arovas

I first met Sol through his daughter Beatrice, probably in 1988 or 1989, at one of his and Bo's magnificent Christmas parties. I took an instant liking to the entire family and in the years that followed attending these events was for me the highest point the holiday season. The other attendees were all fascinating and highly accomplished in various ways — academics, businesspeople, diplomats, artists, authors — I found the company exhilarating. But I was always most drawn to Sol himself and would discreetly follow him around the house to listen in on his conversations and contribute if I could. Sol was a polymath of the highest order, and there was scarcely a subject on which he could not hold forth with a highly informed if not expert opinion. One of the boundary conditions on my marriage in 1996 was that I attend synagogue on Saturdays and Jewish holidays — my wife is Modern Orthodox — and this soon led to a somewhat obsessive fascination in the area of Hebrew Bible and Rabbinic literature. Sol was from a rabbinic family and in his youth was the *baal qore* of his synagogue, but like me he was himself simultaneously secular and yet maintained an ineffable fascination with the texts and traditions of our ancient ancestors. I treasured our discussions of these subjects most of all. It was also great fun to gather around the piano with Sol and some of the other Jews during the Christmas parties and sing *yerushalayim shel zahav*.

I feel privileged to have known Sol. What follows is a brief description I delivered "To Sol, on the occasion of his birthday":

$$\frac{\frac{1}{\pi}\sum_{k=1}^{\infty}\frac{1}{k^2}}{\int_0^{\pi/2} d\theta \frac{1}{1+\tan^{\sqrt{2}}\theta}} \cdot \left\{ \det\begin{pmatrix} 15 & -3 \\ -4 & \sqrt{81} \end{pmatrix} - 6\,\mathrm{Re}\,e^{i\pi/3} \right\} \tag{1.1}$$

In graduate school, he defended six PhD theses simultaneously.
For a summer project, he once counted all the real numbers.
His colleagues surreptitiously attend his classroom lectures.
If he cited your work, even critically, you would put it on your resume.
His beard has been nominated for a Nobel Prize.
He once became confused, just to know how it feels.
As a child, he taught his dog to bark — in Aramaic.

He has solved two Hilbert problems while talking in his sleep. When Wikipedia is down, he is the official mirror site.

He is ⋯ **The Most Interesting Man in the World.**

© 2023 World Scientific Publishing Company
https://doi.org/10.1142/9789811234378_0008

The Issue of *Shift Register Sequences*[*]

Richard E. Blahut

Department of Electrical and Computer Engineering
University of Illinois
Urbana, IL, USA
Department of Electrical and Systems Engineering,
University of Pennsylvania, Philadelphia, PA, USA
blahut@illinois.edu

The book "Shift Register Sequences" by Solomon Golomb has influenced many subsequent developments in the theory and application of sequences. One personal line of such developments is summarized herein.

1. Introduction

I first came across the book *Shift Register Sequences* by Solomon Golomb shortly after it was published in 1967, not knowing then that the subject would enter and re-enter my life many times in different ways in the years to come. Various career assignments and circumstances, both in practice and in academia, continued to bring me back to shift-register sequences and their progeny. I will describe bits and pieces of mathematics as collected on a personal journey that started with my discovery of Sol Golomb's book so many years ago. And so this article has a title that can be read in several ways to acknowledges my debt to Sol.

2. Shift Register Sequences

The Fibonacci sequence is a well-known shift-register sequence in the ring of integers **Z** defined by the recursion

$$V_r = V_{r-1} + V_{r-2},$$

as initialized by two ones. The Fibonacci sequence is not periodic, nor does it eventually become periodic. However, in $\mathbf{Z_7}$, the ring of integers modulo seven, that

---

[*]In Memory of Solomon Wolf Golomb

same recursion produces a periodic sequence of period 16, which is a divisor of $7^2 - 1$.

Other recursions of this form in other arithmetic systems may produce sequences that are periodic or become periodic after an initialization start-up. Sequences in a finite alphabet must always become periodic after the initialization start-up. Only recursions in a finite field $\mathbf{F}_q$ will be considered herein.

A basic shift-register sequence is any sequence of symbols from a finite alphabet generated by a recursion of the form

$$\Lambda_r = -\sum_{j=1}^{L} \Lambda_j V_{r-j}.$$

The right side is written with a minus sign so that, with $\Lambda_0 = 1$, the recursion can be written $\Lambda(x)V(x) = 0$, where $\Lambda(x) = \sum_{j=0}^{L} \Lambda_j x^j$ and $V(x) = \sum_{j=0}^{n-1} V_j x^j$. Moreover, because $V(x)$ is periodic for some period n, this can be written

$$\Lambda(x)V(x) = 0 \mod(x^n - 1),$$

This formulation suggests that the linear recursion can be studied in terms of the zeros of polynomials. This will be discussed herein using the language of the Fourier transform.

3. The Fourier Transform

A Fourier transform of blocklength n in the field $\mathbb{F}$ is an invertible mapping from the set of vectors of blocklength n over the field $\mathbb{F}$ to the set of vectors of blocklength n over the field $\mathbb{F}$ or an extension field of the field $\mathbb{F}$. The Fourier transform of blocklength n exists in the field $\mathbb{F}$ if and only if the field $\mathbb{F}$, or an extension of the field $\mathbb{F}$, contains an element ω of order n. The Fourier transform of the vector $\mathbf{v}$ of blocklength n is the vector $\mathbf{V}$ of blocklength n given by

$$V_k = \sum_{i=0}^{n-1} \omega^{jk} v_j \quad k = 0, \ldots, n-1,$$

The components of $\mathbf{V}$ are in the field of ω, which is possibly an extension of the field $\mathbb{F}$. The Fourier transform in the field $\mathbb{F}$ is inverted by

$$u_j = \frac{1}{n} \sum_{k=0}^{n-1} \omega^{-jk} V_k \quad j = 0, \ldots, n-1,$$

where ω^{-1} also has order n. Thus the Fourier transform and the inverse Fourier transform have the same form except for the choice of the arbitrary ω of order n, and inclusion of the field integer n. In fields of characteristic two, n is odd, and so is always equal to one as an integer of that field. It can be omitted in a field of characteristic two.

The inverse Fourier transform can be described as the evaluation of the polynomial $V(x) = \sum_{j=0}^{n-1} V_k x^k$ at all powers of ω^{-1}. This leads to the expression

$$n^{-1} \Lambda(\omega^{-j}) V(\omega^{-j}) = \lambda_j v_j$$

which is known as the *convolution theorem*.

For an example of a Fourier transform, note that in the finite field $\mathbf{F}_7$, the element 3 has order six. Thus with $\omega = 3$, there is a Fourier transform

$$V_k = \sum_{i=0}^{5} 3^{jk} v_j k \quad k = 0, \dots, 5$$

of blocklength six in $\mathbf{F}_7$. Written explicitly as a matrix-vector product, this is

$$\begin{bmatrix} V_0 \\ V_1 \\ V_2 \\ V_3 \\ V_4 \\ V_5 \end{bmatrix} = \begin{bmatrix} 1 & 1 & 1 & 1 & 1 & 1 \\ 1 & 3 & 2 & 6 & 4 & 5 \\ 1 & 2 & 4 & 1 & 2 & 4 \\ 1 & 6 & 1 & 6 & 1 & 6 \\ 1 & 4 & 2 & 1 & 4 & 2 \\ 1 & 5 & 4 & 6 & 2 & 3 \end{bmatrix} \begin{bmatrix} v_0 \\ v_1 \\ v_2 \\ v_3 \\ v_4 \\ v_5 \end{bmatrix}.$$

Because 6 is composite, the finite field $\mathbf{F}_7$ contains the elements $3^2 = 2$ and $3^3 = 6$ that can form Fourier transforms of blocklength two and three, respectively. Thus a Fourier transform of blocklength six can be regarded as the Kronecker product of two Fourier transforms of blocklengths two and three, respectively.

A Fourier transform of blocklength n exists over $\mathbf{F}_q$ if and only if there is an element of order n in an extension of $\mathbf{F}_q$. Thus, because there are no elements of even order in any extension of $\mathbf{F}_{256}$, there are no Fourier transforms of even blocklength n over $\mathbf{F}_{256}$.

4. Linear Complexity

The linear complexity of a sequence v is defined as the length of the shortest linear recursion (or linear feedback shift register) that will produce that sequence.

The discussion of linear complexity begins with a condition under which two linear recursions that agree up to a certain point must continue to agree thereafter.

Theorem 4.1 (Agreement Theorem). *Two linear recursions Λ and Λ' of lengths L and L', respectively, that both produce the subsequence $V_0, V_1, V_2, \dots, V_{r-1}$ where $r \geq L + L'$ will both produce the same sequence thereafter.*

Proof. It is enough to prove that both linear recursions produce the same V_r. The rest of the sequence would then follow. But

$$V_r = -\sum_{k=1}^{L} \Lambda_k V_{r-k}$$

$$= \sum_{k=1}^{L} \Lambda_k \left[\sum_{j=1}^{L'} \Lambda'_j V_{r-j-k} \right] = \sum_{j=1}^{L'} \Lambda'_j \left[\sum_{k=1}^{L} \Lambda_k V_{r-k-j} \right]$$

$$= -\sum_{j=1}^{L'} \Lambda_j V_{r-j} = V_r.$$

All sums are well-defined because $r \geq L + L'$. The proof is now complete. $\square$

Theorem 4.2 (Massey's Theorem). *Whenever the shortest linear recursion that produces the sequence $(V_0, V_1, \ldots, V_{r-1})$ has length L and does not produce the sequence $(V_0, V_1, \ldots, V_{r-1}, V_r)$, then every linear recursion that produces the sequence $(V_0, V_1, \ldots, V_{r-1}, V_r)$ has length at least $r - L$.*

Proof. Were there a recursion of length L' satisfying $L' \leq r - L$ that produces the sequence $(V_0, V_1, \ldots, V_{r-1})$, then there would be two recursions that produce $(V_0, V_1, \ldots, V_{r-2})$. But $L + L' \geq r - 1$. The agreement theorem would then assert that there are two sequences that produce $(V_0, V_1, \ldots, V_{r-1})$, contrary to the premise of the theorem. $\square$

Theorem 4.3 (Blahut's Theorem). *The linear complexity of a periodic sequence of period n in the field $\mathbb{F}$ is equal to the Hamming weight of its Fourier transform provided F has an element of order n.*

Proof. The linear complexity of the periodic sequence $\boldsymbol{V} = (V_0, V_1, \ldots, V_{n-1})$ of *period* (or *blocklength*) n is the degree of the polynomial $\Lambda(x)$ of least degree such that

$$\Lambda(x)V(x) = 0 \quad \mathrm{mod}(x^n - 1)$$

The inverse Fourier transform of the vector $\boldsymbol{V}$ is the vector $\boldsymbol{v}$. Let the inverse Fourier transform of the vector $\boldsymbol{\Lambda}$ be the vector $\boldsymbol{\lambda}$ of blocklength n. Then

$$\lambda_j v_j = 0 \quad \text{for all } j$$

if and only if $\lambda_j = 0$ for every j at which v_j is nonzero. Then $\Lambda(x)$ must have at least as many zeros as the Hamming weight of the Fourier transform of $\boldsymbol{V}$. There is a $\Lambda(x)$ with zeros only at the nonzero components of $\boldsymbol{v}$, and this polynomial has degree equal to that number. $\square$

5. Linear Block Codes

A linear block code $\mathcal{C}$ of blocklength n can be defined in any field $\mathbb{F}$, including the real field $\mathbb{R}$, though because of the practical issue of numerical precision in the real field, such a code is practical only in a finite field $\mathbb{F}_q$.

A block code $\mathcal{C}$ of code rate $R = k/n$ over $\mathbb{F}_q$ is a set of q^k vectors of blocklength n over $\mathbb{F}_q$, called codewords. The code is a linear code when the componentwise sum of every two codewords is a codeword. The number of nonzero components in any codeword $\mathbf{c}$ of code $\mathcal{C}$ is called the *Hamming weight* of $\mathbf{c}$. The Hamming weight of $\mathbf{c}$ is equal to the Hamming distance between $\mathbf{c}$ and the all-zero codeword. The minimum Hamming distance of a linear block code is equal to the smallest Hamming weight of any nonzero codeword of $\mathcal{C}$. The minimum Hamming weight of a linear code is revealed by an elementary property of a matrix defining that code, as described below.

Let $\mathbf{H}$ be a $n - k$ by n full-rank matrix over the field $\mathbb{F}_q$. The matrix $\mathbf{H}$ is called the *check matrix* of a code C. Define the linear code $\mathcal{C}$ as the set of vectors $\mathcal{C} = \{\mathbf{c} \mid \mathbf{cH}^T = \mathbf{0}\}$. The code $\mathcal{C}$ is the null space of $\boldsymbol{H}$. Each codeword $\mathbf{c}$ of $\mathcal{C}$ corresponds to a linearly-dependent subset of the columns of $\mathbf{H}$. The number of columns in each such subset is equal to the Hamming weight of the corresponding codeword $\mathbf{c}$, and so is equal to $d_{\min}$ for a codeword of minimum Hamming weight. The minimum Hamming weight of code $\mathcal{C}$ is equal to the number of columns in the smallest set of dependent columns of $\mathbf{H}$.

The quality of a linear (n, k) code is traditionaly judged by $d_{\min}$. However, available evidence is that for communication on a discrete memoryless channel, $d_{\min}$ need not always be the best quality measure. For large blocklengths and code rates larger than a rate parameter called R_{crit}, maximizing the Hamming distance between codewords need not be a good way of designing codes, even for a symmetric channel. The *weight profile* of a linear code, which is a list of the number of codewords of each weight, is then a better predictor of the performance of the code under maximum-likelihood decoding. The weight profile, however, has received skant attention in the literature, and so will receive no further attention herein.

6. The Heft of a Matrix

The following definitions of the rank and heft of a matrix are meaningful for matrices in any algebraic field $\mathbb{F}$. The distance properties of linear algebraic block codes in any field can be understood as the interplay between these two general matrix properties.

Definition 6.1. For any matrix $\boldsymbol{A}$ with at least as many columns as rows over the field $\mathbb{F}$:

- The rank of $\boldsymbol{A}$ is the largest value of r such that some set of r columns of $\boldsymbol{A}$ is linearly independent.

- The heft of A is the largest value of r such that every set of r columns of A is linearly independent.

The minimum weight of the code C is equal to the number of columns in the smallest set of dependent columns. It is clear from the definition that heft $\mathbf{H} \leq$ rank $\mathbf{H}$. Therefore,

$$d_{\min} - 1 = \text{heft}\,\mathbf{H} \leq \text{rank}\,\mathbf{H} = n - k$$

because the largest subset of linearly independent columns is equal to the heft of $\mathbf{H}$. Accordingly, $d_{\min} \leq n - k + 1$ for any linear block code, an inequality known as the Singleton bound. The central task of classical algebraic coding theory is the quest for large k by n full-rank matrices $\mathbf{H}$ over the finite field $\mathbb{F}_q$ that satisfy this inequality as tightly as possible. Finding a linear (n, k) algebraic code over the field $\mathbb{F}_q$ with large minimum distance amounts to finding an $n - k$ by n matrix $\mathbf{H}$ over $\mathbb{F}_q$ with large heft.

In general, little is known about matrices with large heft. A vandermonde matrix satisfies the Singleton bound with equality, and so has the largest possible heft for a square matrix of its size. This large heft is inherited by some elementary modifications of a vandermonde matrix over $\mathbb{F}_q$ and certain of its submatrices. For this reason, most known matrices $\mathbf{H}$ of large heft are based on some alteration of a vandermonde matrix. Most of the classical algebraic codes such as the ReedSolomon codes and the BCH codes are based in some way on a vandermonde matrix.

7. The Generator Matrix

The expression $\mathbf{c}\mathbf{H}^T = \mathbf{0}$ defining codewords implicitly can be replaced by an explicit statement suitable for encoding. Any k by n full-rank matrix $\mathbf{G}$ satisfying $\mathbf{G}\mathbf{H}^T = \mathbf{0}$ is called a *generator matrix* for the code C defined by $\mathbf{H}$. Every codeword $\mathbf{c}$ can be written $\mathbf{c} = \mathbf{a}\mathbf{G}$ for some k-vector a. Letting a denote the data vector, this expression describes an encoder.

One way to compute a generator matrix that corresponds to a check matrix is to use column permutations and elementary row operations to rewrite $\mathbf{H}$ in blockpartitioned form as $\mathbf{H} = [\mathbf{I}\,\mathbf{P}]$, where $\mathbf{I}$ is an identity matrix, and $\mathbf{P}$ is a square matrix.. Then $\mathbf{G} = [-\mathbf{P}^T\mathbf{I}]$. An alternative way to compute such a generator matrix is to enlarge $\mathbf{H}$ to an n by n full-rank matrix by adjoining any k additional linearly independent rows in the form of another matrix $\mathbf{B}$. Then

$$\mathbf{M} = \begin{bmatrix} \mathbf{H} \\ \mathbf{B} \end{bmatrix},$$

so that $\mathbf{M}$ is a full-rank n by n matrix with linearly independent rows. Accordingly, $\mathbf{M}$ is invertible. The statement $\mathbf{M}\mathbf{M}^{-1} = \mathbf{I}$ can be block-partitioned as

$$\mathbf{M}\mathbf{M}^{-1} = \begin{bmatrix} \mathbf{I} & \mathbf{0} \\ \mathbf{0} & \mathbf{I} \end{bmatrix} = \begin{bmatrix} \mathbf{H} \\ \mathbf{B} \end{bmatrix} [\mathbf{H}^{-1}\mathbf{B}^{-1}],$$

The Issue of Shift Register Sequences 129

where $\mathbf{M}^{-1}$ is written as $[\mathbf{H}^{-1}\mathbf{B}^{-1}]$. The new submatrices are appropriately denoted $\mathbf{H}^{-1}$ and $\mathbf{B}^{-1}$ because the matrix $\mathbf{H}^{-1}$ is a n by k matrix[1] satisfying $\mathbf{HH}^{-1} = \mathbf{I}$, and the matrix $\mathbf{B}^{-1}$ is an n by $n-k$ matrix, satisfying $\mathbf{BB}^{-1} = \mathbf{I}$. The last $n-k$ columns of the inverse $\mathbf{M}^{-1}$ satisfy $\mathbf{HB}^{-1} = \mathbf{0}$ and so the matrix $\mathbf{B}^{-1}$ is actually the transpose of a generator matrix $\mathbf{G}$ for the code $\mathcal{C}$.

Because $\mathbf{B}$ can be chosen in many ways, there are many such $\mathbf{G}$ corresponding to $\mathbf{H}$. Thus, given either $\mathbf{G}$ or $\mathbf{H}$, there are many ways of choosing the other to satisfy the relationship $\mathbf{GH}^T$.

8. Cyclic Codes and Their Coset Codes

Two classes of cyclic codes are described in terms of the Fourier transform representations of codewords, which is denoted either as $\boldsymbol{C}$ or as $C(x)$. These two classes are the Reed-Solomon codes and the BCH codes. In the first case, $\boldsymbol{C}$ is in the field of the code $\mathbb{F}_q$ (as is ω), which means that the blocklength n is a divisor of $q-1$. In the second case, $\boldsymbol{C}$ is in an extension field of the code $\mathbb{F}_{q^m}$ (as is ω), which means that the blocklength n is a divisor of $q^m - 1$.

For any blocklength n that admits a Fourier transform over $\mathbb{F}_q$, let $C(x)$ be a nonzero polynomial with coefficients in the field $\mathbb{F}_q$ such the $2t$ consecutive coefficients of $C(x)$, viewed cyclically, are equal to zero. Let $\Lambda(x)$ be a nonzero polynomial such that

$$\Lambda(x)C(x) = 0 \mod(x^n - 1).$$

Then $\Lambda(x)$ must have degree at least $2t + 1$. Otherwise, for a degree of $\Lambda(x)$ at most $2t$, let $C_r, \ldots, C_{r+2t-1}$ be the zero coefficients of $C(x)$. Then $\Lambda(x)C(x) = 0$ implies that $C_{r+2t} = 0$, and in turn all $C_j = 0$, contrary to the condition that $C(x)$ is nonzero. The convolution theorem

$$\Lambda(\omega^{-j})C(\omega^{-j}) = \lambda_j c_j$$

then shows that the vector $\boldsymbol{c}$ has at least $2t + 1$ nonzero components because $C(x)$ has $2t$ consecutive zero coefficients.

The set of all $c(x)$ with the same set of $2t$ consecutive zeros is called a Reed-Solomon code when $\boldsymbol{c}$ and $\boldsymbol{C}$ are in the same field. Otherwise, when ω is in an extension field of the field of the code, additional coefficients of the transform $\boldsymbol{C}$ must also be zero so as to satisfy the conjugacy constraints $C_{qk} = C_k^q$. These codes are BCH codes.

Reed-Solomon codes and BCH codes are cyclic codes. Any cyclic shift of a codeword in the code $\mathcal{C}$ is also in the code $\mathcal{C}$. When a long string of concatenated codewords is incorrectly synchronized by a few symbols, the perceived codewords will be cyclic shifts of the actual codewords with a few symbol errors at the end of each codeword. A coset code removes the cyclic structure, thereby preventing this

---

[1]Because $\mathbf{G}$ and $\mathbf{B}$ are not square matrices, $\mathbf{H}^{-1}$ and $\mathbf{B}^{-1}$ are called pseudoinverses. They are not unique.

false decoding, and also enables self-synchronization of the codeword stream when there are only a few error symbols. Simply replace one of the zeros at the edge of the block of $2t$ zeros of C with a one, thereby forming a noncyclic coset code. When the codeword is correctly synchronized, this superfluous one can be removed by the decoder, so the decoding algorithm is unaffected. When the codeword is not properly synchronized, and there are only a few errors, this one in the transform domain will be replaced by a power of ω that can be recovered even in the presence of a few error symbols. The power of ω reveals the slippage. Thus the coset code can correct various combinations of errors, including erasures, and synchronization slippage.

Bibliography

1. S.W. Golomb, *Shift Register Sequences*, Holden-Dey, San Francisco, 1967.
2. R.E. Blahut, *Algebraic Codes for Data Transmission*, Cambridge University Press, 2003.
3. R.E. Blahut, *Cryptography and Secure Communication*, Cambridge University Press, 2014.

© 2023 World Scientific Publishing Company
https://doi.org/10.1142/9789811234378_0009

Solomon Golomb — Some Personal Reflections Plus Random Biographical and Bibliographical Notes

Gary S. Bloom

Computer Science Department, City College of CUNY,
New York, NY 10031, USA

Several years ago I laughed out loud at a recruiting brochure to bring students to the University of Southern California. On the front page in glorious color, captioned by the clause "At USC you'll get much more than a great band and a great football team", there was a picture of our friend and inspiration Sol Golomb in full football uniform (#45) with a football tucked under one arm and his helmet tucked under the other. It occurred to me that with all the roles at USC that Sol has taken on, that he found in this ersatz personna yet another way of broadening his involvement.

I became a student of Sol's at USC in 1974 for several reasons. My thesis advisor at the time was Dick Bellman, who had become seriously ill. He had told me a couple of years earlier: "Take any course that Golomb teaches." I had already followed Dick's instructions and had been delighted with the clarity of insights provided by Sol's probability, information theory, and coding theory courses. As his advisee, I had even more fun. I would visit Sol in his office every week or two, and would listen for an hour or so to his holding forth on his research results and those of other folks. At the time I didn't recognize that I was being steered by Sol in a direction appropriate for my background, but it became obvious when I was able to summarize what I heard, and was able to develop it further in new and unforeseen directions. When new results started appearing, Sol and I got into a cooperative race for problem solutions that was both productive and exhilarating. And I loved every minute of it.

It was also during this period that Sol would sometimes excuse himself from our appointments, because there were conflicts with the substantial blocks of time

---

0898-1221/00/$ — see front matter 2000 Elsevier Science Ltd. All rights reserved. Typeset by
AMS-TEX PII: S0898-1221(00)00100-0

132   *The Wisdom of Solomon*

he was giving to the university for the Faculty Senate. He became its President in 1976–1977. Later, in 1986–1989, he took on the task of serving as USC's Vice Provost for Research. Most recently, he served administratively as Director of Technology at the Annenberg Center for Communication, a post he gave up in 1998.

For all that he has done over his many years for USC, it was fitting that he was given the University's Presidential Medallion in 1985, its most prestigious honor. I sometimes still reflect on what he told me of his reasoning for choosing to join the USC Faculty when he left the Jet Propulsion Laboratory (rather than going to Cal Tech or UCLA): "I wanted to go to the institution where my contributions would make the greatest impact". When we view the University of Southern California that Sol joined with the great university of today, and think of Sol's interactions with it, we find it easy to believe in his success and the satisfaction that his contributions must give him.

Certainly, much of Sol's success in the greater university community was due to his truly being a "scholar" in the broad sense. He belongs to that old school that finds intellectual stimulation over a broad spectrum of topics. I remember my personal delight in working for a person who would not only talk about mathematics, but would hold forth both on Bible scholarship and why "Y sha-ru-tsu" was the Japanese way to write the English phrase "white shirt". I remember also my great pride in almost winning a game of chess with Sol, in which he started by removing his queen from the board.

I do need to mention the real reason for my wanting to work under Sol. He was the only person I knew who could meet my own exacting standards for selecting an advisor. He became for me, not only my academic parent, but also my "anagram-father": nevertheless, I still do suffer some embarrassment when my very existence is questioned by those who notice that the name G. S. Bloom permutes into S. Golomb.

Of course, Sol has been honored in far broader forums than USC. For example, he was the first member of the National Academy of Engineering elected from USC. He is a Fellow of the IEEE and also of the AAAS, he has been elected as a foreign member of the Russian Academy of Natural Sciences, and he has received prizes for the significance of his work including the major recognition of the IEEE Information Theory Group, the Shannon Prize.

Perhaps we should insert Sol's list of publications here, but instead, we'll list a few artifacts of his career that have served as stones tossed in a pond whose ripples have carried his efforts far beyond that point where most of us can see. USC has fortunately been able to reward Sol's intellectual contributions to the scientific community by designating him the first holder of the Andrew and Erna Viterbi Chair in Communications.

Table 3.1 lists Sol's 21 doctoral students who have preserved Sol's influence in their own work for another academic generation.

Sol has written well over 200 journal articles, books, and articles in conference proceedings. The topics are widely varied, but a significant number of these writings have proved seminal. Some have simply been so fundamental in establishing fields of research that the book or paper must be cited. Others have asked such provocative

Table 3.1. Sol's 21 doctoral students.

Ph.D. Date	Doctoral Student	Dissertation Title
1962	Jack Stiffler	"Self-synchronizing binary telemetry codes"
1962	Robert C. Tausworthe	"Correlation properties of cyclic sequences"
1966	Abraham Waksman	"On the performance of a one-dimensional array of finite state machines"
1966	William Hurd	"Statistical properties of speech, music and noise, and the detection of speech in noise"
1968	Harold M. Fredricksen	"Disjoint cycles from the de Bruijn graph"
1968	David Sherman	"Speech spectral measurements—Convergence and applications to language discrimination"
1970	Clarence Fuzak	"On voicing duration in connected English speech"
1970	Herbert M. Trachtenberg	"On the cross-correlation functions of maximal-linear recurring sequences"
1973	Ralph Thoene	"Quasi-polyominoes"
1976	Gary S. Bloom	"Numbered undirected graphs and their uses: A survey of a unifying scientific and engineering concept and its use in developing a theory of nonredundant homometric sets relating to some ambiguities in X-ray diffraction analysis"
1978	Martin J. Cohen	"On the difference $x(x+1)\ldots(x+n-1)-y$"
1981	Herbert Taylor	"Some graph theoretic topics from electrical engineering"
1981	Unjeng Cheng	"Properties of sequences"
1983	Betty Tang	"Comma-free and bounded synchronization delay codes of even word length"
1987	David Rutan	"Difference sets and analysis of the periodic correlation of sequences"
1988	Ning Zhang (Chang)	"Generalized Barker sequences"
1988	Gregory Yovanof	"Homometric structures"
1991	Hong-Yeop Song	"On aspects of Tuscan squares"
1993	C. Wayne Walker	"Solving the error locator polynomial over finite fields in algebraic decoding"
1994	Gregory Mayhew	"Statistical properties of modified de Bruijn sequences"
1999	Peter Gaal	"On the conjectured new families of cyclic Hadamard difference sets"

questions (often stated simply and clearly) that a competitive flurry of activity has resulted. We have checked in the libraries that subscribe to the compact-disk version of the *Science Citation Index* and have noted during the 18 years from 1980 to 1997 inclusive that citations to Sol's work in the papers they reviewed appeared 855 times ... or an average of 48 documented citations per year.

For example, besides producing numerous single papers that dealt with *Shift Register Sequences*, Sol wrote the book in 1967 that is still required reading 32 years later. Not surprisingly, in 1982, Sol wrote an updated revision to the original text, which then became the version to cite. The *Science Citation Index* indicates that in the years from 1980 to 1997 just over half of the refereed papers (434 of the 855) that have cited Sol's work include one of the editions of this book. Among the many application of shift-register technology is the current (IS-95) standard technology for cellular telephony, developed so successfully by Andy Viterbi and implemented by his Qualcomm company.

134   *The Wisdom of Solomon*

Shift-register technology is one kind of *Digital Communications with Space Applications* that Sol wrote about (with portions written by his esteemed colleagues, L. Baumert, M. Easterling, J. Stiffler, and A. Viterbi) in the text of that name in 1964. Because of continuing demand, a reprint appeared in 1981. There have been 58 citations of this book that appeared in refereed papers from 1980 to 1997 according to the *Science Citation Index*.

Few folks are more qualified to write a text on *Information Theory and Coding* than Sol, who did so with Robert Peile and Bob Scholtz. (I might have been able to take fewer notes if it had existed in my student days.) Plenum Publishers indicate that to date well over 1,000 copies of this volume have been purchased.

Perhaps the most read book of Sol's is considered "recreational", because it is accessible to the public that loves intellectual recreations ... puzzles, if you will. *Polyominoes* in its original and revised forms, and its translations, and perhaps by its being described in Martin Gardner's *Scientific American* column, has retained popularity from its appearance in 1965 through its 1996 revised paperback edition until today. Part of its captivating property is that its readers may match their level of mathematical sophistication with the problems on which they work and may even consider a variety of applications to parts of the physical world modeled by polyominoes. This recreational work in applied combinatorial geometry has not only delighted many, but has been cited (in "serious" mathematical, refereed journal articles since 1980) 49 times, as noted by the *Science Citation Index*.

The book *Polyominoes* is only one example of Sol's knowing how to reach people with seductive mathematics. His puzzle columns naturally come to mind, and problems that he brought to Martin Gardner's attention. But there have been more. Log onto the internet at http://members.aol.com/golomb20/intro.html and find a lively and extensive continuing discussion of Golomb Rulers, the shortest ruler with n marks which makes no duplicate measurements. (A measurement is the distance between any pair of the n marks.) Sol calculated rulers through length 11 in 1965. Now, there is an active community determined to find the next Golomb rulers ... at the moment the "next" ruler has a length of 24.

Of course, all of us know of Sol's effectiveness in purely mathematical topics. Among the influential papers cited in the *Science Citation List* are "The information-generating function of a probability distribution" in the January 1966 *IEEE Transactions on Information Theory*, "Formulas for the next prime" in the 1976 *Pacific Journal of Mathematics*, and "An identity for $C(2n, n)$" in the October 1992 issue of *American Mathematical Monthly*. In the *Math Reviews* now there are two subject classification headings pioneered by Sol. One is "Polyominoes", and the other is "Shift Register Sequences".

Another mode of influencing the future of mathematics is to coauthor with other well-known researchers of the present. Table 3.2 lists Sol's 37 co-authors. Perhaps in this Seven-Degrees-of-Freedom-Era, we should refer to these as having "Golomb Number" one. The list naturally includes Sol's students and USC colleagues—many of whom have substantial reputations of their own. In addition, the list contains other major names in coding theory and combinatorics as well as that of

Table 3.2. Sol's 37 co-authors.

Baumert, L.D.	Gaal, P.	Lemple, A.	Selfridge, L.
Bloom, G.S.	Gong, G.	Malling, L.	Song, H.Y.
Cheng, U.	Gordon, B.	Moreno, O.	Tang, B.
Cohen, M.	Gottesman, S.	Mayhew, G.	Taylor, H.
Dahlke, K.A.	Graham, R.L.	No, J.S.	Truong, T.K.
Dai, Z.D.	Grieve, P.	Peile, R.	Weindling, M.N.
Davey, J.R.	Hales, A.W.	Posner, E.C.	Welch, L.R.
Delbrück, M.	Hall, M., Jr.	Reed, I.S.	Yovanof, G.
Etzion, T.	Lee, H.K.	Scholtz, R.A.	Zhang, N.
Even, S.			

Max Delbrück, a Nobel prize winning biologist. Long before such a view had any popular following, Sol saw in the sequencing of DNA and other biological molecules, the coding of life (see for example: "On the plausibility of the RNA code" in the December 1962 issue of *Nature*).

It is also popular to calculate one's Erdős number. (Writing a paper with Erdős, confers on the co-author an Erdős number of one; co-authoring with an Erdős co-author, gives one an Erdős number of two; and so on.) By the standard convention, Sol has a multiple claim to Erdős number two, e.g., through Gordon, Graham, Selfridge, and Taylor. On the other hand, Sol notes that perhaps his Erdős number is negative one, since Erdős wrote a paper on a problem proposed by Sol.

In reflecting about Sol, there is much to say. We've neglected commenting on his commercial ventures and his family ... two more of his successful dealings with life.

It has been an honor and pleasure to help put together *Sol Golomb's 60th Birthday Symposium* and this special issue of *Computers & Mathematics with Applications*. It reminds me that people of outstanding intellect and humanity, of generous spirit and of kind humor not only exist, but can win in the great game of life. And it makes me proud that I have been lucky enough to know and work with one.

Bibliography

1. S.W. Golomb, *Basic Concepts in Information Theory and Coding: The Adventures of Agent 00111,* Plenum Press, New York, 1994.
2. S.W. Golomb, *Digital Communications with Space Applications,* Prentice Hall, Englewood Clifs, NJ, 1964; Reprinted, Peninsula Publishing, Los Altos, CA, 1981.
3. S.W. Golomb, *Polynomials,* Scribner, New York, 1965.
4. S.W. Golomb, *Polynomials: Puzzles, Patterns, Problems, and Packings,* Second Edition, Princeton University Press, Princeton, NJ, 1994.
5. S.W. Golomb, *Shift Register Sequences,* Holden-Day, San Francisco, CA, 1967.
6. S.W. Golomb, *Shift Register Sequences,* Revised Edition, Aegean Park Press, Laguna Hills, CA, 1982.

© 2023 World Scientific Publishing Company
https://doi.org/10.1142/9789811234378_0010

Golomb's Norske Forbindelser*

Tor Bu

IT Director, University of Bergen, Bergen, Norway
tor.bu@it.uib.no

With all his contacts worldwide it is a pleasure for me to report, that in the small mathematical community in Norway, Golomb has personal friends and that his work on number theory, combinatorics, and communication theory has influenced the academic achievements of Norwegian mathematicians.

My own first contact with Sol was on his visit to the University of Bergen in 1970 where he gave a couple of lectures. I recall that one subject was the theory of tiling a 3D Cartesian box with bricks of various dimensions. During the academic year 1979/1980 Sol kindly let me work as a visiting scholar at his department at USC, and I will always remember the hospitality that was bestowed on me and my family by Bo and Sol. I also appreciate how he introduced me to a lot of interesting mathematics and fascinating mathematicians, some of them world famous.

Golomb's first visit to Norway was as early as 1955. I was only ten years of age at that time so I have asked Sol to give his own account of his first meeting with Norway and Norwegian mathematics.

> It was a morning in November, 1954, when I noticed an announcement on a bulletin board at Harvard describing the Fulbright program for overseas study. The deadline for applying was that very day! I went to the graduate school office in Farlow House, and told them I wanted to apply for a Fulbright Grant. "But today is the deadline," was the surprised reply. "That's okay," I said, "I don't have anything else urgent to do today." "You'll have to list four faculty members who have agreed to serve as references." "No problem," I said confidently. "Well, then, what country are you applying for?" "Gee, I hadn't thought about that. What countries have you got?"

---

* Translation: Golomb's Norwegian Connections.

0898-1221/00/$ — see front matter ©2000 Elsevier Science Ltd. All rights reserved. Typeset by AMS-TEX PII: S0898-1221(00)00120-6

138   *The Wisdom of Solomon*

Why did I pick Norway? For one thing, they told me that England and France were always oversubscribed. My thesis was on prime number theory, and I was especially interested in twin primes. I had spent a very fruitful day a year earlier with Atle Selberg, and he was a Norwegian, albeit a permanent member of the Institute for Advanced Study in Princeton. And I had studied Brun's Theorem on twin primes in Landau's *Vorlesungen über Zahlentheorie*, and Brun was a Norwegian, though at that time I had no idea where he was in Norway, or even whether he was still alive. But that was two more pluses for Norway than for any other country on the list!

Much to my amazement, my application was approved. I crossed the Atlantic on the Norwegian America Line's *Stavangerfjord* in mid-June, 1955, with the other 15 or so Fulbright grantees and a couple hundred students bound for the Summer School for American Students at the University of Oslo. I like to tell people I learned Norwegian during the crossing. After all, it was a slow boat. By the time we landed, I could express myself intelligibly in Norwegian in a few limited areas, like asking for directions. Unfortunately, I couldn't understand the replies!

I had written to Viggo Brun. He was very much alive. He turned 70 a few weeks after I arrived, but continued on the faculty of Oslo University for the rest of the year I was there. It was only after I heard Brun's explanation of his "sieve method" that I felt I understood what it really involved. He invited me on several occasions to visit the Brun estate, Tværkjegla, in Drøbak, on the Oslofjord, where he lived with his wife Laura (they had no children). There, he pointed out several local landmarks to me: the house where Leon Trotsky lived when he first fled Russia; and the two cannons ("Moses" and "Aaron") left over from the Napoleonic Wars, which were fired at the invading German fleet on April 9, 1940, scoring a direct hit and sinking the flagship with all hands on board.

Only two years younger than Brun was Thoralf Skolem, whose lectures on group theory I attended, and I was invited to his home for dinner at least once. Two of the other "senior" faculty members were Johansson and Ljunggren, with whom I interacted only slightly. The two youngest members of the Mathematics Department were Lektor Ernst S. Selmer, and Amanuensis Karl-Egil Aubert, both of whom I got to know quite well. Selmer and I had many mathematical interests in common, in prime number theory, sequence generation, combinatorics, etc. He was also the editor of the *Nordisk Matematisk Tidskrift*, which published my second mathematical article, "Properties of Consecutive Integers." I would have submitted it in Norwegian, but I couldn't get the mathematicians at Oslo to agree on a Norwegian word for "consecutive". (One suggestion, "påhinannenfølgende", was branded too *Danish*; another "konsekutive", too *English*; and if the title was to become the word-for-word equivalent of "Properties of Numbers which Follow One After the Other," I felt the whole impact would be lost. Years later, Selmer suggested that "suksessive" would have been acceptable.) A more serious article I wrote the same year, "Sets of Primes with Intermediate Density", was published in *Mathematica Scandinavica*.

Actually, my main task that year in Oslo was to put my doctoral thesis in final form. I mastered material in Hardy's *Divergent Series*, Titchmarsh's *Theory of the Riemann Zeta Function*, and several other works, which played a major role in the ultimate shape of my dissertation. Harvard had very strict rules about the paper, the typing, etc., which were difficult to meet in Norway. I had to find a bulk paper dealer in downtown Oslo to cut large bond sheets of the required quality into 8 (1/2) × 11 pages.

Golomb's Norske Forbindelser

And I found a commercial typing service which assigned a typist who knew no English whatsoever to doing my thesis, formulas and all. (She was very good, but occasionally "of" appeared as "og", the very common Norwegian word for *and.*)*I* mailed my thesis off to Harvard on Norway's National Day, May 17, 1956.

My monthly stipend from the Fulbright Foundation was $120.00, which today would not pay one night's lodgings at any halfway decent Oslo hotel, but which converted to NOK 857.14 at the official rate of 1 norske krone = $.14 U.S., and enabled me to live like an aristocrat in the Oslo of 1955–56. I had brought my "huge" American car over with me, a 1953 Ford Customline, which reinforced my upper-class status, though for travel between the student housing at Sogn Studentbyen, the science campus of the university at Blindern, and Karl Johansgaten in downtown Oslo, I usually went by "trikk" (the electric trains, underground from the downtown Nationalteatret station to peripheral Majorstua, and fanning out from there above ground, though Blindern and Sognsvatn were on the same Ullevål Hageby line).

Oslo was a very provincial small town in those days. To eat in a Chinese restaurant, or to visit a first-class mathematics library, I had to travel the 400 miles to Copenhagen, which I did frequently, by car, by boat, by train, and by air, each on several occasions. I had acquired considerable proficiency in Norwegian when I paid my first visit to Denmark, later in the summer of 1955, and I had read Danish newspapers, which I found scarcely distinguishable from Norwegian. The shock came as soon as I tried to speak with the Danes. It wasn't that I couldn't decode the information, or even that I couldn't demodulate the carrier. I couldn't detect where the signal was! (It took me three decades to realize that speaking Norwegian was actually a handicap for speaking Danish, which had to be approached completely independently.)

I visited Oslo again in the summers of 1960, 1961, 1969, 1976, 1981, and 1984. In January, 1970, I visited the University of Bergen at Selmer's invitation. (At that time, my Ph.D. student Harold Fredricksen was a postdoc in Bergen.) I visited Stavanger (and Oslo) in the summer of 1981; and I visited the University of Bergen again in June 1984 and in June 1990. Starting in 1961, Bo has accompanied me on almost all of these trips, as well as on many trips to Denmark, which have been annually since 1978.

Of the mathematicians Golomb met at that time, only Selmer is still alive, now 80. Skolem and Brun will always be remembered for their outstanding contributions; Skolem in mathematical logic and Brun in number theory. Also, Ljunggren and Aubert have made significant contributions; Ljunggren in the theory of Diophantic equations and Aubert in algebra.

Ernst S. Selmer was at that time active in number theory and won international reputation for his work on Diophantic equations. Recently, the 'Selmer groups' have played a role in the famous solution of Fermat's problem. In 1957 Selmer was appointed as the first professor of pure mathematics at the University of Bergen where he did a great job as a teacher, an administrator, and as an entrepreneur for electronic computing.

What was for a long time known only among a small group within the Norwegian Secret Services was that Selmer was one of Norway's top cryptologic experts. Only recently this has become common knowledge, and Selmer himself gave a vivid account of his "secret life" in his speech to the 1993 EUROCRYPT conference in Lofthus, Norway.

In 1966, Selmer's cryptologic interests surfaced in a series of lectures which led to his widely distributed book *Linear Recurrence Relations over Finite Fields*, in which there are references to Golomb's work on shift register sequences. Golomb's theory was also taught in electrical engineering courses at the Technical University in Trondheim.

In the mid-1960s, Selmer had a group of students working on various problems related to shift register sequences. I and two other students were attacking the 'Norwegian Squares': place the n different translates of a cycle of period n in a square such that the columns also become translates of the same cycle. Even if there is no direct reference to Sol related to this problem I think it is fair to say that it is a puzzle in his spirit. In 1969/1970 Harold Fredricksen was working on properties of the DeBruijn sequences as a visiting scholar in Bergen. Johannes Mykeltveit, one of Selmer's students, through Fredricksen, was introduced to one of Golomb's many conjectures: the number of cycles generated by an arbitrary binary recurrence relation does not exceed the number of cycles generated by the pure cycling register. By an ingenious argument about the mass centers of the cycles of the DeBruijn Graph, Mykkeltveit was able to prove Lempel's conjecture which implies Golomb's conjecture.

Another student of Selmer's, Tor Helleseth, continued to work in the field of error correcting codes and cryptography, and is now an internationally recognized expert in the field. He has spent two sabbatical years with Golomb at USC. One of Helleseth's contributions which is directly inspired by Sol, is a number of theorems on properties of the cross-correlation function between two maximal sequences.

A colleague of Tor Helleseth, Torleiv Kløve, has published a large number of papers on number theory, combinatorics, and cryptography. In three papers, he has generalized the 'Golomb Ruler' to obtain constructions of bounds on sets of integers such that all differences between two numbers are different. In one of these papers, he also applies the "Sonar Sequences" defined by Golomb and Herbert Taylor.

In Golomb's own account of his first visit to Norway he mentions the difference between Norwegian and Danish. I would like to mention the peculiarity that in Norway there are two official written languages, "Nynorsk" and "Bokmaal", and that Sol refers to Bokmaal when he declares that "Danish is scarcely distinguishable from Norwegian". Nynorsk on the other hand is a creation from the period of Norwegian national awakening some 150 years ago, based on dialects from the rural communities and "Old Norwegian" (which is close to the modern language of the Republic of Iceland). As an example of Sol's skills in decoding languages, he seems to have fewer difficulties in understanding Nynorsk than many of my Norwegian friends.

With this I wish Sol a long and happy continuation of his links to Norway.

© 2023 World Scientific Publishing Company
https://doi.org/10.1142/9789811234378_0011

Sol Golomb, My Friend

Leo Buxbaum, MD

I first met Sol Golomb in 1962. The initial introduction to him was via a very dubious real estate investment. A mutual friend of ours, a PhD in Physics from Harvard, and an engineer, got Sol first, and then me, into this misadventure. The investment of course, turned out to be a great fiasco, but it generated an outstanding return for me — I began a pleasant long-term friendship with Sol and his charming wife Bo.

Sol's intellect is well known to everyone, but that had little to do with our friendship. What may not be so well acknowledged is what a pleasant, humorous, very normal, and generous person he was.

I was his part time doctor, as well as his friend, so I got to know him very well. Although I'm no academic, he always invited me to his numerous and interesting social events, exposing me to his very wide range of his friends. I met famous academics, Nobel prize winners, on to people in very ordinary occupations. He treated everyone he met politely and warmly.

He was a great friend to my 2 boys-he wrote them letters of recommendation to their colleges

He was very practical — once during a discussion re a divisive politician-one group thought the fellow was a crook — others that he was very clever — Sol settled the argument in his usual concise mathematical way — stating that "these arguments are not mutually exclusive".

Sol was very generous to me with his time and very patient. I was once fooling around with trying to solve Rubik's cube in 1980. He went to the great trouble of trying to teach me Finite Elementary Group Theory so I was actually able work out an accurate, although a tedious, solution. He did this even after I told him I nearly failed Plane Geometry in high school. That is what I call generous. I realized then what a great professor he must be. Anyone who could teach me abstract math, was an outstanding teacher.

We often met on Saturday AM for brunch at Canter's Deli, a well-known eatery on Fairfax Blvd In West Hollywood (WEHO). A pleasant small group usually turned up, always including my good friend Gus Solomon, another math prodigy. We would solve the world's affairs, sometimes well into the afternoon. Gus was well known in math circles, but was really interested in women and becoming a movie star. He was more successful in mathematics. He co-authored some papers with Sol. This always tickled both of them, since one was G. Solomon, and the other was Solomon G. I guess mathematicians are easily amused.

I'm sure the academic community will greatly miss his outstanding achievements.

I, and my wife and kids, will never forget what a nice fellow he was.

We feel like we've lost a member of our family.

Leo Buxbaum, MD

11/14/2018

© 2023 World Scientific Publishing Company
https://doi.org/10.1142/9789811234378_0012

Ode to Sol

John Dillon

On the occasion of Sol's 70th birthday, which John Dillon was not able to attend, he sent in the following email and the poem.

"I can't make the party, I'm sorry to say, but my thoughts are with you on this special day."

jfd

Through your wondrous discoveries,
inspiring conjectures,
crystal-clear writings,
and sparkling lectures,
you have shown us how beautiful zeros and ones
can be when arranged so correlation's expunged.
And the rich new supply of varying ranks
we'd not have without you,
so to you we give THANKS!

HAPPY BIRTHDAY, SOL!

© 2023 World Scientific Publishing Company
https://doi.org/10.1142/9789811234378_0013

Puzzles and Tilings — A Fascinating Kingdom of Solomon Golomb

Tuvi Etzion

Computer Science Department
Technion, Israel Institute of Technology
Technion City, Haifa 3200003, Israel
etzion@cs.technion.ac.il

Our little world contains quite a few eternal concepts. The human race as we know today does not exist more than a few thousand years. Of course, this is subject to the definition, what is the "human race as we know today"? I would like to think on our human race as we know today from the time that scientists became part of our world. Scientists and not science, since science always exists and it only waited for the human race to explore it and use it. Science has always been combined with technology and engineering. It is not a surprise that the first engineering invention is the wheel. This can be also seen as the first science invention and the first mathematical object used in science as cycle and circle are geometric concepts. Using mathematics in science is something which was used throughout the years. Moreover, our world and its nature hide a lot of mathematical concepts, some of them were revealed and some of them are still there to find. The recognition of mathematics by the Egyptians and the Greeks in the old world is fascinating, even if today it seems to be very simple. They have built the foundations for many simple actions which are done today, e.g., counting and representations of integers. This is the early mathematics that can be dated to 4000 or even 5000 years ago. About the same time, mathematical games and especially board games were born. Such games such as the "royal game of UR" known also as the "game of twenty squares' was played in the Middle East about 4000 to 5000 years ago. Games, puzzles, riddles, and mathematics, were holding hands to each other throughout the years, up to the present. This is part of the wonderful world of recreational mathematics. Surprisingly or not, recreational mathematics have also led to important inventions of our new world.

146   *The Wisdom of Solomon*

University days are always considered as the most beautiful period of someone who wants to join the academic world. An undergraduate student is usually concentrating on the courses that he is taking from semester to semester. When he is getting into the graduate levels he usually concentrates on his specific research. The broader horizon of his research area is usually seen more and more as he progresses in his PhD studies and more than that during his postdoctoral period. The same evolutional process happened to me. As a boy I was most interested in mathematics and games, played chess, solved and composed chess problems, and was always interested in mathematical puzzles. I was very much interested in recreational mathematics which was published by Martin Gradner and appeared in the *Scientific American*. I read his puzzles and riddles frequently in a series of books in which they were collected. This was almost gone during my bachelor studies and completely erased during my graduate studies. During this time I was most fascinating with the de Bruijn graph which was presented to me by Abraham Lempel as I started my PhD studies and shift register sequences were a small step forward. I still consider a lot of problems related to this graph as recreational mathematics. June 1984 was the time of the defence on my PhD thesis. Fortunately, immediately after the exam, between July 1 and July 5, 1984, the Information Theory Workshop took place in Caesarea, Israel, where I met Professor Solomon Golomb for the first time. This meeting led to two extraordinary years of postdoctoral in University of Southern California between September 1985 and August 1987. These two years were the most influential in my academic career. During these two years Sol was the Vice Provost for Research at USC, but even with such a demanding role in USC, I have managed to see him at least 2–3 times a week and each meeting was an important event for itself. I came to Los Angeles towards the end of August 1985. I remember my first meeting with Sol at USC. I came with my wife after we had done the first steps to settle in Los Angeles. Sol asked me if I have been settled in. I replied that the main obstacle is that the check which I brought with me would be cleared only after a week, while I needed cash immediately to rent an apartment. Next, Sol told me about three research problems: Tuscan squares, Sophie Piccard's Theorem, and the middle level's problem. The combination of real life's issues, science, and research problems, were always there during the weekly meetings with Sol. Amazingly, next day the check was cleared thanks to Sol. It took another day and I was completely hooked on the problems related to Tuscan squares. The relations between the three problems that Sol has presented to me and other topics such as shift register sequences, structures with distinct differences, and number theory have been fascinating. This combination between various mathematics concepts and different mathematical areas was one of many lessons I learned during my postdoctoral with Sol. Each meeting I learned something new, and there were no repetitions in the innumerous conversations I had with him, mostly during lunch that we have together every week or two or during the late hours of the day, usually in the office of his longtime colleague and friend Herb Taylor.

I had the honor and the pleasure to hear Sol several times giving lectures in various occasions. In the first talk that I heard he said the sentence "for any given

Puzzles and Tilings 147

beautiful mathematical structure we can find an application". He has demonstrated it on various structures — Latin squares and error-correcting codes; the moves of a Rook in chess and their connection to error-correcting codes; and more. The talk he gave was one of several that I heard in USC, but he also gave talks outside USC and some of them to less expert audience. The one which caught me the most was a two hour talk he gave to teachers of various schools. For this talk, Sol chose the topic of "Polyominoes and Tilings". It was amazing how he had changed the nature of the scientific talk to appeal to this audience who loves mathematics, but with more limited research qualifications.

Two years at USC were the time to learn and to develop as a scientist. I worked on Tuscan squares and related topics such as Costas arrays. Although I stopped working on them after a couple of years, they came back to haunt me twenty years later when I worked on key predistribution for wireless sensor networks, with colleagues in Royal Holloway, University of London. I didn't work on the middle level problem at USC, but twenty five years later I was the first to look on its q-analog problem (a translation from a framework on subsets to a framework on subspaces). On one aspect of Sophie's Piccard Theorem, I worked with Herb Taylor, and I used the foundations of this problem in distinct differences throughout my academic career. Thus, all the problems that Sol presented to me on our first meeting at USC have been influential in my academic career and not just at USC. Finally, I came to USC working on Sol's shift register sequences and especially on m-sequences and de Bruijn sequences. On these I have worked in various applications, such as interconnection networks, single-track Gray codes, complexity of sequences, error-correction for storage devices, robust self-location patterns, coding for non-volatile memories, and coding for DNA storage — many research areas are mostly different in their nature, but all are benefited from the use of shift register sequences.

Combinatorial structures, shift-register sequences, error-correcting codes, space communication, structures with distinct differences, and other topics in coding theory, information theory, and combinatorics, are areas where Sol has a countless number of scientific breakthroughs. But, Sol was known in other areas as well. He was active in recreational linguistics which is no surprise for someone who knew more than ten languages. But, Sol was also known as a recreational mathematician. This was known not only for those who worked with him on related problems, but also all the information theory community knew it. Sol had the regular GOLOMB'S PUZZLE COLUMN in the *IEEE Information Theory Society Newsletter* which appeared every three months. In this column which started on September 1987, exactly when I left USC, Sol has published each time a new set of recreational mathematics problems — each time a new set of intriguing puzzles and problems.

Games and puzzles were always part of recreational mathematics and they have raised many interesting problems in combinatorics and also in coding theory. My interest in combinatorics came in parallel with my interest in games and puzzles. The game in which I was most interested was the game of chess, which was invented about 1500 years ago. The game of chess, in my opinion, hides a rich combinatorics. This combinatorics is inside the game on one hand and combinatorics problems are based on this game on the other hand. One of the greatest, if not the greatest, puzzle-writer and especially chess problems composer was the American Sam Loyd (1841–1911). In the same period lived an English puzzle-writer, Henry Ernest Dudeney (1857–1930), who is considered to be almost as great as Sam Loyd. The two cooperated for a while, but later became bitter enemies. Martin Gardner (1914–2010) has followed them in his *Scientific American* column. These three great recreational mathematicians are known only for their contributions in these areas. Sol has been involved in almost every aspect in which they were involved, but he had also his corner of great contribution to recreational mathematics, the **Polyominoes**. These are geometric shapes on the plane which are formed by joining equal squares edge to edge. One such square is a monomino, two such squares form the well-known domino, three form the tromino, four form the tetromino, five form the pentomino, six form the hexomino, and so on. The name "polyomino" as well as the other names for the specific polyominoes were given by Sol. Sol had a special gift in giving the right names for various objects. He has done so for example, also to the structures on which I worked in USC: Italian squares, Tuscan squares, Florentine squares, and Vatican squares, where before only Latin squares were defined. The definitions for polyominoes were done when Sol was only 22, in his paper from 1954, "Checker board and polyominoes" which was published in *The American Mathematical Monthly*. This first paper of recreational mathematics on polyominoes has motivated lot of research. In this first paper of Sol on polyominoes, the combination of the checker board with the polyomino tiles led the way for many problems on polyominoes by some of the fine scientists of the era. The paper has shaped the way for tiling problems with these shapes and other geometric shapes. Of special interest were the pentominoes for which Sol defined the "game of polyominoes". This is a two player game. There are 12 different free pentominoes and an 8 × 8 checker board. A set of free polyominoes, means that two polyominoes in the set cannot be obtained from each other by any flip or rotation. Each player in his turn places a pentomino on free cells of the board. The last player who put a pentomino on the board is the winner. Tiling of various shapes with given sets of polyominoes is an interesting problem, the one which Sol presented to the math teachers in the talk which I mentioned earlier. Sol has also proved when the tiling problem is solvable and when it is computationally undecidable. After him, others took the helm and the problem also found its applications in various coding theory problems. Furthermore, the polyominoes also motivated the popular game of Tetris for which no introduction is required.

In 2013 US President, Barack Obama, presented Sol with the National Medal of Science for his advances in mathematics and communications, at an awards ceremony held in the White House. In his speech before the audience in the ceremony, President Obama said that in this group of scientists, getting the award, there is someone who inspired the game of Tetris. This sentence was given to the audience, to have the right impression that although the Medal is given to these individuals on complicated inventions and advances of science, their research can also contribute to simple and basic games which were played by most of us.

Puzzles and Tilings 149

The polyominoes has motivated a lot of surprising research in coding theory. Most notable is the research on Lee spheres and codes in the Lee sphere metric. This has been started in another seminal paper of Sol and his longtime colleague from USC, Lloyd Welch. They have considered perfect codes in the Lee metric and their relation to packing of polyominoes. This paper also motivated a lot of research and it demonstrates the connections between games, puzzles, coding theory, and applications. Sol has later demonstrated, in another paper, the connections between sphere packing, coding metrics, and chess puzzles. The tiling with these spheres, which are multidimensional polyominoes, has been a source for extensive combinatorial research. Lots of new techniques taken from combinatorics and algebra were used in the following years, to solve some of the tiling problems, by many researchers. The work on Lee spheres and the related tiling has been used later in various applications such as constrained coding, multidimensional error-correction, coding for flash memories, etc. I have been engaged in all these areas of research and some of my PhD students have been involved in them as well. Some of these students have been the ones who found these interesting connections.

My time with Sol at USC has been inspirational for the next thirty years of my academic career, and hopefully for many years to come. I left USC on September 1987 for my position in the Computer Science Department at the Technion, Haifa, Israel. I returned twice to Los Angeles for professional reasons. The first time was in 1992, when I gave a talk in GolombFest 60, held in Oxnard, on the complexity of de Bruijn sequences. The second time was in 2002 for GolombFest 70, held in the surroundings of USC. A few years earlier I started to work on interleaving schemes for error-correction of multidimensional arrays. Lee spheres and tilings had an important role in the research done in this area and USC days with Sol came back by storm. In GolombFest 70 it was natural for me to give a talk on "Codes, Anticodes, Generalized Lee spheres, and Two-Dimensional Inter-leaving Schemes" which was one of my main research areas at that time. The coming years were also influenced by Lee spheres. After the error-correction for multidimensional arrays, I worked on coding for flash memories. Again, Lee spheres and tilings had an important role in this research area and several of my papers were influenced by this old and good topic which I learned from Sol at USC. The same is true for other research areas in which I was engaged. What I learned from Sol at USC was always of great value and each time a topic which he has advanced, had a role in my research. His guidance, advice, and scientific vision, were with me and will be with me from the day that I left USC until today and in the future as well.

On May 1, 2016, Professor Solomon Wolf Golomb passed away and the world lost one of the last giants in science of our era. On January 31, 2017, a memorial event titled **The Life & Legacy of Sol Golomb** took place in the campus of USC. The event started with a series of talks on Sol Golomb and his research. When I had to decide on the title of my talk it was rather easy. I preferred to talk on his first recreational mathematics, the polyominoes, since everything else is a consequence, even if not directly, from the basics of his recreational mathematics. This is where everything was started from. The title of my talk was **Polyominoes, Lee Spheres, and Tilings**. The slides of the talk will end this chapter.

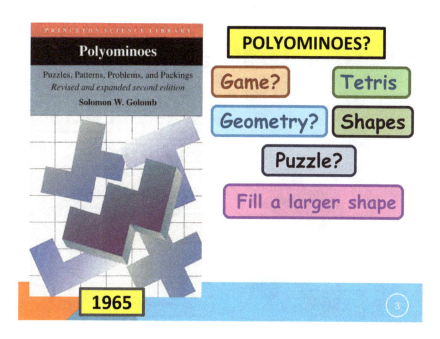

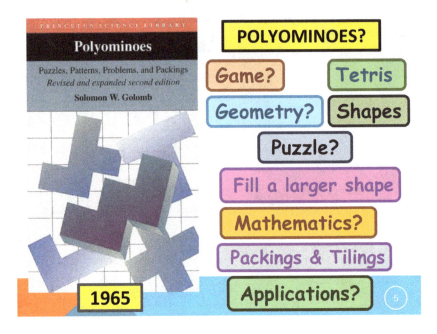

Number of Polyominoes

	name	free	no flip
1	monomino	1	1
2	domino	1	1
3	tromino	2	2
4	tetromino	5	7
5	pentomino	12	18
6	hexomino	35	60
7	heptomino	108	196
8	octomino	369	704

Number of Polyominoes

12 free pentominoes

6 mirror pairs, 18 with no flip

Tiling Rectangles with Pentominoes

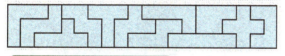

3 × 20 rectangle

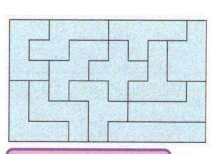

6 × 10 rectangle

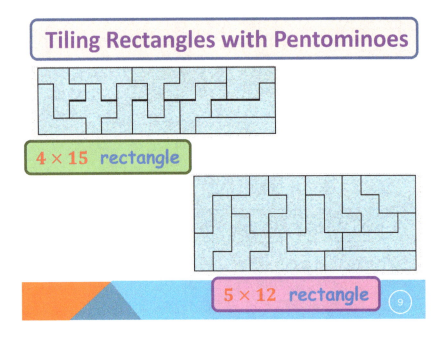

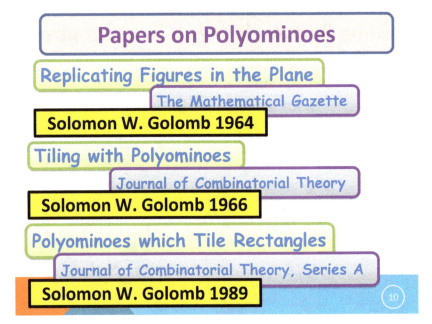

Papers on Polyominoes

Tiling with Sets of Polyominoes

Journal of Combinatorial Theory

Solomon W. Golomb 1970

Theorem

The problem of tiling the infinite plane with replicas of a finite set of polyominoes is computationally undecidable.

Checker board and Polyominoes

Checker boards and Polyominoes

The American Mathematical Monthly

Solomon W. Golomb 1954

Can this board be tiled with dominoes?

Checker board and Polyominoes

Straight tromino and L-tromino.

Checker board and Polyominoes

Tilings with 21 L-trominoes and one monomino?

 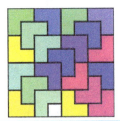 

Checker board and Polyominoes

Can the checker board be tiled with 21 straight trominoes and one monomino?

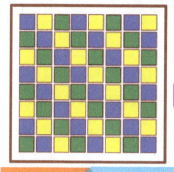

21 yellow 21 green
22 blue

Monomino must be blue

A straight tromino covers one from each color

Checker board and Polyominoes

Can the checker board be tiled with 21 straight trominoes and one monomino?

21 yellow 21 green
22 blue

Monomino must be blue

A straight tromino covers one from each color

use symmetry for the monomino

Checker Board and Polyominoes

GAME OF POLYOMINOES

2 players — A checker board and the 12 free pentominoes

Each player in his turn places a pentomino on free cells on the board. The last player to do so wins.

Applications?

A general Formulation of Error Metrics

IEEE Transactions on Information Theory

Solomon W. Golomb 1969

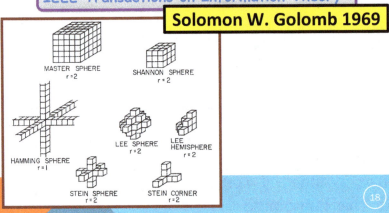

Applications?

Perfect Codes in the Lee Metric and the Packing of Polyominoes

SIAM Journal on Applied Mathematics

Solomon W. Golomb, Lloyd R. Welch 1970

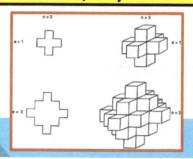

Applications of Lee Spheres/Codes

Constrained and Partial-Response Channels

Roth, Siegel 1994

Two-Dimensional Cluster-Correction

Interleaving Schemes

Blaum, Bruck, Vardy 1998

Etzion, Vardy 2002

Applications of Lee Spheres/Codes

- Coding for Flash Memories
  - Error-Correcting Codes for Rank-Modulation
    - Jiang, Schwartz, Bruck 2008
  - Types of Error-Spheres and their Tiling
    - Schwartz 2012
    - Buzaglo, Etzion 2012

Solomon Wolf Golomb
May 30, 1932–May 1, 2016

Jeremiah and Karen Farrell

Butler University

Sol Golomb, a frequent contributor to *Word Ways,* died on Sunday May 1.

He was a very close friend of Martin Gardner and often was featured in Gardner's *Scientific American Games* column. He also wrote "Golomb's Gambits" for *Johns Hopkin Magazine,* a monthly publication of his undergraduate alma mater.

Jeremiah and Karen Farrell with Sol at a Chicago puzzle conference 2003.

He fully described polyominoes and pentominoes while a graduate student at Harvard where he obtained his PhD in 1957. After a Fulbright year in Norway he became a Senior Research Mathematician at Jet Propulsion Laboratory, later becoming Research Group Supervisor and then Assistant Chief of the Telecommunications Research Section, where he played a key role in formulating the design of deep-space communications for the subsequent lunar and planetary explorations. In 1963 he joined the University of Southern California as a professor.

A partial list of his many awards includes:

(1) USC Presidential Medallion in 1985,
(2) Shannon Award of the Information Theory Society of the IEEE in 1985,
(3) Hamming Medal in 2000,
(4) National Medal of Science 2013,
(5) Benjamin Franklin Medal in Electrical Engineering in April 2016. With this award Golomb joined the ranks of previous Franklin Medal recipients and distinguished laureates, which include Albert Einstein, Marie Curie and Stephen Hawking.

President Barack Obama presents Solomon Golomb with the National Medal of Science at an awards ceremony held at the White House in 2013.

These prestigious awards were well deserved but Golomb's human side was more relished by those of us who had the good fortune to spend time with him. When his health would allow it he attended puzzle and recreational mathematics conferences around the world. At an International Puzzle Party in London we spent a delightful afternoon sightseeing with Sol as an expert tour guide. He attended almost all of the 12 Gatherings for Gardner in Atlanta and sometimes gave entertaining talks on his many recreational topics, pentominoes, word games, geometric novelties, and so on. His memory was prodigious. He told us about an assignment he misunderstood in high school and memorized an entire Shakespearian play in one evening to present to the class the next day. (He could still recite the play decades later!) Sol taught himself over 12 languages primarily because he wanted to lecture in the native tongue when giving a talk in a foreign country.

Obviously he was brilliant. He was one of the first profile professors to attempt the Ronald K. Hoeflin Mega IQ power test which originally appeared in

Omni Magazine. He scored at least an IQ of 176 which represents about 1/1,000,000 of the population.

Sol Golomb was, even so, one of the most humble of men and all who knew him will miss him dearly.

© 2023 World Scientific Publishing Company
https://doi.org/10.1142/9789811234378_0015

Solomon W. Golomb's Enlightening Games

Aviezri S. Fraenkel

Department of Computer Science and Applied Mathematics,
Weizmann Institute of Science,
Rehovot 76100, Israel
fraenkel@wisdom.weizmann.ac.il

We thought that paying true tribute to the memory of Professor Solomon W. Golomb, would be to highlight a gem contribution of his on two-player games [1], which has not received the attention and respect it deserves, even among game experts. In this paper Sol analyzed two-player *take-away* games. We hope that this discourse will do justice to commemorating Sol's fruitful activities in a little-known direction, and, concurrently, illuminating an important yet neglected games corner.

A very simple take-away game is *Nim*. Given a number of positive integers, say 1, 2, 5, 13, the two players alternate in choosing one of the integers and reducing it (i.e. replacing it with a lesser integer), but so that the remaining integer is still nonnegative. The player who first reduced all integers to 0 is called the *winner*, the opponent the *loser*. There is a well-known winning strategy: express each integer in binary, and add them up in binary, but without carry. For the above case,

$$
\begin{array}{cccc|c}
0 & 0 & 0 & 1 & 1 \\
0 & 0 & 1 & 0 & 2 \\
0 & 1 & 0 & 1 & 5 \\
1 & 1 & 0 & 1 & 13 \\
\hline
1 & 0 & 1 & 1 & 11
\end{array}
$$

Alice wants to make this *Nim-sum* 0, by reducing a single row. In this case $13 \to 6$ will do the trick. However Bob now moves (reducing a single row), the resulting Nim-sum will be nonzero, and Alice can always reduce it to 0 by reducing a single row, until eventually arriving at all columns 0. Alice won since she made the last move.

In this missive, we always adopt the convention that the player making the last move wins (*normal* play). There are always precisely two players.

165

Sol began with an even simpler game than Nim, yet succeeded in using it, sleight of hand, as a *catalyst* to generate a class of much more sophisticated games — formulating and solving them. But before formulating the simpler catalyst game, let's jump ahead and present two examples of Sol's achievements in this direction. This can be done without yet disclosing the catalyst.

Notation. Denote by $\mathcal{N}$ the winning positions of the $\mathcal{N}$ext player, the player who moves *from* the current position. Also denote by $\mathcal{P}$ the winning positions of the $\mathcal{P}$revious player, the player who moved to the current position.

Notice that any end position u is in $\mathcal{P}$, since the next player cannot move from u. Every position v that has u as a direct follower is in $\mathcal{N}$, though v may have other direct followers in $\mathcal{N}$ or $\mathcal{P}$. In general, a position w is in $\mathcal{N}$ if and only if it has at least *one* direct follower in $\mathcal{P}$, whereas $w \in \mathcal{P}$ if and only if *all* of its direct followers are in $\mathcal{N}$.

I. Let $V = \{1, 2, 3, 5, 7, 11, 13, 17, 19, 23, 29, 31, 37, \ldots, \}$ the set of primes, with 1 adjoined, be the permitted moves. Sol claims that then $\mathcal{P} = \{4n : n \geq 0\}$. Quite surprising! Usually for take-away games, the move-set is rather regular, such as all of $\mathbb{N}$, offsets of arithmetic sequences, subsets of residues mod T for some integer T, etc, because only for those a winning strategy was found — not the set of primes! Moreover, notice the seemingly unlikely connection between the set of multiples of 4 and the set of all primes.

Illustrating the present case, suppose all positions < 8 are labeled according to Sol's claim. We have to show that $8 \in \mathcal{P}$. It suffices to show that all direct followers of 8 are in $\mathcal{N}$. So suppose that Alice begins to move $8 \to 7$ by subtracting 1. Now $7 \in \mathcal{N}$, since Bob can counter $7 \to 0$, winning. Similarly, if Alice moves to 5, 3, by subtracting 3, 5, respectively, then Bob can counter by moving to 0, winning. But Alice can move to 6 (by subtracting 2), which we have to show is in $\mathcal{N}$. Indeed, Bob moves $6 - 2 = 4 \in \mathcal{P}$. This confirms that $8 \in \mathcal{P}$: Alice lost. Now 9, 10, 11 are in $\mathcal{N}$ by subtracting 1, 2, 3 respectively. We leave it to the reader to show $12 \in \mathcal{P}$.

II. Let $V = \{1, 2, 4, 8, 16, 32, 64, 128, \ldots, \}$ the set of nonnegative powers of 2 be the permitted moves. Sol claims that then $\mathcal{P} = \{3n : n \geq 0\}$. Again rather unexpected!

Illustration. Suppose we already know that all positions < 11 behave according to Sol's claim. We have to show that $11 \in \mathcal{N}$. Indeed, Alice moves $11 \to 3 \in \mathcal{P}$ by subtracting 8. Bob can now only move to either 1 or 2, from each of which Alice can move to 0. Alice won. It is now easy to show $12 \in P, 13, 14 \in \mathcal{N}, 15 \in \mathcal{P}$.

Sol derived these and other results by beginning with the following simple catalyst game, dubbed $\mathbf{G}_k (k \geq 1)$.

Let $T_k = \{1, 2, 3, \ldots, k\}$, where the permitted moves are to diminish the initial position by any positive integer $\leq k$. It is easy to see that $\mathcal{P}(T_k) = \{(k+1)n : n \geq 0\}$.

For example, if $T_k = T_8$ and the initial position is 50, then Alice will move $50 \to 45 = 5 \times 9$. Hereafter, Alice can always maintain a multiple of 9, whereas Bob never can. So eventually Alice will reach $0 \times 9 = 0$, winning.

Sol formulated and proved the following theorem, which demonstrates the connection between the above two sample games and the catalyst $\mathbf{G}_k$.

Theorem. Let $V \subset \mathbb{N}$, and $U_k := T_k \cup V$. Then $\mathcal{P}(U_k) = P(T_k)$ if and only if V is disjoint from $\mathcal{P}(T_k)$.

For the above "Primes" game, V is the set of all primes (and 1), $T_k = T_3 = \{1, 2, 3\}$, so $\mathcal{P}(T_3) = \{4n : n \geq 0\}$, which is disjoint from V. Hence $\mathcal{P}(U_3) = \{4n : n \geq 0\}$. But $U_3 = V$, since $T_3 \subset V$. Thus $\mathcal{P}(V) = \{4n : n \geq 0\}$.

For the "2-powers" game V is the set of all nonnegative powers of 2, $T_k = T_2 = 1, 2$, so $\mathcal{P}(T_2) = \{3n : n \geq 0\}$, which is disjoint from V. Analogously to the primes case, we get $P = \{3n : n \geq 0\}$. The paper [1] contains considerably more sophisticated theorems and examples.

Actually Sol considered *vector* take-away games, where both positions and moves maintain nonnegative integer components throughout. But most of his effort in [1] is devoted to the 1-dimensional case, or to a multi-dimensional game, where only one component can be reduced at every move (such as Nim).

We have neglected to do justice to a very important aspect of [1], namely *shift registers* [3]. Here Sol fused together his talents in both electrical engineering and math to construct $\mathcal{P}$ sets for take-away games. A shift register is a linear sequence of cells connected together electronically, finite or infinite, depending on the size of the set V of permissible moves. Engineers of computer hardware or communication equipment are well-familiar with finite shift registers. The cells typically contain binary bits (0 or 1) and at each time interval are shifted right by one cell. The rightmost bit is lost, and a new bit enters on the left. There are "feedback-taps" on the cells that correspond to V, the set of all moves. The contents of these cells are fed into an appropriate array of gates, whose output determines the bit that enters on the left. This procedure constructs the set $\mathcal{P}$ of V for take-away games.

Incidentally, shift register sequences are used in a broad range of applications, particularly in random number generation, multiple access and polling techniques, secure and privacy communication systems, error detecting and correcting codes, and synchronization pattern generation, as well as in modern cryptographic systems. Many of these were discovered and disseminated by Sol.

We attempted to adhere to Sol's original notation, but replaced just a little of it by modern nomenclature for game concepts, when we felt it would be more comprehensible to current readers.

∗∗∗

Professor Solomon W. Golomb loved puzzles — one person games: **SOL**o games. The most famous of these are polyominoes, which are plane configurations made up of unit squares, joined together by full edge to edge contact. Sol has made them very ubiquitous [2] by analysing them and demonstrating their usefulness in many diverse fields; Martin Gardner popularized them in one of his *Scientific American* columns and Dr Google is full of them. Like many puzzles in recreational mathematics, polyominoes raise many combinatorial problems. The most basic is enumerating polyominoes of a given size. No formula has been found except for special classes of polyominoes. A number of estimates are known, and there are algorithms for calculating them. In statistical physics, the study of polyominoes and their higher-dimensional analogs (which are often referred to as lattice animals in this literature) is applied to problems in physics and chemistry. Polyominoes have been used as models of branched polymers and of percolation clusters.

***

In his talks and writings, Sol succeeded in making just the right compromise between two competing goals: math rigor and reader-friendliness: His statements and proofs are clear-cut rigorous, but he had a natural feeling for when they were beyond the grasp of the average reader; he filled the gap with explicit comprehensible examples, that provided the intuition and demonstrated what was going on, thus contributing huge reader-friendliness.

I encountered Sol first when I had just metamorphosed from an electric engineering PhD to a math PhD student at UCLA, where, *inter alia*, Sol was teaching. His lucid, luminous lectures were important ingredients in providing me with enthusiastic appreciation for mathematics.

Bibliography

1. S. W. Golomb, A mathematical investigation of games of "take-away", *J. Combin Theory* **1** (1966), 443–458.
2. S. W. Golomb, *Polyominoes: Puzzles, Patterns, Problems, and Packings* — Revised and Expanded Second Edition, Paperback, 200 pages Published April 7th 1996 by Princeton University Press (first published September 6th, 1994).
3. S. W. Golomb, Shift Register Sequences, 3rd revised edition, World Scientific, March, 2017.

© 2023 World Scientific Publishing Company
https://doi.org/10.1142/9789811234378_0016

My Life with Sol

Harold Fredricksen

I had just turned 19 and finished my two years at Pasadena City College. The placement office at the college informed me of a possible job opening at the Jet Propulsion Laboratory. Normally this office fielded job offers for machine shop operators and electronic technicians, but this was for a mathematics assistant. The requirement was for a completion of two years of calculus sequence, but calculus had nothing to do with the job.

It was a heady time as the Soviet Union had just launched Sputnik I and the US seemed to be very much in second place in this endeavor. JPL was clearly a player — or player to be — in the space race as it came to be called. In any event, I was invited to appear at JPL to be interviewed as a mathematics aid.

The interview was conducted in an office shared by four MS and PhD members of the staff well versed in mathematics and electrical engineering. My "committee" consisted of Sol Golomb, Stan Lorens, Andy Viterbi and Lloyd Welsh. I, of course, did not know any of them and their various lists of accomplishments, but I soon realized I was in for more than I had anticipated.

Sol was to be the supervisor for the project; so, he naturally took the lead. The job was to have the candidate compute the number of equivalence classes of balanced Boolean functions of 4 variables under the operations of permution and complementation of the variables. This was beyond my understanding at the time and I am not sure that the ultimate goal was ever told to me. I needed to be instructed in the intricacies of GF(2) as even that was something I had never seen in my two college math classes.

I call the group my "committee" as the proceedings quickly took on the form of a candidate oral examination. I have no real memory of how long the examination took but I recall that I felt thoroughly spent by the process. When it was completed, Sol walked me back to the Human Resources Office and surprisingly he said that I would be hired. The money was better than I was making as a supermarket boxboy; so, I was more than happy. My hours were established and I was good to go.

After my interview was completed and I had undergone other preliminary examinations, I then realized that the processes were never intended to embarrass the

170   *The Wisdom of Solomon*

candidate but just to judge his fitness to perform the necessary next activity. Oral exams never really held much dread for me after that experience.

One other point about that day, several months later I was told by Andy Viterbi that there had been several other candidates for the position, many of them who had appeared far better qualified than I, but, Sol had felt that I was the one who was the most likely to take on the work and complete the task.

When I came back for the first day on the job, Sol gave me an office and desk to work at. I felt pretty special as I never had an assigned work area before. He then explained more about what he wanted me to do. Until he got me started, I still wasn't sure about what it was I was doing, but I dug right in.

Sol estimated that the job would probably take me six months to complete and he was surprised and pleased when I announced I had found 153 inequivalent classes of functions in approximately six weeks. Neither of us expected that I would be done so quickly, but then neither of us expected that I would be working on the project whether I was on the job or at home. I supposed my school work suffered, but I found something I really loved to do. I was further gratified when the project was written up and included in Chapter 8 of Sol's book, *Shift Register Sequences*.

As I mentioned, school work studies took a back seat to what I was doing at work and I pretty much dropped out of school. Starting the next summer, my mother, sister and I left for an extended period in Norway. (Sol had been a Fulbright recipient and studied in Norway; so, I was still being close even while far away.) But I was no longer employed at JPL and while away, I took employment at the Ford Motor Company and at General Motors Technical Center. The job was much less satisfying than what I had done at JPL and when I returned in the following September, I applied to return to work with Sol. He wouldn't hire me back unless I was enrolled in a degree program; so, I started back with my studies at LA State College (later absorbed in the California State College system) and returned to my part-time duties in the Math Group of Section 8 at JPL.

I worked on various projects of Sol's design and was always very happy to be engaged in things that were more interesting than I was finding in school. One professor was very attractive at LA State, Dr. Gordon Overholtzer, and I enjoyed his classes, taking several classes with him, but I liked the challenge. One day while we were waiting for a class in Linear Algebra to begin, I chanced to look at another student's paper. He was drawing areas with pentomino pieces and arranging them to fill in rectangular patterns. We did a lot of this kind of recreational mathematics at work and so I engaged my fellow student and asked him if he would like to meet Sol. He was completely engaged with the idea and eventually met and came to work for Sol at JPL. Warren Lushbaugh became a very valuable engineer in the Space Processing Group and was also the illustrator for Sol's book *Polyominoes*.

I spent the next two years completing the requirements for a Bachelor of Science in Mathematics and thought I could then retire to work at JPL. But Sol had other ideas. He wanted me to pursue graduate studies and kicking and screaming, I was bundled off to the University of Wisconsin, Madison to purse a PhD in Mathematics. I returned to JPL for summer studies. When I returned to Madison, Sol also made a move and took a dual appointment at the University of Southern California in Mathematics and Electrical Engineering. He invited me to transfer there; so, I completed my MS in Math at Madison and joined USC in the Electrical Engineering Department as I felt that matched my interest better than continuing with math would.

I had some catching up to do but Sol assured me that EE was really only applied math anyway; so, I buckled down and set to work. I also was working about 20–30 hours a week at JPL with Sol's knowledge as he had kept close contact there and with my supervisor, Edward Posner. I took whatever classes Sol was teaching and whatever other classes I needed to complete to pursue my PhD studies. Sol also kept me on assisting him on some of his projects funded by the Army and other sponsors. I took all my oral exams (again, they weren't as strenuous as what I had already faced at JPL years earlier) and began work on a PhD project, Disjoint Cycles from the de Bruijn Graph. (The report was also accepted as a JPL report so my pleasant relationship continued.)

As I was conferred the PhD degree in EE, I returned to JPL but Sol suggested I might benefit from a Post-Doctoral position abroad and he was able to assist me in getting such a position at the University of Bergen, Norway, where I would study and lecture in the Mathematics Department under the supervision of Ernst Selmer.

It was a wonderful year and I met several people there, who were students and colleagues such as Johannes Mykkeltveit and Tor Helleseth, and learning a great deal more about nonlinear shift register sequences. Both of these became very involved with the community and closely involved with Sol. I am sure that their remembrances will appear somewhere else in this volume. One highlight for me of that year was when Sol came to visit and lecture in Bergen. He and Bodil had been in Denmark and since he was so close, he offered to come up. He announced his intention in a letter to me written in Norwegian. I had a little trouble with reading it, so, I asked Selmer to help me. The letter was in perfect Norwegian. Selmer was very impressed. Except that there was one tiny flaw in his writing. Sol capitalized the words for grandmother and grandfather. That was done properly in Danish, but in Norwegian, these words were never capitalized. Sol also had offered to give his lecture in Norwegian, but Selmer suggested that the audience would understand English perfectly well. I certainly appreciated that as I wouldn't have understood the Norwegian version. All of my lectures were in English. The only Norwegian I knew was that spoken around the coffee room and when I sounded out the words in the newspaper. Interestingly, when I took my language exams in preparation for my degree, Sol suggested I might like to try reading and translating a paper by Axel Thue, and he would be my examiner. I decided it would be easier to pass my exams in French and German as was pretty standard at the time.

A great thing about working for Sol was that everybody seemed to know him and there was an endless string of visitors of the most elite class of mathematicians and engineers. It was wonderful to meet these people, who only were names to me. I assumed by their remarkable celebrity when I read about them that they were surely already deceased. One, who was a real highlight for me while I was finishing

172   *The Wisdom of Solomon*

my dissertation, was N.G. de Bruijn, who came to visit. I took a copy of my thesis to present to him. His remarks, with thanks for the remembrance, were, "So, they're calling these de Bruijn sequences now."

Another visitor to Sol and to JPL in those days was a mathematician from Boston, Neal Zierler. Neal was in the area for quite a while and had, in fact, taken a position as supervisor of the mathematics group at JPL that Sol had left when he went to USC. Working for Neal was wonderful and the best thing for me about it was that he recruited me to work at the Institute for Defense Analyses in Princeton when he returned there. IDA was a shining city on the hill in the mathematical and cryptographic world, and I was very eager to accept the offer. While the clearance process was going on, I happened to be attending a mathematics conference at Madison. The highlights for me as I remember that meeting was one evening when Neal, Sol and I were out for dinner. Usually, these dinners were a free floating exchange of ideas. This one was different. At some point, Sol and Neal decided that since they both spoke Japanese that they would have a conversation in that language. Neal had learned it at the language school in Monterey during the war, and he also had a Japanese wife. Sol picked it up as he did any language and Sol himself was competent in more than 20 languages. I was, of course, at a loss to understand any of what they were saying. But, I had the definite understanding, unstated by either of them, to decide who was the better speaker of the language. I, of course, never said anything.

Sol's proficiency was not limited to languages. At one of the conferences that had been scheduled to specifically celebrate Sol, I heard the Provost of the University of Southern California offer the statement that when he spoke with Sol that when any new topic arose that Sol was able to speak extemporaneously for 15 minutes on that subject. I was never able to refute that claim fully but about some arcane subjects on Norse mythology, he seemed to peter out after only 5–10 minutes of highlights.

In January 1991–1992, some of Sol's students came together to suggest a meeting to honor Sol on his birthday, May 31, 1932. The meeting was to suggest topics that had a typical "Golombic Flavor." Sol was to be 60 on that date, which was also auspiciously the 160th anniversary of the death of the French mathematician Evariste Galois. I suggested the name GF(60) for Golomb-Fest 60. It was also a small joke as GF(60) also appeared to suggest a galois field of 60 elements. The audience would be sure to know that no such field could exist. The meeting was held every five years after that.

At one of these meetings, probably GF(70) in 2002, I gave a short talk. Not really a talk per-se. I merely put up a slide that contained the sequence of numbers as listed below:

12, 18, 24, 48, 110, 112, 133, 140, 152, 190, _____, _____, _____

I asked the audience to tell me what integers come next in the sequence. Of course, this was a trick question because the answer depended upon whether I was talking about "next" in terms of the completed sequence or of "next" in terms of how the sequence was generated. After a few moments, Sol knew the answer to my question though I believe he may have never thought about the question that caused me to ask my question. I leave the answer for the current reader to ponder.

© 2023 World Scientific Publishing Company
https://doi.org/10.1142/9789811234378_0017

Focus and Contrast

Jonathan I. Hall* and Tingyao Xiong[†]

*Department of Mathematics, Michigan State University
East Lansing, MI 48824, USA
halljo@msu.edu
†Department of Mathematics and Statistics, Radford University
Radford, VA 24142, USA
txiong@radford.edu

We present, largely without proof, results motivated by a question in optical microscopy from chemistry. This turns out to be a variation on sequence correlation, a topic of prime interest to Sol Golomb throughout his career.

1. Some Remarks by Jonathan Hall

Solomon Golomb started out as a number theorist, as did my father, Marshall Hall, Jr. Both moved on to other areas and found common ground in combinatorics. Before Sol finished his Harvard thesis on the distribution of sets of primes [13] he had already done the 1954 internship at the Martin Company whose final report was his fundamental study "Sequences with Randomness Properties" [12]. Hall's 1936 Yale thesis [17] provided (as noted by Golomb [15, p. viii]) "an excellent modern treatment" of the algebra associated with modular linear recursions.

Sol went to the Jet Propulsion Laboratory in 1956 while my father arrived at Caltech in 1959. I do not know how they first met, but they may have been introduced by Leonard Baumert — a doctoral student working under my father and also a JPL colleague of Sol's. In 1962 the three wrote a joint paper [1] constructing an Hadamard matrix of order 92, the smallest order for which existence was previously unclear. In any event, Golomb and Hall became close, and Sol's ground-breaking 1967 book *Shift Register Sequences* [15] was dedicated "To Marshall Hall, Jr." One of my most prized possessions is my father's copy of that book, signed both by him and by Sol.

I last spoke with Sol Golomb at the 2012 SETA Conference in Waterloo, hosted by his longtime collaborator, co-author [16], and friend Guang Gong. At that time I was able to remind Sol of my father's respect and admiration for him, and to thank him for the inspiration he has been to me and so many others.

2. Randomness, Correlation, and Merit

Sol Golomb's Martin report [12] (the basis for Chapter 3 of his book [15]) initiated the study of "randomness properties" for sequences. There is a subtlety that Sol described succinctly in his introductory chapter to the volume *Digital Communications with Space Applications*, which he edited [14, pp. 7–8]:

> it is important to realize that the term random applied to a sequence refers to the *a priori* conditions under which the sequence was produced, rather than to the a posteriori consideration of what the sequence looks like, or what properties it exhibits. ... Given a list of tests for the plausibility of randomness, we may define any sequence which passes the tests to be *pseudo-random*.

Golomb's three famous "Randomness Properties" for pseudo-random sequences were first given in [12] and reappear throughout the literature on sequences (in particular [14,15,16]). Although their precise form may vary, they are always consistent with the above quote.

Imagine you produce a binary sequence at random by flipping a fair coin a large number n of times. Before you start (*a priori*), every n-sequence of heads and tails is random, in the sense that each is equally likely to be the result. But it is highly plausible that, when you later (*a posteriori*) examine the particular sequence produced, the number of heads is close to the number of tails. This is at the heart of Golomb's first Randomness Property for a pseudo-random sequence:

Balance Property: The number of heads is nearly equal to the number of tails.[1]

This, of course, only speaks to the (*pseudo*)randomness of a binary sequence — a sequence with alphabet size 2. The natural generalization to arbitrary finite alphabets is that the distribution of letters in the sequence is close to uniform (see [16, p. 123]). It is often algebraically helpful to choose the letters of an alphabet of size ℓ to be the ℓth roots of unity in the complex numbers. Of sole interest in this paper is the binary case $\ell = 2$, where the alphabet is $\{+1, -1\}$, at times abbreviated to $\{+, -\}$.

With this notation, the Balance Property for the binary ± 1-sequence $\mathbf{x} = (x_0, \ldots, x_{n-1})$ says $|\{i \mid x_i = +1\}|$ is nearly equal to $|\{i \mid x_i = -1\}|$. That is, the dot product

$$\mathbf{x} \cdot \mathbf{1} = |\{i \mid x_i = +1\}| - |\{i \mid x_i = -1\}|$$

of $\mathbf{x}$ with the all-1 vector $\mathbf{1} = (+1, +1, \ldots, +1)$ is nearly 0 ; the vectors $\mathbf{x}$ and $\mathbf{1}$ are close to being orthogonal.

Golomb's Randomness Property **R-2** requires the number of runs of heads $(+)$ or tails $(-)$ to drop by a factor of 2 with each increase in length. For instance there should be half as many runs $++$ as there are runs $+$. We will not be concerned with this property here, but it is in the spirit of uniform distribution as described

---

[1] Golomb's **Property R-1** actually requires that the discrepancy be at most 1; see [16, p. 118].

Focus and Contrast 175

above: If all k-tuples are uniformly distributed throughout the sequence, then the particular 3-tuple $-+-$ (a run of length 1) should arise $\frac{1}{8}$ th of the time, while the 4-tuple $-++-$(a run of length 2) should occur $\frac{1}{16}$ th of the time,

Golomb's third Randomness Property — the one most central to this paper- can be motivated by the first. Consider two randomly generated ± 1-sequences $\mathbf{x} = (x_0, \ldots, x_{n-1})$ and $\mathbf{y} = (y_0, \ldots, y_{n-1})$ (produced, say, by coin flipping), and construct from them the third sequence

$$\mathbf{x} \star \mathbf{y} = (x_0 y_0, \ldots, x_i y_i, \ldots, x_{n-1} y_{n-1}).$$

This sequence differentiates the places in which $\mathbf{x}$ and $\mathbf{y}$ agree (where $x_i y_i = +1$) from the places in which they disagree ($x_i y_i = -1$). If the original sequences have been randomly generated, then this sequence is also randomly generated. In particular it is likely that it will satisfy the Balance Property. As the dot product $\mathbf{x} \cdot \mathbf{y}$ is the sum of the entries in $\mathbf{x} \star \mathbf{y}$, we have a fundamental conclusion:

> two ± 1-sequences, both randomly produced, should be nearly orthogonal to each other.

Next, assuming that $\mathbf{x} = (x_0, \ldots, x_{n-1})$ has been randomly produced, it is reasonable to assert that its cyclic shift $(x_1, \ldots, x_{n-1}, x_0)$ is also random. Together with the previous conclusion, we are left with a soft version of Golomb's third Randomness Property:

An Imprecise Statement of Property R-3. The sequence $\mathbf{x}$ is nearly orthogonal to each of its nontrivial cyclic shifts.

2.1. *Correlation*

For real vectors $\mathbf{x} = (x_0, \ldots, x_t, \ldots, x_{n-1})$ and $\mathbf{y} = (y_0, \ldots, y_t, \ldots, y_{n-1})$ the dot product given by

$$\mathbf{x} \cdot \mathbf{y} = \sum_{t=0}^{n-1} x_t y_t$$

is positive definite on Euclidean space[2] $\mathbb{R}^n$. The vectors $\mathbf{x}$ and $\mathbf{y}$ are then *orthogonal* when their inner product is 0.

We have another remark from Golomb's chapter in [14]:

> In statistical terms, orthogonal means *uncorrelated.*

This is consistent with our discussion above, and in the study of $\mathbf{x}$ and $\mathbf{y}$ as sequences we call the inner product $\sum_{t=0}^{n-1} x_t y_t$ their *crosscorrelation*.[3]

---

[2] More generally, we could consider the space $\mathbb{C}^n$ with respect to its positive definite hermitian inner product $\sum_{t=0}^{n-1} x_t y_t^*$. The results of this section would remain true when stated appropriately.

[3] In keeping with the statistical context, the crosscorrelation is at times normalized to be the inner product of the corresponding unit vectors; see [15,16]. For various reasons we use the unnormalized version.

More generally their *aperiodic crosscorrelation function*[4] is defined by

$$C_{\mathbf{x},\mathbf{y}}(k) = \sum_{t=\max(0,-k)}^{\min(n-1,n-1-k)} x_{t+k} y_t,$$

for $-(n-1) \le k \le n-1$.

For $k \ge 0$ this amounts to sliding $\mathbf{y}$ forward k positions before calculating the inner product:

$$x_0 \quad \cdots \quad x_{k-1} \left| \begin{matrix} x_k & \cdots & x_{n-1} \\ y_0 & \cdots & y_{n-1-k} \end{matrix} \right| y_{n-k} \quad \cdots \quad y_{n-1},$$

while for $k \le 0$ the sequence $\mathbf{y}$ slides backwards $|k|$ positions:

$$y_0 \quad \cdots \quad y_{|k|-1} \left| \begin{matrix} x_0 & \cdots & x_{n-1|k|} \\ y_{|k|} & \cdots & y_{n-1} \end{matrix} \right| x_{n+k} \quad \cdots \quad x_{n-1},$$

Clearly $C_{\mathbf{x},\mathbf{y}}(-k) = C_{\mathbf{y},\mathbf{x}}(k)$, so when convenient we may restrict the values k to the range $0 \le k \le n-1$ where

$$C_{\mathbf{x},\mathbf{y}}(k) = \sum_{t=0}^{n-k-1} x_{t+k} y_t.$$

If we pad the sequences $\mathbf{x}$ and $\mathbf{y}$ with sufficiently many 0's at each end, the limits of summation become irrelevant. We can then write the correlation function as

$$C_{\mathbf{x},\mathbf{y}}(k) = \sum_{t} x_{t+k} y_t = \sum_{i-j=k} x_i y_j.$$

The *periodic crosscorrelation* function of $\mathbf{x}$ and $\mathbf{y}$ is

$$\mathbf{P}_{\mathbf{x},\mathbf{y}}(k) = \sum_{t=0}^{n-1} x_{t+k} y_t,$$

for $0 \le k \le n-1$, where the subscripts $t+k$ are to be read modulo n. This is also the crosscorrelation of $\mathbf{y}$ with $\mathbf{x}^{(k)}$, the sequence that results from cyclically shifting $\mathbf{x}$ exactly k times (as in **R-3**). It is easily checked that

$$P_{\mathbf{x},\mathbf{y}}(k) = C_{\mathbf{x},\mathbf{y}}(k) + C_{\mathbf{x},\mathbf{y}}(n-k).$$

Similarly we define the *odd crosscorrelation* function by

$$O_{\mathbf{x},\mathbf{y}}(k) = C_{\mathbf{x},\mathbf{y}}(k) - C_{\mathbf{x},\mathbf{y}}(n-k).$$

---

[4]This is essentially the unnormalized finite crosscorrelation function of [16, p. 8]. The term "aperiodic" has traditionally (for instance [5,28]) been used to distinguish this from the "periodic" crosscorrelation function, soon to be defined. The aperiodic version is the most primitive and so should probably be styled as the "crosscorrelation function" without further modification, as in [8,9]. But, as we have seen with Golomb's **R-3**, the periodic version has been studied much more extensively. Among other things it exhibits nicer algebraic properties.

This is also the crosscorrelation of $\mathbf{y}$ with the sequence that results from k *negacyclic shifts*[5] of $\mathbf{x}$ and so is also called the *negacyclic crosscorrelation*.

If we let $\epsilon \in \{\pm 1\}$, then we can combine the periodic and odd correlation functions by writing

$$C^{\epsilon}_{\mathbf{x},\mathbf{y}}(k) = C_{\mathbf{x},\mathbf{y}}(k) + \epsilon C_{\mathbf{x},\mathbf{y}}(n-k)$$

so that

$$P_{\mathbf{x},\mathbf{y}}(k) = C^{+}_{\mathbf{x},\mathbf{y}}(k) \text{ and } O_{\mathbf{x},\mathbf{y}}(k) = C^{-}_{\mathbf{x},\mathbf{y}}(k).$$

Lemma 2.1.

(a) $C_{\mathbf{x},\mathbf{y}}(-k) = C_{\mathbf{y},\mathbf{x}}(k)$.
(b) $C^{\epsilon}_{\mathbf{x},\mathbf{y}}(n-k) = \epsilon C^{\epsilon}_{\mathbf{x},\mathbf{y}}(k)$.

For a single sequence $\mathbf{x}$ the aperiodic autocorrelation function $C_{\mathbf{x}}(k)$ of $\mathbf{x}$ is then defined as $C_{\mathbf{x},\mathbf{x}}(k)$ (by the lemma we need only consider $0 \le k \le n-1$). We also define the *periodic autocorrelation* and *odd autocorrelation* functions of $\mathbf{x}$ via

$$C^{+}_{\mathbf{x}}(k) = P_{\mathbf{x}}(k) = P_{\mathbf{x},\mathbf{x}}(k)$$

and

$$C^{-}_{\mathbf{x}}(k) = O_{\mathbf{x}}(k) = O_{\mathbf{x},\mathbf{x}}(k).$$

Over the binary alphabet $\{\pm 1\}$ we have

$$C_{\mathbf{x}}(0) = P_{\mathbf{x}}(0) = O_{\mathbf{x}}(0) = n.$$

2.2. *Merit*

The merit factor, introduced by Golay [8,9], is a single parameter that measures the (aperiodic) autocorrelation properties of the binary sequence $\mathbf{x} \pm 1^n$. The *base energy* of $\mathbf{x}$ is $C_{\mathbf{x}}(0)^2 = n^2$; the *off-base energy* is $\sum_{k \ne 0} C_{\mathbf{x}}(k)^2 = 2\sum_{k=1}^{n-1} C_{\mathbf{x}}(k)^2$; and the *merit factor* of $\mathbf{x}$ is the ratio of these two energies:

$$F_{\mathbf{x}} = \frac{C_{\mathbf{x}}(0)^2}{\sum_{k \ne 0} C_{\mathbf{x}}(k)^2} = \frac{n^2}{2\sum_{k=1}^{n-1} C_{\mathbf{x}}(k)^2}.$$

The integer $C_{\mathbf{x}}(k)$ is always congruent to $n-k$ modulo 2, so the best (lowest energy) sequence one can hope for has $C_{\mathbf{x}}(k) \in \{-1, 0, +1\}$ for all $k \ne 0$. Such a sequence is called a *Barker sequence*. These sequences are only known for $n \in \{2, 3, 4, 5, 7, 11, 13\}$. The Barker sequences of length 13 have merit factor 14.08, the largest known value for any sequence. An example:

$$+ + + + - - + + - + - +$$

Let

$$S = \{\mathbf{s}^1, \mathbf{s}^2, \ldots, \mathbf{s}^i, \ldots\},$$

---

[5] A single negacyclic shift of $\mathbf{x}$ yields $(x_1, \ldots, x_{n-1}, -x_0)$.

be a family of sequences, where each $\mathbf{s}^i$ is a binary sequence of length n_i, an increasing function of i. If the limit of $F_{\mathbf{s}^i}$ exists as i approaches infinity, we call

$$F_S = \lim_{i \to \infty} F_{\mathbf{s}^i},$$

the *asymptotic merit factor* of the sequence family S.

Basic goals are to find sequences with large merit factor and families with large asymptotic merit factor.

The merit factors and asymptotic merit factor have been calculated for several important families of sequences; see the excellent survey by Jedwab, Katz, and Schmidt [21]. We focus here on two of the most fundamental examples.

For each integer i, let $\mathbf{m}^i$ be (the ± 1-version of) an *m-sequence*[6] having length $2^i - 1$. This is a sequence of period $2^i - 1$ generated by an i-stage linear feedback shift register, the maximal possible period for such a sequence. These were central topics in Golomb's Martin report [12] and his book [15] on shift register sequences (where they were called *PN*-sequences, for "pseudonoise"). Jensen, Jensen, and Høholdt [22] (following Sarwate [27]) proved:

Theorem 2.2. *The family* $M = \{\mathbf{m}^1, \mathbf{m}^2, \ldots, \mathbf{m}^i, \ldots\}$ *of m-sequences has asymptotic merit factor* 3.

For p an odd prime, the *Legendre sequence* of length p is defined by the Legendre symbols

$$\alpha_j = \left(\frac{j}{p}\right), \quad j = 0, \ldots, p - 1,$$

$$\text{where } \left(\frac{j}{p}\right) = \begin{cases} 1, & \text{if } j \text{ is } 0 \text{ or a square modulo } p \\ -1, & \text{otherwise.} \end{cases}$$

Golay [11] gave arguments suggesting the next theorem, which was proven by Høholdt and Jensen [18].

Theorem 2.3. *The family* L_f *of Legendre sequences of odd prime length* p, *each cyclically rotated by the fraction* f *of its length, has asymptotic merit factor* F_f *given by*

$$\frac{1}{F_f} = \frac{2}{3} - 4|f| + 8f^2, \quad |f| \leq \frac{1}{2}.$$

In particular, the family $L_{\frac{1}{4}}$ *has asymptotic merit factor* 6.

For many years, the value 6 of the theorem was the best know. The current record for proven asymptotic merit factor is $6.34\ldots$, achieved by a family of appended and rotated Legendre sequences [20,21]. Golay [10] gave arguments pointing to 12.32 as an absolute upper bound for all families. This remains unproven; indeed it is not known whether any upper bound exists.

---

[6] The "m" stands for "maximal" as in "maximal length".

3. Binary Phase Shaping

There is a longstanding relationship between optics, particularly spectroscopy, and the mathematics of sequences and codes. In the late 1940's Fellgett and Golay [6,7] introduced multiplex spectrometry, whereby radiation passes through grids with multiple slits (rather than the traditional single slit) and then its spectrum is calculated by transform methods. By the late 1960 's m-sequences (and the related Hadamard transform techniques) were being used; see, for instance, Sloane, Fine, Phillips, and Harwit [29]. These advances in optics were in part motivated by Sol Golomb's work [14] on sequences and radar ranging (according to Harwit [4]).

In December 2003, JIH was approached by Marcos Dantus, MSU Foundation Professor and University Distinguished Professor of Chemistry, with a question about pseudorandom sequences that was related to recent work of the Dantus Group on optical microscopy. They had improved upon previous instrumentation by using a ± 1-sequence determined by prime numbers. Dantus wondered whether there were better sequence choices. It was clear that the relevant sequence property was small autocorrelation, but there was an interesting variation from the usual.

Professor Dantus and his group were designing instruments that drive chemical reactions by using very short (femosecond) laser pulses [3]. A laser pulse is highly coherent, but the wish was to shape the pulse further so that (ideally) a single frequency became isolated. The basic idea is very simple: use appropriately arranged filters and mirrors so that the beam interacts with a delayed version of itself. Energy can be concentrated on a narrower spectrum by tuning the resultant interference.

Given a laser pulse, described by having intensity $\mathbf{E}(\omega)$ at frequency ω, they were concerned with two types of nonlinear, second order interference. For impulsive stimulated Raman scattering (**SRS**) the interference is negative, and the resultant intensity at frequency ω is

$$\mathbf{E}^{\mathbf{SRS}}(\omega) = \int_{\alpha - \beta = \omega} \mathbf{E}(\alpha)\mathbf{E}(\beta)*d\alpha.$$

On the other hand, for second harmonic generation (**SHG**) the interference is positive with intensity given by

$$\mathbf{E}^{\mathbf{SHG}}(\omega) = \int_{\alpha + \beta = \omega} \mathbf{E}(\alpha)\mathbf{E}(\beta)d\alpha.$$

Careful shaping of the pulse $\mathbf{E}$ yields $\mathbf{E}^{\mathbf{SRS}}$ and $\mathbf{E}^{\mathbf{SHG}}$ with improved spectral properties.

In binary phase shaping (**BPS**) the situation is simplified considerably; see [2,23,24,30] The initial pulse $\mathbf{E}$ consists of equally spaced frequencies ω_k for $0 \leq k \leq n - 1$. At a given frequency the intensity can be masked to 0 but otherwise is at a fixed amplitude, the only choice being whether its phase is left unchanged or is switched by π. The shaping of $\mathbf{E}(\omega)$ is thus encoded by the shaping sequence $\mathbf{e} = (e_0, \ldots, e_i, \ldots, e_{n-1})$ where $e_i = 0$ if frequency level i has been masked, and otherwise e_i is $+1$ or -1, depending upon the phase change 0 or π. (For all i out of the interval, set $e_i = 0$.)

The intensity measures are then determined by

$$E_{\mathbf{e}}^{\mathbf{SRS}}(k) = \sum_{a-b=k} e_a e_b$$

and

$$E_{\mathbf{e}}^{\mathbf{SHG}}(k) = \sum_{a+b=k} e_a e_b.$$

The goal is to find shaping sequences $\mathbf{e}$ that produce focus at a chosen frequency.

That is, for **SRS** we seek sequences for which the peak energy $|E_{\mathbf{e}}^{\mathbf{SRS}}(h)|^2$ is large at the desired focal frequency h while the remaining (off-peak) energies $|E_{\mathbf{e}}^{\mathbf{SRS}}(k)|^2$ are small for all $k \neq h$. (Unavoidably the base energy $|E_{\mathbf{e}}^{\mathbf{SRS}}(0)|^2$ is relatively large.) Similarly we seek sequences that admit focusing for the energy function $|E_{\mathbf{e}}^{\mathbf{SHG}}(k)|^2$.

Set $T_{\mathbf{e}}^{\mathbf{SRS}} = \sum_{k=1}^{n-1} |E_{\mathbf{e}}^{\mathbf{SRS}}(k)|^2$, the total off-base **SRS** energy. Then for a chosen focal frequency h we wish to choose a shaping sequence $\mathbf{e}$ to accomplish various things:

(i) maximize the peak energy $|E_{\mathbf{e}}^{\mathbf{SRS}}(h)|^2$ at the focal level h;
(ii) minimize the total off-peak energy $B_{\mathbf{e}}^{\mathbf{SRS}}(h) = T_{\mathbf{e}}^{\mathbf{SRS}} - |E_{\mathbf{e}}^{\mathbf{SRS}}(h)|^2$;
(iii) maximize the contrast ratio at level h:

$$R_{\mathbf{e}}^{\mathbf{SRS}}(h) = \frac{|E_{\mathbf{e}}^{\mathbf{SRS}}(h)|^2}{B_{\mathbf{e}}^{\mathbf{SRS}}(h)}.$$

Similar issues arise for second harmonic generation. Let the **SHG** total energy be $T_{\mathbf{e}}^{\mathbf{SHG}} = \sum_{k=0}^{2n-2} |E_{\mathbf{e}}^{\mathbf{SHG}}(k)|^2$. Then for the focal frequency h we wish to choose $\mathbf{e}$ in order to:

(i) maximize the peak energy $|E_{\mathbf{e}}^{\mathbf{SHG}}(h)|^2$;
(ii) minimize the total off-peak energy $B_{\mathbf{e}}^{\mathbf{SHG}}(h) = T_{\mathbf{e}}^{\mathbf{SRS}} - |E_{\mathbf{e}}^{\mathbf{SHG}}(h)|^2$;
(iii) maximize the contrast ratio:

$$R_{\mathbf{e}}^{\mathbf{SHG}}(h) = \frac{|E_{\mathbf{e}}^{\mathbf{SHG}}(h)|^2}{B_{\mathbf{e}}^{\mathbf{SHG}}(h)}.$$

4. Focus and Contrast

The contrast ratio problems are very similar to the traditional correlation and merit factor problems, but the desire to focus at a fixed off-base level introduces a new aspect.[7]

Let $\mathbf{x}$ be a ± 1-sequence of length n. As discussed above, for optimal focusing, the autocorrelation value $C_{\mathbf{x}}(h)$ at the desired focal level h should be large (yielding a peak) while all other off-peak values $C_{\mathbf{x}}(k)$ should be small. Again, a single

---

[7]Here we only treat the **SRS**- rather than the **SHG**-motivated contrast ratio problem. That also has a natural setting in terms of crosscorrelation; see Section 7 for a brief discussion.

Focus and Contrast 181

parameter provides a gauge for this information. Define the *contrast ratio function* to be

$$R_{\mathbf{x}}(h) = \frac{C_{\mathbf{x}}(h)^2}{\sum_{k \neq 0, h} C_{\mathbf{x}}(k)^2}.$$

for $0 \leq h \leq n-1$. (Notice that $R_{\mathbf{x}}(0)$ is equal to $2F_{\mathbf{x}}$, twice the merit factor.)

There is an easy upper bound on the peak focal energy $C_{\mathbf{x}}(h)^2$.

Lemma 4.1. *Let $\mathbf{x} \in \{\pm 1\}^n$. Then for $0 \leq h \leq n-1$ we have $|C_x(h)| \leq n-h$ with equality if and only if we have $x_{t+h} = \epsilon x_t$ for a fixed $\epsilon \in \{\pm 1\}$.*

We say that the sequence $\mathbf{x}$ has an optimal peak at h if $C_{\mathbf{x}}(h)^2 = (n-h)^2$; that is, if its autocorrelation function has the maximum possible energy at level h.

We will study those sequences that have an optimal peak at a chosen level h. Of particular interest will be sequences of the form

$$(c_0, c_1, \ldots, c_{m-1}, \epsilon c_0, \epsilon c_1, \ldots, \epsilon c_{m-1})$$

of length $n = 2m$; these have an optimal peak with energy m^2 at the central level $h = m$.

5. Focus on the Center

The concatenation of the sequences $\mathbf{a} = (a_0, \ldots, a_r)$ and $\mathbf{b} = (b_0, \ldots, b_s)$ will be written

$$(a_0, \ldots, a_r, b_0, \ldots, b_s) = (\mathbf{a} : \mathbf{b}).$$

Thus the sequence displayed above is

$$(\mathbf{c} : \epsilon\mathbf{c}) = (c_0, c_1, \ldots, c_{m-1}, \epsilon c_0, \epsilon c_1, \ldots, \epsilon c_{m-1})$$

for $(c_0, \ldots, c_m) \in \pm 1^m$ and $\epsilon \in \{\pm 1\}$.

Most of the analysis of focus and contrast for ± 1-sequences [2,23,24,30,32] restricts attention to sequences $(\mathbf{c} : \epsilon\mathbf{c})$ of length $2m$ with the chosen focus level $h = m$ at the center. In this section we offer some reasons for this.

We begin with a quote from Lozovoy, Xu, Shane, and Dantus [24]:

When the signal to be maximized is not in the center of the spectrum, we form the [sequence] from the longest possible symmetrized sequence. The leftover [sequence symbols] must be set to 0.

What does this mean? First suppose that the focus level is h with $2h \leq n$, the length of the sequence $\mathbf{x}$. Then to have an optimal peak at h, we must (by Lemma 4.1) have

$$\mathbf{x} = (\mathbf{a} : \epsilon\mathbf{a} : \mathbf{t})$$

where $\mathbf{a} = (x_0, \ldots, x_{h-1}) \in \pm 1^h$ is the initial segment of $\mathbf{x}$. The above stricture sets the leftover subsequence $\mathbf{t}$ to the zero vector $\mathbf{0}$, masking out the final $n-2h$ frequencies. In doing this the peak energy at h remains large while the off-peak energy is minimized for this choice of $\mathbf{a}$. Thereby, the problem of focusing at level

h and length $n(\geq 2h)$ is replaced with the question of focusing at the center h for length $2h$.

A similar situation arises for $2h \geq n$. In that case, set $l = n - h$ and let $(\mathbf{b} : \epsilon\mathbf{b}) \in \pm 1^{2l}$, optimal at its center l. Then

$$\mathbf{x} = (\mathbf{b} : \mathbf{0} : \epsilon\mathbf{b})$$

has an optimal peak at $h = n - l$ with its middle, leftover section set to $\mathbf{0}$, as directed by [24] in the above quote.

The message is that good solutions for all levels (at a given length) can be found by finding solutions at the center (for a different length). This is the topic of the next section.

But these good solutions may not be the best solutions. For instance, a computer search reported in [23] states that, for length $n = 16$ and level $h = 6$, the sequence

$$+ + + - + + \,|\, + + - + + + + -$$

is a better solution than

$$+ + + - + + \,|\, + + + - + + \, 0\,0\,0\,0.$$

Consider the argument of Lemma 4.1 regarding an optimal peak at level h in $\mathbf{x} \in \pm 1^n$. Let $\mathbf{a} = (a_0, \ldots, a_{h-1})$ be the initial segment of $\mathbf{x}$.

We calculate $C_{\mathbf{x}}(h)$ as the dot product of the following two lines:

$$
\begin{array}{cc|cc|cc|c}
x_0 & \cdots \; x_{h-1} & x_h & \cdots \; x_{2h-1} & x_{2h} & \cdots \; x_{3h-1} & x_{3h}\cdots \\
0 & \cdots \quad 0 & x_0 & \cdots \; x_{h-1} & x_h & \cdots \; x_{2h-1} & x_{2h}\cdots
\end{array}
$$

That is,

$$
\begin{array}{cc|cc|cc|c}
a_0 & \cdots \; a_{h-1} & x_h & \cdots \; x_{2h-1} & x_{2h} & \cdots \; x_{3h-1} & x_{3h}\cdots \\
 & & a_0 & \cdots \; a_{h-1} & x_h & \cdots \; x_{2h-1} & x_{2h}\cdots
\end{array}
$$

For an optimal peak at h, this must be (for some $\epsilon \in \{\pm 1\}$)

$$
\begin{array}{cc|cc|cc|c}
a_0 & \cdots \; a_{h-1} & \epsilon a_0 & \cdots \; \epsilon a_{h-1} & x_{2h} & \cdots \; x_{3h-1} & x_{3h}\cdots \\
 & & a_0 & \cdots \; a_{h-1} & x_h & \cdots \; x_{2h-1} & x_{2h}\cdots
\end{array}
$$

Continue:

$$
\begin{array}{cc|cc|cc|c}
a_0 & \cdots \; a_{h-1} & \epsilon a_0 & \cdots \; \epsilon a_{h-1} & x_{2h} & \cdots \; x_{3h-1} & x_{3h}\cdots \\
 & & a_0 & \cdots \; a_{h-1} & \epsilon a_0 & \cdots \; \epsilon a_{h-1} & x_{2h}\cdots
\end{array}
$$

and so (as $\epsilon^2 = 1$)

$$
\begin{array}{cc|cc|cc|c}
a_0 & \cdots \; a_{h-1} & \epsilon a_0 & \cdots \; \epsilon a_{h-1} & a_0 & \cdots \; a_{h-1} & x_{3h}\cdots \\
 & & a_0 & \cdots \; a_{h-1} & \epsilon a_0 & \cdots \; \epsilon a_{h-1} & x_{2h}\cdots
\end{array}
$$

then

$$
\begin{array}{cc|cc|cc|c}
a_0 & \cdots \; a_{h-1} & \epsilon a_0 & \cdots \; \epsilon a_{h-1} & a_0 & \cdots \; a_{h-1} & x_{3h}\cdots \\
 & & a_0 & \cdots \; a_{h-1} & \epsilon a_0 & \cdots \; \epsilon a_{h-1} & a_0\cdots
\end{array}
$$

and

$$a_0 \cdots a_{h-1} \left| \begin{array}{c} \epsilon a_0 \cdots \epsilon a_{h-1} \\ a_0 \cdots a_{h-1} \end{array} \right| \begin{array}{c} a_0 \cdots a_{h-1} \\ \epsilon a_0 \cdots \epsilon a_{h-1} \end{array} \left| \begin{array}{c} \epsilon a_0 \cdots \\ a_0 \cdots \end{array} \right.$$

and so on.

This leads to a refinement of Lemma 4.1:

Proposition 5.1. *The sequence $\boldsymbol{x} \in \pm 1^n$ has an optimal peak at h if and only if there is a sequence $(a_0, \ldots, a_{h-1}) \in \pm 1^h$ and a sign $\epsilon \in \{\pm 1\}$ for which $\boldsymbol{x}$ is the truncation at length n of*

$$\mathbf{a} : \epsilon\mathbf{a} : \mathbf{a} : \epsilon\mathbf{a} : \mathbf{a} : \epsilon\mathbf{a} : \mathbf{a} : \epsilon\mathbf{a} : \mathbf{a} : \epsilon\mathbf{a} : \mathbf{a} : \epsilon\mathbf{a} : \cdots$$

So, for instance, with $n = 16$ and $h = 6$ one possibility is the sequence

$$+ + + - + + \,|\, + + + - + \,|\, + + + -$$

found in [23] (and given above) in which two symbols have been dropped from the end of the 18-tuple $(+ + + - + +)^3$.

A corollary emphasizes that the center is a more fruitful focal point than the extremes.

Corollary 5.2.

(a) *Every sequence $\boldsymbol{x} \in \pm 1^n$ has an optimal peak at $h = n - 1$.*
(b) *The only four sequences $\boldsymbol{x} \in \pm 1^n$ with an optimal peak at $h = 1$ are $\mathbf{1}$, $((-1)^0, \ldots, (-1)^k, \ldots, (-1)^{n-1})$, and their negatives.*

The sequences of the proposition also have large peaks at other multiples of h. As suggested above, this can be handled by zeroing out the tail. But for practical purposes, focus at h can be achieved provided none of the $|C_{\boldsymbol{x}}(k)|$ are large for those k close to h. This can be realized provided h itself is not small. The second part of the corollary shows that small values of h will always be a problem.

6. Focus at the Center

As discussed in the previous section, length $2m$ sequences $(\mathbf{c} : \epsilon\mathbf{c})$ can be used to construct sequences of length $n \geq 2m$ with good focusing properties and optimal peaks at m or $n - m$.

In this section we give some specific results and then some asymptotic results about central focus for $(\mathbf{c} : \epsilon\mathbf{c})$.

184   *The Wisdom of Solomon*

6.1. *Contrast and merit*

The (ordered) *energy spectrum* of the ± 1-sequence $\mathbf{x} = (x_0, \ldots, x_{n-1})$ is the sequence

$$(C_{\mathbf{x}}(0)^2, \ldots, C_{\mathbf{x}}(h)^2, \ldots, C_{\mathbf{x}}(n-1)^2).$$

In particular the contrast ratio function

$$R_{\mathbf{x}}(h) = \frac{C_{\mathbf{x}}(h)^2}{\sum_{k \neq 0, h} C_{\mathbf{x}}(k)^2}.$$

and so the merit factor $F_{\mathbf{x}} = R_{\mathbf{x}}(0)/2$ of x are entirely determined by its energy spectrum. It is therefore helpful that the energy spectrum is left invariant by the group G (typically of order 8) generated by the three transformations $T : \mathbf{x} \longrightarrow \mathbf{x}'$ given by

(i) Negation: $N : x_i' = -x_i$;
(ii) Reversal: $R : x_i' = x_{n-1-i}$;
(iii) Alternation: $A : x_i' = (-1)^i x_i$.

For instance, starting with the Barker sequence

$$+ + + + + - - + + - + - +$$

given earlier, there are exactly four Barker sequences of length 13 : the above, its reverse,

$$+ - + - + + - - + + + + +$$

and their negatives. (These two sequences are skew: They are left fixed by $RA = AR$.)

Our calculations at the center are therefore aided by:

Lemma 6.1.

(a) $(N\boldsymbol{c} :\epsilon N\boldsymbol{c}) = N(\boldsymbol{c} : \epsilon \boldsymbol{c})$;
(b) $(R\boldsymbol{c} :\epsilon R\boldsymbol{c}) = \epsilon R(\boldsymbol{c} : \epsilon \boldsymbol{c})$;
(c) $(A\boldsymbol{c} : \epsilon A\boldsymbol{c}) = A(\boldsymbol{c} :\epsilon(-1)^{m+1}\boldsymbol{c})$.

For instance

$$R(c_0, c_1, c_2, c_3, c_4 : -c_0, -c_1, -c_2, -c_3, -c_4)$$
$$= (-c_4, -c_3, -c_2, -c_1, -c_0 : c_4, c_3, c_2, c_1, c_0)$$
$$= -(c_4, c_3, c_2, c_1, c_0 : -c_4, -c_3, -c_2, -c_1, -c_0)$$
$$= -(R(c_0, c_1, c_2, c_3, c_4) : -R(c_0, c_1, c_2, c_3, c_4))$$

and

$$A(c_0, c_1, c_2, c_3, c_4 : c_0, c_1, c_2, c_3, c_4)$$
$$= (c_0, -c_1, c_2, -c_3, c_4 : -c_0, c_1, -c_2, c_3, -c_4)$$
$$= (A(c_0, c_1, c_2, c_3, c_4) : -A(c_0, c_1, c_2, c_3, c_4)).$$

Therefore in calculating central contrast ratio we only need consider $(\mathbf{c}{:}\mathbf{c})$ and $(\mathbf{c}:-\mathbf{c})$ to include the calculation for all the G-images of $\mathbf{c}$.

Arguments similar to those of the Section 5 yield

Proposition 6.2. *For a positive integer m, sign $\epsilon \in \{\pm 1\}$, and ± 1-sequence $\mathbf{c} = (c_0, c_1, \ldots, c_{m-1})$, for the sequence $\mathbf{x} = (\mathbf{c}:\epsilon\mathbf{c})$*

$$
C_{\mathbf{x}}(k) = \begin{cases} C_c^\epsilon(k) + C_c(k), & \text{if } 1 \le k < m; \\ \epsilon m, & \text{if } k = m; \\ \epsilon C_c(k-m), & \text{if } m < k < 2m. \end{cases}
$$

For $k = h < m$ this can be seen as $C_{\mathbf{c}}(h) + \epsilon C_{\mathbf{c}}(m-h) + C_{\mathbf{c}}(h)$ from

$$
\begin{array}{cccc}
c_0 \cdots c_{h-1} & \begin{array}{cc} c_h & \cdots & c_{m-1} \\ c_0 & \cdots & c_{m-1-h} \end{array} & \begin{array}{cc} \epsilon c_0 & \cdots & \epsilon c_{h-1} \\ c_{m-h} & \cdots & c_{m-1} \end{array} & \begin{array}{cc} \epsilon c_h & \cdots & \epsilon c_{m-1} \\ \epsilon c_0 & \cdots & \epsilon c_{m-1-h} \end{array} \cdots
\end{array}
$$

while for $m \le k$ with $h = k - m$ we have the simpler

$$
\begin{array}{ccc}
c_0 \cdots c_{m-1} & \begin{array}{cc} \epsilon c_0 \cdots \epsilon c_{h-1} \end{array} & \begin{array}{cc} \epsilon c_h & \cdots & \epsilon c_{m-1} \\ c_0 & \cdots & c_{m-1-h} \end{array} \begin{array}{c} \\ c_{m-h} \cdots \end{array} & .
\end{array}
$$

We have implied that sequences with good merit factor can be used to construct sequences with good contrast ratio. At the center, the precise relationship is given by:

Proposition 6.3. *For a positive integer m, sign $\epsilon \in \{\pm 1\}$, ± 1-sequence $\mathbf{c} = (c_0, c_1, \ldots, c_{m-1})$, and $\mathbf{x} = (\mathbf{c}:\epsilon\mathbf{c})$ we have*

$$
\frac{1}{R_{\mathbf{x}}(m)} = \frac{1}{F_c} + \frac{\sum_{k=1}^{m-1} C_c^\epsilon(k)^2}{m^2} + 2 \times \frac{\sum_{j=1}^{m-1} C_c^\epsilon(k) C_c(k)}{m^2} \tag{11.1}
$$

where F_c is the merit factor of sequence $\mathbf{c}$.

Indeed, by Proposition 6.2

$$
\begin{aligned}
\frac{1}{R_{\mathbf{x}}(m)} &= \frac{\sum_{k=1}^{m-1} (C_c^\epsilon(k) + C_c(k))^2 + \sum_{k=m+1}^{2m-1} C_c(k-m)^2}{m^2} \\
&= \frac{\sum_{k=1}^{m-1} C_c^\epsilon(k)^2 + 2\sum_{k=1}^{m-1} C_c^\epsilon(k) C_c(k) + \sum_{k=1}^{m-1} C_c(k)^2 + \sum_{k=m+1}^{2m-1} C_c(k-m)^2}{m^2} \\
&= \frac{2\sum_{k=1}^{m-1} C_c(k)^2 + \sum_{k=1}^{m-1} C_c^\epsilon(k)^2 + 2\sum_{k=1}^{m-1} C_c^\epsilon(k) C_c(k)}{m^2} \\
&= \frac{1}{F_c} + \frac{\sum_{k=1}^{m-1} C_c^\epsilon(k)^2}{m^2} + 2 \times \frac{\sum_{j=1}^{m-1} C_c^\epsilon(k) C_c(k)}{m^2}.
\end{aligned}
$$

6.2. *Contrast ratios*

The following table contains the central contrast ratios of both $(\mathbf{c}{:}\ \mathbf{c})$ and $(\mathbf{c}:-\mathbf{c})$ (and so also for all G-conjugates) for certain interesting sequences $\mathbf{c}$ of length at most 13.

The Wisdom of Solomon

Specifically, we consider:

1. all optimal merit sequences **c** of length at most 13 (including all known Barker sequences), as listed in [25];
2. the sequences **c** used by Lovozoy and Dantus [23, p. 1986] (given in the table in **boldface**);
3. the Legendre sequence of length 13 (given in the table in italics). (Shorter Legendre sequences and short m-sequences are Barker.)

In our list we use the "run notation" of [25], whereby the sequence is assumed to begin with $+$ and the lengths of the consecutive runs are given. For instance, our Barker sequence

$$+ + + + + - - + + - + - +$$

is, in run notation,

$$5221111.$$

Length m	Sequence **c**	ϵ	$R_{(\mathbf{c}:\epsilon\mathbf{c})}(m)$	Length m	Sequence **c**	ϵ	$R_{(\mathbf{c}:\epsilon\mathbf{c})}(m)$
2	2	$+$	0.4	9	42111	$+$	1.125
	2	$-$	2		42111	$-$	1.125
3	21	$+$	1.5		32211	$+$	1.125
	21	$-$	1.5		32211	$-$	1.125
4	112	$+$	4		31122	$+$	1.125
	112	$-$	0.8		31122	$-$	1.125
5	311	$+$	2.08333	10	42211	$+$	1.02041
	311	$-$	2.08333		42211	$-$	1.72414
6	141	$+$	0.666667		52111	$+$	1.02041
	141	$-$	1.2		52111	$-$	1.72414
	123	$+$	0.666667		311122	$+$	1.02041
	123	$-$	1.2		311122	$-$	1.72414
	312	$+$	0.666667		41122	$+$	1.02041
	312	$-$	1.2		41122	$-$	1.72414
	1113	$+$	0.666667		33121	$+$	0.409836
	1113	$-$	1.2		33121	$-$	1.08696
	411	$+$	0.666667	11	112133	$+$	4.03333
	411	$-$	1.2		112133	$-$	4.03333
7	1123	$+$	2.7222	12	4221111	$+$	2.76923
	1123	$-$	2.7222		4221111	$-$	2.11765
	151	$+$	2.742424		4111221	$+$	2.76923
	151	$-$	2.742424		4111221	$-$	2.11765
8	32111	$+$	1.333333		**222114**	$+$	1.56522
	32111	$-$	1.333333		**222114**	$-$	1.89474
	31121	$+$	1.333333		**151131**	$+$	0.349515

	31121	–	1.333333		**151131**	–	1.02857
	341	+	0.195122	13	5221111	+	4.69444
	341	–	0.615303		5221111	–	4.69444
9	311121	+	0.632812		*2124211*	+	0.692623
	311121	–	0.50625		*2124211*	–	0.338

The patterns to be found within the table may only reflect the smallness of the sequence lengths considered. For instance, in the table it is nearly true that contrast ratio is a function of merit factor and ϵ, but that is false in a small number of cases.

The most interesting entries in the table may be at length $m = 7$ where Lovozoy and Dantus' sequence 341 has better contrast ratio than the Barker sequence 1123, which is also a Legendre sequence and an m-sequence. Thus, although they are related, the contrast ratio and merit factor problems are not equivalent.

6.3. *Asymptotic contrast ratios*

For the asymptotic case, we present some results from [32] without proof.

Proposition 6.4. *Let*

$$S = \{\mathbf{s}^1, \mathbf{s}^2, \ldots, \mathbf{s}^i, \ldots\},$$

be a family of ± 1-sequences of increasing length n_i, and assume that its asymptotic merit factor

$$\mathrm{F}_S = \lim_{i \to \infty} \mathrm{F}_{\mathbf{s}^i}$$

exits.

Choose an $\epsilon \in \{\pm 1\}$ and assume that $\mathrm{C}^\epsilon_{\mathbf{s}^i}(k)$ is $O(1)$ for all $1 \leq k \leq n_i - 1$.

Then the family SS^ϵ, consisting of the sequences $(\mathbf{s}^i : \epsilon \mathbf{S}^i)$ for all $\mathbf{s}^i \in S$ has asymptotic central contrast ratio

$$\mathrm{R}_{SS^\epsilon} = \lim_{i \to \infty} R_{\mathbf{s}^i}(n_i)$$

equal to the asymptotic merit factor F_S.

The proposition then leads to

Theorem 6.5. *Let $M = \{\mathbf{m}^1, \mathbf{m}^2, \ldots, \mathbf{m}^i, \ldots\}$ be a family of m-sequences as in Theorem 2.2. The family MM^+ has asymptotic central contrast ratio 3.0.*

Theorem 6.6. *Let $L_{\frac{1}{4}}$ be the family of $\frac{1}{4}$-rotated Legendre sequences as in Theorem 2.3. The family $L_{\frac{1}{4}} L_{\frac{1}{4}^+}$ has asymptotic central contrast ratio 6.0.*

Similar results hold for other families with known asymptotic merit factor. For instance, if $\mathbf{r^i}$ is the Legendre sequence for the ith odd prime p_i, set

$$\mathbf{t^i} = (\mathbf{r^i} : \mathbf{r^i}) = (t_0^i, \ldots t^i p_{i-1}, t^i p_i, \ldots t_{2p_i-1}^i) = (t_0^i, \ldots, t_j^i, \ldots, t_2^i p_i - 1).$$

Then

$$\mathbf{s^i} = (\ldots, (-1)^{j(j-1)/2} t_j^i, \ldots)$$

$$= \mathbf{t^i}{\star}(+1, +1, -1, -1, +1, +1, \ldots)$$

is a *Parker sequence* of length $2p_i$. The family $S = \{\mathbf{s^1}, \mathbf{s^2}, \ldots, \mathbf{s^i}, \ldots\}$, has asymptotic merit factor 6, as predicted by Parker [26] and proven by Xiong and Hall [31]. This leads to

Theorem 6.7. *For the family* $S = \{s^1, s^2, \ldots, s^i, \ldots\}$ *of Parker sequences, the family* SS^- *has asymptotic central contrast ratio 6.0.*

7.   The Second Harmonic and Convolution

Some work on the contrast ratio problem motivated by **SHG** has been done (for instance [2,23,24,30,32]) in a similar manner to that discussed above.

The problem here can be more complicated, since instead of the autocorrelation function $C_\mathbf{x}(k) = C_{\mathbf{x},\mathbf{x}}(k)$ studied before, we must consider the convolution function

$$V_\mathbf{x}(k) = \sum_{i+j=k} x_i x_j = C_{\mathbf{x}, R\mathbf{x}}(n - 1 - k),$$

for all $0 \leq k \leq 2n - 2$, the crosscorrelation function between the sequence $\mathbf{x}$ (of length n) and its reverse $R\mathbf{x}$ given by $Rx_i = x_{n-1-i}$.

Bibliography

1. L. D. Baumert, S. W. Golomb, and M. Hall, Jr., Discovery of an Hadamard matrix of order 92, *Bull. Amer. Math. Soc.* **68** (1962) 237–238.
2. M. Comstock, V. V. Lozovoy, I. Pastirk, and M. Dantus, Multiphoton intrapulse interference 6; binary phase shaping, *Opt. Exp.* **12** (2004) 1061–1066.
3. M. Dantus, *Femtosecond Laser Shaping: from Laboratory to Industry*, CRC Press, Taylor and Francis Group, Boca Raton, 2018.
4. M. Harwit, Chapter 7. Hadamard transform analytical systems, in: *Transform Techniques in Chemistry*, editor P.R. Griffiths, Plenum Press, New York (1978).
5. T. Helleseth and P. V. Kumar, Chapter 8. Pseudonoise sequences, in: *Mobile Communications Handbook*, editor S.S. Suthersan, CRC Press, Boca Raton, 1999.
6. M. J. E. Golay, Multi-slit spectrometry, *J. Opt. Soc. Am.* **39** (1949) 437–444.
7. M. J. E. Golay Static multislit spectrometry and its application to panoramic display of infared spectra, *J. Opt. Soc. Am.* **41** (1951) 468–472.
8. M. J. E. Golay, A class of finite binary sequences with alternate autocorrelation vaiues equal to zero, *IEEE Trans. Inform. Th* **18** (1972) 449–450.
9. M. J. E. Golay, Sieves for low autocorrelation binary sequences, *IEEE Trans. Info. Th.* **23** (1977) 43–51.

Focus and Contrast

10. M. J. E. Golay, The merit factor of long low autocorrelation binary sequences, *IEEE Trans. Info. Th.* **28** (1982) 543–549.

11. M. J. E. Golay, The merit factor of Legendre sequences, *IEEE Trans. Info. Th.* **29** (1983) 934–936.

12. S. W. Golomb, *Sequences with Randomness Properties*, Martin Company, 1955.

13. S. W. Golomb, Sets of primes with intermediate density, *Math. Scand.* **3** (1955) 264–274.

14. S. W. Golomb, editor (with contributions by L.D. Baument, M.F. Easterling, J.J. Stiffler, and A.J. Viterbi), *Digital Communications with Space Applications*, Prentice Hall International Series in Electrical Engineering, Prentice-Hall, Inc. Englewood Cliffs, New Jersey, 1964.

15. S. W. Golomb (with portions co-authored by L.R. Welch, R.M. Goldstein, and A. W. Hales), *Shift Register Sequences*, Holden-Day, Inc., San Francisco, Calif.-Cambridge-Amsterdam, 1967

16. S. W. Golomb and G. Gong, *Signal Design for Good Correlation. For Wireless Communication, Cryptography, and Radar*, Cambridge University Press, Cambridge, 2005.

17. M. Hall, An isomorphism between linear recurring sequences and algebraic rimgs, *Trans. Amer. Math. Soc.* **44**(1938) 196–218.

18. T. Høholdt and H. E. Jensen, Determination of the merit factor of Legendre sequences, *IEEE Trans. Info. Th.* **34** (1988), 161–164.

19. T. Høholdt, H. E. Jensen, and J. Justesen, Aperiodic correlations and the merit factor of a class of binary sequences, *IEEE Trans. Info. Th.* **31** (1985) 549–552.

20. J. Jedwab, D. J. Katz, and K-U. Schmidt [Littlewood polynomials with small L^4 norm, *Adv. Math.* **241** (2013) 127–136.

21. J. Jedwab, D. J. Katz, and K-U. Schmidt, Advances in the merit factor problem for binary sequences, *J. Combin. Th (Series A)* **120** (2013) 882–906.

22. J. M. Jensen, H. E. Jensen, and T. Høholdt, The merit factor of binary sequences related to difference sets, *IEEE Trans. Info. Th.* **37** (1991) 617–625.

23. V. V. Lozovoy and M. Dantus, Systematic control of nonlinear optical processes using optimally shaped femotosecond pulses, *Chem. Phys. Chem.* **6** (2005) 1970–2000.

24. V. V. Lozovoy, B. Xu, Janelle C. Shane, and M. Dantus, Selective nonlinear optical excitation with pulses shaped by pseudorandom Galois fields, *Phy. Rev, A* **74** (2006) 041805(4).

25. T. Packebusch and S. Mertens, Low autocorrelation binary sequences, *J. Phys. A: Math. Theor.* **49** (2016) 165001 (18 pages).

26. M. G. Parker, Even length binary sequence families with low negaperiodic autocorrelation, in: *Applied Algebra, Algebraic Algorithms and Error-Correcting Codes. AAECC 2001,* Lecture Notes in Comput. Sci. 2227, Springer, Berlin, 2001, pp. 200–209.

27. D. V. Sarwate, An upper bound on the aperiodic autocorrelation function for a maximal-length sequence, *IEEE Trans. Info. Th.* **30** (1984) 685–687.

28. D. V. Sarwate and M.B. Pursley, Crosscorrelation properties of pseudorandom and related sequences, *Proc. IEEE* **68** (1980) 593–619.

29. N. J. A. Sloane, T. Fine, P. G. Phillips, and M. Harwit, Codes for multiplex spectrometry, *Appl. Optics* 8 (1969) 2103–2106.

30. P. J. Wrzesinski, D. Pestov, V. V. Lozovoy, B. Zu, S. Roy, J. R. Gord, M. Dantus. Binary phase shaping for selective single-beam CARS spectroscopy and imaging of gas-phase molecules, *J. Raman Spectrosc.* **42** (2011) 393–398.

31. T. Xiong and J. I. Hall, Construction of even length binary sequences with asymptotic merit factor 6, *IEEE Trans. Inform. Th* **54** (2008) 931–935.

32. T. Xiong and J. I. Hall, "Sieves for Binary Sequences with large SRS and SHG Contrast Ratios," 2021 IEEE International Symposium on Information Theory (ISIT), 2021, pp. 1659–1664, doi: 10.1109/ISIT45174.2021.9518247.

© 2023 World Scientific Publishing Company
https://doi.org/10.1142/9789811234378_0018

Taped Conversation re Sol and JPL

Jon Hamkins

Conversation: Jon Hamkins (JPL), Beatrice Golomb, Al Hales (11-15-2018)

BG: We are now recording a tape on November 15, 2018, with Jon Hamkins, Beatrice Golomb and Al Hales

AH: Are you at JPL now?

JH: Yes

AH: I see. When did you first come to JPL?

JH: 1996

AH: Oh, so you were a very recent arrival back then by my standards — I was a student working under Sol Golomb at JPL during the summers of 1958–1962.

BG: So how did you get to know Sol?

JH: Well I think first at an IEEE conference. I knew his reputation already from his contributions to many aspects of communications, especially shift registers. I think it was at a symposium in Germany in the early or mid 90s. He had entered a chess "tournament" where they had hired an expert to play 30 simultaneous games against the participants. So all these 30 people would be sitting trying to win their one game and Sol was one of those people. Unfortunately he lost — as did all of them.

Sol was a very good chess player — if you look back at the early JPL newspaper, the "Lab Oratory", you see that he was winning the JPL tournaments being put on by the lab at that time.

Anyway people like Vera Pless were walking up to talk to Sol during that chess event and he kept saying "stop talking, stop talking".

AH: This reminds me of the recent book "Rise of the Rocket Girls" where, in 1958 while the lab workers were waiting to see if the Explorer satellite launch would succeed, Sol started playing chess with a woman named Margie Behrens to kill time during the suspense. She was a calculator at JPL and did not know who Sol was — except that he was an engineer involved with the project, and was a chess expert.

BG: Go ahead.

JH: But later on what got me interested in Sol was this. Early on in my career I was asked to be supervisor of the information processing group at JPL, and I was supervisor of this group for 15 years — then promoted in 2015. Sol was really the creator of that initial group in April 1960 and it only had a handful of supervisors, including Bob McEliece shortly after Sol. In digging into the history of that research group I saw how a lot of real great work had come out of it and how it was led by Sol Golomb while he was a young man at JPL. That set the stage for the kind of research that would be done in that group. In fact that group still has the name "Information Processing Group", as it has been from the very beginning. I still have the organizational charts going all the way back and a lot of prominent people have gone through that group, including a number of Shannon Award winners. It has been an honor, not just to be in that group but to lead it.

I was interested enough in the history that at one point since Sol was living in La Canada to invite him to lunch where we reminisced over what was going on at JPL during his time and my time. That was kind of fun.

AH: Who was your predecessor as head of that group?

JH: Fabrizio Pollara

AH: The names I knew from earlier on included Ed Posner and Laif Swanson. Were they each group heads at one time?

JH: Laif is still at the lab, and yes she was a long term supervisor of that group. At one point I put together the full list of people but I'm not sure if I can find it $\cdots$

AH: It would be nice to have the list to put in the book we are coediting.

JH: OK, here it is:

Sol Golomb 1960–62
Neal Zierler 1962–67
Ed Posner 1968–70
Bob McEliece 1971–78
Richard Lipes 1978–81
Joseph Yuen 1982–85
Stephen Townes 1985–86
Laif Swanson 1987–95
Fabrizio Pollara 1995–2001 (acting 1995–97)
Jon Hamkins 2001–2015
Sam Dolinar 2015–present

Sol came here in 1956 and left in 1963, well before my time. I was born in 1968, five years after he left. He did a lot in his time here. He often talked about his work on the Venus radar ranging project, where they bounced a signal off of Venus and from the response could measure distance, etc. In a sense this was the first human interaction with another planet.

But his major legacy was the development of convolutional codes, which have been a work horse here at JPL. There were two stages for this. At first very long codes were required. Then, after the work of Andy Viterbi, much shorter codes could be used. By now, of course, there have been many further advances. But the whole program, based from the start on the JPL work —

Taped Conversation re Sol and JPL

shift register encoding, etc. — has been super-important as a critical element in the NASA space program and is enshrined in international space program standards and the IEEE standards as well.

AH: I have a question about the Venus radar ranging project and its implications for one of Einstein's predictions — Sol was given credit for this. There was a monograph which mentioned JPL's involvement with this. I had my hands on it a couple of years ago — I think it was someone's Ph.D. thesis. Do you know anything about what I'm referring to?

JH: What was the specific title or topic?

AH: I can't remember now but it covered what was going on at JPL in the 50s and early 60s and mentioned both Sol's work and the work of Dick Goldstein, along with many other things.

JH: There are a few books. There is a book called *JPL and the American Space Program* which has a lot of early information. I don't think I read that with Sol specifically in mind. Then there is another called *Uplink-Downlink*. I don't have a copy of that. But the first book has a lot of material going back to the 40s and 50s, with many pictures of Pickering and other early folk at JPL.

AH: Maybe that is what I am thinking of.

JH: The author of that is Koppes.

AH: Right. Thank you.

BG: Does JPL have a historian?

JH: Yes. Let me see if I can dig out her name. Julie Cooper. She is the archivist at JPL. She identifies and processes the corrections to the JPL archives and helps researchers find information.

BG: Very well. Thank you.

AH: You mentioned Bob McEliece several times. Do you know him personally?

JH: Yes. I was a student at Caltech when he was a professor there in the middle 80s. He was such a phenomenal teacher that he really started me on my path to becoming a communications research engineer. He taught Math 2 and a coding theory class and was an inspiration. Then later I went to grad school at the University of Illinois and he was there — I was trying to follow his path in some sense! He always maintained a connection with Caltech and JPL from the very beginning, from the day he set foot on the Caltech campus — he always bragged that Marshall Hall was his freshman advisor there and remained his advisor all the way through graduate school.

AH: I was a student at Caltech and Marshall Hall was one of my first professors there.

Bob McEliece is back in the Pasadena area now but is in very bad shape.

JH: Yes, I know. I had a friend who went to visit him. He's really in a bad way, I understand.

AH: I knew him quite well, and I have tried several times to contact him. I talked to his nurse and then she put him on, but it was very hard. He has trouble communicating and the phone connection was not very good. It was clear he wanted very much to communicate and produce something as a contribution

194   *The Wisdom of Solomon*

to our book. But it was very hard to get anything useful. Maybe trying to do it in person or have his nurse take dictation might work but I doubt it.

***N.B. Bob passed away in 2019 before any useful input could be obtained from him, so we omit further discussion here about him.

BG: Do you know who still at JPL, if anyone, might have been there when Sol was there?

JH: No. There is nobody left here who was here when Sol was here.

BG: OK.

JH: He left in 1963. The earliest person I can think of in our entire section came in 1971 many years later.

I do have access to the lab newspaper so I can see here your birth announcement!

And Astrid's as well. Astrid actually dated my brother when I was in college.

BG: Oh.

JH: Yes, I see in this issue various pictures of displays of the Pioneer Four Probe that Sol had in his office.

BG: I might know what it looks like, but I didn't know what it was. I have a picture of Sol with something in his office and it would be great if you could send anything like that on.

JH: I appreciate that. It's a kind of a grid with a trajectory of the probe and then there's another picture in the paper with him developing a trajectory, a tool that's used to help predict the orbit.

BG: Oh that would be great.

JH: Anyway I'll send you these. There's one from 1957, one from 1958 and one from 1959 that mention various aspects of what's going on with Sol. Either a birth or a chess tournament or other little items.

BG: Great

AH: Are these from the JPL Newsletter or the *Star News* or ???

JH: Yes. JPL. It's called the *Lab Oratory.*

AH: Oh right. You mentioned that earlier.

JH: So this came out monthly. It was eventually changed to the *JPL Universe* and went on for many years. But then about 5–10 years ago they stopped the print version — just e-mailed a pdf version — and now even that has stopped.

BG: Oh that is too bad.

JH: But there's a lot of interesting stuff that the paper had.

BG: Another queston. Is there someone who has access to names of all the people at different times?

JH: They don't. I've asked about that in the past. There is not a comprehensive documentation on that point which is unfortunate. I have at least maintained the organizational charts going back to the 50s for my own area, my own section. The people when they retire they give their pile of stuff to the next person. So I have organizational charts going back for where Sol was in the organization for the few years that he was here.

AH: Does it give the names of the people working under him at the time?

Taped Conversation re Sol and JPL

JH: Yeah. In fact, let me get out my list. Just a minute.

AH: OK.

JH: So let me go back to the beginning, back in the old days. I'm the deputy section manager now which has about 100 people and you've got six groups in the section and each group has ten people. When Sol was appointed the supervisor, at that time the section manager was called the "Chief" instead of section manager. Anyway the earliest one I have here is for April 1960 and it shows him as head of the Information Processing Group, with members Baugh, Baumert, Easterling, Jewett, Stiffler and three technicians. You're one of the technicians.

AH: I am?

JH: Yes. Part time.

AH: OK. I thought I might be there but I was not sure if part timers were listed.

JH: Yes, along with Branfield and Fredricksen. There were other people in other sections, not Sol's, that you'd remember like Goldstein, Tausworthe, Viterbi, and Welch.

AH: All of these have been asked to contribute to the book we are working on.

JH: OK, great. They were all in Lorens' group which was called the Communications Research Group in April. 1960. And then sometime after that Posner became the supervisor and Sol moved on to be the Assistant Chief of the bigger section and then Andy Viterbi became the supervisor of the Communications Research Group. Then there were other people like Tom Kailath. I know that Sol was very proud of the people he had hired, and the great achievements that JPL was able to make. When I had lunch with him he was particularly proud of the people he brought in.

BG: Sol told me long ago about the idea he had for testing Einstein's general relativity, based on the effect of gravity on radar signals. He had explained this at JPL in a room with several other people. One of them, Duane Muhleman, had later gone off to MIT and gotten into a priority dispute with someone there (Irwin Shapiro) over the idea, not remembering (apparently) that he himself had gotten the idea from Sol. But there was another junior person in the room, a man with first name Dennis (maybe a PCC student?) who could vouch for Sol's priority. Might it be possible to track down this person?

JH: Too little for me to go on. If you knew the surname it would give me something to go on. Then I could try to do some historical exploration on this. Oh, by the way, this Julie Cooper can also point you to photographic records.

BG: Great.

JH: I have already done that for some historic records I was looking up. I do have a couple of early pictures of Sol and he has been in a group picture from that time period as well — I think you have already seen the one about the work with Len Baumert and Marshall Hall. I talked to Bob McEliece about that and he said it was based on work Len did while a grad student at Caltech, working with Marshall Hall and Sol. Sol put out a great press release for that — I think I have a copy of the original.

BG: OK, great. Any sort of memories or humorous things from JPL in those days?

JH: Let's see. I guess I'm not the best person. You probably know that JPL itself was created because once the Lab got kind of blown up a little — had a "little accident". Caltech trustees eventually told them "You know, you guys, you gotta stop, you gotta go off campus if you are going to work on rocket propulsion" — so they sent us out to what at the time was a very barren area — there was no city around then — and what eventually became JPL kind of grew from there, with the war effort, etc.

AH: There are a whole lot of interesting stories connected to that, including the one about Jack Parsons. Have you ever read anything about him? He was an early pioneer and did blow himself up, but on Orange Grove after he had stopped working on the project.

BG: Wow — too bad!

AH: One thing I think you wanted to ask about, Beatrice, was the "Miss Guided Missile Contest"!

***Here the tape breaks off — the last few minutes are missing (See Tausworthe contribution.)

But one item from an earlier partial transcription remains:

AH: We were at some point in contact with Dick Goldstein. Did you ever at some point actually talk to Dick Goldstein, Beatrice?

BG: Yes, I did. The first thing he said was that Sol was the smartest person he ever met. Then he pretty much said he had Alzheimer's and could no longer provide details.

© 2023 World Scientific Publishing Company
https://doi.org/10.1142/9789811234378_0019

Board Game to Board Room

Robert Hanisee

rhanissee38@gmail.com

I first met Sol Golomb in the 1970s. At that time, I was a principal in a small Los Angeles based investment banking firm that was on the prowl for banking opportunities with defense related companies. Though small, we did possess a deep reservoir of knowledge of the aerospace and defense industries. Our search led us to Technology Service Corp (TSC), a small Santa Monica based defense consulting firm that our due diligence discovered was badly undercapitalized and so in need of a capital infusion.

1. Technology Service Corp.

Technology Service Corp., was founded in 1966 by Peter Swerling to provide various types of engineering services and consulting to the US Military and the US Defense Department. A core focus and competency of TSC was advanced Radar Systems. At its founding, Dr. Swerling invited two colleagues from USC to join his board of directors (BOD), Professor Solomon Golomb and Professor Irving Reed. In his autobiography *Alaska to Algorithms*, Reed described Peter Swerling as "the classic boy wonder; Swerling had enrolled at Caltech at age 15 and graduated in three years". Following his death in 2000, Sol Golomb called Swerling "the leading radar theoretician of the second half of the 20th century, certainly in the United States and probably in the entire world". Having access to Irv Reed through his BOD membership added strength to TSC credentials in radar technology. Dr. Reed had been extensively involved in radar development at both Lincoln Labs and the Rand Corporation earlier in his career.

Having concluded that TSC was a strong viable company with a leadership role in radar system engineering and research, our small firm Seidler Amdec Securities began competing actively to help TSC resolve its underfunded status and to provide capital that would permit TSC to resume a growth trajectory. However, there was competition for this banking assignment. That competition came from a large highly regarded national firm by the name of Drexel Bernham Lambert. Drexel Bernham was the leading US advocate of high-yield bonds (aka junk bonds) and had been enormously successful raising capital for both US and foreign companies through the issuance of high-yield bonds. Our small firm was challenged to devise a more viable strategy in order to compete with this investment banking behemoth. After our detailed analysis of TSC's historic capital raising efforts, but more importantly its then current balance sheet, debt and equity relationships, it became clear that TSC's most pressing need was an infusion of new equity capital, not more debt. We were fortunate to be given a hearing before several members of senior management and several members of their Board of Directors. After making our pitch, part of which required our convincing them that Seidler Amdec Securities was capable of successfully completing an equity financing for the company, we then got to the Q & A part of our pitch. The questioning of one Director, Dr. Solomon Golomb (at the time known to us as a Professor of Mathematics and Communications Engineering at USC) indicated that he clearly understood the merits of our pitch that TSC needed on equity infusion, not more debt. I later learned that Professor Golomb, within the inner council at TSC had become an advocate for our approach and took the lead in convincing management and the Board that our proposal was superior to the alternative. Our small firm, Seidler Amdec Securities was selected by TSC, management to lead a public offering of common stock. That financing, completed in 1983, was a success. With its new capital, TSC's growth resumed and within three years the company was acquired by Westinghouse Electric Corp. It was a happy outcome for the original investors, the Board of Directors and management and certainly for those institutional investors that had bought stock in our offering. For me it was the beginning of a 40-year relationship with Sol on a friendship level, but always centered around business investments and relationships. From the TSC episode, I learned that Sol Golomb had a deep and abiding interest in business and especially in helping young companies and young entrepreneurs (often start-ups) succeed. Most often, entrepreneur and business managers were anxious to attract Dr. Golomb into their orbit, as a consultant or a member of their Board of Directors, for his technical expertise. They quickly learned that Sol's interest and contributions extended to most other domains of business challenges, from finance to marketing to production or provision of services to management structure, succession and ultimately liquidity for investors. Another observation worth making here is that Sol had an extensive core of respected colleagues, scientists, engineers and mathematicians from his time and activities at JPL (Jet Propulsion Laboratory) and at USC. Many of these colleagues also had entrepreneur impulses and started their own companies.

They all wanted Sol in their corner, often as a member of their Board of Directors or alternatively as a consultant. Sol's early business involvements and the colleagues involved include:

Technology Service Corp. — Peter Swerling, Irv Reed
Space Computer Corp. — Bill Kendall, Irv Reed, Elwyn Berlekamp
Swerling, Manassey and Smith — Pete Swerling, Elywn Berlekamp
LinCom Corp. — Bill Lindsey
Cyclotomics Corp. — Elwyn Berlekamp

Illgen Simulation Technology — John Illgen
Stoneage Equipment — Charles Cole

Except for the last two, all of these business relationships sprang from Sol's time at JPL and USC.

2. Mark Resources

After the sale of TSC, several engineers left to start a new company, Mark Resources. Founded in the late 1990s by Bill Kendall and an associate with strong support from Irv Reed and Sol Golomb, Mark Resources was another "beltway bandit", a consulting firm set up to provide a variety of technical services to the military and DOD. The core competency of Mark Resources was studies on US Air Force Radar Systems. Sol's involvement was not as a Board member but as a consultant for matters both technical and financial.

3. Space Computer Corp.

About 12 years later, Bill Kendall left Mark Resources following a dispute. With the advent of the Reagan Administration "Strategic Defense Initiative" aka "Star Wars", Irv Reed and Sol saw a market opening for computing in space. They recruited Bill Kendall (known from TSC and Mark Resources) and Bill Jacobi, an engineer with a Princeton PhD, known to Sol, to start a company to capitalize on this perceived new market opportunity. Both Sol and Irv Reed served on the Board of the new company Space Computer Corp. Here is Bill Kendall's description of the beginnings of Space Computer Corp., its successes and its ultimate sale to Exelis.

3.1. *Space Computer Corporation (SCC) and Sol Golomb*

By William Kendall

> **Background.** The associations that led to the formation of SCC and Sol's role in it go back to Peter Swerling's Technology Service Corporation (TSC). In 1968 Sol and Irv Reed were on the Board of Directors of TSC, and Lee Petillon did legal work for TSC. Since Irv Reed and I had worked together at the Rand Corporation in the early 1960's, he recommended to Swerling that he hire me away from JPL to join TSC. At that time I was ready for a change and became the fifth employee of TSC. It was during my time at TSC that I came to know Lee Petillon.
>
> **SCC Incorporation.** In 1986 Bill Jacobi wanted to form a corporation to take advantage of some opportunities he envisioned becoming available under the Strategic Defense Initiative (SDI), or 'Star Wars' program, started by the Reagan administration. In particular, he noted the huge amount of computation that almost all potential components of the SDI

would require, and the limited capabilities of then available space-qualified computers. So Jacobi put together a business plan and went to Lee Petillon to convert his existing Jacobi Engineering Corporation into Space Computer Corporation. On the Board of the renamed corporation were Sol Golomb and Irv Reed.

At that time in mid 1986 after over 12 years as president of Mark Resources Inc. I was looking for a new opportunity. Lee Petillon heard (probably from Sol, who had been a consultant to the Board at Mark Resources and knew my situation) that I was looking for something new, and he recommended to Jacobi and me that we explore forming the new SCC together. Over one lunch meeting Jacobi and I agreed to team up to form and operate SCC The holders of SCC's founders' stock were Jacobi and I, Sol and Irv, and Frank Moothart (a friend of Jacobi's who never participated in SCCs).

Early SCC Success. Of course, developing any computer, and especially a space qualified one, is an extremely expensive undertaking, and we at SCC did not have and could not get the necessary investment. However, there were many opportunities to do paper studies for SDI's computing requirements and potential processing architectures, and we aggressively pursued those. For about six months we used Jacobi's phenomenal writing ability and my technical ability to grind out one proposal after another. As a result, at the end of 1986 we received a $50k subcontract from Rockwell International to support their work on one of the large proposed SDI systems.

In early 1987 in short order we hired Alan Stocker, rented office space, and hired Kelley Aaron as our office manager. We continued our proposal writing, had some marketing successes, expanded our staff, and were off and running as a viable company doing high-end research, and all but forgot about producing any space-based computers.

Sol's Role. From SCC's founding in mid 1986 until its sale to Exelis in mid 2012, Sol was a consultant to SCC management, and an active member of the Board of Directors. However, in spite of his extraordinary intelligence and deep understanding of mathematics, his contributions to SCC were almost exclusively in the areas of finance and management. He met with the Board quarterly, and met with Bill Jacobi and me more or less monthly. That way we benefited from his experience in small-business dealings with the Department of Defense and he maintained a good understanding of our operations and accomplishments.

By the middle of 1987 SCC needed a capital infusion to fund its rapid growth, and so we sold new shares to Golomb, Elwyn Berlekamp, Peter Swerling, John Jamieson, and Douglas Buddenhagen. Sol was instrumental in getting Berlekamp and Swerling to invest, and between the three of them provided over three-fourths of the new capital raised. Jacobi convinced Jamieson (an expert in infrared sensor design) to invest and I convinced Buddenhagen (a friend and serial entrepreneur) to invest. Berlekamp and Swerling did not know Jacobi and me well, and I am sure that they invested because of Sol's recommendation, and the knowledge that Sol would give them an objective view of the company's progress.

The next major requirement for an infusion of cash into SCC came in the Fall of 1991 as a result of the collapse of the Soviet Union and the resulting disarray in the research initiatives of the Department of Defense. New stock was sold to Golomb, Berlekamp, Moothart, and Jamieson. Sol played a major role in making that happen.

By 2005, SCC was flourishing and we began to think about an exit strategy for the existing investors. In support of this possibility, Sol brought Robert

Board Game to Board Room 201

Hanisee onto the Board because Hanisee had experience in finance and had participated in the sale of several small companies in which Sol had had an interest. The influence of Golomp and Hanisee on the SCC Board led eventually in 2009 to the selection of Philpott Ball & Werner to manage the sale of SCC, and of Bill Carr to conduct negotiations with the eventual buyer. The sale was originally to ITT Corporation, but while negotiations were ongoing in 2011, ITT spun off its defense business into a company named Exelis Inc., and the sale was consummated on July 1, 2012 as a sale of SCC to Exelis. That brought the 26-year long association of Sol Golomb with Space Computer Corporation to a successful conclusion.

As noted by Bill Kendall's notes above, when I joined the Board in 2005, it was top heavy with PhDs: Sol Golomb, Irv Reed and Elwyn Berlekamp (who had joined the Board following his investment in SCC) and of course Dr. Kendall. Also, as a gesture to one of their founders, Dorothea Jacobi, wife of Bill Jacobi who had passed away in 2004 and who was a significant shareholder, joined the Board at that time. SCC had emerged as a technical leader of a somewhat arcane technology known as hyperspectral image analysis. This involved automated spectroscopic analysis of remotely-sensed imagery acquired in hundreds of narrow spectral bands for "color" in the optical and/or infrared wavelength regions. By employing sophisticated algorithms to process this data in an on-board fusion computer using software developed in-house by the team led by Alan Stocker, it was possible to detect objects that would ordinarily be hidden to the naked eye, and to identify specific targets or materials based on fundamental properties of their reflection or emission of light. One example application would be an enemy tank hidden under camouflage netting which rendered it invisible to an observer in an aircraft passing overhead, but which could be readily found using SCC's hyperspectral analysis of sensor data in nonvisible wavelengths. When I joined the Board, SCC was under contract by the Civil Air Patrol (a US auxiliary government funded entity) to develop a hyperspectral system to automatically locate aircraft that had crashed in remote areas. SCC had also been working with DOE National Laboratories to develop infrared hyperspectral capabilities for detecting and identifying chemical effluents associated with the manufacture of weapons of mass destruction (WMD).

The company under Bill Kendall's and Alan Stocker's leadership was thriving; it was consistently profitable with positive cash flow. Nonetheless, after joining the Board it became clear that the technology and market was limited and the company small such that a path to a sale of the company and liquidity for the owners and investors was at best problematic. Then a change occurred. At the height of the war in Afghanistan, at a time when improvised explosive devices (IED's) were causing hundreds of casualties, the company was asked by the US Military to come to the theatre to test and demonstrate whether SCC's hyperspectral analysis could identify specific materials being used to fabricate homemade explosives. At this point in time, counter IED products and techniques were among the highest priority of US military. The SCC team went to Afghanistan, loaded their sensors and computers onto aircraft and flew multiple passes over remote villages. The tests were successful; orders and backlog expanded rapidly and SCC suddenly became a prime take-over

202   *The Wisdom of Solomon*

candidate for larger companies focusing on counter IED technology. Thus, as noted
in Bill's account, a sale to Excelis was accomplished, with Prescott, Ball and Warner
brought in to structure the deal and to represent SCC's interests. Another deal guy,
Bill Carr who was known to Sol back in the TSC days, proved effective in negotiating
the final deal terms. It was a happy ending for the long suffering outside investors,
for the Board members and for management (the Board and senior management
had been generous in granting stock awards deep into the employee base so they
were happy also).

4.   Swerling, Manasse and Smith

As the Space Computer saga was unfolding, Sol Golomb was busily engaged in helping other friends and colleagues in their own business endeavors. In no particular
order, we will begin with his good friend Peter Swerling who after being ousted from
his leadership position at TSC founded a new venture, Swerling, Manasse and Smith
in 1983. As noted earlier, Pete Swerling was a boy wonder who matriculated at the
California Institute of Technology at age 15, where he completing his Bachelor of
Science degree in three years. He then earned a second Bachelor degree in Economics
from Cornell University in 1949. Pete then went on to earn his PhD in Mathematics
from UCLA in 1955. At the time he founded SMS, Dr. Swerling was regarded as one
of the top radar theoreticians in the world. Given his national reputation, SMS had
little trouble winning studies contracts from the US Defense Department. During
its 12 years of life, SMS made notable contributions to national defense including
vulnerability studies of such major defense systems as AWACS, the Patriot Missile
and electronic counter measures. Most notably, he developed advanced techniques
(doppler radar) for tracking Stealth targets. He was appointed adjunct Professor of
Electrical Engineering at USC where he taught advanced seminars in communications theory. In 1997, Sol asked me to take a look at SMS to see if there were any
prospects for a sale. At that point, both Manasse and Smith had drifted off (back
to Washington and the Defense Department) so that SMS was a one man show and
far too small for viability. After completing existing contracts, Pete Swerling put
SMS to sleep (this was a new experience for me) and retired in 1998. Through the
whole 12 years of life Sol was a consultant to SMS and a member of the Board of
Directors, but more importantly, a close friend and confidant to Pete Swerling. Pete
died of cancer in August 2002.

5.   LinCom Corp.

Next up was LinCom Corp. a company founded by Dr. William Lindsey. Bill, despite
his well crafted "Good Old Boy" from Arkansas persona was in fact a highly skilled
engineer and scientist. He earned both a Masters Degree and his PhD in Electrical
Engineering from Purdue University. In the summer of 1962, Bill joined the Eb
Rechtin group at JPL that was attacking the problem of deep space communication
(where he and Sol worked together). He made numerous contributions in the field of

Board Game to Board Room

communication science and was a member of the JPL team that developed modern digital communications. Bill is a life member of the IEEE (Institute of Electrical and Electronics Engineers) and was named to the NAE (National Academy of Engineering) in 1997. In 1968, Dr. Lindsey joined the Electrical Engineering faculty at USC as a full Professor, recruited to USC by Sol. In 1972, he published a highly regarded and widely used textbook on digital communications.

Like the other polymaths in this account, Bill had a desire to convert academic knowledge into practical application that would be of service to mankind. So, in 1972, he started LinCom Corp. a communications technology company, to achieve those goals. LinCom did DOD studies, and provided engineering support to its customers (other companies focusing on communications as well as the US Defense Department). One signal achievement was the development of the algorithm that made the walk around telephone (untethered from the base station) both a success and ubiquitous in the world today. Sol Golomb was a paid consultant to LinCom as well as a member of its Board of Directors.

In my ripening relationship with Sol, I think that I had become Sol's go to guy for both financing of undercapitalized start-up companies as well as the agent provocateur for arranging the sale of those companies and thereby liquidity for the entrepreneurs and investors. So when, in 2000, Sol asked me to meet Bill Lindsey and take a look at LinCom, it was only another episode in a repeating pattern. In our meetings at the company's offices, I quickly ascertained that: (1) LinCom was a technology rich company; (2) it was a well run and profitable company; and (3) Dr. Lindsey felt that it had gone about as for as he could or was willing to take it given his multiple other engagements and activities. I was then a member of the Board of Directors of Titan Corp., a much larger New York Stock Exchange listed company headed by Dr. Gene Ray, located in La Jolla, California. Titan was a mature engineering, consulting and services company (mostly defense) that had embraced a growth strategy based on the acquisitions of smaller, technology rich companies with high growth potential that were looking for liquidity. LinCom was a natural fit with Titan; in what was probably the easiest transaction of my career, one phone call with Dr. Ray was all it took. LinCom was acquired by Titan in 2000. It was another happy ending.

6. Cyclotomics

While Sol was at JPL in the late 1950s and early 1960s, he met and worked closely with another mathematician and engineer, Dr. Elwyn Berlekamp. The mind weld between Sol and Elwyn had to be almost instantaneous. Both had deep and abiding interests in applied mathematics, coding theory and error correction coding and mathematical games. Dr. Berlekamp completed all of his degrees at MIT earning his PhD in Electrical Engineering in 1964. In that same year, he joined the electrical engineering faculty at U.C. Berkeley. Several years later, he left Berkeley to work at Bell Labs, but returned to U.C. Berkeley in 1971 as a full Professor in both the mathematics and computer science departments. For much of his professional life, Elwyn portioned a slice of his brain power on combinatorial games, publishing on the game Dots and Boxes and the Asian game of GO. Also, like Sol, he was elected a member of the NAE (National Academy of Engineers) the NAS (National Academy of Science) and the American Academy of Arts and Sciences. His additional awards and honors are far too numerous to list here, but worth mentioning is his contribution to error correction decoding during his time at JPL. His contributions made possible the transmission of accurate and detailed images from spacecraft such as Voyager and the Hubble Space Telescope. Indeed, in his autobiography, Irv Reed credited a decoding algorithm developed by Elwyn Berlekamp for "paving the way for greater usage of the Reed-Solomon codes in communications."

Any listing of Elwyn Berlekamp's interests and accomplishments would be remiss if it failed to note his short but highly successful foray into the investment world. First, Elwyn was a co-founder of Cylink, a successful start-up developing commercial cryptography. He served on the Board with a number of notable people such as former Secretary of Defense Bill Perry, and mathematician Jim Simons, founder of Medallion Quant Fund, one of Wall Street's most successful hedge funds. Elwyn later became involved in improving the trading algorithms for a smaller hedge fund which he ended up acquiring. This in turn led Jim Simons to ask Elwyn to run part of the Medallion Funds from Berkeley. In 1990, Elwyn led the Fund to a return of 55.99% net of fees, a phenomenal result. After several more years running hedge funds, Elwyn returned to teaching at U.C. Berkeley.

In 1973, in a Business and Finance biography penned by Dr. Berkekamp, he noted that in 1973, he, his wife Jennifer along with Solomon Golomb co-founded Cyclotomics. Cyclotomics was a research and engineering firm specializing in developing error correction coding and decoding algorithms for digital communication and mass data storing applications. My professional involvement with Cyclotomics was minimal although Sol invited me to visit the company in Berkeley in early 1985 in order to consider its viability as an acquisition candidate. After meeting Dr. Berlekamp and learning as much as possible in a short time, I concluded that the technology was overly arcane and lacking in scaleability such that any potential buyer would have to be a company that needed the technology but which had limited interest in Cyclotomics as a revenue and earnings generating business. Some months after my original visit, such a buyer showed up. Under contract from Eastman Kodak, Cyclotomics had developed a sound encoding/decoding system for EK's Digital Sound System that later (1995) had won an award for scientific and technical achievement. So, in 1985, Eastman Kodak showed up with an open checkbook. They offered a price and valuation of one-time revenue which equated to a shockingly high multiple of earnings. Both Sol's and my advice to Elwyn was to take the money and quickly. Clearly, Eastman Kodak wanted control of the technology and the price offered was only a rounding error for a company of EK's size. For all involved, it was another happy ending.

7. Illgen Simulation Technology, Inc.

In 1988, I had been involved in attempting to raise money for a new defense focused company in Santa Barbara that saw an opportunity to provide Modeling and Simulation services to the government as a new tool to prevent the outsized cost overruns on large systems being developed by the largest defense contractors. At that time, I partnered with Dr. William Perry who, being between US Defense Dept. postings, was employed by a San Francisco based investment bank and venture capital firm. (Bill Perry later was appointed Secretay of Defense in the President Bill Clinton administration.) The founding entrepreneur at the time was John Illgen who had established himself as one of the leading experts and proponents of Modeling and Simulation to address some of the wasteful cost overruns that plagued the defense industry. Another founder/director, Paul Kaminski, noting that the Illgen name was well known and highly regarded within the simulation world, suggested the company be named Illgen Simulation Technology and so it was.

Here, I will let John Illgen describe the capabilities of ISTI:

> ISTI was founded March 1988 and headquartered in Santa Barbara, CA. ISTI's technology focus was Modeling and Simulation (M&S). ISTI was one of the first small businesses to focus on M&S. ITSI's M&S capabilities are listed below:
>
> (1) Developed Models and Simulations for Defense FAA and NASA applications.
> (2) Applied M&S to system, subsystem and platform design which allowed the clients to determine whether their products would meet performance requirements before spending millions or billions of dollars constructing systems (communications, radar, intelligence sensors, antennas, aircraft, vessels, ground vehicles and associated systems and more).
> (3) Mathematical Modeling of the telecommunications including wireless, networks and data links (narrow and wide band), navigation/ position location (GPS, LORAN-C, inertial, Distance Measuring Equipment) and sensors (imaging, infra-red, laser and radio frequency) and many others.
> (4) Software Architecture development to automatically link and integrate multiple, dissimilar models, simulations, and databases in an interactive process over a geographically distributed system (real-time, fast-time, and non-real-time). This architecture was also scalable in terms of entity throughput and entity interactions in real-time where an entity is a tank, aircraft, sensor, data link, human and so much more. Number of entities in a simulation can be 100s of thousands.
> (5) Verification and Validation of models and simulations and systems using unique static and dynamic software diagnostic tools causing efficiency, quality, and cost savings.
>
> ISTI provided M&S support to US military forces in South Korea under contract with the Army Military Command. This included using M&S to evaluate Command and Control Systems in South Korea and evaluating area along the border that would be important to Mission Level Simulations. This work required several trips to South Korea requiring interaction with US Forces and the Korea Institute for Defense Analysis (KIDA). This effort resulted in improving database inputs for low and high-fidelity models and simulations. Command and Control Systems were greatly improved as a result of this work.

206 *The Wisdom of Solomon*

Despite the rich array of capabilities and a wide customer base, the company ran into a rough patch in the mid-1990s. At that point in time the Board of Directors consisted of Larry Roberts, who was representing the San Francisco financial interest, Jarvis Gant, a business man from Texas with a strong defense background, John Illgen and me (both Larry Roberts and Jarvis were steady and good contributors). I suggested to John that we add Sol Golomb to our Board. John had never met Sol but was familiar with his work in coding and his seminal work, *Shift Register Sequences.* The Board quickly voted to invite Sol to join the Board. For the next seven or so years, Sol was a consistent and broad band contributor to board deliberations. Here, I will insert John Illgen's comments and evaluations of Sol's contributions to him (as CEO) and to the company:

7.1. *Sol's impact*

(1) I was using Viterbi's error correction decoding scheme for a defense project. I needed to verify my results. I called Sol to ask for his verification assistance. Sol got Dr. Andy Viterbi (the developer of the decoding scheme) and me together, where I gave Andy my decoding application. Andy double checked my work with his coding. The customer and the Defense Department were delighted with the resulting work product; it is a tribute to Sol. There were many contributions from Sol like this one.

(2) Sol checked and several times corrected me or gave me stronger mathematical formulas for use in my M&S applications. This occurred a lot as well.

(3) Sol gave me guidance for a contract with SAAB in Stockholm. It focused on improving ISTI's architecture for integrating and interoperating disparate models that have not worked together before. His suggestion resulted in a broader use of the architecture for SAAB. His suggestion resulted in a huge success with SAAB. Even today I still interact with SAAB because of Sol's wisdom.

(4) Sol would give me hours of his time to review Key Note talks I would give at different M&S Conferences. His suggestions were always used and worked in an outstanding fashion.

(5) Sol invited me for lunch at USC where we would discuss ISTI Technical and Business areas on a routine basis. I have lost track on how many times but it was a lot. Sol would have me in his office for 1 to 2 hour discussions then another hour for lunch.

(6) Sol's contributions on the ISTI Board were amazing for both business and technology. Sol had us hire Bill Carr to assist from a business point in selling ISTI. This was instrumental in selling ISTI. Sol had worked with Bill before. Bill and Sol guided me carefully during this process and Bill conducted the financial negotiations. Sol also provided me careful advisement on presentations I gave to possible buyers and most importantly when it came down to the wire with Northrop Grumman. Sol was always on target. This acquisition occurred because of Sol's wisdom.

Board Game to Board Room

By the early 2000s, the company was thriving and growing. Though small, ISTI had emerged as the leading company in the M&S space and as such, was being noticed by the large Defense Aerospace companies. Suitors began knocking on our door: first Lockheed who made an inadequate offer. Next, along came Northrop Grumman Corp. who made a very attractive offer. Sol brought in his old pal Bill Carr who represented the company in negotiations with Northrop Grumman, a deal was completed. After a very long slog, the result was a very happy ending for the original investors, for management and for John Illgen. John continued with Northrop Grumman for the next 15 plus years and thrived as a technology leader in the M&S space and a leader (director for M&S throughout Northrop Grumman) for an important business unit of Northrop Grumman. Both Sol and I were very happy for John's success.

8. Stone Age Equipment

The last chapter in the long and successful business career of Sol Golomb was his engagement with Stone Age Equipment Co., aka Five-Ten. Up to this point, every one of Sol's business activities revolved around his many associates in the academic, scientific, and an engineering communities and were, in every case technical in their core competencies. Five-Ten then is a headscrather; the company's products were all in the outdoor recreational arena. The most important and sustaining product was a line of rock climbing shoes (favored by the community of high risk athletes who for sport, enjoyed scaling near vertical rock walls) where the main requirement was not slipping. As I had heard the story, Sol's daughter, Beatrice, during her undergraduate years at USC, had become friends with fellow student Charles Cole. Together they decided to try their hand at rock climbing, a popular outdoor sport in Southern California at that time. In time, Beatrice moved on to graduate school (the rest of which story we all know); Charles became hooked on rock climbing and migrated to Yosemite Valley where he spent time climbing sheer rock faces. Over time, he became sufficiently skilled that he pioneered several routes up the face of El Capitan (an iconic rock face in Yosemite). It is worth noting here that sandwiched around his climbing vocation, Charles completed his Bachelors degree in Engineering at USC and then a MBA at Notre Dame University. Later, when faced with the necessity of earning a living, Charles began making and selling various articles of equipment used in rock climbing. As he told the story, climbing shoes at that time were primitive; little more than upgraded tennis shoes. The rubber soles of the available shoes did not hold well enough on the rock faces then being scaled. Using his engineering training and high intelligence, Charles taught himself rubber compounding, and after multiple tries came up with a compound that had a higher coefficient of friction then anything on the market at that time. Named "Stealth Rubber" (which was copywrited) the Five-Ten shoes with sticky soles moved up the ranks and in time emerged a leader in the small but prestigious rock climbing shoe industry. Among his other talents, Charles Cole was an absolute genius at marketing, consistently coming up with product names such as "Stealth Rubber" and Five-Ten (the brand name that describes the degree of difficulty of an advanced climb) and slogans such as "Brand of the Brave" that captured the imagination of rock climbing aficionados. The family of rock climbing shoes was a smashing success; the business not so much.

I had met Charles Cole several times at the Golomb home during their annual Christmas party, and was vaguely aware that Charles had a company that manufactured rock climbing shoes and other outdoor paraphernalia but was otherwise unfamiliar with his business. One day in 2009, Sol called and asked me to take a look at Charles' company Five-Ten. What was quickly apparent was that Five-Ten was critically undercapitalized; indeed, the company had been bootstrapped from its inception and had never had a real infusion of operating capital. My concern about getting involved was that Charles, despite all of his good qualities and brilliant product strategies, tended to be erratic in his business decision-making (damn the torpedoes and all that). (Five-Ten had come close to collapse several times in its history.) Nonetheless, and with some prodding from Sol, we decided to step-in and help. My own calculus was that "Stealth Rubber" and the company Five-Ten had such a strong position and mind share in the climbing shoe market that it would eventually become a desirable acquisition candidate by a larger shoe manufacturer.

In 2009, Sol and I put together a syndicate of friends, associates and family and raised $1 million in new capital for the company in exchange for a 20% interest in the company, a pre-money valuation of $5 million. Both Sol and I joined Charles on the Five-Ten board from which position we could help Charles while looking after our investors' money. Two years later, at a trade show in Germany, Charles was approached by executives of Adidas, one of the largest outdoor shoe companies, expressing an interest in acquiring Fine-Ten. After agreeing in principle to a deal, Sol immediately brought in his favorite deal maker and negotiator, Bill Carr. Bill did his usual magic in negotiating very favorable terms. At that time, Five-Ten's annual revenue was approximately $10 million. The agreed upon price was $20 million or two times sales. Within the shoe industry there had been numerous takeovers of small, nichey startup companies, none of which sold for even one times sales. So, the final price of $20 million plus an earn out (not achieved) was a real success; another and last happy ending.

At this point my chronicling of Sol Golomb's business dealings and successes would be complete. Earlier in 2020, I asked Bill Carr to send me an account of his recollections of business dealings with Sol. What Bill crafted was in some instances a revelation to me (early TSC history and why Peter Swerling was tossed out) but was so informative and amusing (in a Bill Carr sort of way), and I will end this chapter with Bill's account in his own words:

> I met Sol in 1978 when TSC was in financial difficulty due to acquisition of Karex and Ktronix businesses from Rohm & Haas. Karex and Ktronix were magnetic media businesses in Sunnyvale, California making mag-cards and floppy discs.

A lawsuit was being prosecuted with TSC claiming breach of warranty and R&H making claims for payment of deferred purchase price. The allegations were that R&H had failed to disclose the depth of the "black holes" that were Karex and Ktronix. So deep were holes that TSC was on the verge of sinking while its "base business" was successful. The potential acquisition of TSC by ATI provided the leverage and fresh faces to accomplish a "walk away" settlement with R&H. That settlement was the foundation of Sol and my friendship and our attorney/client relationship that lasted 40 years.

Allied Technology Inc. was a contract manufacturer of circuit boards and small electronic components. Headquartered in Dayton, Ohio, it was a "step-child" or "tar baby" of H. Talbot Mead, son of George H. Mead – founder of The Mead Corporation (Mead/Westvaco now West Rock).

Allied Technology was a company without a business that had "struck it rich" by being the contract manufacturer of the Fuzz Buster, the first police radar detector. The Fuzz Buster was invented and marketed by Dale Smith in response to a Georgia speed trap. ATI did not own the Fuzz Buster, but prospered with the product's success.

Fresh with a pile of cash and new found intelligence, ATI went looking for a business or product to call its own. Somehow, the paths of ATI and TSC converged. ATI viewed TSC as a real radar based business with two "product-based" businesses in Karex and Ktronix. Management of ATI believed it could tolerate the scientists at TSC and grow the acquired products. Confident it could understand radar and anxious to have products of its own, ATI found its merger partner. The combined companies had a brief success. Peter Swerling was "edged aside" by ATI, but Glen Gray was allowed to continue to ably direct the Santa Monica based government contracting business. 'Bruce Emmeluth was found' to manage and achieve temporary success at Sunnyvale/San Jose.

I met Sol in my capacity as outside legal counsel for ATI. Sol's role with respect to the ATI/TSC transaction can best be described as "subtle". It was never clear what Sol's exact role was or would be, but "things" were always better when he was present. The same is true with respect to the next 40 years of our friendship. Our relationship was centered on a series of business transactions. He normally said little. He generally made a brief statement at the beginning of a negotiation and at the time of critical decisions. While he was present "things" always seemed to be "on track". He never "over played" a hand.

The life cycles of the products made by Karex and Ktronics and that of the Fuzz Buster were nearly coterminous. ATI like TSC before it, went broke. Sol, Glen Gray and I were left standing with a successful TSC and a bankrupt parent-ATI. We decided to pursue a bankruptcy reorganization to preserve TSC.

The secondary objective of a Chapter 11 proceeding was to protect the "Insiders" from claims of general creditor of ATI. Its shareholders and the Bank (Winters National Bank which was acquired by Bank One which was acquired by Bank of America). The strategy was to negotiate a slow note with the Bank, divide the stock of the surviving company among TSC management, ATI creditors and former shareholders of ATI/TSC. We did not allow the directors of the former ATI to participate in the new company.

TSC Corporation emerged from the Chapter 11. TSC's management continued in place with a new Board including Glen Gray, Irv Reed, Sol and me. The base business was intact and growing. The "then novel" reverse subsidiary merger saved a large tax loss carry-forward. All was well. Sol served as a director of the new TSC and as a consultant to Glen Gray. The extent of their dealings are unknown to me. I was with a law firm in Ohio. The client and Sol were in Santa Monica.

A few years later Siedler Amdec Securities emerged from Sol's doings. A possible public offering transaction appeared. Bruce Emmeluth, a banker at SAS, was assigned the deal. Sol also mentioned his friend Bob Hanisee as being President of the securities firm.

Sol and I "tasked" ourselves with causing the Registration Statement to be written and the public offering to proceed without a mention of the ATI bankruptcy. Neither the Securities Exchange Commission nor the Underwriter caused us to air the past Rohm & Haas/Allied Technology saga. The public offering proceeded with TSC as the registrant to a successful closing. It was time for dinner at the Capital Grill. The first of a handful.

New TSC Corporation (now well-funded) prospered until sold to Westinghouse in 1985 for $23,000,000. Again, the cause for a dinner with an attendance too large for the usual table at the Capital Grill. I remember being accompanied to this dinner with my future wife Maryanne and her daughter Britany. Although they only met Sol on this one occasion, Maryanne and Britany "knew Sol" until his death and were saddened by his passing. Sol provided instant and lasting credibility and stature to me and his other friends. Maryanne thought that even I must have a "redeeming social value" because of my friendship with Sol.

While TSC was our first deal together, Stone Age was our last. We started with radar and ended with rocks. In between were others (four of which I remember enough to comment upon). The first deal began with litigation and the last deal ended with a two week trial. Both disputes concerned purchase price.

Part of the strategy for the Stone Age versus Adidas trial was to color the too colorful genius Charles Cole as credible rather than eccentric. It was decided to sandwich Charles Cole's testimony between that of Sol and Bob Hanisee. Sol's testimony was intended to show that Charles was a bona fide genius. David Greer was our trial counsel. Part of David's fame was that he and Alan Dershowitz shared the head of their Yale Law School class. David questioned Sol about a Wikipedia reference to Sol's I.Q. as being 175. Sol's response was priceless. "Yes it has been measured that low. It has been measured between 175 and 200". Sol was not boasting. It was a "matter of fact". Dave continued with respect to Sol's interest in chess. Sol admitted studying the game and being quite an expert, but that Charles usually prevailed 2 out of 3 matches. Instant credibility was obtained. Charles was in fact a genius and Stone Age (aka 5-10) was a technology company with the World's most unique high friction rubber.

Sol's friend Bob Hanisee was our other book end. We used Bob's testimony to normalize Charles as a businessman.

Unlike the TSC transaction, Stone Age was not a total success. Charles failed to follow Sol's advice which resulted in business and contractual flaws. Charles trusted his "new hire" to manage his business and the deal rather than follow Sol's direction. A potential earn-out of up to $65 million in additional price was lost. Charles and his new hire agreed to an ill-drafted earn-out provision which was frustrated by Adidas freezing Five-Ten out of the market for running shoes. No assurance was provided that the China

Board Game to Board Room

market would be pursued with Adidas' marketing power. The right to a trial by jury to settle disputes was also sacrificed.

With the above deficiencies, the Adidas/Stone Age deal closed in 2012. The initial payment was substantial and the investors accumulated by Hanisee received a pay-day. It was time for dinner. The Capital Grill was re-visited; however, without Charles and his then protégé. One irony was that the protégé turned on Charles prior to trial and was a witness for the "other guy" Adidas.

At dinner during a pause in the conversation, Sol whispered that he was to receive a high honor, but he was pledged by the White House to only tell family. My high honor was being told by Sol that he regarded me as family. Sol's high honor was President Obama presenting to him at the White House the National Medal of Science.

My conversations with Sol almost always were business related but politics and life crept into most discussions. Sol was unflappable, but I saw him upset at our last Capital Grill dinner. Sol was bothered by (and did not see a clear path to deal with) the plagiarism of one of the articles he had written 20 years before. A "scholar" had just plagiarized a paper Sol had written on encryption. I failed to respond intelligently — amazed that then "state of the art work" on encryption was Sol's from 20 years previous. Unfortunately, much of Sol's early work remains classified and the author will always be anonymous.

When asked Sol would also share opinions on Global Warming and its "money trail" or the birth rate (or lack thereof) within the American Jewish community. His opinions always added a "different but thoughtful" insight to the conversation.

The common denominator in the "string of deals" was Sol being involved with the founding of a company using "wireless" technology. It was never clear whether the first roots evolved from Sol's work on wireless communications at the Jet Propulsion Lab or was a spin off from TSC.

A very early deal was the sale of William C. Lindsey's company to Titan Industries. Lindsey had a business but was selling wireless capability. Bob Hanisee found a home for what was a financially challenged, but technology rich, enterprise.

Next Sol arranged for me to travel to Berkeley to meet and visit with Elwyn Berlekamp. The exact year is lost to me but the rivalry between the USA and the USSR was in "full swing". This being my first and only time on campus at Cal Berkeley, I was particularly surprised to see students hauling their bicycles with them into the free standing restroom. The next strange encounter was Elwyn himself. He waved me into his office while he was on the telephone. When his conversation ended he announced that the score was USSR 9 U.C. Berkeley 8 (exact #s unclear). Elwyn then explained that the USSR had 9 Nobel Prizes and U of C had 8. He would have one except one was not awarded in his field. What had Sol "got me into"?

After the initial shock, Elwyn and I became friends. He even visited my home in Springboro, Ohio. However, I believe the main destination of his trip was Wright Patterson Air Force Base.

Elwyn's company was Cyclotomics. Wikipedia defines a "cyclotomic field" in number theory as a number field obtained by adjoining a complex primitive root of unity to Q, the field of rational numbers. In other words, I have no idea what Cyclotomic means. Elwyn gave me a copy of the book he authored on game theory. I was lost in the first paragraph of the introduction, Elwyn and Sol were from another world.

According to Elwyn, Cyclotomics could send 32 pieces of data to a satellite with 16 errors and receive back a correct or corrected message.

It was never clear what use Eastman Kodak had for the technology or if they even understood it.

Sol was more involved and visible in the sale of Cyclotomics to Eastman Kodak than in the other transactions. He and Elwyn structured the earn-out provision which was an "up-side kicker" to the cash price. Sol and Elwyn laughed at Kodak's willingness to pay a premium price in part because Elwyn had employed a general manager type person who had no understanding whatsoever of the technology or business. A company which never made money was successfully sold and the entire earnout was obtained. Kodak's checks cleared and Elwyn was off to teach at M.I.T. and Cal Berkeley and to write Algorithms with which to trade securities.

After Elwyn Berlekamp and Cyclotomics, Sol introduced me to John Illgen and Illgen Simulation Technology Inc. John operated a profitable modeling and simulation business and served as a $500,000 per year consultant to General Paul Kern (then a 4 star in charge of Army procurement). John and Sol explained that when the military procured a major arms system, they required modeling and simulation as part of the bid package and award process. Illgen Simulation could provide a competitive advantage to major defense contractors concerning significant proposals. Hence, we were able to trigger a bidding war between Lockheed Martin and Northrop Grumman.

Sol's role, among others, was to "keep too many cooks out of the kitchen".

After the usual back and forth, a transaction was closed with Northrop Grumman. I recall Sol counseling John to the effect that he was a relatively young man and that it was OK to consider where his future would be most comfortable. The last dollar need not be outcome determinative. However, Northrop made life easy by being the best fit and with the largest check.

Sol, John and I were willing to host a dog and pony show occasionally. Northrop Grumman was initially greeted at Illgen by a video of General Alexander Haag and John Illgen hosting a TV show. John, Sol and General Haag provided instant credibility for which the buyer would pay.

While alternative purchasers for Stone Age were percolating, Sol called me from Mexico (where I had become a manufacturer) to meet William Kendall and Space Computer Corporation. Sol advised that unlike many of our past adventures, this company made money and was not "on the edge" hanging from a rock.

Space Computer, among other things, could locate land mines in Afghanistan from sensors on an air craft or satellite. It had real technology with real management. The shareholders were ageing but the next generation was in place. A sale was necessary.

The Space Computer transaction with ITT Exelis provided substantial opportunities for deal making "value creation". SCC was a subchapter-S corporation using the cash basis of accounting. Its untaxed and relatively large accounts receivable allowed for "double dipping". We were successful in selling a tax election that was favorable to the buyer, but then create a pro-forma G.A.A.P. liability owed to the shareholders resulting in the Buyer permitting a substantial dividend after the price was set. Sol and I shared a laugh and the shareholders benefited substantially. Sol even caused me to receive a $100,000 bonus. The Buyer and its Big 5 accounting firm were happy. Bill Kendall, Bob Hanisee, Sol and I returned to the Capital Grill. This time, due to Sol's induced generosity, I picked up the check.

From his narrative I think that you will quickly see the reason that Bill Carr was held in such high esteem by Sol and why Sol wanted Bill on his team in every negotiation. One last point, the dinners referenced by Bill in his comments were in keeping with a well-established Wall Street tradition. On successful completion of a financing or transaction, the banker would host a "closing dinner" with most of the principals involved and usually their spouses. They were fun events and Bo Golomb, Sol's wife enjoyed them immensely.

To conclude this narrative on the business engagements of Sol Golomb, one is led to conclude that it is a remarkable record of success. Of the eight companies noted, six ended with successful sales: TSC, Space Computer, LinCom, Cyclotomics, Illgen Simulation, and Five-Ten. One, Mark Resources was indeterminate and one, SMS ended without success but also without liabilities or loss. Of the six successes, four had episodes of extreme difficulty during their life cycles. It was during those troubled times that Sol's understanding and skillset shone most brightly. His calm demeanor and perspicacity were critical in identifying the core problem and then crafting a strategy that led to overcoming those problems. It was my pleasure to have accompanied Sol on this journey.

© 2023 World Scientific Publishing Company
https://doi.org/10.1142/9789811234378_0020

The Wisdom of Solomon

Lawrence Hueston Harper

1. In the Beginning

The summer of 1962 was a watershed in my life. I had just completed my first year in graduate school in the Mathematics Department at the University of Oregon. I needed to work during the summer to make ends meet. In 1960 and 1961, a physics major at Berkeley, I had trudged up the hill to put in my application for summer employment at the Rad Lab. To be in that scientific environment I was willing to do anything; empty wastebaskets, sharpen pencils, count specks on photographic emulsions. But I never heard back from them and ended up in menial jobs at minimum wage. For graduate school I switched to mathematics for the purest of reasons: I loved solving hard problems, studying the beautiful theories of mathematics and sharing it with like-minded people. At the University of Oregon in the Fall of 1961, another math grad student, Bob Jewett, told me about having a summer job at the Jet Propulsion Laboratory in Pasadena and said they were looking for people to solve combinatorial problems. I had never taken a class in combinatorics but relished problem-solving so I applied to JPL for a summer job. I was shocked when my application was accepted and immediately began to feel like a fraud. What will they think when they find out that I know nothing about combinatorics?

After finishing exams in June, I prepared to leave Eugene for Pasadena. In order to not be completely ignorant, I went to the U of O bookstore to look for books on combinatorics. They only had one, *Introduction to Combinatorial Analysis* by John Riordan, which I bought. Riordan's book is limited to enumeration (counting) and difficult to read, but there was one technique that I had seen before in my course on probability theory at Berkeley: The Principle of Inclusion-Exclusion. "Well", I thought, "I am not completely ignorant in combinatorics. At least I know the Principle of Inclusion-Exclusion".

Reporting for work at JPL Section 331 (Communications), my boss, Ed Posner, assigned me to assist engineers who were designing and building the communications systems for space probes. The first two weeks I did pedestrian calculations. Then one morning Posner called me on the phone, "Come to my office", he said, "Golomb has

215

216   *The Wisdom of Solomon*

a combinatorial problem for you". This was the moment I had been dreading, being exposed as a fraud. Golomb, the Section Chief, was the only person in Section 331 I had heard of before starting work there. He had been a regular contributor to Martin Gardner's "Mathematical Games" column in the *Scientific American* of which I was an avid reader.

I went to Posner's office and we waited for Dr. Golomb. Upon entering, Golomb went straight to the board to explain his problem: "You have a deck of 64 cards, 3 suits with denominations 1 to 20 in each suit plus 4 jokers. If 37 cards are dealt at random, what is the probability that all 20 denominations will be represented?". Upon finishing he dropped the chalk in the tray, turned and left the room. Posner looked at me and said, "Well, there you have your combinatorial problem!"

I left Posner's office feeling like a condemned man. Back in my office I took stock of the situation. "I only know one thing in combinatorics, the Principle of Inclusion-Exclusion. If that doesn't work, I'm dead." So, I reached up to the shelf above my desk, pulled down my copy of Riordan's book, opened it up to the section on Inclusion-Exclusion and sought to interpret the formula there in terms of Golomb's problem. Then a miracle occured: It worked!! As I was scribbling down the formula for Golomb's probability, my phone rang again. It was Posner, "Come back to my office, Golomb has an idea for you to solve that problem". In Posner's office, we waited for Golomb. He came in, went directly to the board and outlined a process of successive approximations for his probability. Upon finishing he turned and left the room. I said to Posner, "I don't know if I understand Dr. Golomb's idea well enough to carry it out, but I have this formula" (pulling the paper from my pocket). Posner's eyes widened and he exclaimed, "That's it! Why didn't you say something?" That was the moment I became a combinatorist."[*]

I found out later that Golomb's problem was not just an academic exercise: Sol was working on one of the great scientific challenges of the 20th century: Decoding DNA. A few years earlier at Cambridge University, Crick, Watson, Wilkins & Franklin had shown that genetic material was composed of long strings of deoxyribonucleic acid (DNA) which is made up of 4 bases, adenine, thymine, cytosine and guanine. These can be regarded as 4 letters, A,T,C,G, in a message (the DNA string comprising a chromosome) that somehow specifies the structure and function of its organism. The 4 letters occur in "words" of length 3 so there are $4^3 = 64$ possible words. Each word may specify one of the 20 amino acids, the building blocks of all living organisms. The challenge was to "break the code", i.e. sort out which words stand for each amino acid. Because there are more words, 64, than amino acids, 20, some redundancy could be expected. The people working on this problem were attempting to reverse engineer the work of the Creator (or Evolution, depending on

---

[*]Editors' Note: Inclusion-exclusion (which Sol well knew – and taught as the second topic in a combinatorics course at JPL, taken by editor AH, in 1959 – three years before Harper's arrival) does provide an exact formula, but not one from which the value could be readily computed. What Sol wanted was the actual probability (or estimate thereof), to provide evidence for (or against) a proposed genetic code model – hence Sol's provision of an approach to secure precisely that.

The Wisdom of Solomon 217

your religion), so guessing at the design principles behind the system was an awesome undertaking. Sol was actually dealing with RNA, the relatively short molecule that would transcribe a segment of DNA and transport it to a ribosome where the corresponding sequence of amino acids could be assembled into a protein. In RNA the thymine (T) is slightly altered to uracyl (U). Sol's hypothesis about the code was nondiscriminatory, in that each amino acid (denomination 1 to 20) had 3 words (the suites) representing it and the remaining 4 words stood for nothing (jokers). There are $3^3 = 27$ words with no U in them so $64 - 27 = 37$ do have a U. At the time (1962) that Sol wrote his paper, every word that had been decoded, i.e. the corresponding amino acid identified, had a U in it. So if the assignment of DNA words to amino acids is random, what is the probability that all 20 amino acids are specified by at least one codeword with a U in it?

Aside from its impact on my career, I had not thought much about the calculation for Sol until 1993. At Sol's 60th birthday symposium that year (in Santa Barbara), Ed Posner opined that Sol and I should have gotten part of the Nobel prize for solving the DNA coding problem (awarded to Nirenberg, Khorana and Holley in 1968). I was astounded and seriously questioned Ed's grasp on reality. But he seemed fine otherwise and gave a wonderful talk about his work on the physiology of olfactory glands entitled, "A Code in the Nose".

As recent books on the history of the DNA decoding process make clear, there were two very different approaches to the problem: Mathematical modeling and mucking around in the lab. Sol's work with information and coding theory put him in the right place at the right time to theorize about possibilities for the genetic code. The significance of that theorizing by Sol and others (such as George Gamow with his comma free and diamond codes) has been controversial: The first recounting of the process, *The Eighth Day of Creation* by Horace Freeland Judson (1978), gives the mathematicians no credit at all and does not even mention Sol. However, Judson also pointed out the difficulty of the task he (Judson) was undertaking: To reconstruct the process that produced one of mankind's greatest intellectual achievements, a process of immense complexity involving interactions between hundreds of sophisticated scientists with diverse personalities scattered across America & Europe over a period of 22 years (1944–1966). It makes Humpty Dumpty look easy by comparison.

A later account, *Who Wrote the Book of Life* by Lily E. Kay (2000), does mention Sol's work on the comma free coding hypothesis, but still affords the mathematicians and coding theorists little credit. The most recent history, *Life's Greatest Secret: The Race to Crack the Genetic Code* by Mathew Cobb (2015), was much more positive about the contribution of mathematicians and specifically mentions Sol's paper, "Plausibility of the Ribonucleic Acid Code". on page 202, "The real answer to the conundrum of the predominance of U nucleotides in the cell-free data was inadvertently provided by Solomon Golomb. He performed various calculations and concluded it was not possible to deduce anything about the role of non-U sequences without doing an experiment. Which is what the biochemists did...". I was again amazed and gratified to read that. Maybe Ed Posner was not delusional after all.

2. Kautz's Conjecture

A month into the summer of 1962, Ed Posner had to leave for a week to attend the International Congress of Mathematicians (ICM) in Stockholm. Before he left he mentioned several combinatorial problems that were relevant to the mission of JPL-331, to design and build the communications systems for space probes. One particularly caught my attention because it involved one of my favorite mathematical structures, the n-dimensional cube: The vidicon tubes on space probes (such as the one that took the first closeup pictures of the surface of the moon two years later) register 64 shades of gray, from white (0) to black (63). These outputs are encoded as 6-tuples of 0s & 1s for transmission back to earth. The obvious way to encode the numbers 0 to 63 as 6-tuples is to use their base 2 representation, 000000 for 0, 000001 for 1, 000010 for 2, 000011 for 3, ..., and 111111 for 63. In general, $a(5)a(4)\cdots a(0)$ codes for $a(0)2^\wedge 0 + a(1)2^\wedge 1 + \cdots + a(5)2^\wedge 5$. Since the transmitter is very small and low powered (every ounce sent to the moon or Mars is costly) sometimes a 0 will be changed to a 1 or a 1 to a 0 in transmission, causing an error at that pixel in the picture. If 3 is transmitted by the base 2 code, so 000011 is sent but 001011 is received (a mistake in a (3), changing 0 to 1) it would be decoded as $8 + 2 + 1 = 11$, an error of 8. Is it possible that another code could be better than the natural (base 2) code? Better in the sense that its error, averaged over all possible mistakes in transmission, is less. William Kautz, an engineer at Stanford, had generalized the problem to arbitrary n and conjectured that the natural binary code always minimizes the average error, so no other code could do better.

In the week that Posner was in Stockholm, I was on my own at JPL and started to think about Kautz's conjecture. After several days and a false start, another miracle occurred and I could see a way to prove Kautz's conjecture. Showing my ideas to Posner upon his return was tremendously gratifying. It took most of the next year to work out all the details and even then I missed a case. That paper was the beginning of my research career. Fifty-seven years later I am still working on followups to it.

The following summer ('63) I was able to parlay my success in solving combinatorial problems at JPL into a job at Bell Labs in Murray Hill, New Jersey. That was another 3 months of exciting experiences (Starting off with my first transcontinental flight, LAX to JFK, taking a helicopter shuttle to New Jersey that landed on the southern tip of Manhattan where I had my first view of the Statue of Liberty and the New York skyline). I did write another paper that summer. But it was not the same quality as the paper I had written the previous summer. Nor was my experience of the same quality. The summer after that ('64) I elected to return to JPL-331.

3. The Summer of 1964

The Summer of '62 had been wonderful, but the Summer of '64 was beyond my wildest dreams. Another of the combinatorial problems that Ed Posner had mentioned before he left for Stockholm in July of '62 had been buzzing around in my head. It was closely related to the one I had solved that first summer and it seemed

The Wisdom of Solomon 219

to me that a variant of the solution to the first problem might solve the second. It was also a coding problem on the n-dimensional cube. The new challenge was to minimize the maximum difference (as opposed to the sum of the differences) between the numbers assigned to neighboring n-tuples (those that differ in exactly one coordinate). Al Hales and Bob Jewett had worked on the problem three summers before and Al had conjectured a solution, which I called the Hales code. My insight into the original problem had been that the sum of (absolute) differences between the numbers applied to adjacent vertices could be recast as the sum of values for a natural functional on the initial segments of the code (the vertices numbered 0, 1, 2, ..., k for any $k < 2^{\wedge}n$). That functional I now understood, was the analog of the boundary of a set (I called it the edge-boundary) and minimizing it (given the size of the set, $k + 1$) was an analog of the classical isoperimetric problem in the plane. For minimizing the maximum difference there is a similar functional on k-sets of vertices, the vertex-boundary, whose maximum value over all k bounds the maximum difference below. Utilizing the same logic as before, I solved the vertex-isoperimetric problem on the n-cube, showed that Hales code achieved the resulting lower bound and therefore had to be optimal.

I rode my bicycle to work at JPL every day that summer. The last mile before the main gate was very tranquil, through a grove of live oak trees and by a horse stable. On the morning of July 30, 1964, however, that last mile was lined with large vans from all the TV networks. After six failures, JPL's Ranger 7 had succeeded in taking the first closeup pictures of the moon. For a recounting of that historical moonshot, link to https://en.wikipedia.org/wiki/Ranger_7 and see those closeup pictures of the moon (transmitted by "my" binary code).

4. Sol's Memo on the Partitions of n

As I mentioned before, Sol had been a contributor to Martin Gardiner's column, "Mathematical Games," in the *Scientific American*. He would also pass around "columns" on combinatorial puzzles to Section 331. One of these caught my attention that summer: "Over all partitions of n, $n = p(1) + p(2) + \cdots + p(k)$, what is the maximum value of the product, $p(1)p(2)\cdots p(k)$?" Sol gave a hint: The optimal partition should contain as many 3s as possible. It may also contain one or two 2s or one 4 but no more. If it contains more, or any larger number, such as 5, then it may be further partitioned ($5 = 2 + 3$) so that its contribution to the product $2(*3 = 6)$ is greater.

That memo of Sol's stimulated me to ask, "Why 3?". Was it a sign from the Holy Trinity or was there some more mundane explanation? Having long been fascinated by the interaction between discrete and continuous structures, I examined the continuous analog of Sol's problem on integers: Over all representations of a positive (real) number, x, as a sum of positive numbers, $x = y(1) + y(2) + \cdots + y(k)$, what is the maximum value of the product, $P = y(1)y(2)\cdots y(k)$? If $y(i) \text{not} = y(j)$, then $y(i)$ and $y(j)$ may each be replaced with

$$(y(i) + y(j))/2$$

since y(i)+y(j) = (y(i)+y(j))/2+(y(i)+y(j))/2 and y(i)*y(j) < ((y(i)+y(j))/2)^2 by simple algebra. Thus we may assume P = (x/k)^k and need only determine which k is optimal. Letting k be continuous and applying differential calculus, k = x/e is optimal and P = e^(x/e) where e = 2.71828... is Euler's constant, the base of the natural logarithm function. So the reason that 3 predominates in the product derived from partitioning an integer is that 3 is the integer nearest to e!

The use of this hidden connection between discrete and continuous structures to explain a mysterious property of the integers further intrigued me. I decided to look at closely related combinatorial structures to see if similar questions about them could be illuminated by continuous methods. My first question was about those partitions of n: Euler had defined $p(n, k)$ to be the number of different partitions of n into k positive summands (without regard to order, so we may assume $p(1) =>$ $p(2) => \ldots => p(k)$.

For example if n = 5, then

$$5 = 5$$
$$= 4 + 1$$
$$= 3 + 2$$
$$= 3 + 1 + 1$$
$$= 2 + 2 + 1$$
$$= 2 + 1 + 1 + 1$$
$$= 1 + 1 + 1 + 1 + 1.$$

$$\text{So } p(5, 1) = 1,$$
$$p(5, 2) = 2$$
$$p(5, 3) = 2$$
$$p(5, 4) = 1$$
$$p(5, 5) = 1.$$

By calculating $p(n, k)$ for larger n, $1 <= k <= n$, one observes that $p(n, 1) = 1 = p(n, n)$ and for $1 < k < n$, $p(n, k)$ increases to a peak value and then decreases. But unlike the binomial coefficients it is not symmetric ($p(n, k)$not $= p(n, n-k+1)$) so its maximum value does not appear at $k = n/2$. So my question (my first independent research question) was: At what value of k does the maximum value of $p(n, k)$ occur?

5. Sol's Departure from JPL

Sol left JPL in the fall of 1964 to join the faculty at the University of Southern California. In retrospect I see that as the beginning of the end of a golden era for Section 331. In the fall of 1965 Gus Solomon took a job at Hughes Aircraft. Several years later Ed Posner left to join the faculty at Cal Tech. Ed could talk to engineers and mathematicians. He was a master at mathematizing engineering problems and

The Wisdom of Solomon 221

passing them on to mathematicians able to solve them. Gus was everybody's best friend and personal counselor. He made it fun to come to work every day. I had little personal contact with Sol, but my impression is that he was the overseer who set the tone and created the environment that was so productive. A couple of incidents clinch this idea in my mind:

(1) Late in the summer of 1965, just before Gus went to work at Hughes, we had dinner in Pasadena and drove up to Sol's beautiful vintage Spanisn-style home in La Canada. Gus had some business with Sol, I was just along for the ride. At some point they got to reminiscing about being students together at Johns Hopkins (I think Sol was an undergrad and Gus was a postdoc). The Hopkins math department had been dominated by Aurel Wintner, a man whom the biography at http://www-history.mcs.st-and.ac.uk/Biographies/Wintner.html characterizes as inflexible, bitter and impatient with people not as accomplished as himself. Gus quoted Wintner as saying, "Who are you, with your fingernails of clay, to think that you can scratch the hide of the great rhinoceros of mathematics?!!", and they both laughed. I realized that they had not only survived such a negative environment, thrived professionally in spite of it, were able to laugh about it, but they were now creating much more positive and humane environments. What wonderful role models.

(2) In reading Sol's memoir about those years, I was struck by his reason for going to USC: "That was where I felt I could do the most good". In his memorial blog for Sol (https://blog.stephenwolfram.com/2016/05/solomon-golomb-19322016/), Stephen Wolfram, enlarges on Sol's statement, pointing out that Sol also had offers from CalTech and UCLA, as well as USC. At that time (1964), USC was not in the same league as CalTech and UCLA in electrical engineering but is now nationally prominent. This attests to Sol's effectiveness as a mathematician, scientist, engineer and administrator, but most of all as a leader.

6. The Summer of 1965

Sol was no longer at JPL, having departed to USC. He was replaced by Richard Goldstein, a radar astronomer. Otherwise, the personnel and ambiance were essentially the same. Plus I was still occupied with my first independent research question which had developed out of Sol's memo of the summer before (for which value of k (Kmax), does $p(n, k)$ take its maximum value?). My idea for computing Kmax, at least asymptotically, was to show that $p(n, k)/p(n)$ is asymptotically Gaussian. If so then Kmax would be near the mean of the distribution. Despite effort, I was unable to show that the distribution was Gaussian or even compute its mean. Not wanting a promising idea to go to waste, I tried it on the Stirling numbers of the second kind, $S(n, k)$, the number of partitions of an n-set. $S(n, k)$ has a nice recurrence which I used to show that $S(n, k)/B(n), B(n) = Sum\{S(n, k) : 1 <= k <= n\}$, is asymptotically Gaussian and compute the mean (see https://www.researchgate.net/publication/38365311_Stirling_Behavior_is_Asymptotically_Normal). This showed that for $S(n, k)$, Kmax is asymptotic to $n/\ln(n)$. I could not have done this without the assistance of Ed Posner and Adriano Garsia (an academic consultant for JPL-331). Ed pointed out Leo Moser's asymptotic formula for $B(n)$ in the literature. Adriano helped me overcome a technical problem: If $B(n; x) = \text{Sum}\{S(n, k)x^\wedge k : 1 <= k <= n\}$, the generating function for the nth order Stirling numbers, I was guessing that the n roots of $B(n; x)$ were all real, distinct and nonpositive so

$$B(n; x) = (x - (-r\_1))(x - (-r\_2))(\cdots)(x - (-r\_n))$$

$$= (x + r\_1)(x + r\_2)(\cdots)(x + r\_n)$$

and the distribution, $\{S(n, k)/B(n) : 1 <= k <= n\}$, would be the distribution of the sum of n independent, 0-1 valued random variables. Goncarov's extension of the deMoivre-Laplace theorem of probability theory would then imply that $\{S(n, k)/B(n) : 1 <= k <= n\}$ is asymptotically Gaussian. My plan was to prove it by induction on n using Rolle's theorem (between any two zeros of a continuously differentiable function there is a point where the derivative is zero) but it wasn't quite coming together. I mentioned it to Adriano and he replied, "Try multiplying by $e^\wedge x$". I did and found that

$$d((e^\wedge x)B(n; x))/dx = (e^\wedge x)B(n; x) + (e^\wedge x)d(B(n; x))/dx, \text{ by the product rule}$$

$$= (e^\wedge x)\text{Sum}\{(S(n, k-1) + kS(n, k))x^\wedge k : 1 <= k <= n+1\}$$

$$= (e^\wedge x)\text{Sum}\{(S(n+1, k)x^\wedge k : 1 <= k <= n+1\},$$

$$\text{by the recurrence for } S(n, k),$$

$$= (e^\wedge x)B(n+1; x)$$

which was just what I needed.

7. A Gift

My best illustration of Sol's humanity was the time he arranged for me to spend with Robert Vivian, Emeritus Dean of Engineering at USC (in 1978). Sol had somehow found out that my grandfather, John Fay Wilson, had been a Professor of Electrical Engineering at USC (1918−22). I never knew John Fay because he died of diabetes in 1922 (just before insulin became available), when my mom was 8 years old. His premature death was a central tragedy in our family. Aside from my mom and grandmother, who said little about him, I had never met anyone who had known John Fay. Dean Vivian had been on the USC campus almost continuously since 1914 when he started as an undergraduate student in chemical engineering but he had graduated in 1918, just before John Fay joined the USC faculty. Vivian then left to work in the oil fields of Bakersfield until after John Fay's death, so he had never met John Fay. However, he had written a book, *The USC Engineering Story*, that had several entries on my grandfather. After sharing what he did know about John

The Wisdom of Solomon 223

Fay and his time at USC, he presented me with a copy of his book and inscribed it. After my mother's death, I found John Fay's copy of the textbook he authored, *Essentials of Electrical Engineering* (1915) among her personal possessions. In it were my grandfather's marginal notes for corrections (he must have been planning a second edition) and the names of those enrolled in his last class on a sheet of his USC letterhead stationary. I treasure those two books and the interview with Dean Vivian as the closest I ever got to knowing my grandfather.

8. The Technological Accomplishments of JPL-331 Under Sol's Leadership

My impression, though I lack the expertise to make a definitive assessment, is that the digital revolution in communications and computing that we find ourselves so immersed in today, had three principal sources: Bell Labs at Murray Hill, NJ, which pioneered transistors, information theory & fiber optic cables, Lincoln Labs (at MIT) which pioneered error correcting codes & artificial intelligence and JPL which pioneered extra-terrestrial communications (requiring error correcting codes and other hi tech refinements). Gus Solomon, the principal coding theorist in Section 331 during the early 1960s was tickled when a Reed-Solomon code was adopted as the standard for compact discs. He told me, "It means that you can take a CD, scratch it with a needle and it will still play flawlessly". Andy Viterbi, an engineer in Section 331, followed Sol to USC get his Ph.D. Subsequently he and Irwin Jacobs started two companies based on their expertise in digital communications. The second and most successful was Qualcomm which created our cell phone system. Another colleague at JPL-331 was Len Kleinrock. Len was a Professor at UCLA, part time at JPL. He and a Professor at Stanford set up the first internet between their two campuses. And then there was Sol's own work on linear-feedback shift register (LFSR) sequence generation. Being simple, fundamental and having nice autocorrelation properties, the maximal length LFSR sequences that Sol laid out an elegant theory for have been incorporated into just about every aspect of digital communication and computation. In Wolfram's blog he estimates that Sol's algorithm for generating maximal length LFSRs has been executed an octillion (10^27) times. In Wolfram's opinion, this would make it the algorithm that has been executed more times than any other in human history.

Sol's work, applying Galois' theory of field extensions to electrical engineering problems, also had the effect of exorcising the ghost of G. H. Hardy. Hardy was a renowned mathematics professor at Cambridge University. His specialty was analytic number theory. In 1940 he published a book, *A Mathematician's Apology*, praising "pure mathematics" (that with no applications) as beautiful and uplifting of the human spirit. On the other hand he characterized "applied mathematics" as ugly and destructive. These opinions were evidently based on his experience of WWI, when the flower of his generation was mowed down by the hi tech weapons; machine guns, howitzers, tanks and explosives produced with the aid of mathematics. WWII evidently reinforced Hardy's ideas for a lot of mathematicians

224   *The Wisdom of Solomon*

because they were very prevalent in the postwar period. Sol told Wolfram that at Harvard (in the early 1950s) "the question of whether anything that was taught or studied in the mathematics department had any practical applications could not even be asked, let alone discussed". For Hardy of course, number theory was the ideal of pure mathematics. So to have Galois theory, the pristine heart of algebraic number theory, become a tool for electrical engineers was a fatal blow to Hardy's thesis.

9.   Why was JPL-331 Such a Special Place?

The movie, "Hidden Figures", is set at NASA's compound in Langley, West Virginia in 1961–62, but captures the feeling of the work environment at JPL at that same time. The story focuses on Katherine Johnson, a mathematician specializing in flight trajectory calculations. In the movie's dramatic high point, an IBM computer fails and she provides hand calculated data that brings orbiting astronaut John Glenn safely back to earth. Both JPL and Langley were involved in setting the stage for the fulfillment of John Kennedy's promise (1961) to land a man on the moon.in 10 years. The political pressure and technological challenges facing NASA at that time were tremendous. They had the effect of welding people together and making them feel part of something larger than themselves. Status meant relatively little at work: Everybody had a job. If you did your job, you were part of the team. If not, you were out. Socializing, as at parties or going out to lunch, was egalitarian. We were all on the same team and a person's worth was measured by what he (or she) could contribute toward the goal of the team. If one could help a coworker with his task, it was done.

This contrasted significantly with the environment at Bell Labs. There, the hierarchy was much more obtrusive. Status symbols such as title, group affiliation, the location of your office (a corner office was prestigious), were very significant. Socializing was a fraction of what it was at JPL. The same for mutual assistance and exchange of ideas. Competition was more significant than cooperation. Those differences drew me back to JPL in the ensuing summers.

So why was there such a difference? I think there were two reasons. One I already mentioned, the sense of mission that permeated NASA because of JFK's promise to land a man on the moon in ten years. The other was the quality of the people, particularly the leadership. The only serious conversation I recall having with Sol in those two summers that we were both at JPL, was him telling me about another member of the group, a PhD engineer, who had been assigned a piece of Section 331's project. The man had made regular progress reports over several months, but when it came time to turn in the completed work, it turned out he had done nothing. I think Sol allowed him to clean out his desk and had him escorted to the gate. But why would he tell me about it? Many (maybe most) administrators faced with such incompetence or malfeasance would attempt to cover it up (it could damage the image of the organization, as well as the reputation of the administrator for omnipotence and control). But Sol was open about it with one of the lowest status

The Wisdom of Solomon 225

members of the team. Why? I think he was educating me about the foibles of humanity and sharing the misery of being in a position of responsibility like his.

10. A Grand Synthesis

After receiving my PhD from the University of Oregon (June 1965), I landed my dream postdoctoral position at the Rockefeller University in New York with big assists from Ed Posner and Gus Solomon, I spent the next two summers ('66 and '67) attending combinatorics conferences in Italy & Germany and traveling around. By 1968 I was married and looking for a more settled life. So I went back to JPL-331.

One day about the middle of that summer I was sitting in my office working on what I called "Combinatorial Coding Theory", a project to pull together and develop the ideas that I had used to prove the conjectures of Kautz and Hales. Warren Lushbaugh, one of our engineers came into my office (I suspect that Ed sent him). Warren wanted help with a problem. He was working to incorporate a Viterbi decoder into the communication system for Mariner probes (destined for Mars and beyond). The wiring diagram of the decoder had 16 electronic components corresponding to the 4-tuples of 0s and 1s. Those components then, were essentially the same as vertices of the 4-cube, but the edges (wires) of the graph were different: An edge was determined by eliminating a digit (0 or 1) on the left end of a 4-tuple and replacing it by a 0 or 1 on the right end. This of course corresponded to the "shift" operation of a shift register. Lushbaugh's problem was how to lay out that wiring diagram on a linear chassis so as to minimize the total length of all the wires. Such a "layout" would minimize the crosstalk (mutual inductance) in the decoder, making it more efficient and reliable. After a few minutes of discussion, it dawned on me that Lushbaugh's problem was essentially the same as Kautz's: To number the vertices of the graph so that the sum of their differences over the edges is minimized. The only difference between the two problems was in the edges of the graph, so the same analysis could be applied, reducing it to the edge-isoperimetric problem on the graph.

With the tools I had developed for solving the edge-isoperimetric problem on a graph, solving it on the Lushbaugh-Viterbi-Golomb-deBruijn[1] graph was straightforward. Two of its vertices, 0000, and its complement, 1111, have loops (loops do not contribute to the edge-boundary). All the other vertices have 4 edges incident to them, so {0000} & {1111} constitute the sets of size 1 that minimize edge-boundary. Also there are two triangles, 0001−>0000−>1000 and its complement that minimize the edge-boundary for all the sets of size 3. The graph (disregarding directions of

---

[1] Nicholas deBruijn was a Dutch mathematician who first studied this graph (c. 1950). DeBruijn showed the existence of "pseudonoise" sequences by applying Leonhard Euler's celebrated solution of the Bridges of Konigsberg problem to it. Sol's maximal length LFSRs were essentially the pseudonoise sequences that deBruijn had shown to exist but generated constructively in a very simple way.

edges) has two symmetries, complementation of components and reversal of 4-tuples that can be used to systematically simplify its edge-isoperimetric problem.

In 2004 I published a monograph, *Global Methods for Combinatorial Isoperimetric Problems* (GMCIP). It was the ultimate product of the project I was working on when Warren Lushbaugh walked into my office. In the early 1970s, while I was making a presentation of my solution of the edge-isoperimetric problem for the dodecahedron to my postdoctoral mentor, Gian-Carlo Rota, he observed that my method reminded him of Steiner symmetrization. Rota was a profound scholar of mathematics. His personal quest was to find connections between combinatorics and classical mathematics. Sol had accomplished that by connecting linear shift register sequences with Galois theory. What I found in following up on Rota's observation was a strong and useful analogy between the transformations that those of us who had worked on combinatorial isoperimetric problems had developed and symmetrization, the transformation that the great Swiss mathematician Jakob Steiner had invented to give the first rigorous proof (1840) of the classical isoperimetric theorem in the plane (The circle is the simple closed curve of a given length in the plane that surrounds the maximum area). I called those transformations *Steiner operations*. GMCIP is an indepth study of Steiner operations and their applications in combinatorics. For the reader who would like more detail, the full solution of Lushbaugh's problem is on pages 12–14 of GMCIP.

11. In Retrospect

Looking back (with that long perspective of 50+ years) those two summers, '62 and '64, when I worked in JPL-331 under Sol's leadership, were foundational for my career and had such a positive influence on the rest of my life. There I was, making twice as much money as I had ever made in any previous job, working on challenging and meaningful math problems (which, if I had been wealthy, I would have done for nothing) in a technological and scientific environment (unmanned space exploration) that was so stimulating and with leadership that was so supportive. It was as though Sol and Ed were working for me, helping me achieve my dream of a career in math and science. I remember sitting in my office at JPL one day and saying to myself, "Amazing to think that only 6 years ago my greatest academic accomplishment was playing Junior College basketball!" And I anticipate that this volume of memoirs by those fortunate enough to have been associated with Sol will contain many more such stories of fulfillment.

Thanks, Sol, for jumpstarting my career in combinatorics and providing me the opportunity to light my candle from the Promethean fires of DNA and space exploration.

© 2023 World Scientific Publishing Company
https://doi.org/10.1142/9789811234378_0021

The Norwegian Connections of Solomon W. Golomb

Tor Helleseth

Department of Informatics, University of Bergen
N-5020 Bergen, Norway
tor.helleseth@uib.no

Solomon W. Golomb has had a significant influence on Norwegian researchers within coding theory, cryptography, sequence designs, number theory and combinatorial designs. In his earlier days as a PhD student he spent one year on a Fulbright Overseas Grant at the University in Oslo during 1955–1956. During this year he built strong contacts with the excellent group in number theory at the University of Oslo. One of the young researchers there at this time was Ernst S. Selmer who, a few years later, became a professor at the University of Bergen. On one of his later visits to Norway Golomb visited the University of Bergen for the first time in January 1970. This was the first time that the author of this article met him and this was the beginning of a life-long friendship and the start of Golomb's long contact with the research group in coding and cryptography in Bergen. This is an account of his influence and inspiration of Norwegian researchers.

1. Introduction

Solomon W. Golomb, known as Sol among friends, was an international giant in applying mathematics to practical problems leading to major consequences for modern communication systems. I am happy to say that Sol during his lifetime built strong relations and long term friendships with Norwegian researchers for more than 60 years. This article will present some of the connections between Sol and Norwegian researchers

In 2000, Tor Bu, who was then the IT Director of the University of Bergen, published a paper [1] with the Norwegian title: "Golomb's Norske Forbindelser" that translates to "Golomb's Norwegian Connections".

This paper will give an updated and expanded version of Sol's influence in Norway. There may be some overlapping facts about Sol's first visit to Norway described in the earlier paper by Tor Bu that are repeated here for the sake of completeness. The interested reader is still strongly recommended to read the paper by Tor Bu, in particular, since in part of this paper Sol himself describes in his own

228   *The Wisdom of Solomon*

words a very readable account of his life having a Fulbright Overseas Grant in Oslo during the academic year of 1955–1956.

1.1. *Golomb's early education and activities*

Solomon W. Golomb was born in 1932 and obtained his bachelor degree from John Hopkins University in 1951 and his master degree from Harvard University in 1953. He obtained his PhD degree from Harvard University in 1957.

In 1954–1955 Golomb worked part time for the Glenn Martin Company. He wrote an important technical report on shift register sequences with the title *Sequences with Randomness Properties*. This report was later published as a chapter in his groundbreaking book — *Shift Register Sequences* — [4] that was first published in 1967. This book was a landmark in the theory of linear and nonlinear shift registers and had a tremendous influence on researchers in these areas. Sol later stated that his shift register work has had the broadest impact to cell phone signals, the GPS system, error-correcting codes, radar and cryptography. The book appeared in a second edition in 1982 and its third edition in 2017 was completed by Sol a short time before he passed away.

1.2. *Golomb's first visit to Norway*

Sol's plan to come to Norway took shape in 1954 when he applied for a Fulbright Overseas Grant to visit the University of Oslo. The reason for choosing Norway was that his PhD thesis that he was working on was on number theory and Norway had many excellent number theorists. In particular, he received a positive impression after meeting the Norwegian number theorist Atle Selberg at the Institute of Advanced Studies at Princeton.

Golomb's Fulbright application was successful and he arrived in Oslo in June 1955 after crossing the Atlantic Ocean by boat. The University of Oslo had several senior number theorists, including Viggo Brun and Thoralf Skolem.

The younger number theorists at the University of Oslo included Ernst S. Selmer, whom Golomb called a distinguished young mathematician that he learned to know quite well. Ernst S. Selmer and Sol Golomb had many interests in common and in particular Sol mentioned that Selmer got interested in linear shift registers from him. Sol completed his PhD thesis when he was in Oslo and submitted his PhD thesis to Harvard on May 17, 1956, Norway's national day.

Two facts were important for the strong connections developing between the University of Bergen and Sol Golomb. The first was that Selmer in 1957 became one of the first professors in Mathematics at the University of Bergen. Furthermore, in 1964–1965 Selmer had a sabbatical year in Cambridge, England where he gave a series of lectures on shift register sequences. The lecture notes were entitled *Linear Recurrence Relations over Finite Fields* [12]. These lecture notes were not published but in spite of this they were frequently in demand and were ordered by mathematicians all over the world. In fact, at Eurocrypt 1993, Selmer gave an invited lecture and he sold all his 200 remaining copies of his lecture notes during 20 minutes immediately after his lecture.

1.3. Golomb's connections to the University of Bergen

During the 1960's Selmer regularly supervised a group of master students on linear shift registers. This group included Tor Bu who wrote the aforementioned paper in 2000 about Golomb's Norwegian connections, and Sverre Spildo who for many years was the vice-director of the University of Bergen. Two of Selmer's master students, Johannes Mykkeltveit and myself, went on to complete doctoral degrees on shift register sequences in the 1970's.

I started as a master student with my supervisor Selmer during the academic year 1969–1970. Harold Fredricksen from the Jet Propulsion Laboratory (JPL), a former PhD student of Golomb, was visiting Bergen for one year to give a series of lectures on nonlinear shift registers. The theory of linear shift registers was well known by the audience in Bergen due to their knowledge of Selmer's lecture notes. We were very eager to learn about the new topic of nonlinear shift registers. The audience in Bergen attending Hal's lectures consisted of Professor Ernst Selmer and Svein Mossige who was an associate professor and the computer expert in the group. The two students were the PhD student Johannes Mykkeltveit and myself, the only master student in the weekly seminar. Furthermore, Kjell Kjeldsen, a researcher from the Chief Headquarter of Defense, Norway came from Oslo just to attend the lectures of Hal Fredricksen. Occasionally, Tor Bu came from Stavanger to follow these activities.

The lectures given by Hal were very interesting and contained the results from Sol's book on *Shift Register Sequences* in addition to recent results including problems from his own PhD thesis. In particular, he came frequently up with many challenging open research problems on nonlinear shift registers that the doctoral student Johannes Mykkeltveit spent a large amount of time on.

1.4. Nonlinear shift registers and Golomb's conjecture

One of the problems that Hal Fredricksen brought to our attention was Golomb's conjecture on the de Bruijn graph. This was a fascinating and apparently a difficult research problem. To briefly describe the problem we need some basic definitions of nonlinear shift registers.

A Boolean function in n Boolean variables is a mapping f from $B_n = \{0,1\}^n$ to $\{0,1\}$ of the form,

$$f(x_0, x_1, \ldots, x_{n-1}) = \sum_{1 \leq i_1 < i_2 < \cdots < i_k < n} a_{i_1 i_2 \cdots i_k} x_{i_1} x_{i_2} \cdots x_{i_k}.$$

There are 2^{2^n} different Boolean functions $f(x_0, x_1, \ldots, x_{n-1})$ of degree n. To construct a shift register sequence let $(s_0, s_1, \ldots, s_{n-1})$ be the initial state and define a binary sequence (s_t) using the Boolean function to generate the next bit as a function of the n previous bits,

$$s_{t+n} = f(s_{t+n-1}, s_{t+n-2}, \ldots, s_t).$$

The mapping F maps the set of n-dimensional vectors B_n into itself by

$$F(x_0, x_1, \ldots, x_{n-1}) = (x_1, x_2, \ldots, x_{n-1}, f(x_0, x_1, \ldots, x_{n-1})).$$

The mapping F is a permutation (the case of interest) of B_n if and only if

$$f(x_0, x_1, \ldots, x_{n-1}) = x_0 + g(x_1, x_2, \ldots, x_{n-1}).$$

This gives $2^{2^{n-1}}$ permutations of B_n. The de Bruijn graph is a directed graph with nodes B_n and an edge from $(x_0, x_1, \ldots, x_{n-1})$ to $(y_0, y_1, \ldots, y_{n-1})$ if and only if $y_i = x_{i+1}$ for $i = 0, 1, \ldots, n-2$.

A simple example is the mapping F corresponding to $g = 0$. In this case the function $F(x_0, x_1, \ldots, x_n - 1) = (x_1, x_2, \ldots, x_{n-1}, x_0)$ corresponds to a cyclic shift of the nodes in B_n. This is called the Pure Cycling Register of order n (PCR_n).

For $n = 4$ the PCR4 decomposes B4 into disjoint cycles: (0000) and (1111) of period 1, (0101) of period 2, (0011), (0111) and (1011) of period 4. PCR_4 has 6 cycles. Each cycle in PCR_n consists of all the different cyclic shifts of a node.

It is a well known fact that the number of cycles in the pure cycling register of order n is given by

$$Z(n) = \frac{1}{n} \sum_{d|n} \varphi(d) 2^{\frac{n}{d}}.$$

One can try any other permutation F and decompose B_n into cycles. For $n = 9$ there are more of these functions than the atoms in the universe. The conjecture can now be stated as follows.

1.5. Golomb's conjecture

The number of cycles generated by an arbitrary Boolean recurrence relation $f(x_0, x_1, \ldots, x_{n-1})$ does not exceed $Z(n)$. Thus the maximum number of cycles is obtained from the PCR_n.

This is an amazing conjecture that holds for so many Boolean functions. The conjecture was based on a complete computer search for $n = 5$, so Golomb certainly must have had a strong intuition for making the conjecture.

Golomb's conjecture became a challenging problem that Johannes Mykkeltveit worked on for more than one year before he came up with a surprising proof using the concept of the mass center of the nodes of length n.

There was progress in proving the conjecture for $n < 10$ by Hal and Abraham Lempel [10] who made a conjecture that implied Golomb's conjecture.

The Norwegian Connections of Solomon W. Golomb 231

1.6. Lempel's conjecture

The minimum number of vertices which, if removed by Bn will leave a graph without cycles are $Z(n)$.

Lempel's idea was to select *exactly one node* from *each cycle* in the PCR_n (i.e., to select $Z(n)$ nodes from B_n) and to show that any cycle in the de Bruijn graph B_n contained at least one of these $Z(n)$ selected nodes. Mykkeltveit invented an ingenious scheme using the concept of a mass center for a PCR_n cycle. After one year of hard work he finally succeeded.

Given a node on a PCR cycle $(x_0, x_1, \ldots, x_{n-1})$ of length n. Place the bits along on the nth-roots-of-unity (x_i corresponds to w^i where w is the canonical primitive nth root of unity) and compute the mass center of the nodes (where you may consider the 1's to weigh 1 kilo and the 0's to be weightless).

For example, to select one node from the cycle (0111) in PCR_4 consisting of the four nodes (1110), (1101), (1011), and (0111). Then place the bits on the 4th roots of unity and compute the mass center. Computing the mass center of a cyclic shift of node is just to rotate the mass center 90 degrees.

The mass center is, on the y-axis, to the right of the y-axis, or to the left of the y-axis. Then, if the mass center is not in origin, you select the unique node corresponding to the cyclic shift such that the mass center is on one fixed (say on the right) side of the y-axis and its previous shift is not. In the case that a cycle on PCR_n has mass center in origin you can select any node on this cycle.

Based on this selection of $Z(n)$ nodes it is very simple to prove Golomb's conjecture. The full paper [11] published in 1972 is less than 6 pages including the full description of the problem and its solution. This was essentially the first problem Mykkeltveit solved in his PhD thesis. We were worried that with such a short paper his thesis would be too meager so maybe he would later regret to have given such a marvelous simple and short proof. However, there was no reason to worry, since he completed his thesis by solving many other problems in coding theory and on nonlinear shift registers.

For me, as a master student, this was an inspiring learning process and I enjoyed watching Golomb's conjecture being solved in such an elegant way.

1.7. Cross correlation of m-sequences

In the Christmas break at the university in early January 1970 Professor Selmer called me at home and told there would be a guest lecture by Professor Solomon Golomb that I should attend. I was very enthusiastic for the opportunity to meet such a famous professor and the author of the celebrated book on shift register sequences. This visit by Golomb significantly changed my life.

The excellent lecture given by Sol was related to a combinatorial problem on tiling with a 3D Cartesian box with bricks of different sizes. After the lecture Sol discussed more informally a new research problem on the cross correlation of m-sequences of period $2^n - 1$ that they had been working on at USC. He explained that in a recent groundbreaking paper [3] by Robert Gold in 1968 he had shown that the cross correlation between two m-sequences differing by a decimation $d = 2^k + 1$, where n is odd and $\gcd(k, n) = 1$, has only three values $-1, -1 - 2^{(n+1)/2}$ and $-1 + 2^{(n+1)/2}$. This observation was the main idea behind the family of Gold sequences that probably are among the most used sequence family in communication systems and many other applications, for example used in the Global Positioning System (GPS). It is well known that the autocorrelation of an m-sequence only takes on two different values, -1 out-of-phase and $2^n - 1$ in-phase.

Golomb mentioned that his colleague Lloyd Welch had generalized Gold's result to $d = 2^k + 1$, where k and $n/\gcd(n, k)$ are odd, and also proved that the decimation $d = 2^{2k} - 2^k + 1$ had the same correlation distribution for the same conditions for k and n. Welch did not publish his proof. The first published proof was by Kasami a few years later.

Originally my supervisor Selmer intended me to write a master thesis on nonlinear shift registers but he found that the nonlinear topic is very difficult and suggested instead that I should write my master thesis on the cross correlation of m-sequences. Selmer gave me the thesis by Trachtenberg [14] written at USC. This thesis had generalized the Gold and Kasami-Welch decimation to non binary m-sequences of period p^n - 1 for any odd prime p to $d = (p^{2k} + 1)/2$ and $d = p^2 - p^k + 1$ and proved a three-valued cross correlation.

I worked on the cross correlation problem for more than a year and completed my master thesis in 1971. I enjoyed the problem and the combination of computer search and theoretical results. I came up with several new results and made several tables that produced some interesting and longstanding conjectures. Some of these conjectures are still open. The best known of these conjectures is the "Helleseth -1 conjecture" [7] that states that -1 is always one of the values that appear among the cross correlation values between any two binary m-sequences (for $p > 2$ the conjecture needs an extra condition on d).

I remember Selmer told me that he found the conjecture interesting and my grade would be improved if I could solve it. In spite of working hard I was unable to solve it. However, my comfort is that many good mathematicians have also tried to solve the conjecture during almost 50 years without success.

In 1969 Sol published [5] where he also mentioned a conjecture by Welch that the decimation $d = 2^{(n+1)/2} + 3$ gives a three-valued cross correlation for all odd n. This problem found its solution more than 30 years later by Canteaut, Charpin and Dobbertin [2]. A simpler observation mentioned in this paper was that the cross correlation between m-sequences always had at least three values. Sol mentioned the result without proof (that he did not have). The proof of this observation was one of my first results in my master thesis and was a nice problem on Diophantine equations.

1.8. Sabbatical at USC 1977–1978

During my Dr. Philos. studies I applied to the Research Council of Norway for a sabbatical year at USC. This was an excellent place to stay because the only

The Norwegian Connections of Solomon W. Golomb

two known PhD thesis on the cross correlation of m-sequences had been written at USC, but also because of the climate was significantly better than in rainy Bergen. Furthermore, Sol encouraged me to come to USC and was willing to be my host and help me with the practical formalities and paperwork needed.

I had a wonderful experience staying at USC during the academic year 1977–1978. I met and worked with (later) Shannon award winners including Sol Golomb, Irving Reed and Lloyd Welch. I had frequently lunches with all of them and they often asked me about Norway. I sometimes remember my feeling a little embarrassed when I realized that Sol knew more about Norway than I did and could supplement me with information that I did not know. Sol was a walking encyclopedia and he constantly impressed me and others with his great knowledge about basically everything including sports, knowledge of languages and even discussing Chinese characters with my Chinese PhD students.

Sol treated me very friendly and invited me, my wife and little daughter frequently to his home. Since we were from Norway, Bo (Sol's wife) liked to talk with us in her own Danish language. One of our favorite moments were when we were invited to celebrate Christmas Eve in their home in 1977. I remember that we were walking around their Christmas tree singing Christmas songs and also holding hands with two large toy dolls to be able to reach around the Christmas tree. This Christmas celebration we will always remember. I always appreciated the nice treatment Sol always gave me and my family and that he had time to take so good care of such a young researcher as I was then.

Me and my family enjoyed the first sabbatical at USC so much that we came back in 1992–1993 and also in the autumn of 1999. In recent years I visited USC frequently and when I was in Los Angeles I always found time to meet Sol for a nice chat and a good meal.

Other Norwegian researchers influenced by Sol Golomb included Torleiv Kløve who generalized the "Golomb ruler" and obtained bounds on the constructions of integers such that all differences between the integers are all distinct. He also applied the concept of sonar sequences used frequently by Herbert Taylor and Sol Golomb. Furthermore, we also solved a conjecture on a nonlinear shift register problem [8]. My PhD student Erik Hauge wrote his PhD thesis on nonlinear shift registers. Hal Fredriksen was an opponent during his defense. This thesis provided several publications in the 1990s, including [6].

Kjell Kjeldsen wrote a very elegant paper [9] in 1976 on nonlinear shift register that were fundamental in the analysis of determining the periods of nonlinear shift registers with symmetric feedback polynomials. This inspired further work on this topic by another young Norwegian researcher Søreng [13].

Sol often asked me about conferences in Scandinavia in combinatorial theory, information theory or related areas. He and his wife Bo used to visit Scandinavia almost annually and I could update him so he could combine his visit with a relevant conference.

The last time Sol was in Bergen before he passed away was in 2014. He planned to go to a conference in London and he wondered if I knew of a suitable conference in Scandinavia. I told him I did not know of any at that time but I invited him to Bergen for a week. I remember he gave an excellent talk on some very recent new results on the twin-prime conjecture that he was very excited about. It is strange to think that his last lecture in Norway was on number theory, the topic that actually brought him to Norway for the very first time too.

Sol received the prestigious Franklin Medal in Philadelphia in April 2016. I was honored to be able to attend the ceremony when he received his well deserved medal. But I was also very honored to be one of the five invited speakers to give a talk at a seminar presenting Sol's work the day before the award ceremony. The other four speakers were Sol himself, Guang Gong, Alfred Hales and Andrew Viterbi.

During the 46 years I knew Sol we had many scientific discussions that I learned from but only a handful of joint publications. I consider Sol as a very good friend and mentor and I am very happy for being lucky to experience our long term friendship. Sol knew so many things that impressed me in addition to having a very good sense of humor and always making funny observations in his way of looking at the world. One of the favorite observations and formulations he mentioned to me was when we walked at USC and I realized that a Nobel Laureate at USC had a free parking spot. Then Sol said dryly: "You should observe that *first* a Nobel Prize *then* you may get a free parking at USC."

In my 60th birthday celebration in Norway with many of my friends and former colleagues Sol and Bo came and I had reserved a table for myself, Sol and my colleague Torleiv Kløve and our wives. The last time I visited USC quite recently someone told me "life is not the same here without Sol." I think that summarizes what I feel after he passed away that something in my life is missing without him. He was a fantastic person that enriched my life.

Acknowledgement

Supported by The Research Council of Norway under project no. 247742/O70.

Bibliography

1. T. Bu, Golombs Norske Forbindelser, *Comput Math Appl* **39** (11) (2003) 135–138.
2. A. Canteaut, P. Charpin and H. Dobbertin, Binary m-sequences with threevalued crosscorrelation: A proof of Welch's conjecture, *IEEE Trans Inform Th* **46** (1) (2000) 4–7.
3. R. Gold, Maximal recursive sequences with 3-valued cross-correlation functions, *IEEE Trans Inform Th* **14** (1968) 154–156.
4. S.W. Golomb, *Shift Register Sequences,* San Fransisco, CA, USA, Holden-Day, 1967.
5. S.W. Golomb, Theory of transformation groups of polynomials over GF(2) with applications to linear shift registers, *Inform Sci.* **1** (1968) 87–109.
6. E. Hauge and T. Helleseth, DeBruijn sequences, irreducible codes and cyclotomy , *Discr Mathe* **159** (1-3) (1996) 143–154.
7. T. Helleseth, Some results about the cross-correlation between two maximal linear sequences, *Discr Mathe* **16** (3) (1976) 209–232.

8. T. Helleseth and T. Kløve, The number of cross-join pairs in maximal length linear sequences, *IEEE Trans Inform Th* **37** (6) (1991) 1731–1733.

9. K. Kjeldsen, On the cycle structure of a set of nonlinear shift registers, *J Combin Th (A)* **20** (2) (1976) 154–169.

10. A. Lempel, On the extremal factors of the de Bruijn graph, *J Combin Th (B)* **11**(1971) 17.

11. J. Mykkelveit, A proof of Golombs's conjecture for the de Bruijn graph, *J Combinl Th (B)* **13** (1972) 40–44.

12. E.S Selmer, *Linear Recurrence Relations of Finite Fields*, Lecture Notes, University of Bergen, 1966.

13. J. Søreng, Symmetric shift registers, *Pacific J Math* **65** (1) (1979) 201–229.

14. H.M. Trachtenberg, *On the Cross-Correlation of Maximal Linear Recurring Sequences*, PhD Thesis, University of Southern California, 1970.

© 2023 World Scientific Publishing Company
https://doi.org/10.1142/9789811234378_0022

Sol Golomb and a Twice-in-a-Lifetime Celestial Event

Jonathan Jedwab

Department of Mathematics, Simon Fraser University
Vancouver, BC V5A 1S6, Canada
jed@sfu.ca

When I was a high school student in the United Kingdom in the 1980s, one of my teachers generously ran a mathematics enrichment club for enthusiasts. In one of these sessions, he fascinated us with the topic of polyominoes [1], a rich source of geometric puzzles that is nowadays recognised as the inspiration for the wildly popular video game Tetris. The first time I can recall hearing the name Solomon Golomb was when I asked the teacher: who proposed the concept of polyominoes?

Some thirty years later, it was my honour to be the co-chair with Tor Helleseth of the Technical Program Committee for the 2012 SETA (SEquences and Their Applications) conference at the University of Waterloo [2]. By this time I had met Sol in person several times at conferences, and knew that he was a legendary figure in information theory and discrete mathematics. Guang Gong, the conference local organiser, proposed holding the conference in early June in order to coincide with Sol's 80th birthday. Tor and I readily agreed, and set aside 5 June for a full-day special session of the conference in his honour, followed by an evening banquet celebration. When we heard that Sol had been awarded the 2012 Proctor Prize of the Sigma Xi Scientific Research Society, we invited the director of the society to present Sol with the prize at the banquet. The day of 5 June 2012 was shaping up to be memorable, but I didn't yet realize there was another delight awaiting us.

Five weeks before the conference, Alex Pott emailed me from Germany: "Do you know that SETA is a venus transit conference? Well done! It is easier to watch in Waterloo than in Germany." I have to say that I had no idea what Alex was talking about! After some online browsing, I discovered that a transit of Venus occurs when the planet Venus passes directly between the Sun and Earth. It is similar to a solar eclipse of the Moon, but differs in two important ways. Firstly, because Venus is much further away from Earth than the Moon is, it cannot obscure the entire disc of the Sun but instead appears as a tiny black dot moving across the Sun's surface. Secondly, a transit of Venus is extremely rare, occurring only twice within eight years and then not again for more than a century. The previous transit had occurred in June 2004, and the next one was due on 5 June 2012: the very

238   *The Wisdom of Solomon*

day we had dedicated to Sol's 80th birthday celebration! This really was a twice-in-a-lifetime celestial event: the next transit will not occur until December 2117, so although we knew that the conference participants might one day attend another SETA conference, they surely would never again have the opportunity to witness a transit of Venus.

We therefore adjusted the conference schedule to incorporate an observation of the transit of Venus. I found out that the local branch of the Royal Astronomical Society of Canada [3] was holding a "star party" on the campus of the University of Waterloo, and their representative graciously told me we were all welcome to drop by to view the transit through large telescopes on the evening of 5 June. We provided each conference attendee with a pair of viewing goggles so that they could observe the transit safely just before the start of the banquet. (See the photos in Figs. 1 and 2 for Sol Golomb observing the transit of Venus.)

I had learned that transits of Venus were historically important for estimating the Astronomical Unit (the average distance between the Earth and the Sun) using the principle of parallax to combine information from observations of the transit made from different points on the Earth's surface. In fact, Captain James Cook's first voyage was to Tahiti, in order to observe the 1769 transit. A marvellous poster [4] gave this colourful background:

> Much of the mystique of the ToVs [transits of Venus] stems from the incredible stories of observers' astronomical endurance and fortitude. Le Gentil missed one ToV and hung around the far east for another 8 yrs. to see the next, only he couldn't (clouded out; and people at home thought he had perished — perhaps he wished he had). Chappe d'Auteroche successfully observed his ToV, but died from yellow fever almost immediately afterwards (saving the trouble of a return journey). Mason & Dixon's expedition got beat-up by unfriendly cannonballs, Winthrop was harried by Newfoundland's finest insects, and Nevil Maskelyne before he got home had to make his way through a lot of expensive drink (doubtless assuaging the hurt — he too was cheated by cloud).

On the first day of the conference, we played a five-minute NASA educational video [5] to provide some basic scientific and historical context for the following day's transit viewing. Afterwards Sol came up to me and, with a gleam in his eye, said "I actually know a little about Venus". I suspected this actually meant that Sol knew considerably more than "a little", and asked him to elaborate. Without hesitation, Sol began to recount historical events and technical details from fifty years earlier as if they had happened the previous week. After listening to his answer with increasing excitement, I arranged a special speaking slot shortly before our transit viewing for Sol to repeat what he had told me to all the conference attendees.

Here is the riveting story Sol told us (supplemented by documents and personal recollections he later kindly sent me). In 1962, the NASA space probe Mariner 2 became the first spacecraft to send back information from another planet, passing within 21,000 miles of Venus after a journey of 36 million miles from Earth. The key to successfully directing the probe to reach the vicinity of Venus was having a highly accurate estimate for the distance from Earth to Venus (or, equivalently, for the

Fig. 1. Sol Golomb observing the transit of Venus, 5 June 2012 (photo courtesy of Kelly Boothby).

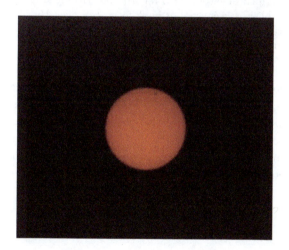

Fig. 2. The transit of Venus observed from Waterloo, Ontario, 5 June 2012 (photo courtesy of Krystal Guo).

Astronomical Unit). The problem for the space probe designers was that previous estimates of the Astronomical Unit, based on the two 19th century observations of the transit of Venus, were actually in error by one part in a thousand. If Mariner 2 had depended on these estimates, it would not have come remotely close to observing Venus. But by the time Mariner 2 was launched, the error in the estimate of the distance to Venus had been reduced to less than one part in a million! How was this critical improvement achieved?

In 1961, a team at NASA's Jet Propulsion Laboratory (JPL) used a radar signal to measure the distance between the Earth and Venus with enormous accuracy. This was the first successful radar contact with another planet of the Solar System. The

240   *The Wisdom of Solomon*

team determined the distance by measuring the time for a radar signal transmitted from Earth to be reflected from Venus and detected back on Earth. Sol was the project manager of the Venus radar project at JPL, and had already conducted pioneering investigations into linear feedback shift register sequences (having the 2-level-autocorrelation property) that he would later present in his classic book [6]. In a critical insight, Sol realised how to combine many such sequences advantageously. In Sol's words:

> the "ranging system" I designed at JPL combined a number of much shorter 2-level-autocorrelation sequences of relatively prime periods, using a Boolean function such as the "maximum decision" function on an odd number of inputs, which is positively correlated with each individual input ... The result was a random-looking binary sequence of sufficiently long period that its "gross ambiguity" would exceed the diameter of the solar system!; but because it was positively correlated with each of the component subsequences, our correlation (over hours at a time) was sensitive enough to determine "where we were" in each of the subsequences, and this information was combined, using the Chinese Remainder Theorem, to specify where we were in the overall system ... The details of my JPL ranging system were described in several JPL reports, but (to my subsequent regret) were never published in journal or book form. However, the above description summarizes the mathematics of it. I had some very talented engineers working for me to reduce everything to hardware that worked.

I still marvel at the amazing coincidences that unfolded during the week of the SETA 2012 conference. Not only had the organisers unknowingly fixed the conference schedule to include the last day in more than a century that would include a major astronomical event involving the planet Venus; but our guest of honour turned out to have played a pivotal role in combining discrete mathematics with communications engineering five decades earlier to determine the position of this celestial body — and thus the astronomical unit — with unprecedented accuracy. Thank you, Sol!

Bibliography

1. S.W. Golomb, *Polyominoes: The Fascinating New Recreation in Mathematics*, Charles Scribner's Sons, New York, 1965.
2. SEquences and Their Applications (SETA 2012), University of Waterloo, ON, 4–8 June 2012 (http://seta2012.uwaterloo.ca).
3. Kitchener-Waterloo Centre, The Royal Astronomical Society of Canada (http://www.kw.rasc.ca/).
4. R.A. Rosenfeld, Transit of Venus 2012 June 5, Poster.
5. The Transit of Venus Part One (2012), https://sunearthday.nasa.gov/2012/materials/transit1.m4v.
6. S.W. Golomb, *Shift Register Sequences* Holden-Day, Inc., San Francisco, CA, 1967.

Solomon Golomb in Pentominoes

Scott Kim[*]

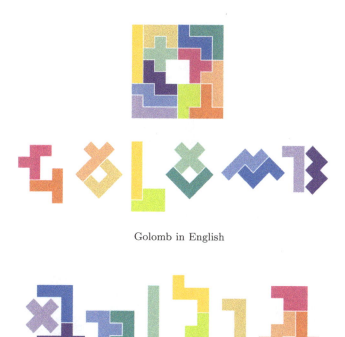

Golomb in English

Golomb in Hebrew

Both versions, English and Hebrew, used each of the 12 distinct pentominoes

---

[*]© Scott Kim, scottkim.com. Reprinted with permission.

© 2023 World Scientific Publishing Company
https://doi.org/10.1142/9789811234378_0024

Solomon Wolf Golomb: A Man of Humility, Integrity and Compassion

Rochelle Kronzek

Is the measure of a man his intellect and all that he accomplishes during his lifetime or in how he treats his fellow human beings and those less fortunate or with less opportunities? Sol Golomb was a polymath and a man of humility and integrity. And so he excelled in both measures of greatness.

It is hard to believe that it has been four years now. May 1, 2016 marked the passing of a brilliant, accomplished applied mathematician and the kindest of souls — Dr. Solomon Wolf Golomb, Distinguished Professor of Mathematics and Electrical Engineering at the University of Southern California (USC) where he had been an active member of the faculty for more than fifty years. He specialized in combinatorial analysis, number theory, coding theory and communications and became famous for the mathematical puzzles and games he created (including Polyominoes which he named in 1953, was popularized by Martin Gardner in his *Scientific American* column in 1960 and inspired the game Tetris).

Only ten days prior to his passing, Dr. Golomb had been awarded the prestigious Franklin Medal 2016 in Electrical Engineering for his revolutionary work on *Shift Register Sequences* and their applications to space communications, satellite communications and cellular communications. Marie Curie, Thomas Edison, Nikola Tesla, Bill Gates, Albert Einstein and Stephen Hawking are but a few of the visionaries whose work has been honored by the Franklin Institute previously.

Dr. Golomb had accumulated many such honors and awards during his career including the 2011 National Medal of Science given by former US President Barack Obama in 2013.

What I Remember Most about Solomon Golomb was his Deep Humanity and Compassion for Others

I first met Dr. Golomb in March 2010 during a **Gathering for Gardner** recreational mathematics Conference held in Atlanta. Dr. Golomb had become a fixture at the Gathering bi-annual meetings. Many people know of Sol's passion for recreational mathematics. Sol had been a puzzle columnist for several publications over the course of many years. He was a frequent contributor to *Scientific American's* "Mathematical Games" column and also wrote "Golomb's Puzzle Column" for the IEEE Newsletter over many years. He also did a column "Golomb's Gambits" containing puzzles that ran in the *Johns Hopkins Magazine* for more than twenty-five years. Dr Golomb also ran a recreational mathematics puzzle class at the USC campus many times.

I had heard of Sol Golomb and he was high on my editorial list of people to approach for potential book projects. I sought him out and felt honored to meet him after one of the general sessions during the Ninth Gathering for Gardner (G4G9). I came up to introduce myself and found him to be warm and friendly.

During that same conference, I was walking down a busy downtown street returning to the conference hotel from lunch and found Dr. Golomb kneeling outside to give a homeless person the hot lunch he had purchased for them at a nearby restaurant.

Solomon Wolf Golomb had similarly walked across the USC campus, greeting many students and colleagues along the way each day for the more than five decades that he was a beloved professor.

I spoke with Sol Golomb several times over the years. He called me to suggest several classic works that needed to be back in print, including one of his own on *Shift Register Sequences*. Another of the classic works that Sol insisted that I reprint is the now late Elwyn Berlekamp's book on *Algebraic Coding Theory*. Both of those books are bestsellers for us today and have stood the continued test of time.

Sol and I once visited together in his USC office where we spent hours talking about books and puzzles. His banter was comfortable and relaxed as we brainstormed together. Sol had a way of making whomever he was speaking to feel important and a priority — me included. I asked Dr. Golomb to become a series consulting editor for me at World Scientific on classic reprints. He was very willing until his wife Bo became ill. Life came to a screeching halt when Bo took ill.

Sol authored or coauthored more than 200 journal articles and seven books and was particularly well known for his pioneering and extensive work since 1953 on *Shift Register Sequences* and their applications to cryptography, radar & coded, spread-spectrum and wireless communications. Originally published in 1967, World Scientific has produced a reprint of this classic work © 2017.

Several tributes were held in Dr. Golomb's honor during the latter half of 2016 and in 2017. This book is an attempt to give voice to some of the many people that were impacted by the life and legacy of Sol Golomb.

SEquences and Their Applications (SETA) is the leading international conference in the areas of sequences and their applications to communication and cryptography. It has been held in various locations around the world starting in 1998 biannually.

Solomon Wolf Golomb: A Man of Humility, Integrity and Compassion 245

A special feature of SETA 2016, which was held in Chengdu (09–14 October 2016), China, was a memorial session in honor of Solomon W. Golomb. The featured speakers for the SETA memorial and their talks were as follows:

1. **Tor Helleseth**, University of Bergen, Norway, *Cross-correlation of Sequences and Golomb's Norwegian Connection.*
2. **Steven Wolfram** (video talk), **Mathematica, GB, Cellular Automata and Shift Registers, or** *How Sol Golomb Almost Made My Favorite Discovery Before I Was Born.*
3. **Hong-Yeop Song**, Yonsei University, Korea, *Prof. Golomb was my Advisor.*
4. **Pingzhi Fan**, Southwest Jiao Tong University, China, *Solomon W Golomb and Sequence Design at SWJTU.*
5. **Vijay Kumar** (video talk), IISC, India, *Some Facets of Sol Golomb.*
6. **Guang Gong**, University of Waterloo, Canada, *Golomb's Shift Register Sequences for Wireless Communication, Cryptography and Radar — Work with a Great Mind.*

On 31 January 2017 a day-long event entitled "**The Life & Legacy of Sol Golomb**" celebrated Professor Golomb's great contributions to the electrical engineering department, the University of Southern California (USC) and the campus Hillel.

I Attended the Day and Events

The morning portion was held in the Viterbi School of Engineering within the department of electrical engineering that Sol called home for more than five decades. Dr. Andrew Viterbi, a longtime friend of Dr. Golomb attended and spoke at the memorial event. Viterbi, co-founder of Qualcomm Inc. and the namesake for the Viterbi School of Engineering at USC met Golomb on his very first day at JPL, in 1957. Together they worked on NASA's earliest satellite programs.

"My friendship with Sol lasted nearly six decades, during which time he grew from the brightest kid in the room to the wisest man on campus," said Viterbi.

Several of Dr. Golomb's former PhD students and frequent research collaborators attended the USC symposium. Colleagues flew in from Norway, Canada, India, Korea and Israel to participate. Albums and posters containing pictures of Dr. Golomb were shared by his two daughters Beatrice and Astrid.

After the morning symposium, the attendees then gathered in the Hughes Aircraft Electrical Engineering Center lobby, where Dean Yannis Yortsos spoke and a bronze relief of Golomb was unveiled

Being the granddaughter of a rabbi myself, I appreciate that Dr. Golomb was both the son and grandson of rabbis. A portion of the afternoon's memorial service was given by the campus Hillel rabbi, Hebrew songs were sung by children and Bailey London, Executive Director of USC Hillel also paid tribute to Dr Golomb.

The office of the President of USC co-sponsored the afternoon tribute and held a private reception in Dr. Golomb's honor.

UC San Diego's **Information Theory and Applications** (ITA) Center, at its 2017 Workshop in February, devoted an afternoon memorial session to "Remembering Sol Golomb". It was organized by Tuvi Etzion (Israel), with the following speakers:

(1) Guang Gong (Canada), "**Golomb's Invariants and Modern Cryptology**";
(2) Tor Helleseth (Norway), "*Shift Register Sequences* **and Golomb's Norwegian Connections**";
(3) Hong-Yeop Song (Korea), "**Existence of cyclic Hadamard difference sets and some memories of Prof. S.W. Golomb**"; and
(4) Andrew Viterbi (USA), "**Celebrating the Life of Solomon Wolf Golomb**".

That evening an interactive puzzle session (with prizes) in Sol's honor was also held, organized by Joe Buhler, Paul Cuff, Al Hales and Richard Stong.

Dr. Golomb will be missed, but his legacy lives through his 50 years of teaching at USC, how he revolutionized digital communications with his book on **Shift Register Sequences** and his kindness that touched me deeply after knowing him just seven years. I was honored to have been his editor.

"Wisdom rests in the heart of the intelligent..." Proverbs 14:33

"The tongue of the wise uses knowledge rightly..." Proverbs 15:2

Rochelle J. Kronzek
Executive Editor
World Scientific Publishing Company

© 2023 World Scientific Publishing Company
https://doi.org/10.1142/9789811234378_0025

Sol Golomb — My Hero and My Idol

Abraham Lempel

The contribution #1 includes the letters I wrote to Beatrice and Mitra as well as a letter I've received from the President of the Technion about Sol's contribution to the Technion. Contribution #2 includes a few short e-mails exchanged between Sol and me over several years.

1. Contribution #1

Email to Beatrice

Dear Beatrice,

Your father was a dear friend of mine and I was shocked by the sad news and am very sorry for your loss. Your father was very proud of you. I hope you remember me, being a close friend of your father for nearly 50 years. I first met him in 1967, at an IT workshop at the Technion, when I presented him with a proof of one of his conjectures re shift register sequences. At the time I was a fresh PhD at the Technion EE department and Sol invited me for my post-doc at USC which I was happy to accept. I arrived at USC in July 1968 and worked very closely with Sol for a year. Under Sol's influence and guidance I have traded my interest in network theory for the rich realm of digital sequences. Since my year at USC, I've met with Sol at least once a year for lengthy discussions of a wide variety of topics. One of my happiest meetings with Sol was at the USC ceremony honoring another dear friend of mine, Andy Viterbi, with the announcement of the Viterbi School of Electrical Engineering. I remember also meeting you at that event. During the many years Sol was on the BoG of the Technion, we had an annual lunch appointment in Haifa. Actually, on my calendar for next month, I have a sad reminder: A June 4, 12:30, lunch appointment with Sol in Haifa. Please share this note with your mother and sister.

Best regards,
Abraham Lempel

Email to Mitra

Dear Professor Mitra,

Sol Golomb was my idol and dear friend for almost 50 years. I first met him in 1967, at an IT workshop at the Technion, when I presented him with a proof of one of his conjectures re shift register sequences. At the time I was a fresh PhD at the Technion EE department and Sol invited me for my post-doc at USC which I was happy to accept. I arrived at USC in July 1968, shared an office with Lloyd Welch, and worked very closely with Sol for a year. Under Sol's influence and guidance I have traded my interest in network theory for the rich realm of digital sequences. Later on, Tuvi did his PhD under my guidance in this area. Since my year at USC, I've met with Sol at least once a year for lengthy discussions of a wide variety of topics. One of my happiest meetings with Sol was at the USC ceremony honoring another dear friend of mine, Andy Viterbi, with the announcement of the Viterbi School of Electrical Engineering. During the many years Sol was on the BoG of the Technion, we had an annual lunch appointment in Haifa. Actually, on my calendar for next month, I have a sad reminder: A June 4 lunch appointment with Sol in Haifa. Professor Mitra, I could go on for ever describing my memories of meetings with Sol, but I'm aware of the constraint of a short quote at this time. Please choose anything appropriate to this end from the above segment.

Best regards,
Abraham

Email from President of Technion to Abraham Lempel

President@technion.ac.il on 10/30/16 to Abraham Lempel

Prof. Solomon Golomb was a member of the American Technion Society (ATS) Southern California Board of Directors and the Technion Board of Governors: he was a member of the Academic Committee from 2003–2009, and a regular member from 2010 till he passed away in 2016. He was a frequent Technion visiting professor and a loyal board member. Golomb had a close relationship with both the Technion and the Los Angeles Chapter of the ATS. He also worked to strengthen the relationship between USC and Technion. Golomb was awarded a Technion Honorary Doctorate in 2011, "In tribute to your innovative contributions to the theory and practice of digital sequences, space communications and spread spectrum communications; in recognition of your long service in university teaching and research; and in gratitude for your many years of commitment to the State of Israel and your support for and cooperation with the Technion." Golomb was also a donor to several Technion funds.

2.   Contribution #2

On May 2, 2004, at 8:59 AM, Lempel, Abraham wrote:

Sol, Thank you very much for initiating my nomination for Associate Member of the National Academy of Engineering. I will pass this time. It is quite impossible to prepare a case for a nomination like this on such short notice. I'm looking forward to seeing you here in June.

Best Regards,
Abraham

Abraham, I got your message. Let's get an early start for next year. — Sol

Solomon Golomb on 4/23/14 to me Terrific. Let me suggest 12:15 p.m. I'll wait in the lobby of the Dan Carmel. Sol

On Apr 21, 2014, at 11:37 PM, Abraham Lempel wrote: Dear Sol, Great — we have a date, Saturday, June 14, for lunch. What time should I pick you up from Dan Carmel?

On Sat, May 23, 2015 at 4:51 AM, Solomon Golomb wrote:

Dear Avraham, I will attend the Technion Board of Governors meeting again this year. As before, I would like to have lunch Saturday, June 13, if you are available. Let me know. Shabbat Shalom, Chag Shavuot Sameach! Sol Golomb

Abraham Lempel on 5/23/15 to Solomon Golomb Dear Sol, Great! And, of course, I'm available. I'll pick you up at 12:30 at the Dan Carmel. Chag Sameach, Avraham

On Apr 5, 2016, at 10:39 PM, Abraham Lempel wrote: Dear Sol, Great! Looking forward to our lunch together on Saturday June 4. Is 12:30 a good time to pick you up from the hotel? Best wishes and Chag Pesach Sameach, Avraham

On Wed, Apr 6, 2016 at 3:06 AM, Solomon Golomb wrote: Dear Avraham, I plan to attend the Technion Board of Governors meeting again this year, arriving on June 2. Can you join me for lunch on Saturday, June 4? I expect to stay at the Dan Carmel again. All best wishes, and Chag Pesach Sameach, Sol Golomb

© 2023 World Scientific Publishing Company
https://doi.org/10.1142/9789811234378_0026

Golomb and Radar Waveforms

Nadav Levanon

1. Costas Arrays

The acquaintance between Solomon W. Golomb and John P. Costas probably happened thanks to their major contributions in Communications. Golomb with his Shift Register Sequences and Costas with his Costas Loop. This can explain why Costas approached Golomb sometime in 1980–1981 with a description of his sonar and radar waveforms with "... nearly ideal range-Doppler ambiguity properties."

Golomb grasped the challenge. It matched his vast expertise in number theory and his special interest in two-dimensional puzzles (e.g., polyominoes). Enlisting his colleagues at USC, Golomb responded very quickly. The two papers — Costas: "A study of a class of detection waveforms having nearly ideal range-Doppler ambiguity properties" and Golomb and Taylor: "Construction and properties of Costas arrays", appeared in two consecutive issues of the *Proceedings of the IEEE* [1, 2].

The radar community took its time, but the math community embraced the subject. A sort of competition evolved whose subject was enumeration of Costas arrays through construction algorithms and exhaustive search. The present state of the art appears in an IEEE dataport [3].

What quenched potential enthusiasm among the radar engineers was the complete contrast between Costas waveforms and the prevailing pulse-compression radar waveform — linear frequency modulation (LFM). In both signals the envelope is frequency modulated, but in a completely different way. In LFM, the frequency evolution is a single line connecting the frequency vs. time points (Fig. 1, red). In a Costas waveform it is the furthest from a single line (Fig. 1, black). Costas requires that each line connecting two consecutive points must be unique in slope, or length or both.

Note that in practice the frequency evolution of LFM is smooth and not in steps, as shown in Fig. 1.

Both waveforms produce pulse-compression, namely a narrow autocorrelation peak (recall that the autocorrelation is the response of a matched filter — the processor that yields the highest signal-to-noise ratio (SNR) on receipt). However,

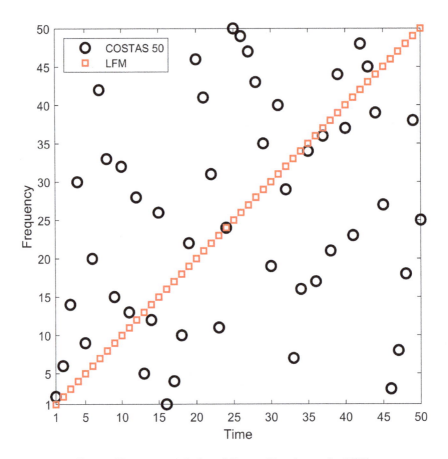

Fig. 1. Frequency evolution of Costas 50 and stepwise LFM.

the two signals differ drastically in what happens when the signal is reflected from a moving target that adds Doppler shift. The resulting response can be described by Woodward's ambiguity function (AF) [4]. AF is a two-dimensional function: t(delay) and v(Doppler). We usually are interested in its absolute value $|(t,v)|$. It has several important properties: (a) Constant volume under $|\chi(\tau,v)|^2$, (b) a peak at the origin, (c) symmetry with respect to the origin.

Because of property (c) it is enough to display only two quadrants of the function, typically the ones that represent positive Doppler.

To create a radar waveform based on Costas arrays it is necessary to define t_b the transmission duration of a frequency element. That automatically describes the frequency spacing Δf between neighboring frequency values, since $\Delta f = 1/t_b$. The interest from the math side is usually in the values of the AF at the grid points, namely at integer multiples of t_b and Δf. Costas waveforms of order N exhibit the unnormalized (left) or normalized (right) AF:

$$|\chi(mt_b, n/t_b)| = \begin{cases} N & m=0, n=0 \\ 0,1 & m \neq 0, n \neq 0 \end{cases} \text{ or } |\chi(mt_b, n/t_b)| = \begin{cases} 1 & m=0, n=0 \\ 0, 1/N & m \neq 0, n \neq 0. \end{cases}$$

(1)

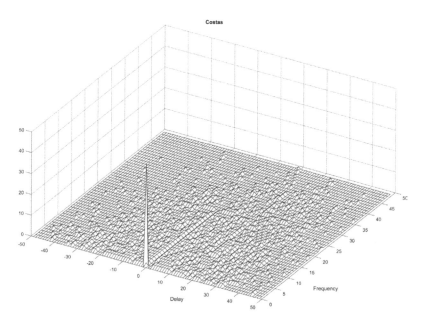

Fig. 2. DAF (unnormalized) of a Costas 50.

Equation (1) is called the digital ambiguity function (DAF). An unnormalized DAF of a Costas array of order 50 appears in Fig. 2.

For comparison, Fig. 3 displays the DAF of the LFM waveform shown also in Fig. 1. The "thumbtack" shape of the Costas 50 DAF was assumed to provide good resolution in both delay and Doppler. The LFM waveform provides the same delay resolution but poor Doppler resolution (known as Doppler tolerance). However, in modern radar, it is possible to maintain coherence over the duration of a long pulse train. Using the long coherent processing interval (CPI) is what provides the Doppler resolution.

Delays τ of target returns do not occur only at discrete values, like $\tau = mt_b$. Hence, the radar community is interested in the AF values between grid points; and there the AF sidelobes do not obey the results in (1). Figure 4 displays the zero-Doppler cut (i.e., autocorrelation) in dB, of the AF of the Costas 50 waveform. The delay span zooms on $0 < \tau/t_b \leq 2$ (the first two bits out of 50). Figure 4 shows that the autocorrelation sidelobes between the grid points cannot be ignored. The 1st null at $\tau/t_b = 0.02$ implies pulse compression of $N^2 = 2500$.

What keeps LFM the most common radar waveform, much ahead of Costas signal, is the option to use stretch processing in the receiver. In stretch processing the received signal is mixed with the transmitted signal, resulting in a narrow-band beat signal that allows analog-to-digital conversion at low sampling rate. However, as the civilian use of radar increases (e.g., in automotive), the interest in Costas signals may rise. The reason is diversity, needed to minimize mutual interference between many radars operating in close physical proximity. There is very little diversity in LFM (up slope vs. down slope). On the other hand, in some orders N the number of different Costas waveforms is in the hundreds.

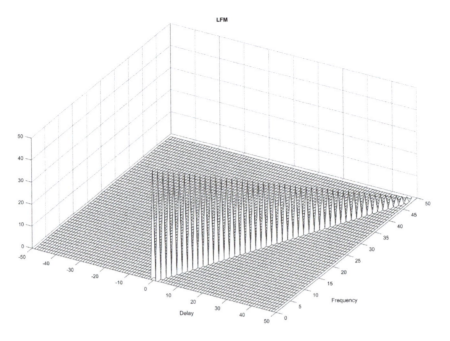

Fig. 3. DAF (unnormalized) of a stepwise LFM array of order 50.

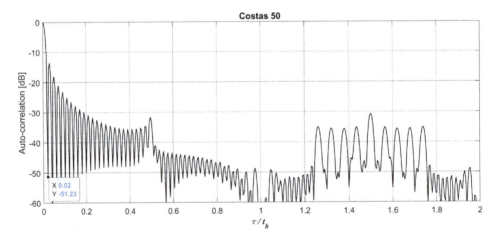

Fig. 4. Autocorrelation [dB] of a Costas 50 (zoom).

1.1. *Personal note*

Costas arrays prompted my acquaintance with Sol Golomb. I was among the first radar engineers who got interested in Costas arrays as radar waveforms [5,6,7]. So when I introduced myself to Sol in a seminar at the Technion in Haifa, he recognized the name and a discussion evolved followed by some correspondence. Sol's sister Edna Sharoni and her husband (who was her commander in the 1948

Fig. 5. Solomon Golomb in Jerusalem, 1991.

Israel's war of independence) lived in Tel Aviv, surrounded by a clan of children and grandchildren. That was an important motive for Sol to attend the annual meetings of the Technion's Board and probably also why he accepted Tel Aviv University's invitation to come for a short sabbatical visit in 1991. In the weekends we took trips, one of them to Jerusalem, where the photo in Fig. 5 was taken.

During that short sabbatical visit I proposed a radar research topic that he grabbed enthusiastically. Within one month he developed a complete theory of two-valued sequences that yield perfect periodic autocorrelation. Perfect periodic autocorrelation can be the key to a continuous wave (CW) radar waveform, free of delay (range) sidelobes. With my Ph.D. student A. Freedman, we studied the periodic ambiguity function (PAF) of that family of waveforms. Sol's paper and ours appeared, next to each other in 1992 [8, 9]. Correspondence on that subject continued after the sabbatical. An example appears in the insert below. More details on that contribution, will follow.

During Desert Storm (1991) I sent Sol a rather long letter describing our daily life under the rocket attacks from Iraq. Sol distributed it to some of his friends, including Andrew Viterbi. That started a brief correspondence with Viterbi, which led to a sabbatical at Qualcomm in 1995 and a few summers afterward. Staying in San Diego gave Sol and me more opportunities to meet. I especially recall a dinner at Sol's home, in which, for my sake, he invited Peter Swerling and his wife. Swerling is a radar legend, and talking with him on his early radar work at the RAND Corporation was fascinating. In my radar course, to this day, when I come to the chapter on the Swerling target models, I show my students a picture from that dinner (Fig. 6).

8 January 1992

Dear Sol,

In your paper on the "Two Valued Sequences" you referred to the 1989 paper by Rohling and Plagge

"Mismatched Filter ...", which I brought to your attention. Well, I may have tripped you. In the November 1991 issue of the IEEE Trans. on AES, there is a letter by the Russian scientist V.P. Ipatov, concerning Rohling's paper, and claiming that he did the same work back in 1977. Ipatov gives references to his papers in journals that are translated to English. I was able to find in Israel only the first one, and indeed he has done it before Rohling. Ipatov's letter, and a copy of his first paper, are enclosed. The paper is more general than Rohling's, and more in line with your paper. It does not preempt your much more comprehensive paper. As a matter of fact, had we known it before, the only effect would have been a reference to Ipatov instead of (or in addition to) Rohling.

If you will be able to find Ipatov's other 3 papers, I will appreciate it if you could send me copies. Last week I received the proofs of my paper, so I assume that yours, came and went back already. Regarding my later paper on the subject: "CW alternatives ..." of which I sent you a copy some time ago, It has been tentatively accepted with the condition that it be shortened to fit as a correspondence. One reviewer was critical about the practicality of a CW signal, which he interpreted as infinitely long. In the revision I show that when the reference is N periods long, the dwell time in any given antenna direction need not be longer than $N + 2$ periods, when there is no range ambiguity (i.e., the delay t to the furthest target is shorter than the period T), and $N + 2 + p$ periods, when there is range ambiguity of order p (i.e., the delay to the furthest target is pT $< t <$ $(p + 1)T$).

I am still very anxious to implement these signals in hardware. May be the appearance of the papers will get the attention of those who can afford it.

I hope you have enjoyed the holidays. Here we are very busy keeping our two boys happy despite their service in the army. That entails being ready, within a few hours' notice, with meals, laundry, and transportation. The fact that they managed without us during our three weeks in China, is completely forgotten.

Sincerely,
Nadav Levanon

Sol Golomb kept visiting Israel at least once a year, usually for the annual Technion board meetings. We always used his few days stop with his sister's family in Tel Aviv, to meet, chat and tour interesting places in town. One irregular visit took place in February 2012, when he was a member of a delegation from USC to Tel Aviv University. The picture in Fig. 7 was taken on that visit.

Fig. 6. Dinner at the Golombs' house. From left: Bo and Sol Golomb, P. Swerling, unknown, N. Levanon.

Fig. 7. Golomb visiting Tel Aviv University, February 2012.

2. Waveforms for Continuous Wave Radar

In radar, the probability of detection is a function of the average power, while the transmitter's hardware must support the peak power. In pulse radar the ratio between the peak and the average is usually very large. Pulse compression was developed to lower the peak-to-average ratio. In CW that ratio reaches the minimum attainable value of a unit. This is why most modern automotive radars use CW waveforms. To get range resolution from a signal that transmits continuously, requires modulation or coding. Surprisingly, periodic CW waveforms can achieve what a compressed pulse cannot — range response that is free of sidelobes. This valued property cannot be obtained by frequency modulation (e.g., periodic LFM) but only by phase modulation. For simplicity binary phase coding $(0^\circ, 180^\circ)$ would be preferred over polyphase coding. Unfortunately, only a Barker code of length 4 is a binary code that has this property. Such a short code does not provide useful pulse compression.

Golomb in his comprehensive 1992 paper [8] showed that his shift-register sequences, Legendre sequences, and sequences of a few other pseudo-noise binary families, can reach the property of periodic autocorrelation with zero sidelobes (perfect periodic auto-correlation, PPAC), by using two-valued, non-antipodal, phase coding $(0, \beta)$. The phase value of β (in radians) is determined by the code length N according to the simple expression

$$\beta = \cos^{-1}\left(-\frac{N-1}{N+1}\right). \tag{2}$$

Figure 8 displays the phase evolution of one period of a Legendre sequence of length 19, with the non-antipodal two-valued phase coding (top) and the resulted perfect periodic autocorrelation (bottom). Figure 9 displays the periodic ambiguity function resulting from processing coherently 16 periods of this waveform. The ambiguous Doppler ridge appears at a Doppler shift equal to the inverse of the duration of one code period. Its height drops as the code length increases. This is a property unique to this waveform family. In CW-LFM, a copy of the high peaks at zero Doppler will replace the low ridge.

To demonstrate the decrease in the ambiguous Doppler ridge for long codes we resort to a logarithmic [dB] response scale. Figure 10 displays the delay-Doppler response for a shift-register code of length 1023 with coherent processing interval (CPI) of 64 periods. For a code length of $N = 1023$ the ambiguous Doppler ridge typical height is $20\log_{10}(1/N) = -30.1\,\mathrm{dB}$. We used the phrase Periodic delay-Doppler response rather than periodic ambiguity function because the reference signal in the receiver was not identical to the transmitted signal, but was amplitude weighted by a Hamming window, in order to lower the Doppler sidelobes. Note also that the Doppler resolution improves as the number of code periods in the CPI increase. The 1st null in the Doppler cut of the PAF happens at the inverse of the CPI. It widens a little if amplitude weighting is used on receipt.

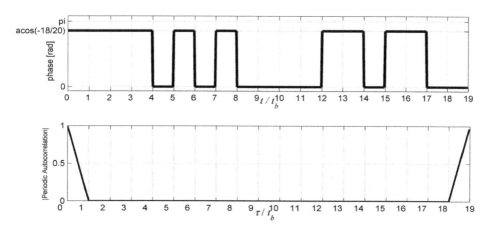

Fig. 8. Legendre sequence of length 19. (Top) Non-antipodal two-valued phase coding. (Bottom) The resulting perfect periodic autocorrelation.

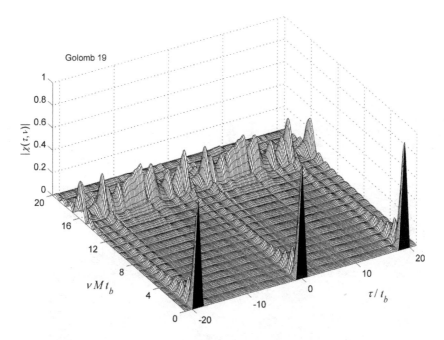

Fig. 9. Periodic ambiguity function of 16 periods of the signal in Fig. 8.

Field trials with a 1023 Golomb waveform were described in [10]. To obtain good velocity (Doppler) resolution the CPI contained 2048 code periods. The observed scene (Fig. 11, top) shows four cars approaching on a road. A guardrail obscured the wheels of the more distant cars (# 2, 3 and 4), but the wheels of car #1 were visible to the radar. The radar's range vs. range-rate display of the scene (Fig. 11, bottom) shows the reflections from the four cars, demonstrating sidelobe-free, high resolution, range response. The excellent Doppler resolution is revealed in the returns from car #1, where the micro-Doppler returns from the rotating wheels

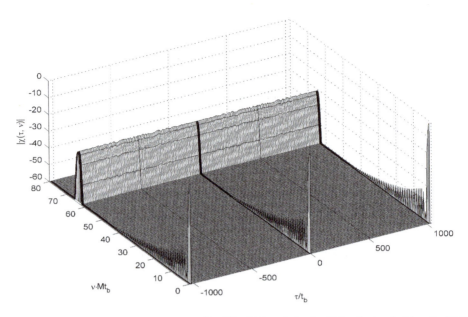

Fig. 10. Periodic delay-Doppler response (in dB) of 64 periods of a Golomb signal of length 1023. Hamming weighted reference.

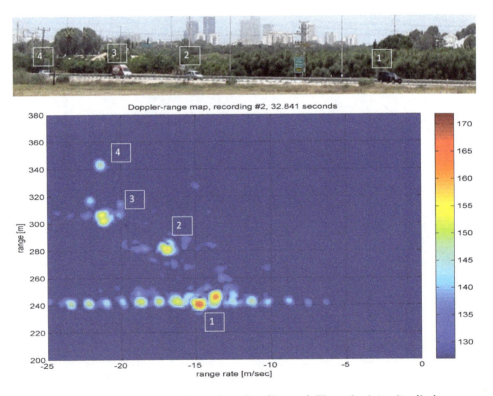

Fig. 11. (Top) The scene observed from the radar. (Bottom) The radar intensity display, range vs. range-rate.

are clearly seen on both sides of the car body returns (two strong scaterrers). In a moving car the circumference of the wheel exhibits forward velocities stretching from zero velocity to twice the car speed.

Golomb's shift register sequences are available at lenghs $N = 2^n - 1$, $n = 1, 2, \ldots$ These lengths are rather sparse, but at each of these lengths, there are several distinct codes. Distinct excludes trivial variations like flipped direction or flipped polarity. A more dense availability is found in Legendre sequences of any length N that is a prime number that obeys $N = 4k - 1$, $k = 1, 2, \ldots$ However, at each of these lengths there is only one distinct Legendre code.

The competition between LFM and pseudo-noise (PN) phase or frequency-coded signals exists also in CW radar. Here too, LFM allows simpler stretch processing while PN coded waveforms provide more diversity. However, in CW, LFM has one more advantage. LFM is Doppler tolerant. Range performances do not degrade in high range-rate cases. On the other hand, processing PN phase-coded waveforms in high Doppler situations becomes more difficult. In low duty-cycle pulse waveforms, matching Doppler-shifted returns are performed by adding inter-pulse phase ramps. This is known as slow-time Doppler compensation. In high duty-cycle and especially in CW (duty cycle = 100%) Doppler compensation is required also within each period (fast time Doppler compensation). If the need for diversity increases and the cost of signal processing drops, CW pseudo-noise phase coding will gain popularity.

Conclusions

Golomb's two most major contributions to radar waveforms are (a) the comprehensive study and development of construction algorithms for Costas arrays and (b) the adaptation of non-antipodal two-valued phase coding to pseudonoise sequences like shift register sequences (his own) and Legendre sequences. Contribution (a) provides diversity and high compression ratio to pulse radar systems. Contribution (b) creates many diverse periodic CW radar waveforms with zero range sidelobes. Presently, the dominance of LFM in both pulse and CW radars obscure the practical value of these contributions. However, the need for diversity in radar waveforms may bring them up front.

Bibliography

1. Costas, J.P. A study of a class of detection waveforms having nearly ideal range-Doppler ambiguity properties, *Proceedings of the IEEE*, **72** (8) (1984) 996–1009.
2. Golomb, S.W. and Taylor, H. Construction and properties of Costas arrays, *Proceedings of the IEEE*, **72** (9) (1984) 1143–1163.
3. Beard, J.K. Costas arrays and enumeration to order 1030, IEEE Dataport, 2017 [Online].
4. Woodward, P.M. (1964) *Probability and Information Theory, with Applications to Radar*, 2nd edition, Oxford, Pergamon Press.
5. Freedman, A. and Levanon, N. Any two N*N Costas signals must have at least one common ambiguity sidelobe if $N > 3$ — a proof, *Proceedings of the IEEE*, **73** (10) 1530–1531.

6. Freedman, A. and Levanon, N. Staggered Costas signals, *IEEE Transactions on Aerospace and Electronic Systems*, **22** (6) (1986) 695–702.
7. Levanon, N. *Radar Principles*, New York, J. Wiley and Sons, 1988, Ch. 8, 145–152.
8. Golomb, S.W. Two-valued sequences with perfect periodic autocorrelation, *IEEE Transactions on Aerospace and Electronic Systems*, **28** (2) (1992) 38x–386.
9. Freedman, A. and Levanon, N. Periodic ambiguity function of CW signals with perfect periodic autocorrelation. *IEEE Transactions on Aerospace and Electronic Systems*, **28** (2) (1992) 387–395.
10. Cohen, I., Elster, R., and Levanon, N. Good practical continuous waveform for active bistatic radar, *IET Radar Sonar Navigation*, **10** (4) (2016) 798–806.

© 2023 World Scientific Publishing Company
https://doi.org/10.1142/9789811234378_0027

Funeral Oration — Sol; and A Tribute to Dr. Solomon Golomb*

by C.L. Max Nikias

Today we mourn the passing of one of the most brilliant lights in the American academy. One of the intellectual giants of the past generation. And one of the most cherished and beloved members of our professoriate. Today, we grieve the loss of a man who had not only an impact on the scientific community as a whole, but a profound effect on the academic community that he called home for so long.

For more than 50 years Solomon Golomb cared for the University of Southern California as deeply as it is possible to care. And for more than 50 years he served it as magnificently as it is possible to serve. And yes, he loved Trojan football. He expended every atom of his immense genius and every drop of his enormous kindness towards lifting up his university and the students and faculty within it. Sol embodied academic excellence and he lived each day for the purpose of building up such academic excellence in others. For the past quarter-century Sol was one of my own most beloved and inspiring mentors and advisors. While serving on the Viterbi School faculty and then in each one of my university leadership positions. But today as USC president I salute him, not simply as a mentor or friend, not simply as one of our very best faculty, but as one of the most distinguished and consequential scholars in the 136 year history of our University. We can most fully appreciate the career of Sol Golomb when we see how dramatically it intersected with the rise of USC and the advent of a new era of communications unlike anything ever witnessed before by humanity.

Solomon Wolf Golomb's arrival at USC in 1963 heralded USC's arrival as a leading American research university. This intellectual titan was a mathematician at heart. But more than that, he was a true polymath, dazzlingly gifted in engineering and

---

*The "Funeral Oration — Sol" was originally delivered by USC President C.L. Max Nikias at the funeral of Solomon Golomb at Mount Sinai Memorial Parks and Mortuaries in Hollywood Hills on May 4, 2016. "A Tribute to Dr. Solomon Golomb" was delivered by C.L. Max Nikias at the Celebration of Life of Solomon W. Golomb held at the Town and Gown ballroom at the University of Southern California in Los Angeles on January 31, 2017.

263

communications while also grounded solidly in the humanities, languages, and religion. He had an exceptional ability to see connections between fields of study and an intense need to apply the solutions of one, to the problems of the other. And from his first days at USC, he was instrumental in drawing other top scorers and innovators into USC's orbit. He led the recruiting of scores of world class faculty who swiftly made USC a leader in global communications and the digital media revolution. He not only helped bring the best talents to USC, he helped them to shine on the national stage. He was pivotal in successfully promoting dozens of his faculty colleagues for election into the National Academies of Science and Engineering. This was yet another way in which he placed USC on the map in perpetuity as a world-class institution.

He loved teaching, especially undergraduate courses. He did it until his last breath. "Teaching doesn't really exist," Sol once said. "What exists is learning." And we all learned so much because of him. Not the least of which was his book on the polyominoes which laid the foundation for the wildly popular online game Tetris. Born with a superhuman ability to memorize numbers, Sol was using obscure mathematical number patterns that he called "shift register sequences". The concept opened the world of digital communications unleashing explosive social change over the next 50 years. Take a moment to consider how our world has changed since 1963. The cell phone, the internet, GPS, all use shift register sequences. For the tens of thousands of students and faculty who came to know him, he showed us how staying curious was the key to life. But all of humanity owe him a debt too, with his ideas as a base, today we have Twitter, Facebook, keyless ignitions, and remote home security.

With his bushy white beard, barrel chest, and penetrating eyes, he was one of the most recognizable faculty members on our campus. I remember a group of us walking together near engineering and Sol was talking about something when he suddenly said, "that's it, that's the answer." I can assure you, none of us had any idea what he was trying to solve, but that was Sol. He was always dressed in a suit, it seemed, and he was always thinking. In 2011 when President Obama awarded him the National Medal of Science, Sol reflected a little on his career. "I'm proud that I lived long enough to see so many of the things I have worked on being so widely adopted that no one even thinks about where they came from," he said.

Throughout his life Solomon Golomb proudly drew upon the values of his Jewish heritage allowing him to care in the most profound ways for his academic community, for America, for Israel, and for the global community. That heritage and those values fueled his desire to promote tolerance, human dignity, diversity, and a universal respect for people from all walks of life. I will always remember the spirit of passion and the spirit of play that drove this extraordinary man and his superhuman intellect. You could raise any subject under the sun and Sol could hold forth on it with authority for 20 minutes or more. And once, I honestly tested his knowledge, perhaps unfairly, in an area in which I presumed to have an advantage. On the genealogy of the royal family of Greece. But not only did Sol know the names and the genealogy of the kings, he had a perfect recall of even such details

Funeral Oration — Sol; and A Tribute to Dr. Solomon Golomb

as the dates of their weddings. Yet, yet, we can all be lastingly grateful that this commanding intellect was devoted to a purpose to the life of the academy and to the betterment of our university. His genius, his curiosity, and his character, cast new intellectual sparks in every direction over this past half-century, helping pave the way along numerous new academic frontiers.

For virtually his entire career, Solomon Golomb gave his very best to USC and his very best is almost beyond description. But as the beneficiary of his very best for more than half-a-century our university is infinitely better and it is enduringly richer. That is why, I have the privilege of announcing to you today the creation and establishment of the Solomon W Golomb Chair in Communications at the Viterbi School of Engineering, to celebrate his memory in perpetuity. And so while we grieve the departure of our dear friend, this most beloved member of the Trojan family, we know with certainty that his legacy will endure. And we know it will nourish this university and higher education, for decades and generations to come. Thank you.

Funeral Oration — Sol; and A Tribute to Dr. Solomon Golomb

by C.L. Max Nikias

Solomon Golomb
Celebration of Life
January 31, 2017
By C. L. Max Nikias
USC President

Good afternoon, everyone.

It's a privilege to be here as we pay tribute to the legendary life and lasting legacy of University and Distinguished Professor Solomon Golomb.

We are here to celebrate the extraordinary impact of a pioneering professor, a cherished colleague, and a magnificent mentor who meant so much to so many members of our academic community.

Solomon Golomb was much more than simply a member of USC's faculty.

He was one of the most incredible intellects in the history of the American academy.

He was one of the great geniuses of his generation, a titan in the world of mathematics and technology, a courageous explorer who ventured beyond the boundaries of engineering, mathematics, and digital communications. He was a genuine polymath.

And he was also one of the most treasured members of the Trojan Family. A little secret – he loved the USC football team.

A revered researcher who advanced the great body of human knowledge, who enhanced our academic excellence, and who enriched our global society.

*** PAUSE ***

For more than half a century, Sol stood as a symbol of profound passion for the life of the mind and the life of our university.

The entire trajectory of our history changed the moment he chose to bring his brilliance to USC.

Over the decades, he dedicated his energy and every moment of his time to enhancing and elevating our academic excellence, helping so many of our faculty get elected into national academies.

Funeral Oration — Sol; and A Tribute to Dr. Solomon Golomb

His deep dedication earned him the highest esteem from his faculty colleagues, and the greatest respect from students and scholars from diverse disciplines.

In this way, his impact stretched across our campuses and reached around the world.

I know how important Sol was to USC because he also played a pivotal role in my own life.

When we served together as colleagues at the Viterbi School of Engineering, he was one of the first to welcome me, generously offering his astute advice and his wealth of wisdom. I still remember my interview with him in 1990!

He quickly became one of my most trusted and valued advisors, serving as a memorable mentor and a faithful friend for 25 years.

*** PAUSE ***

When I became president, I gained a new perspective on Sol's privileged place in our academic community.

When he joined our faculty in the early 1960s, USC was still in the early stages of its transition from regional eminence to national prominence.

Sol arrived on campus six years before USC joined the prestigious Association of American Universities.

And he appeared at the dawn of a new day for digital communications, an industry where his imagination and innovation would lead the field into exciting new frontiers of possibility and discovery.

*** PAUSE ***

Beyond the confines of our campus, his brilliant innovations shaped our lives and swept the world.

His concept of shift register sequences unleashed a revolution in digital communications.

His work formed the foundation for the cell phone, the Internet, and GPS.

His ideas led to popular products that have dominated popular culture, such as Tetris and Twitter.

And his research set the stage for the first clear images that streamed through space from a mission to Mars.

*** PAUSE ***

With all of his extraordinary accomplishments, Sol naturally earned considerable acclaim.

He received so many honors and awards that I will only hit on the highlights.

At USC, he was presented with our highest honor, the Presidential Medallion, as well as the rare distinction of being named both a Distinguished and University Professor.

Outside of our university, he was awarded the Benjamin Franklin Medal in Electrical Engineering. He was very proud of that.

And, of course, we were filled with pride when the President of the United States presented him with our nation's highest honor for scientific innovation – the National Medal of Science – in a festive ceremony at the White House.

Although Sol is no longer with us in person, he is still with us in spirit.

You can see his impact and his influence all around us, across the university.

You can see it in the countless careers here at USC and across the academy.

You can see it in the scores of the finest faculty he helped recruit to USC, acting as the architect of a global revolution in communications and digital media.

You can see it in the pioneering professors he guided and mentored, encouraging them to stretch their minds and shape their fields of study.

You can see it in the students and scholars he taught and trained, whose lives were improved by his profound wisdom and his great generosity.

While we are saddened Sol's extraordinary journey at USC has come to an end, we are very grateful that we gained so much.

His incredible impact as a professor and as a person will always be felt. . . and his legacy will always endure.

And we will carry the memories of his legacy in our hearts and our minds.

Thanks to his immense intellect and his boundless energy, USC's academic excellence has forever been elevated.

Thanks to his kindness and compassion, our lives have been greatly enriched.

And we will always remember the gentle man with the brilliant mind and the big heart, the true polymath, who will stand proudly as one of the great intellectual giants in the history of our beloved university.

USC President C. L. Max Nikias honors 2013 National Medal of Science recipient Solomon Golomb

President Nikias, Solomon Golomb, and USC Viterbi School of Engineering Dean Yannis Yortsos

President Nikias and Solomon Golomb at the USC President's annual Faculty Address in 2015

President Nikias honors Professor Golomb with a book at the 2013 National Medal of tScience event.

© 2023 World Scientific Publishing Company
https://doi.org/10.1142/9789811234378_0028

Memories on 40th Anniversary of Bo and Sol: May 18, 1996

Robert Rosenstein

Below is something written by now-deceased "Uncle Bob" Rosenstein, who was with Sol when he met Bo in Copenhagen, in 1955. It was written for Sol's and Bo's (May 18) 40th anniversary.[1]

Well, here goes:

The first time I visited Sol and Bo they had already moved into their house in La Cañada. It sat on the side of a hill opposite a mountain. I gathered that it had been built by a newspaper publisher named Descanso who had donated a large public garden to the city of Pasadena, and I seem to recall a rose qarden at the front of house sloping down towards the road. I do remember thinking that it wouldn't take too much longer for them to have been in every room in the house.

Apart from two beautiful, intelligent, active, trusting, and affectionate daughters, there must have been a small menagerie of pets, among them a huge hairy dog called Blue Boy for his blue eyes, a family of cats, and perhaps a chicken. Blue Boy, whose name I thought somehow would have been more appropriate for a pig, seemed content to being wrestled with by the children from time to time when he wasn't otherwise reigning over the rugs and basking in the sun.

Bo had found a local artist to paint some airy murals based on snapshots, recognizably portraying the family and their environs, and in the spirit of certain early painters and manuscript illuminators, repeating the same subjects obliviously posing in different parts of the same picture.

One day, I doubt if it was that particular visit, I came upon Bo and Sol in front of their house laboring over a boulder, converting it into a rune stone in the spirit of their Nordic ties. (Now I really begin to ramble on.) While in Sweden, which of course has a good supply of genuine rune stones, I had dabbled in the subject instead of concentrating on my degree, and so was acquainted with some of the more elementary aspects, for example, that the letters of the runic alphabet, the FUTHARK, were in a different order than ABC (or even QWERTY, not to speak of ETOIN SHRDLU), and that their shapes in Scandinavia were not the same as in England, as well as depending on when they were cut.

272   *The Wisdom of Solomon*

The most relevant precedent for their enterprise was obviously the rune stone found in Minnesota during the last century, the era of P.T. Barnum. (This has nothing to do with Garrison Keilor.) I had read a few papers about the dispute over its authenticity, one by a favorite professor I had studied French with at Cornell University, Robert A. Hall, Jr, another by the author of the Norwegian textbook used at the Summer School at Oslo University when I was there together with Sol, Einar Haugen. There were of course other critics whose names escape me. The stone, called the Kensington Runestone (see the Web site, which has a nice color picture) had been unearthed in a farmer's field, like the Cardiff Giant, and had been accepted as genuine by as many believers as UFO sightings are today.

The rune stone purported to commemorate an expedition by some Vikings who had run afoul of hostile Red Men. It is evidently now a monument, and Hall vouched for it, although it was distinctly out of his field of expertise. (Sol knew the author of a monograph supporting the claim, and was willing to side with him. As he is usually, nay, invariably right, that too should be part of the story.)

Haugen refuted the claim with a volley of arguments, casting doubt on the consistency and appropriateness of the language, and accusing the farmer whose land the artifact was found on of outright forgery and concocting a fraud. This was a Swedish immigrant who dabbled in antiquities, although he had stayed discreetly clear of the find, deferring to his son, the actual discoverer. Haugen even scoffed at the premise that Vikings were inherently explorers. Most appositely, he queried the softness of the stone, an unusual choice for a Viking marker.

(Now back to the subject.)

This objection was especially evident in the rock Sol and Bo were toiling on, which was satisfactorily hard and resistant. With the tools they had at hand they had barely scratched the surface when they realized it was better to postpone the job until they could apply "more power", as Tim the Toolman would put it. Sol told me that they ultimately did so, and so their rune stone now unquestionably does have a verifiable provenance, wherever it wound up, another roadmark on the information highway.

(Sorry for the delay, and it doesn't even rhyme: 3rd slightly revised edition.)

Note

1. 'Uncle Bob' (not actually related) is seen in the background in the photo of Sol and Stephen Wolfram on page 8 of the Part II, Golomb "A Career in Engineering." The photo was taken at the home of Sol's daughter Beatrice Golomb and son-in-law Terrence Sejnowski in Solana Beach. Uncle Bob, who was also undertaking studies in Norway at the time, was with Sol in Copenhagen when Sol first met Bo, and Bob remained a lifelong friend of both. Bob died on Feb 19, 2017 at age 94.)

© 2023 World Scientific Publishing Company
https://doi.org/10.1142/9789811234378_0029

From Professor Solomon Golomb's Math PhD Advisee to His Stock Market Guru; & Orbits and Calendrics*

Stephen Schloss

Beatrice Golomb: I knew of Stephen Schloss as Sol's longtime Merrill Lynch stock-broker — but it turns out that he had first been Sol's PhD student. He had had a master's degree in mathematics from Harvard, and had finished a PhD with Sol, after his previous advisor, Bellman, became too ill. Schloss requested that he contribute by phone, and that I write this out; he has graciously indulged my insertion of occasional comments.

1. Sol was my PhD mentor

My PhD with Sol was the best time in my life.

Professor Richard Bellman (at USC in biomathematics), for whom I was a Research Associate, had told me about Sol. Sol was the one professor that Bellman highly respected at USC. He asserted that you should always take any course given by Sol Golomb. I followed his advice.[1]

Four years later I decided I might want to do a PhD, if possible, with Sol. In 1974, I made an appointment to meet with Sol to explore studying with him for a doctoral degree. What a wonderful beginning for a fascinating, wide-ranging intellectual adventure that became.

---

*This is based on conversations on 10-17-2019, 12-17-2019, 12-30-2019, and subsequently revised and edited including by Dr. Schloss.

[1]Comment from Beatrice Golomb: Prof. Robert Gray, Electrical Engineering, Stanford, shared a similar reflection after Sol's death. "When I went to USC in 1966 for my PhD, my MS thesis advisor (Irwin Jacobs) told me that Sol was a genius and he suggested I take every course Sol taught, regardless of the topic. He also advised me that it would be a good idea to learn the material first. It was great advice and for three years I learned about nonlinear shift registers, pseudo-random noise generators, the Möbius mu and Euler phi functions, the Möbius inversion formula, counting the number of distinct necklaces with n beads in k colors, polyominoes, synchronized lossless codes, and other gems that still hold an honored place in my few class notebooks that remain in my bookshelves along with his books (shift registers, polyominoes, information theory)."

He asked me about my mathematical background. As I was talking, I could tell Sol was taking mental notes. Three years later when I was preparing to take my PhD orals, he told me that at the beginning of the exam he will ask me to tell the committee about my mathematical background, namely, with whom I had studied and what were the subjects.

Schloss's fascinating educational path, culminating in a PhD with Sol and Lloyd Welch

I had a bachelor's degree in mathematics from the University of Minnesota. Then I went to Harvard. Because I was not sure I had such a good background, if I wanted to get into a PhD program in math, I determined to start in a different department, and take math courses to show I could do the work. I started in the Department of German Literature. (I was born in Germany, and went to the US at age 5.) (*BG: So, this would have been ∼ 1936, in time for his nuclear family to escape extermination at the hands of Hitler.*) (Schloss was born one year before Sol.)

I was in the Department at Harvard for a few months and they had a meeting to recommend people for fellowships to study in Germany. They had recommended me for a Fulbright but I wasn't that interested. There was a von Humboldt Scholarship, but they gave out about 20 of them. And a nomination for the fellowship of the Germanistic Society; if I get that I'll have to go, there's only one of them given out in the country. (I got it.)

These, I told Sol, are the noted mathematicians I studied under:

- Emil Artin (Modern Algebra) at Göttingen University. His lectures were immortalized in the classic *Moderne Algebra* by B. L. van der Waerden published in the Springer Verlag's famous series on the foundation of mathematical science, dubbed the "yellow devils" by math students.[2]
- Max Deuring (Group Theory) at Göttingen University.

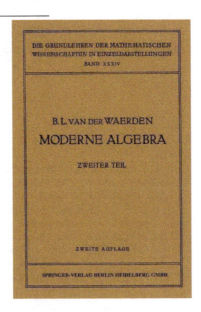

From Professor Solomon Golomb's Math PhD Advisee to His Stock Market Guru 275

- Carl Ludwig Siegel (a "mathematician's mathematician" lured back after the war to Göttingen, where he had been until 1940, from Princeton) at Göttingen University. André Weil, see below, named Siegel as the "greatest mathematician of the first half of the 20th century."
- Oscar Zariski (Algebraic Geometry) at Harvard University. Has been called "One of the most influential algebraic geometers of the 20th century."
- Richard Brauer, (Non-associative Algebra): I did a reading course under him personally, at Harvard University.
- Professor Lars Ahlfors' (Riemann Surfaces) reading course at Harvard University.
- Then I was a grader for Professor David Widder at Harvard University.
- S. S. Chern (Seminar in Topology) at University of Chicago, a world-renowned topologist, who has been called the "father of modern differential geometry."
- André Weil (Siegel Spaces), founder of (and the organizer of the first meeting of) the Bourbaki School of Mathematics. (On Bourbaki see below.)
- Herbert Busemann at USC (Differential Geometry).

Aside on the Bourbaki School: This was a group of primarily French mathematicians named after a fictitious person Nicolas Bourbaki, initial members of which were associated with the École Normale Supérieure in Paris, who took it upon themselves to attempt a systematic revision of all of mathematics. Although, in fact, a number of fields — like probability and combinatorics, the latter dear to Sol — were omitted from their treatment.

(According to Wikipedia, the name "Bourbaki" refers to a French general, Charles Denis Bourbaki; it was adopted by the group as a reference to a student anecdote about a hoax mathematical lecture, and also possibly to a statue. It is said that André Weil's wife Evelyne suggested using "Nicolas" as the first name instead of Charles. This story was more or less confirmed by Robert Mainard.)

Sol and I had a number of hilarious discussions about the Bourbaki movement. Their books had no illustrations. They believed in abstract intuition and spurned concrete illustration. If they discussed a geometrical construction, they would not show it. While Sol and I were at Harvard, it was the peak of the movement, which had limited success, as no Professor would adopt their texts.

As Sol noted, all this abstract math had its origins in application, via people like Carl Friedrich Gauss, a mathematician and physicist, who found their inspiration for mathematical topics from physics. I told him that while I was a graduate student in Mathematics at Harvard, I took a course in Solid State Physics. I didn't dare mention it to anyone in the math department, because they looked down on studying anything applied at that time!

It was those trained in the Bourbaki school that were the motive force behind the "new math" so ably parodied by Tom Lehrer. (*BG: Lehrer was a song humorist and Harvard mathematics lecturer. "It's so simple, so very simple, that only a child can do it."*)

Its tenets were diametrically opposed to those of Sol, who understood that concrete examples aided development of abstract intuition, and who understood and valued the fact that even "beautiful" mathematics could (and in Sol's case, would) be found to have applications. Furthermore, problems in physics, communications, biology, etc. provided motivations to solve the associated mathematical problems.

I took Algebraic Topology from Professor Andrew Gleason, who at that time displayed no interest in applications and certainly not in the use of computers. However, years later when my son went to Harvard, Professor Andrew Gleason was the one who suggested that all undergraduate students at Harvard should learn programming. This shows how far the pendulum had swung.

Fast forward, all of this broad mathematical background aided me in my PhD research but also in easily defending my thesis in the PhD oral exam. I became a PhD student of Sol in 1974. When it came to my PhD orals, Sol figured how I could quickly get through it without a single question from the Committee. It is standard to prepare a printed folder that you hand out to the committee at the beginning of the exam that contains a title page with your name, a page with a summary of the dissertation and a biography. Sol said on the second page list each of my graduate school professors and the courses I took from them. The Committee will recognize who they are.

Sure enough, when I went through whom I studied under and their significance outside of EE (the PhD was in EE Communications under Sol and Lloyd Welch — though it was all theorems and proofs and mathematics, Sol had his office in EE), not a single member of the Committee asked me a question. Other than Lloyd Welch and Sol, none knew what algebraic topology was — one advantage of having studied graduate mathematics for many years before I finished up, that proved very useful.

Back to my meeting in 1974 with Sol, he asked about my background, who I'd studied under. Sol asked, "In what area would you like to get your PhD?" In advising students what area they should pursue, Sol would say, "Do what you would enjoy the most." This made working wonderful, because one was excited to go to work.

I said, "Sol, you know my background, what would I be suited for?"

Sol replied that there is a new type of error correction code by a Russian named Valerii Goppa. He has written five papers and they are all in Russian but this circumstance should not pose a problem.

(*BG: As Schloss told the tale, he paused, then laughed, to let this sink in.*)

Since, Sol himself had such facility in languages, I supposed that Sol underrated the obstacle for me to read Russian. But not really. Sol had such confidence in one that it was very inspiring, and I got a Russian dictionary — and of course, mathematics has its own stylized language. It was so nice, he showed such confidence in me, and this whole adventure, I had such a pleasurable experience working under Sol and Lloyd Welch.

At that time in 1974, USC had added remote classrooms in several locations with television links to the main campus. One was close to my office near the LA airport. USC was very early to do remote courses.

My first course to prepare myself for my thesis research in information theory was a course in algebraic coding theory. Dr. Robert Scholtz was teaching the course. I took that course, remotely, over television. That was very interesting. (*BG: I mentioned that Scholtz retired this year — as we had this discussion — last of the "magnificent seven.")*

I went to the USC classroom building by the airport, which had several rooms for different courses. I was the only one taking the course on algebraic coding theory taught by Scholtz (at that location). The text for the course was an early draft for a book Sol was writing with Bob Scholtz (two-volume book on information theory). This course was the first chapter of the book.

I came into the room, there was the TV, I'm trying to figure out how to operate everything. Dr. Scholtz says "Can anybody tell me what a mathematical group has to have." So, I find the microphone, nobody in the classroom on campus is answering. I press the button, my voice booms out: "It has to have at least a left inverse." Dr. Scholtz said: "Yes, but lower your voice. What else?" I said "And it needs to have a right identity." I'd taken group theory at Göttingen and a reading course under Richard Brauer at Harvard. I gave the weakest, the minimal requirements for a group. If it has a left inverse, we can prove it has a right inverse and can then show they are equal. Similarly, if it has a left identity, we can prove it has a right identity and that the two are equal. A typical textbook, simply assumes an identity and an inverse.

The above approach is from *Lehrbuch der Gruppentheorie* by H. Zassenhaus 1937.

Over the next week, I started making some notes on my copy of the draft of the chapter, where something was unclear etc. I gave Sol my annotations. Sol said: "If you do this for the whole book, I'll give you full credit in Information Theory, and pay you." It was a nice, fun experience for me.

So, all the years of graduate school were paying off.

By the way, someone had told me early in my graduate career: "Seek the master, seek the best, study with the teacher not the student."

So, I did.

2. Schloss Meets Sol in 1970

In fact, however, I'd met Sol four years earlier — and our paths and interests had crossed before that...

I first met Sol at the home of a mutual friend, hematologist Dr. Michael Rubinstein and his wife Vera. Sol arrived toward the end of dinner. It is a Jewish tradition that if there are at least three men at a meal, they should say grace at the end of the meal (*"Birkat Hamazon"*). Rubinstein asked Sol to lead, which he did all by heart without looking at the book — which impressed me. This was around 1970. That evening was like a real European salon type meeting, starting with Dr. Rubinstein's son playing classical piano on their Steinway Grand in the living room followed by good conversation. Sol asked what I did. I said I'd gotten a master's degree in mathematics from Harvard, then went to North American Aviation, which at that time was the prime contractor for the Apollo Lunar program.

278 *The Wisdom of Solomon*

I was the Supervisor for Guidance and Navigation Analysis and applied Kalman filtering (see below) to determine, among other things, the fuel requirements for the Delta V corrections for the flight to and from the moon.

As I mentioned earlier, Professor Richard Bellman had told me I should take any course that Sol teaches — so I had heard of Sol before that first meeting.

We discovered that evening that our educational and career paths crossed many times over the previous 15 years — first at the Mathematics Department of Harvard University where we both did our graduate work and became associated with the same professors. We also talked about Professor David Widder, who was Sol's PhD advisor and for whom I was a grader for his Advanced Calculus course. Thus, began a unique, magical for me, friendship, rich in intellectual interchange that lasted 46 years!

Schloss becomes Supervisor Level II of Apollo Lunar Mission Guidance and Navigation

Before continuing to describe the discussions, I had with Sol, let me tell how it came about that I suddenly began to play an important role during the infancy of the manned space program. In 1962, I received a telegram from Curt Zoller, the Manager of the Guidance and Control Department of North American Aviation, the Prime Contractor of the Apollo Lunar Mission offering me a fine position. As soon as I came on board, Curt called me in his office and said that there are plans to apply Kalman filtering to the navigation in the various phases of the lunar mission: the orbital phase, the translunar phase, the lunar orbital phase, etc. This work was being done by Stanley Schmidt at NASA Ames and Richard Battin at MIT Instrumentation Lab.

First who was Kalman? He was Rudolph (Rudy) E. Kálmán, a Hungarian-American electrical engineer and mathematician at that time at the Research Institute for Advanced Study (RIAS), the research part of Glenn L. Martin Company in Baltimore, where Solomon Golomb spent some time each year during his Harvard Graduate School years 1951–1955 culminating in his producing the classic book on the theory of shift register sequences.[3]

Kalman had written his main paper, "A New Approach to Linear Filtering and Prediction Problems," *Transactions of the ASME-Journal of Basic Engineering*, 82 (Series D), pp. 35–45, 1960. This paper applied the work of Norbert Wiener to dynamical systems with a state space formulation.

---

[3]Another prominent member of RIAS at that time was Professor Solomon Lefschetz, a founder of the mathematical disciplines of Algebraic Topology and Algebraic Geometry. He was recently retired from Princeton. I had met him in 1959 at the American Mathematical Society meeting in Chicago. Lefschetz results in Algebraic Geometry, which I had studied under Oscar Zariski at Harvard in 1957, proved useful years later for my dissertation on error correcting codes for Sol Golomb. All in all, the many courses in very abstract mathematical topics which I had studied at Harvard, Göttingen, Chicago and USC, curiously found application in much of my engineering work as well as for my dissertation, which was highly abstract mathematically with theorems and proofs, characteristic of a mathematical dissertation, and yet has many applications in electrical engineering, particularly in communications, computer processors, and even in the creation of high quality musical recordings.

Sol noted at the time that Kalman filtering iterative estimation process on n measurements is equivalent to using the escalator method of inverting the n by n information matrix, which is done one row and one column at a time in determining the Maximum Likelihood Estimate of the state vector from n observations.

A Kalman filter is a minimum variance estimator. In the space navigation application, you begin with an estimate of the six-dimensional state vector of the spacecraft at time t which is the deviation from the reference trajectory, namely, delta $x(t) =$ (delta position, delta velocity) and then apply the transition matrix to find a preliminary delta $x(t+1)$. Then incorporating a radar measurement consisting of range, range rate, and angles and an updated covariance error matrix as a weighting factor you determine an optimal estimate of the delta $x(t+1)$.

When we discussed this subject, Sol pointed out to me that this iterative process of incorporating a single measurement at a time has an interesting relationship to the traditional estimation process of taking n measurements $(t, t+1, t+2, \ldots t+n)$ and producing a minimum variance estimator at time $(t+n+1)$. In the latter case, you invert the n by n information matrix. Sol said that the Kalman estimation process is equivalent to inverting the n by n information matrix of the traditional estimation process by using the escalator method which operates on one row and one column of the matrix at a time!

Kalman's concepts of observability and controllability were among the subjects which Sol and I discussed often.[4]

I visited the Research Department (by then called RIAS) of Martin Company in Baltimore where Sol developed the theory of shift register sequences during the summers of 1951–1955.

Details for this section are contained in a footnote. The footnote also introduces Herbert Taylor, a mathematician in my group, who later worked at Jet Propulsion Laboratory, and subsequently also received his PhD under Sol Golomb. (*Note*: In order that I could get up-to-date on the work that had been done so far in celestial navigation for the Apollo mission, as well as proposed plans for the future, I traveled first to Baltimore to consult with Rudy Kalman to get some background on minimum variance techniques from one of the leaders in this field. Martin Company treated their noted researchers well and RIAS was housed in a mansion in a wooded area of Baltimore, instead of a dingy office park. Rudy Kalman was a delightful person, not only full of wonderful ideas which he was very willing to share, but he also gave me details of where his research was heading. For example, generally research on linear filtering assumes that the error distribution is Gaussian, also referred to as "white noise." He was researching how one would do optimal filtering in the presence of non-Gaussian noise.)

Next, I went to MIT Instrumentation Lab (MIT/IL) which was an associate contractor on the Apollo project of North American Aviation. MIT/IL was tasked

---

[4]When I was a research associate for Professor Richard Bellman, these factors proved useful when I applied Kalman filtering. They were useful in many applications, for example, in mathematical biosciences, specifically computer-aided analysis of vector-cardiograms and the determination of optimal drug dosages.

by NASA to develop the onboard navigation system, which included a sextant and the Apollo Guidance Computer. I was referred to Dr. James Potter, the brilliant mathematician on MIT/IL's staff, a meeting that turned out to be the first of many meetings at MIT in Cambridge and North American in Downey, CA. Incidentally, he later founded and was the first chairman of MIT's Department of Aerospace Engineering where my granddaughter, Barbara Schloss, earned her bachelor's degree in Aerospace Engineering. The Apollo Guidance Computer had a very small amount of memory: there were only 2,048 words (1 word is 15 bits plus 1-bit of parity) of RAM (magnetic core memory) and 36,864 words of ROM (core rope memory), equivalent to the amount of data in an average email of today. Dr. Potter was tasked with the problem of reducing the memory requirements for the celestial navigation equations. He had noticed that the matrices employed in the celestial navigation equations were, for the most part, symmetric and hence he could use a square root method which would save on limited memory space. When he would visit me at North American Aviation, I would invite Herbert Taylor, a brilliant mathematician in my group who was assisting me in developing my simulation program of the Apollo Navigation System. This program simulated both the onboard backup navigation system and the primary navigation system using the JPL Manned Spaceflight Network with their 26-m antennas at Goldstone, Madrid and Canberra. Herbert Taylor later left for Cal Tech/JPL where he met Sol and became a doctoral student of Sol's at USC and we met again years later at USC in Sol's office!

Then I went to NASA Ames Research Center to meet with Stanley F. Schmidt, who in 1960 (after a visit from Rudy Kalman) had adapted his linear filtering methodology to the Apollo Navigation System. When I returned to North American, I wrote a series of memos which presented the mathematical and statistical background needed to understand the celestial navigation process. With another memo, I answered Rudy Kalman's question to me: Under what conditions would increasing the number of tracking measurements lead to the true value of the state vector (the spacecraft position and velocity)? Next, I wrote a memo presenting the derivation of the Kalman filter and its application to the Apollo Navigation System. Finally, I created a flow chart of the statistical celestial navigation process, which, as far as I am aware, was the first time a flow chart of the process appeared in print. In 1963, North American Aviation decided to publish these memos as a booklet, Schloss, Stephen E., "Introduction to Statistical Celestial Navigation", SID 63-1018, March 1963. Hundreds of copies were printed, and thousands were distributed on microfiche to NASA Centers and contractors.

I consulted for Bissett-Berman evaluating the performance of S-band Radar Stations in JPL's Manned Spaceflight Network: After North American Aviation, I consulted with Bissett-Berman, a company founded by two very bright people, Thomas B. Bissett and Howard L. Engle. Bissett-Berman had a contract to perform analysis of the performance of the radar tracking stations in the JPL Manned Spaceflight Network. In addition, the company was tasked to make enhancements to the precision orbit determination program, a massive trajectory determination program that included models of the gravitational effects of the planets. This was the work horse program, not only for Apollo but also for the numerous unmanned interplanetary missions. We worked directly for Christopher (Chris) Kraft, the legendary director of the manned Mission Control Center in Houston. On one of our interactions with JPL, I had noted that a tracking station on an island in the Pacific was suddenly providing poor data. I called the manager and asked, if by chance, they had moved the antenna. He said that they had, by several hundred feet. I recommended that we remove this station from the MSFN and the station was removed!

I had also worked at TRW from 1972 until 1992. From 1974 to 1977 I was engaged in my PhD studies with Sol. During this period Gus Solomon was there at TRW. By that time, the academics at TRW were more isolated and in their silos, and it was considered to be difficult to get a meeting with Gus. I knew that Gus had been coauthor on the Reed-Solomon code. I phoned Gus around 1975 and told him I was working on the codes of this Russian Goppa, and added that his codes were supposed to be more powerful than the Reed-Solomon code. This produced an immediate invitation from Gus to come see him! There might have been some issues related to the Nixon presidency, and Gus Solomon's possible involvement in some protests.

2.1. *Schloss discusses Apollo Program, MIT/IL, JPL, North American, etc. with Sol in 1970*

That evening — of our first meeting at the Rubenstein's home — we found that we both had concluded in the early 1960s that the onboard navigation instrument which was comprised of a sextant, together with the limited computational capability of the Apollo Guidance Computer, despite the fine mathematical skills of the MIT/IL team, would not provide the accuracy and reliability needed for the space program. NASA had very early on concluded the same thing, namely that all the navigation would be done with radar tracking data provided by the JPL Manned Spaceflight Network, a range and range rate network. This system was the only one that could provide the accuracy and reliability for a successful mission. The onboard system would not even be useful as a backup.

At North American Aviation we had dealings with Sol's Communications Department (and other Departments) at JPL, throughout the Apollo program. As one of only a few mathematicians at North American Aviation on the Apollo Program at the beginning I was given a series of fascinating tasks. Study the problem of the likelihood of significant solar flares in the late 1960s which was the planned period for the manned lunar flights. I was asked to report on the JPL determination of the Earth's gravitational field from the calculations of the orbits of earth satellites. I told Sol, this led me some time later in 1967, when I was doing research with Professor Bellman in biomathematics, to see an analogy between the problem of finding myocardial infarctions from dips in the vectorcardiogram to the dips in an earth satellite when it is above a gravitational dense area of the Earth. At that time, JPL had found four such dense mass regions on the Earth (one, I recall, was off the coast of Chile and another about 100 degrees east near the equator). This conversation on this subject led to an important discovery about Sol. I learned Sol was a master at seeing analogies like this so that one could apply mathematical tools in many apparently totally unrelated areas. Would you believe the Rubik's cube! Sol found that the eight corners of a Rubik's cube provide a model for many aspects of quark behavior (Solomon W. Golomb, *American Scientist*, Volume 70, No. 3, May-June, 1982, pp. 257–259).

3. Schloss is Sol's stockbroker

After I moved to Merrill Lynch, Sol and I had lunches every other month at the Beverly Wilshire next door to my offices. These lunches usually followed a characteristic pattern, in which I would raise a question (or it would arise in some other way) on any subject whatsoever, and Sol would spend the next 2 hours answering it. Sol's knowledge on the subject was never superficial, with further questions revealing the breathtaking depth and formidable breadth of his understanding.

In 1948, Sol was 16 and I was 17. It turned out, as I discovered about 6–7 years ago, when looking through my papers, that Sol and I were both finalists for the National Honor Society competition. Sol was at Baltimore City College, and I was in Sioux Falls, South Dakota at Washington High School.

3.1. *Sol Frequently Touched on Mathematics, Visualizations, Education and Art*

My opening remark on one of our lunches was that I was planning to visit Spain that summer. Little did I expect that this casual remark would cause Sol to reveal to me the mathematics that is revealed in art! His face lit up and he urged me while visiting the Alhambra to notice that in the magnificent tiles exhibited there, one can find sixteen out a possible seventeen symmetry groups! Sol also noted that these tile patterns inspired the artist M.C. Escher. Related to this, Sol also mentioned his tilings of the plane, such as rep-tiles.

This led to my noting that I encountered art in math texts for the first time when I studied at Göttingen University. A text on group theory included illustrations of wallpaper patterns that exhibited symmetry groups and similarly such patterns in nature. In the main hall of the Mathematics Institute of Göttingen University there were numerous mathematical models. So, for example when Professor Max Deuring gave his introductory lecture on group theory, he had a wire model of a dodecahedron with various colored strings displaying the symmetry axes which he handed around for us to examine.

We talked about the educational path Sol followed, which encompassed the European classical phases of *Bildung*: *Lehrjahren*, *Wanderjahren*, and *Meisterjahren*. To understand this we need to note that prior to World War II the centers of development of mathematics were at European Universities in such cities as Göttingen, Hamburg, Berlin, Zurich, etc. While a math student would spend his first 2 or 3 years at one university to learn Calculus, Real Analysis, Complex Variables, Abstract Algebra, etc., the student was encouraged to travel to study under a master in his chosen field. Sol told me while he was at Harvard for his initial graduate studies, he found that there was no one in Number Theory, his chosen field of endeavor. Indeed, in American Universities at the time there was little interest in Number Theory and he needed to travel to Europe to study under the leaders in the field. Sol had his Lehrjahren at Johns Hopkins and Harvard then his Wanderjahren meeting with mathematicians in Norway and Denmark. Then came his *Meisterjahren* in California.

With this background, we had interesting discussions about an aspect of beauty in mathematical proofs that were elegant. A good way to understand this aspect of a proof is to show a proof that Sol said was not elegant, although a very, very important proof, namely the proof given in September 19, 1994, over a hundred pages in length, of Fermat's Last Theorem. (As an aside, he told me in 1993, when this proof was initially given, that an error had been found in the proof and then a year later the corrected proof was published.) In contrast, the proof of countability of rational numbers which is given in a handful of lines is certainly both elegant and very important.

Another aspect of beauty of mathematical proof lies in the duality of proofs. There are many proofs in projective geometry in which one can replace lines with points and points with lines in the theorem and in the proof, and thus as you prove the theorem you also are proving the dual of the theorem is true.

Some of the most elegant proofs lie in the field which the great mathematician Georg Cantor invented, namely Set Theory. Sol, half-jokingly, told me that mathematicians resent that Georg Cantor not only created the wonderful field of Set Theory, but he then proceeded to prove all the major theorems in the field, leaving very little for them to add. There is, however, a very important problem from Set Theory that has occupied mathematicians for the last 100 years. First, we need to define the cardinality of a countable set, such as the set of integers or, as can be proven by an elegant proof, the set of rational numbers, designated as $\aleph_0$ (pronounced aleph null). A set is countable if it has one-to-one correspondence with the set of natural numbers. The higher cardinality of the set points on a line, or a segment of a line, which is uncountable, is designated as $\aleph_1$. The question is: Is there a cardinality between aleph null and aleph one. What became known as the Continuum Hypothesis states that there is no set whose cardinality is between aleph null and aleph one.

Sol and I also had many discussions about Rabbi Gamliel, Halley's Comet, the Hebrew Calendar, and Jewish astronomers — that you probably won't get from anybody else.

It was something that inspired me to really delve into the study of ancient, over 2,000 year old, scientific knowledge.

I first learned about the calculations needed for the Hebrew Calendar from my research for the Apollo program on how to calculate the orbit of the moon. More about this in the next installment.

June 1958 (written on back in Bo's handwriting)

Sol was My Mentor, Part 2: Ancients' Knowledge of Orbits and Calendrics

Stephen Schloss

My search over the years for a mentor in research of "the scientific knowledge the Jews had for over 2000 years" came to a successful conclusion when I met Sol. Sol not only knew the traditions in the Written Law, the Torah, but also the Oral Law which was passed down over generations and then transcribed in the Mishnah and the Talmud. In addition, he had an encyclopedic knowledge of science AND the history of science. For example, just as we find that much of the scientific knowledge of Jews of ancient times are being rediscovered in modern times, it is similar to what I had heard about concepts introduced by Euclid. Sol agreed and said that while Euclid had in mind the concept of mappings in his definition of congruence, scholars in later generations did not understand it and this sophisticated notion lay dormant through the dark ages and well into modern times!

At one of our meetings the topic focused on precisely this subject, namely "the scientific knowledge the Jews had for over 2000 years." My interest in the subject intensified when, beginning in 1962, I researched how to calculate the orbit of the Moon about the Earth as part of my role as Supervisor of Guidance and Navigation Analysis at North American Aviation on the Apollo Program. The Moon's orbit is very complex. The Earth is not a sphere, it's an oblate spheroid and if that isn't enough complexity its mass is not uniform and there is even plasticity so the mass distribution changes, etc. During my research in the 1960's, the gravitational field of the Earth was being updated periodically by researchers at JPL from measurements obtained from orbiting satellites. I remember being startled to learn that they had found four regions of mass concentration on the Earth. This proved very useful in determining the optimal position of communications satellites which are in 24-hour orbits. This analysis is beyond the scope of this discussion. However, due to these findings, the moon's orbit about the Earth, while elliptical overall, has deviations from the ellipse because the Earth is not a sphere; there are variations in the gravitational field; there is plasticity, etc.

The biggest accomplishment that I talked to Sol about was the knowledge of celestial mechanics from over 2000 years ago (!), especially as evidenced by the determination of the orbit of the moon to very fine accuracy (details to follow below). The particulars are summarized in the paper I gave in 1992 at the three-day celebration of Sol's 60th birthday at the Radisson in Oxnard. The proceedings were published by Pergamon Press of Elsevier, Netherlands. The article is "Accuracy of the Hebrew Calendar", *Computers and Mathematics*, 39, 2000, pp 23-24 H. S. Schloss. This entire issue was a *Festschrift* (a collection of writings published in honor of a scholar) in celebration of Sol's 60th birthday and included the papers that were presented on that occasion. Sol and the authors made corrections and edited the papers before this issue was published eight years after the event.

How I learned that the knowledge of the average lunar orbit was known over 2000 years ago

In the course of my endeavors in my role as supervisor of Guidance and Navigation Analysis on the Apollo Lunar Program at North American Aviation, I started studying various texts on celestial mechanics, such as the American Ephemeris and Nautical Almanac Explanatory Supplement, first developed in the 1930s and then updated every ten years. In the Great Depression, the US Government sponsored many scientists to compute mathematical tables, also develop the coordinate transformations, etc. needed for celestial mechanics. In the Supplement there is a discussion of calendars, including the Hebrew calendar, the Moslem calendar and the Gregorian calendar. Of these, the most accurate is the Hebrew calendar. It is a "lunar/solar" calendar; the months are lunar months, but fitted to the solar year. Passover occurs in the spring, and that's a solar event.

The present Hebrew (or Jewish) calendar was put into effect by Hillel II in the year 358 C.E. (see *Bircas Hachammah,* J. D. Bleich, Mesorah Publications, New York 1980; text is in English). Prior to that time, various sages imparted fundamental astronomical data with regard to the lengths of the lunation (synodic lunar month). The elders of the ruling judicial system, the Sanhedrin, established the calculations upon which (with corroborations by witnesses) declarations of the new moon were based. The destruction wrought by the Romans and dispersion of the people produced the need for a permanent calendar whose basis could be taught to skilled laymen. It still is complex, since the calendar months are lunar and yet, for example, Passover must occur in the Spring, which is a solar event. Thus, the Hebrew calendar must be in harmony with both the lunar and solar cycles. The fact that 19 mean solar years contain almost precisely 235 lunar months (the *Metonic cycle*) makes such a lunar/solar calendar possible.

Sol said to me that while this cycle is termed the Metonic cycle (named after the Greek mathematician Meton who presented this in 432 BCE) this 19-year cycle was conceived by Babylonian scholars hundreds of years earlier. The details are given in *The Exact Sciences in Antiquity* by Otto Neugebauer. The accuracy of this calendar is examined. It is shown that despite the approximations necessary to provide a fixed calendar, the Hebrew calendar's mean lunar month duration discrepancy from current astronomical values (a small positive number) amounts to only one day in 14,000 years! Therefore, this calendar is the most accurate one in existence. Since the lunar month is the basis of the Hebrew calendar and this discrepancy is so small

Sol was My Mentor, Part 2: Ancients' Knowledge of Orbits and Calendrics 287

and since it is a crude simplification of that calendar possessed by the Sanhedrin, we can only marvel at the knowledge and skill of these sages.

Prior to Hillel II developing a fixed calendar with a 19-year cycle, the manner in which the time for the New Moon was determined was as follows: The Sanhedrin, who were the Elders of the Jewish judicial system, selected because they were scholarly, established the calculations for the determination of the new moon, which then had to be corroborated by two witnesses to the sighting of the new moon. Sol noted that then the announcement of the new moon was communicated throughout Israel by visual sightings of smoke signals from mountain top to mountain top.

In order to understand how the average lunar month in the Hebrew calendar compares with the actual average length of the lunar month we need to define some terms: The *lunar month* is the time to complete a cycle of lunar phases, which we know to be 29.530588 days. This is also termed a *synodic lunar month*. Twelve synodic lunar months equals 354.36706 days, almost eleven days shorter than the *tropical year*. The *tropical year* is the time between vernal equinoxes (late March). This is known to be 365.242199 *mean solar days* according to *The Astronomical Almanac for the year 1990* issued jointly by the US Naval Observatory in Washington D.C and the Royal Greenwich Observatory, and Her Majesty's Stationery Office in Great Britain, 1989. The *mean solar day* equals the average of two passages of the sun across the meridian. This is 24 hours, 3 minutes, 56.55 seconds or 24.0657083 hours. Furthermore, a *sidereal* day (with respect to the stars) is 23 hours 56 minutes 4.10 seconds, of mean solar time, or 23.9344721 hours.

The solar day is longer than the sidereal day because of the earth's motion about the sun. That is, it has to turn more than a whole revolution to bring the sun back to the meridian. The secular calendar year is 365.2425 days which is 365.25 minus three days out of 400 years. That is, 3 out of 4 one hundred-year cycles have no leap year.

As noted above, the Hebrew calendar is lunar-solar. The lunar part has a "small cycle" of nineteen years. The solar part has a "large cycle" of twenty-eight years. Thus, the unique Jewish holiday of *Bircas Hachammah* (Blessing of the Sun), which is based on the large solar cycle of twenty-eight years, was only celebrated twice in forty-six years of discussions between me and Sol. This occurred on Wednesday, April 8, 1981 and Wednesday, April 8, 2009. It is a tradition that when the Sun completes the twenty-eight-year cycle, it returns to the position it was in at the time of its creation, on the fourth day, which is Wednesday in the Jewish tradition. Per *Bircas Hachammah*, Mesorah Publications, New York, 1980 J. D. Bleich (text is in English) p. 39, "...astronomy provides eloquent testimony to the intricacy and precision of creation. Knowledge of what astronomers call celestial mechanics is a key to a traditional understanding what lies behind creation."

While discussing this subject with Sol, a number of questions and/or issues arose. For example: What measurements were made or even possible at the time (2000 years ago)? What technical aids were available? What measurements were made and how were they recorded? How was knowledge transmitted?

288   *The Wisdom of Solomon*

I got an indication of how the knowledge might have been handed down from generation to generation from Sol, who told me that it is known that astronomical data were stored on Hebrew scrolls in the buildings constituting the library of Alexandria, which were subsequently lost in fires over many years. The Romans burned some of the buildings and the remaining buildings were later destroyed by others. It's a shame. Yet we read about the accomplishments of the Great Roman Empire and are reminded that history is written by the victors.

In order to achieve a calendar with an accuracy of one day in 14,000 years, which is equivalent to one hour in 583 years or one-half hour in 291 years, it is clear that data had to have been recorded for, perhaps, a hundred years or more.

Sol referred me to George Sines who had made recent discoveries of ancient convex and concave lenses in the Middle East that are intriguing. See "Lenses in Antiquity", *American Journal of Archeology* **91**, 191–196 (1987). I talked to George Sines about these lenses he discovered. Many had holes in the middle, some thought these were merely jewelry, but if you put them together you had a telescope. No one has found how they were utilized except possibly as magnifying lenses and/or in the manufacturing of jewelry. Furthermore, there is evidence that Jewish scholars had knowledge of the orbit of a comet which only could be calculated from data recorded over a number of generations. For example, there is a story in the Talmud which Sol described to me in which is indicated that almost 2000 years ago Rabbi Yehoshua knew about the orbit of what 1700 years later became to be known as Halley's comet. This story is about Rabbi Yehoshua and his knowledge of the "wandering star" and its orbit and takes place in the year 66 CE (Common Era). In 66 CE, Rabbi Gamaliel and Rabbi Yehoshua were on a voyage at sea, probably on the Mediterranean traveling from Israel to Rome. (Rabbi Gamaliel later rose to prominence when he became head of the Sanhedrin.) Rabbi Gamaliel took provisions on the trip, but ran out of food. Rabbi Yehoshua had taken double the standard amount of provisions. So, Rabbi Gamaliel asked his companion Rabbi Yehoshua how he knew to take the extra provisions. Around this time, said Rabbi Yehoshua, there is a "wandering star" appearing in the sky. He gave the orbit as 76 years. When it appears, the navigators can get confused by this *wandering star*, and the trip will take longer, because the sailors use the stars to navigate. The Chinese and Babylonians also knew of this comet, but only Rabbi Yehoshua has been recorded as knowing the period of the orbit of the *wandering star* that 1700 years later became known as Halley's comet.

In summary, clearly measurements were taken and recorded over long periods of time. Perhaps someday scrolls will be found which will show how such refined calendars that preceded the fixed calendar were achieved.

© 2023 World Scientific Publishing Company
https://doi.org/10.1142/9789811234378_0030

Bob Scholtz's Memories of Sol

Robert Scholtz

I first met Sol early in 1963 when he visited my thesis advisor Norman Abramson at Stanford on a recruiting visit for USC. I remember accompanying Sol as he walked to his rental car. He had noticed some familiar signal correlation properties in my description of my thesis, and was questioning me with interest. We stopped and spread some writing paper on the rental's trunk lid and talked about the mathematical details. As it turned out, based on Sol's advocacy, I soon received an offer of employment from USC. And I accepted.

Back in those days the full-time faculty in electrical engineering was a close-knit group — we lunched together at the faculty center, took coffee breaks together at the Kite diner on Vermont, and conducted research and university business on these occasions. On many such lunches, Sol and I would talk about sequence design with the objective being to achieve good correlation properties. The paper placemats on our table soon became filled with diagrams, equations, hypotheses, partial proofs, etc. Often at the end of the meal, Sol would tear off a useful part and take it back to his office. These early lunch mats were the genesis of my first paper, published with Sol, on "Generalized Barker Sequences".

We kept up these lunch discussions through the 1960s. During that time, Sol had an interest in tiling problems, and in particular games that could be played with pentominoes, the twelve distinct shapes constructed from five adjacent squares. One game, now known as "Golomb's game", is played on an 8X8 board, with players taking turns placing pentominoes on the board, without duplication. The last player to put a pentomino on the board is the winner. Our placemats were filled with 8x8 cross-hatch game boards as Sol searched for a winning strategy and I attempted to foil him.

Parties at the Golombs were equally memorable. I remember one garden party at Sol's and Bo's Linda Vista home that Lolly and our two sons Michael and Paul attended. Mike and Paul were in grade school and were somewhat at loose ends at this gathering, so Sol suggested that I take the boys for a spin in a rickshaw (!) that he happened to have available. I was a little self-conscious at the time, having just vacated a lawn chair in which the webbing had given way, and I replied "Maybe later." Undeterred, Sol decided to get between the rails and give Mike and Paul a rickshaw ride with him as the driver. Indeed Sol gave them a ride around the lawn, but on the second circle he took a turn too quickly and the rickshaw overset. Not at a loss for words, Sol remarked that the centrifugal force and the centripetal force must have been out of balance!

You always met interesting people at a Golomb party. On one occasion my son Paul, then in junior high, found another likeminded older person with which to discuss science fiction. They huddled on the sofa for a considerable portion of the evening discussing stories and plots. Not knowing the person, I asked Sol who the person was. "Oh, that's John Pierce. He writes science fiction under the name J. J. Coupling!"

A meeting with Sol was always an educational experience!

Solomon Golomb: Wise Man, Biblical Scholar and Father-in-Law

Terrence J. Sejnowski

Salk Institute
terry@salk.edu

My first visit to the Golomb residence in La Canada left a strong impression. It was like stepping into the past when great houses really were great and high ceilings really were vaulting. A chess set was set up ready to play in the living room. The house itself was a work of art, with outside and indoor walls painted with pastoral scenes of the family at a picnic, when Beatrice and Astrid were young girls, and at different stages of the family history. I had heard much about Bo and Sol but nothing could have prepared me for their warm welcome and fascinating stories that ranged from famous mathematicians to the Alexandrian library. After I married Beatrice and joined their family, I was told that Sol had located my Curriculum Vitae and had warned Beatrice that in one year I had given 55 talks around the world. This is truly due diligence. I was fortunate that this did not deter Beatrice from marrying me. I was later added to a mural by the pool, holding up a glass of wine.

Sol had season tickets to Trojan football games at the Coliseum and I would sometimes accompany him. The Coliseum is next door to the University of Southern California and cheering fans would fill the enormous stadium. Sol was not a cheering fan, but he was a loyal one. I grew up with football and thought I understood the game, but Sol had a Talmudic knowledge of football strategy that would illuminate each play. In a famous fund-raising photo, Sol is sitting in the Hoose library dressed in a Trojan football uniform looking convincingly like a heavyset tackle. It is clear from his expression that he was enjoying the spoof.

It did not matter where Sol was sitting: He could seemingly tell you more about the background and history sitting at a Shakespeare play than if you were sitting next to Shakespeare himself. In a discussion about a biblical episode at dinner he explained when it was written and why, as well as the likelihood that it reflected an actual historical event. I only realized later that he knew the story behind *every* biblical episode. His knowledge base was broad, deep and profound.

But more important to me than all of his encyclopedic knowledge was his wisdom in all matters of the world and the heart. I would consult with him when I had a difficult decision to make.

His advice would put everything into perspective. This was like having access to an all-knowing god, a benign god who had your best interest in mind and could frame your problem in the wider context of a complex world. Many have benefitted greatly from his wise counsel.

I was with the family when Sol was awarded the National Medal of Science in the East room of the White House. This is the highest honor a scientist can achieve in our country. President Obama announced that there had not been more brain power in the room since the last time the medals were awarded. Had he known Sol, he would have had to go back a lot further in history.

I was also with Sol in 2016, when he was awarded the Benjamin Franklin Medal "for pioneering work in space communications and the design of digital spread spectrum signals, transmissions that provide security, interference suppression, and precise location for cryptography; missile guidance; defense, space, and cellular communications; radar; sonar; and GPS." It was a wonderful celebration of his life and I felt he was as pleased by this award as the National Medal of Science.

Solomon Golomb is dead; long live Sol in our memories and in our hearts.

© 2023 World Scientific Publishing Company
https://doi.org/10.1142/9789811234378_0032

Professor S. W. Golomb was My Advisor and Mentor

Hong-Yeop Song

School of Electrical and Electronic Engineering
Yonsei University, Seoul, 03722 Korea
hysong@yonsei.ac.kr

Professor S. W. Golomb was my life-time mentor ever since he became my PhD advisor at USC from 1986 to 1991. I would like to recollect some good memories I have with him in various times in my life. This article is based mainly on my talks in memory of him at various occasions in 2016–2017: *Prof. Golomb was my advisor*, at SETA 2016, Chengdu, China, October 2016 (organized by Profs. Gong and Helleseth); *Existence of cyclic Hadamard difference sets and some memories of Prof. Golomb*, at Life and Legacy of Solomon W. Golomb, USC, January 2017 (organized by Profs. Gong and Willner) as well as at ITA UCSD, February 2017 (organized by Prof. Etzion).

I remember clearly on a summer day in 2015 when I thought it had been too long since I had seen Prof. Golomb. I had been so overwhelmed by all kinds of things from my department after two years as the department chair and it took me another some years to recover the shape of my lab in Yonsei University. I was just finishing a revision of the manuscript on perfect sequences using Fermat-quotients.[1] It turned out that this paper has become the last joint paper with him. I had some email correspondences recently but not had a chance to see him face to face or talk with him for a long time. In 2009, I remember when I needed his advice because I was asked to serve as a department chair for two years. I visited New York for a few days, but I stopped in Los Angeles for one day to try to get his advice. I remember he described some complicated system of faculty evaluation, which gave me a clear direction that I should pursue in my department (Figs. 1 and 2).

---

[1]Ki-Hyeon Park, Hong-Yeop Song, Dae San Kim and Solomon W. Golomb, Optimal families of perfect polyphase sequences from the array structure of Fermat-quotient sequences," *IEEE Transactions on Information Theory*, 62(2) (2016) pp. 1076–1086.

Fig. 1. Prof. Golomb, myself, and Dr. Herbert Taylor, January 28, 2016.

Fig. 2. Prof. Golomb's lecture after the lunch, January 28, 2016.

Professor S. W. Golomb was My Advisor and Mentor 295

My plan was to visit him and have lunch together and possibly with Dr. Herbert Taylor at USC campus while I visited Los Angeles in January 2016 for participating in Information Theory and Application Workshop (ITA) at UCSD. I called him on the phone and he replied that he would invite Dr. Taylor for lunch as well. That is how we had lunch together after such a long time, specifically since 1995 when I left Southern California for my new job at Yonsei University in Korea. As Dr. Taylor and I had done in the past while we all were there, we asked Prof. Golomb if it would be fine with him if we went to his class on Combinatorial Mathematics right after the lunch. He was teaching the course in Spring 2016. His response was that it could be boring. We said it will not be, and we followed him to the classroom full of math majors. His style of lecture and his style of using the black board and chalk did not change. Dr. Taylor and I had a good time at the back of the classroom, while most of the students probably wondered who these old guys could be. This has become the last moment with him forever.

I first came to USC in 1984 and took three classes from him from my second semester in the MS program: Linear Algebra, Information Theory A for conventional theory of entropy, etc., and Information Theory B for m-sequences and various other sequences with good correlation properties. After taking these courses with pretty good and impressive grades (if I remember correctly), and passing the screening exam, I asked him to be my advisor. I was so impressed by his three lectures and thought that I would like to be his student in whatever area I have to study in the future. I still believe that he is the best lecturer that I have ever met because his style of giving a lecture is truly exceptional. All three-unit major classes at USC were given twice a week lasting one and half hours in the studio for on campus and remote location students. He made about 30 lectures as a series of beautiful talks with every talk consisting of a short introduction and toy examples, a main topic of serious discussion and development, and sometimes the current status of a subject, including his own contribution to the subject. He has done all these without any memo or material in his hand! I still have most of his hand-written notes, all three courses mentioned above and more, and I would like to share one page of this, saying the famous open problem on the existence of cyclic Hadamard matrices of size $4t \times 4t$, as in Fig. 3. It is still wide open as of 2019. Some of my research activities for the last 20 years or so have been somehow related to this problem also.

My first technical paper was a joint paper with him and Dr. Herbert Taylor. Dr. Taylor was a math expert, especially in discrete math and combinatorics. After 20 or more years later since he got an MS in mathematics at UC Berkeley, he wrote a joint paper with Prof. Golomb in the early 1980s while he was working at JPL, and then moved to USC to work as his research associate. His concentration on math problems or any open problems is really exceptional. Probably that is why he was very good at playing the game of Go. In fact, he had a title of North American Go Player Champion in the late 1970s.

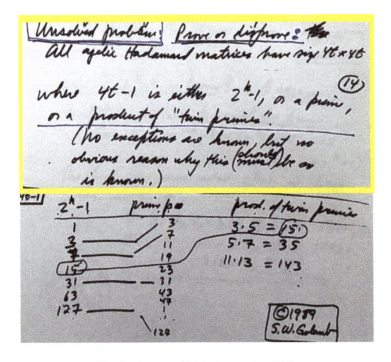

Fig. 3. A page of his lecture notes, 1989.

The paper[2] contains some answer to the question about the number $G(n)$ initially proposed by Prof. Golomb: given a positive integer n, the number $G(n)$ is the largest integer k such that no matter how Z_n is two-colored, some progression $a, a+d, a+2d, \ldots, a+(k-1)d$ of k distinct elements of Z_n will appear in one color. Major results in this paper were obtained in the Commons restaurant at USC with Dr. Taylor. He trained me seriously also in writing some technical results.

I spent two years as a post-doc research associate of Prof. Golomb after my PhD defence in Dec 1990, and then moved to Qualcomm, San Diego, for my first job in California in 1993. Finally in the summer of 1995, I moved to Yonsei University, in Seoul, Korea. Before coming back to Korea, I visited him for advice. I left Korea for graduate study and spent 11 years in the states, so my home country seemed to be a bit strange to me. I do not remember exact wording but he gave me an advice as something similar to this:

> Be careful of everything there. Korea must be a foreign country to you, except that everybody speaks your mother tongue. But do not worry too much since there are old friends and family members too.

Sequences and their Applications (SETA) is an international conference initially organized by C. Ding and T. Helleseth in 1998 in Singapore. The first time I went to SETA in 2001 in Bergen, Norway, was chaired and organized by Prof. Helleseth.

---

[2]Hong-Yeop Song, Solomon W. Golomb, and Herbert Taylor, Progressions in Every Two-coloration of Z_n, Journal of Combinatorial Theory, Series A 61, Nov (1992).

Fig. 4. Professors Gong, Yang, Song, Golomb, Scholtz, Welch, Kumar, No, and Chung in SETA 2001, Bergen, Norway.

Some USC-CSI graduates and faculty members, including Prof. G. Gong took a picture together here, which is Fig. 4.

I should talk about his visit to Korea in the fall of 2004. It was SETA organized by Prof. Jong-Seon No at SNU who was a PhD from Prof. Vijay Kumar at USC. I was working as a member of the Organizing Committee. Also, there was a big conference for USC alumni from China, Korea and Japan in Seoul. Professor Golomb came to Korea to attend both the Alumni Conference and SETA since these two events were held back-to-back in the same week. After the alumni conference was over, I picked him up every morning and drove him back every evening to his hotel. Figure 5 is a picture taken at the dinner of SETA 2004 with my PhD students (Fig. 6). He was pretty much happy with his PhD (myself) and my grad students together. You will also see Profs. Scholtz, Welch and Vijay Kumar here.

Professor Golomb was not only a great mind in various technical areas but also in various language and culture all over the world. During his short visit to Seoul, Korea, with only driving along the streets of Seoul in my car, he understood how to use the Korean alphabet. He later told me that he learned from all the signs on the streets with both English and Korean. On his last day in Seoul, I asked him to deliver an hour lecture at Yonsei University, where I have been teaching a class on digital communication. He selected the topic of "shift register sequences" and gave a lecture to about 300 undergrad students in my department. Some of them were from my class, and the remaining attendants were from other classes, recognizing the fame of Prof. Golomb. The first page of his slides is shown in Fig. 7.

Fig. 5. A scene at the dinner at SETA 2004, Seoul, Korea.

Fig. 6. Group photo of SETA 2004.

Professor S. W. Golomb was My Advisor and Mentor

PROPERTIES OF SHIFT
REGISTER SEQUENCES
SOLOMON GOLOMB
쌀어먼 걸엄브
UNIVERSITY of SOUTHERN CALIFORNIA
南加州大學 (남 카 대 학 교)
29 OCTOBER 2004
十月 二十九日 (二千四 年)
YONSEI UNIVERSITY
연 세 대 학 교

Fig. 7. Title page of his lecture slides at Yonsei University, 2004.

After spending only a few days in Seoul, he completely understood and correctly used the Korean alphabet in this title page! He even used some Chinese characters on this slide for USC, the date and year. I vividly remember his explanation on how some Chinese characters have changed their shapes in various parts of the world using these, especially, in China, in Korea and then in Japan. He argued that Korea is the country where the old and traditional shapes of most Chinese characters have been well-preserved, much better than in any other parts of the world including China itself. This is the true and the well-accepted situation in the region of East Asian countries.

I brought him to a Korean Traditional Folk Village for a day tour and we had a good time there (Fig. 8). He seemed to be very much interested in the old traditional shape of society, cloths, materials, writings, buildings and houses, living area, etc. Figure 9 is a collection of some pictures taken there with me.

Fig. 8. In the classroom at Yonsei University, 2004.

On the last morning when I drove him to the airport in Korea, he gave me a hand-written Thank-you letter. I would like to share the first few words here in Fig. 10.

In 1991, I finished my PhD defence. In 1992, I was working as his research associate at CSI-USC; we had his 60th birthday. Dr. H. Taylor from CSI-USC and Prof. G. Bloom from New York prepared for most of the conference in Oxnard, CA. I just brought my wife and new born baby (my first daughter, Rachel) in a stroller. I was so impressed by all the attendants, including Dr. A. Lempel from Israel, Dr. McEleice from Caltech, Dr. I. Reed and Dr. G. Solomon, the inventor of RS codes, Dr. Posner from JPL, etc. The list of some attendants and title of their talks are shown in Fig. 11. The famous late M. Gardner did not come to Oxnard but sent his contribution to the special issue celebrating Prof. Golomb's 60th birthday.

Well, that makes the year 2002 his 70th birth year, and this time, the conference was organized by CSI, and the program was prepared by Profs. Vijay Kumar, Tor Helleseth, Jong-Seon No, and myself, at the campus of USC. I remember that we all had a great time. Jong-Seon and I were responsible for collecting all the presentations as manuscripts and for publishing then together with Kluwer Academic Publishers in 2003, titled "Mathematical Properties of Sequences and Other Combinatorial Structures." In 2007, Prof. Golomb had another gathering to celebrate his 75th birthday called GF(75) and now the conference was organized by himself, Profs G. Gong and T. Helleseth, and myself at USC campus again. We again had

Fig. 9. Visiting Korean Folk Village, 2004.

a great time with some subsets of participants of GF(70). Mrs. Golomb held my arm after the party and told me "Be happy!" many times, and that was my last moment with her. Prof. Golomb took an emotional picture with all his great old-time buddies at JPL, which is shown in Fig. 12. Prof. McEleice gave a talk with a beautiful voice with shaking hands to change the slides, and he has just recently passed away. Dr. B. Gordon did not appear, and I guess he had passed away already. Dr. H. Taylor gave an impressive talk on queens problem, but he looked so old and had some difficulty in changing the slides. Dr. Berlakamp has written a beautiful

31 Oct. 2004

Dear Hong-z

There is no way I can possibly thank you enough for all your care, thoughtfulness and hospitality during my visit to Korea this past week. Your generosity with your time, not even to mention the expense, was far beyond anything I could possibly have expected.

As I meant to

Fig. 10. Beginning of the letter, 2004.

Special Issue of GolombFest60, may 1992, Oxnard, California

Table of Contents for
Computers and Mathematics with Applications
Volume 39, Issue 11, June 2000 (Special Issue of GolombFest60, May 92, Oxnard, CA)

Te-Won Lee, M. Girolami, T.J. Sejnowski and A.J. Bell
A Unifying Information-Theoretic Framework for Independent Component Analysis — 1–21

H.S. Schloss
Accuracy of the Hebrew Calendar — 23–24

D.R. Estes and C. Waid
An Algebraic Analysis of Bore Hole Samples — 25–30

Hong-Yeop Song
The Existence of Circular Florentine Arrays — 31–35

G.S. Yovanof and H. Taylor
B_2 –Sequences and the Distinct Distance Constant — 37–42

B. Tang
Evaluation of Some DNA Cloning Strategies — 43–48

B. Gordon
Multiple Tilings of Euclidean Space by Unit Cubes — 49–53

M. Gardner
Dominono — 55–56

G.L. Mayhew
Clues to the Hidden Nature of de Bruijn Sequences — 57–65

L.M. Butler and A.W. Hales
Generalized Flags in p-Groups — 67–76

E. Berlekamp
Unimodular Arrays — 77–88

I.S. Reed
A Brief History of the Development of Error Correcting Codes — 89–93

M. Buck and N. Zierler
Decimations of Linear Recurring Sequences — 95–102

G. Solomon
Golay Encoding/Decoding Via BCH–Hamming — 103–108

R.E. Peile and H. Taylor
Sets of Points with Pairwise Distinct Slopes — 109–115

A.L. Whiteman
Some Balanced Incomplete Block Designs — 117–119

W.J. Hurd
Maximum Likelihood Global Positioning System Receiver — 121–125

E.C. Posner
A Code in the Nose — 127

Y.S. Abu-Mostafa and R.J. McEliece
Maximal Codeword Lengths in Huffman Codes — 129–134

T. Bu
Golombs Norske Forbindelser — 135–138

M.J. Cohen
The Difference Between the Product of n Consecutive Integers and the nth Power of an Integer — 139–157

N. Hamada and T. Helleseth
Arcs, Blocking Sets, and Minihypers — 159–168

R.K. Guy Edited by
Problems Sessions 92–05–29 and 92–05–31 — 169–171

Fig. 11. The list of participants of GF(60), Oxnard, California.

Fig. 12. A picture at GF(75), USC, 2007.

and memorable Preface for the conference proceedings, and he too has passed away recently. I could not describe explicitly but I figured in my mind that time was coming. All these great minds, including Prof. Golomb, cannot resist the time. That was so sad and I cried in my hotel room for long time after the dinner party that night.

It was my great fortune that he was my PhD advisor. He was not only a PhD advisor but also a great mentor to me. My life has changed completely ever since I met him in 1985. He introduced me to the world of information theory and discrete mathematics, was a role model of how an academic advisor should be, and showed me how an hour lecture to students must look like. Additionally, he taught me how beautiful and thrilling a fundamental research could be and how exciting an application research could be. He was always there to give me advice. He will live forever in my heart.

© 2023 World Scientific Publishing Company
https://doi.org/10.1142/9789811234378_0033

Solomon Golomb in the Faye Zlok Days at JPL

Robert C. Tausworthe

Foreword

Mind you, I am recreating memories of things that happened 60 years ago, principally from 1958 to 1963 when Sol was at JPL. There are other treatises on Sol's contributions to mathematics, coding, communications, and information theory. Here, I will attempt to describe instead, from my perspective, the rich intellectual research environment that Sol created at JPL, the intellectual camaraderie of the people that thrived in it, and some of the events occurring in those years. It was a fun place to work. I also will include a few remembrances relating to his legacy, the continuing influence he had on JPL that prevailed even after his departure for USC.

In telling this story, I will be describing things that happened to me or within my purview that were related to Sol's action or influence. It is thus a personal narrative not so much about Sol, as it is about the influence he had on those of us who were fortunate enough to know and work with him.

Sol's Coming to JPL

Sol came to the California Institute of Technology Jet Propulsion Laboratory (JPL) sometime in 1956 as I recall being told, just before having taken his final PhD exam at Harvard University. At that time, JPL was a guided-missile research and development organization, a division of Caltech, under contract to the U.S. Army as a Federally Funded Research and Development Corporation (FFRDC). Sol was assigned to the Communications Research Section, managed by Walter K. Victor, as a Senior Research Engineer.

One of his first tasks was to use spacecraft radar data to determine the orbit of a missile or satellite during its launch phase of operations. He then left for a very short time to completed his PhD orals. On return he contributed significantly to the development of the Coded Doppler Radar Communications system, or CODORAC,

306   *The Wisdom of Solomon*

which was intended to be part of the Sergeant surface-to-surface missile guidance system. Not surprisingly, it employed pseudo-noise techniques to make the missile guidance unjammable.

I had become aware of JPL as a Caltech student in October 1957, when the Soviet Sputnik was launched and in orbit. Newscasts were that JPL was in hot pursuit; it launched its Explorer I and the era of U.S. space exploration in early February 1958. It was a time of great change around JPL and in the space program. In May 1958 a new JPL Information Processing Group was formed, and Sol was put in charge.

In July of 1958 President Eisenhower signed the National Aeronautics and Space Act, establishing NASA. This new agency was to conduct all non-military space-related activities. Earlier, in February, the Advanced Research Projects Agency (ARPA) had been created to develop space technology for military application. JPL was not immediately inducted into NASA, but JPL's employees knew it was coming.

My Coming to JPL

I had come to Caltech in 1957, just after graduating from New Mexico State University, to pursue my master's degree. At NMSU, I was enrolled in its ROTC program and was commissioned as a first lieutenant, slated to report for duty in July, 1958. So I had a year to complete my MSEE degree, which I did, before entering active duty in the Army. In early 1958, however, I had injured my back, and showed up at the Aberdeen Proving Ground Ordinance Depot in Maryland, ready for service, still wearing the back brace that my doctor had prescribed. After a short physical examination and a longer administrative grind, altogether lasting 19 days, I was released from active duty for medical reasons.

I was aware, after flunking the physical, that I would be released. What was I to do? I pondered. It was too late in the year to apply to Caltech for entry into a PhD program. I needed to find a job right away, as funds were already meager from my being in school so long. I asked myself, "Where can I find a job in Pasadena for a year so that I can reenter Caltech in the fall of 1959?" I had not concerned myself with this before, as I had planned to be in the Army for two years.

The only place I could think of that might hire me in a technical job was JPL. I sent them a letter early in the 19-day mustering-out period, to which they responded without delay by telegram, requesting me to come in for interviews immediately on my return to Pasadena. I was interviewed by three different sections at the time, two of which I do not recall. The third was conducted by previously mentioned Walter Victor, with C.S. (Stan) Lorens and Solomon Golomb, who were group supervisors within his section, present and doing most of the questioning. I was trying to say "Whatever you have, I'll take it! I'm desperate! I need a job for a year!" But I didn't put it that way, and I certainly didn't say I was only going to be there a year. Finally, Walt Victor came right out and said "Well, do you want to build stuff or write papers?"

Solomon Golomb in the Faye Zlok Days at JPL 307

Still hedging, I said that I preferred doing research and writing papers, but I was no stranger to building stuff, being employed as a Co-op Student Trainee at White Sands Missile Range for the tenure of my undergraduate degree, which involved, among other things, building radio and power supply equipment for one of its field tracking stations.

Victor offered me a job in his Communications Research Group, led by Stan Lorens, an ScD graduate from MIT a few years earlier. This, they said, was probably the best fit for what they judged my talents to be towards the kinds of work the section performed.

I reported for work in September, 1958. My paycheck would come from Caltech, but my desk belonged to the Army. Walt Victor's organization was then called "Section 8, Communications Research." I remember being amused by this numbering, because, at the time, the Army used the "Section 8" nomenclature to categorize recruits found to be mentally ill or otherwise unfit for service. I asked myself, "What am I getting into here?"

Section 8 was one member of what was called the Telecommunications Division, led by Eberhardt Rechtin. Other sections involved new circuits, and spacecraft and ground communications systems. The ground system, just under construction, was known as the Deep Space Instrumentation Facility (DSIF), later retitled The Deep Space Network (DSN).

Sol was first my colleague and in a few days, a fast friend. He was unpretentious, quick witted, a gifted raconteur, and a phenomenal linguist. I was amazed to learn of his photographic memory and exceptional intelligence that similarly astounded all who would ever know him. He later became my mentor, Deputy Section Manager, and my Caltech thesis advisor, but I will cover these encounters in subsequent narrative.

The Communications Research Environment

Office space at that time at JPL was sparse. I was given a desk in a four-person office that included Stan Lorens, Andrew Viterbi., and Lloyd Welch. The rest of Stan's group occupied adjoining offices to the south. Sol Golomb and members of his group shared offices just to the north.

The two groups were tightly collocated and had practically the same charters. Looking back, I can't remember exactly the makeup of each group, but Stan's included Mahlon Easterling, Richard Epstein, and Richard Goldstein. Sol's group would soon be joined by Ed Posner, Neal Zierler, Jack Stiffler, and Gus Solomon. Almost everyone there was either a recent graduate or part-time student.

There were always a number of part-time or summer-hire Caltech students throughout Sol's tenure at JPL. Dick Goldstein enrolled at Caltech's PhD program, as I did, in 1959. Jack Stiffler had completed his class work and was looking for a thesis topic. Al Hales and Bob Jewett were grad students at Caltech at the time working part time or summers. Leonard Baumert and Bob McEliece came a little later, still undergrads at Caltech. McEliece became full-time in 1964 and Baumert

308   *The Wisdom of Solomon*

in 1965. Other summer hires of note include Larry Harper, Charles Greenhall, Hal Fredericksen, and Rick Greene. Those of you reading this who are familiar with communications, coding, and information theory will recognize these names from their outstanding accomplishments that began under Sol's tutelage.

Although the Lorens group was termed the Communications Research Group and Sol's was called the Information Processing Group, there was nevertheless a free exchange of ideas among the two. There appeared to be no specific charter for the groups other than ones implied by their titles. It was clear to me after only being there a short time that Sol was the intellectual leader of the two groups. Collaborations were free and encouraged.

In the years I knew Sol, there always seemed to be a quiet intensity within him. He was insightful, funny, a raconteur, and off-the-chart intelligent. He seemed to be involved in all things that tweaked his curiosity, but he didn't usually say much on any particular subject if you didn't seem to share his curiosity in it. Occasionally I would talk to him about things outside science and mathematics; usually he responded with detailed stories, often very erudite, about the personalities and organizations involved.

Learning the Communications Science

Since I was the new kid on the block, because I was new to communication and information theory, and since Davenport and Root, whom Stan and Andy knew from their MIT days, had just published their book, *An Introduction to the Theory of Random Signals and Noise*, I was given the task of leading a seminar with that book as text, one chapter each Friday afternoon. All the members of the two research groups were to participate. The mandate was that we all had to work all the problems given in each chapter. I certainly learned a lot about communications very quickly!

JPL was transferred to NASA in December 1958, becoming the agency's primary planetary spacecraft center. As a part of the transition process, all the research that had been conducted for the Army needed to be collected, documented, and reported. Working for the Army at that time required secret clearances, and the material we produced for that document was considered a national secret. As the junior member of the team, I was tasked with being the editor of the Final Report On (title redacted). Although I cannot say what was in that report, I can relate that I began the introduction with the text

"The ether is free and the enemy is smart. This is the environment the secure communicator must face as a natural condition of this chosen profession. This report documents the works in this field by the Communications Research Section of JPL." Or words to that effect.

That was the last I ever saw of the report. It is still classified to this day. Curiously, some of the mathematics that Sol had created and recorded, after many years, was rediscovered by another author, who claimed it as his original work.

Sol sought to seek out the report and ask that it be declassified, as proof that he was the true originator. But that never happened. Apparently, no one in the government could be found who had the authority for declassification and was deemed competent enough to judge whether the material it contained was still sensitive.

Decoding Lloyd Welch

Lloyd Welch came to JPL in 1956, got his PhD from Caltech in 1958, and left JPL in 1959 to work at the Institute for Defense Analysis, and later, the University of Southern California. I immediately was aware that Lloyd had a brilliant, formidable mathematical ability. To further my training in the current state of communication theory, I was given an internal Section document that he had written and was asked to read, understand, and supply the missing logical chain of detailed reasoning between the implied statement of a problem and what seemed to be a very complicated mathematical result. Lloyd's intellect was such that he deemed the result as being rather trivial and obvious; but to me at that early stage of my learning, the steps leading to the answer were practically opaque.

The problem was to formulate the spectral density of a random series of binary intervals of constant length modulating a carrier frequency. I was able to supply the steps he had left out and to extend the theory to include the spectral computation of modulation by a pseudorandom sequence of a given arbitrary period. I was also able to characterize asymptotic spectral behavior in regions far away from the carrier frequency. The results were documented as a JPL external report available to other NASA centers and their contractors (or anyone else who learned of it).

Lloyd later published a paper with a co-author who had also co-authored a paper with the famous and revered Hungarian mathematician, Paul Erdős. Erdős, who had published over 1500 papers in journals during his lifetime, was so respected that mathematicians instituted a distance scale in his name, measured as the distance, in co-authorship between Erdős and another mathematician.

Lloyd therefore is endowed with an Erdős number of two. That work I did with him earned me an Erdős number of three! (Although my name is not listed among those in the current listing of such distances.)

Sol Golomb similarly has an Erdős number of two, but he claimed that he deserved to be rated at *minus one*, since Erdős had published a paper on a subject previously treated by Sol.

Likewise, Al Hales has an Erdős number of two. And, by the way, Bob McEliece has an Erdős number of one, having co-authored a paper with Erdős himself. He may well have been the link between Sol's and Al Hales's number of two.

Because there were so many collaborations with Sol, Al, and Lloyd, the Communications Research Staff was loaded with low-numbered Erdős scholars.

310   *The Wisdom of Solomon*

ISOCAIT

Sol, Stan Lorens, and Dick Epstein had decided that JPL should convene the prominent authorities in the new and evolving field of digital communications together in what would become the First International Symposium on Coding and Information Theory, or ISOCAIT. They particularly wanted participation from experts in Russia. They were able to get the necessary JPL and US Government approvals to proceed, but telephone contact with the Russians involved was deemed necessary. They needed an interpreter.

Henry Richter, who was an engineer on Eb Rectin's division staff, had been learning Russian from a Polish emigrant working at that time in the JPL machine shop named Joseph Zygielbaum.

It appeared that Zygielbaum had completed a mechanical engineering degree and was a reserve officer in the Polish Army in 1939, when Germany invaded Poland, launching World War II. Remarkably, he fled and joined the Red Army, fighting courageously in far-flung places across Europe, and decorated for his service. Here he had learned to speak fluent Russian and learned much about the Soviet military and its leaders. After the war he emigrated to the US, became a US citizen, and entered the JPL workforce.

In preparation for the upcoming communications, Sol, Stan Lorens, and Dick Epstein solicited Joe to convene a Russian class that several others at JPL also participated in. When the time came for the telephone conversations, Joe translated the interchanges between the Russian and JPL sides. Those in attendance, especially Sol, had picked up enough Russian as to recognize some of the technical content of the exchange, but Joe supplied the details.

The Symposium was held in the late spring of 1959, as I recall. I have been unable to find references to works presented at this time. After this, Zygielbaum was moved into JPL's technical library to begin translation and analysis of technical publications and materials on Soviet space and military programs. His son, Arthur, joined JPL in 1968, was employed there almost 30 years before he left to extend JPL Remote Sensing technology to the University of Nebraska.

Norbert Wiener Class

In the summer of 1959, MIT Prof. Norbert Weiner, coiner of the term *cybernetics*, presented a series of lectures at UCLA, entitled *Nonlinear Problems in Random Theory*. A number of people in the Communication Research groups, especially those who had graduated from MIT, expressed interest in attending. I don't remember the exact number of us who made the daily one-hour trek over to UCLA that summer, but I do know it included Sol, Andy Viterbi, Stan Lorens, Dick Epstein, and me. I recall filling out the enrollment form, which asked, "Why are you taking this course?" Stan Lorenz responded "for entertainment."

And entertaining it was. Everyone who has ever attended one of Wiener's lectures agrees that he was well accomplished with keeping the listener's attention.

At MIT he became the subject of dozens of absent-minded, nearly blind, brilliant professor tales, jokes, and anecdotes. In this class, he walked in each morning with a fresh cigar, lit it, and began the presentation. He wore very thick, heavy glasses, which slipped down over his nose. In order to maintain eye-to-eye contact with his audience, he would tilt his head backward. He held the cigar gingerly, in an almost stationary position, moving his head over to draw upon it, carefully, not to disturb the ash, which got longer, and longer, and longer. After a while, all student attention was focused on the ash rather than the content of the lecture, in anticipation of the impending drop. Near the end of each lecture, the ash fell, the students sighed in unison, and Wiener dismissed the class.

Following each class several teaching assistants convened smaller groups for two-hour discussions of the material that had been covered. It was in these smaller groups that the extremely complex and opaque mathematical subject was revealed.

Sol and Max Delbruck

Caltech biology Prof. Max Delbruck had done work that influenced the Francis Crick and James Watson discovery of the helical structure of DNA, for which they had won the Nobel Prize in 1953. In fact, he had been Watson's post-doc advisor at Caltech. He was thus very familiar with their work on DNA encoding, but much remained a mystery about how the four base amino acids, designated by A, T, C, and G, and visualized using color coded as beads in simulated DNA molecules, were arranged within the helix. It wasn't yet clear just how sequences of the four base pairs encoded the 20 amino acids making up the helix.

In 1956 he asked people he knew at JPL if anyone there could help him figure it out. Naturally, he found Sol, who was only too eager to study the problem. He and members of his group analyzed an idea espoused by Francis Crick's that led them to invent self-synchronizing "comma-free codes." Here they sought for overlapping triples of the four base pairs that could encode amino acids but constrained so that no overlaps of symbols could be valid triplets. Their analysis showed that exactly 20 amino acids could be encoded this way.

Sometime between 1958 and 1962, Sol sent Max a cryptographic message that he had carefully encoded using the amino acid alphabet using all 20 of the amino acid code symbols. I do not recall in what form it was conveyed to Max, as I only saw a copy of it written down on paper in base form with its deciphered form alongside just before he sent it. It ended with the demand "Decipher this message or give back Nobel Price [sic]." This part stuck in my memory because his amino acid cipher alphabet did not have room for a Z in it. I remember being a bit confused, too, because Max had not won the Nobel Prize, yet. He did win it however in 1969.

I also got to know Max through another channel. My first wife taught his children piano lessons, and I was the tenor soloist at Throop Memorial Church in Pasadena. Max often convened, and may have played in himself, a string quartet that came to his home a few times a month. Whenever they would decide to do a

312   *The Wisdom of Solomon*

Bach cantata or something that required a piano or vocal participation, they would invite us over to perform with other musicians they would pull out of the Caltech woodwork. It was great fun. After the great divorce of 1969, I no longer was asked to participate.

Under Sol's guidance, the subject of comma-free codes, in the guise of self-synchronizing binary telemetry codes, became the topic of Jack Stiffler's 1962 PhD thesis. Jack had thus just become Sol's first PhD student.

Miss Guided Missile and Miss Faye Zlok

Beginning in 1952, JPL held "Miss Guided Missile" pageants. In 1959, the title changed to the "Queen of Outer Space"—mimicking the name of a 1958 Zsa Zsa Gabor film of the same name. These continued throughout the 1960s, halting in 1970. Billie Jean Gunter in 1959, the secretary servicing our two groups, decided to enter the competition. She was very pretty, outgoing, friendly, and popular.

But Sol was concerned by the categorization of women as objectified by the contest and decided to enter a humorous ersatz candidate of his own creation. Andy Viterbi and I had been studying the nonlinear lock-in characteristics of phase-locked loops using phase plane trajectory analysis methods. Andy had produced several depictions of the behavior via an analog computer capability that JPL had at the time.

Sol selected one of the phase plane portraits, dubbed it Faye Zlok, placed on placards, and distributed them about the Lab in advertisement of his candidate. Faye, he surmised in his description of her, was a modest girl of Polish extraction, who was "so sinusoidal, she don't need a goidle." Needless to say, she did not win, but everyone in our gang had a lot of fun in promoting the fictitious model. A portrait of Faye, much like the one actually used on posters, appears in Fig. 1.

Billie Jean Gunter wound up winning that year over Faye. How close the race was, was never announced.

The Random Golfer, Theory and Validation

In the fall of 1959 I reentered Caltech to work on my PhD degree. I was only allowed by Caltech to work 3 hours a week at JPL, which I wanted very much to do, because I wanted to continue my employment there after getting the degree, and also, to keep tabs on what my friends and colleagues were up to, to keep updated on their latest achievements. The next summer, and thereafter, I was allowed to work 20 hours a week until my PhD work completed. But every Friday afternoon of that academic year, I spent on Lab.

It was during this period, in the spring of 1960, that I became aware of rather focused discussions concerning the hypothetical performance of a randomly chosen, first-time golfer, or RFTG. In the short time slots of these weekly windows, I could

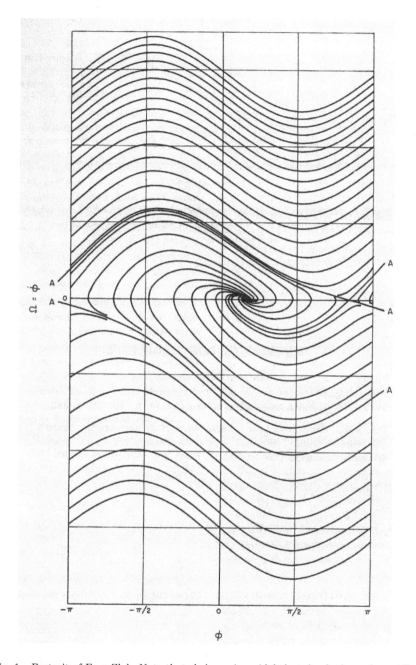

Fig. 1. Portrait of Faye Zlok. Note that she's so sinusoidal that she don't need a goidle.

only piece together little slices of the evolving mathematical theories that were being developed in support of the various conjectures that were being made.

The debates supposedly came about as the result of a casual conversation between Dick Epstein and Dick Goldstein, both of whom were very good golfers.

Epstein claimed that a randomly chosen, first-time golfer should be able to post a score somewhere around 150. Goldstein countered that this figure was too low. Sol and others within earshot offered their opinions, and so began the controversy. The friendly, jocular, and spirited debate over the estimated score carried forth for weeks. It seemed to be the favorite Friday afternoon coffee break topic of discussion.

Sol suggested that there was an important more general mathematical question here: how to make the best intelligent guess of the outcome of a single random event without any *a priori* statistical knowledge of the ensemble of events. Sol was aware of, and brought into the discussions, the work of Emil Gumbel a few years earlier, on the *"Statistical theory of extreme values and some practical application"*.

I referred to their endeavors at the time as "distribution-free" parameter estimation theory, which was a reference not to the underlying population as being without a statistical distribution, but to the estimation basis being free of the assumptions about what those statistics might be. For while there are surely underlying distributions that characterize such phenomena, these, in this case, depended on a range of unknown complex factors, such as past athletic involvements and innate athletic ability. Without access to Google in those days, we had no way of even guessing what the statistics might be.

Others called it by its more accurate designation, a Scientific Wild-Ass Guess, or SWAG.

Finally, the theory, such as it was, was ready for field test, a friendly game among the opposing advocates and an anonymous test subject was scheduled, preparations were made, and wagers on which theory was correct were laid. The payoff was to be a dollar a stroke from the agreed-to median score.

I was not present when the "random" selection of the test subject was made, but I rather think he was more likely pressed into service than willingly volunteered for the experiment. His name shall remain anonymous so as to retain a semblance of stochastic selection.

Luckily for me, the field test occurred in the latter part of a Friday afternoon at the Brookside golf course, next to the Rose Bowl, just across I-210 from JPL, so I was able to attend and bear witness to the performance of the RFTG relative to the estimated behavior.

Almost immediately, at about the second or third hole, it was realized that a significant adjustment, right there, on the fly, or more accurately, on the whiff, was necessary to the theory. It was recognized that the forecasting model did not adequately address the statistical problem known as "gamblers ruin". Without much theorization about when it would occur, a liability limit was set, quickly reached, and the test terminated.

I doubt that the test subject ever boasts that he was the first and only RFTG of forecasting theory. But this event so early in his professional career certainly may have aroused his interest and directed his intellect toward applying statistical methods to those other venues for which he now is rightly famous throughout the world.

Editor comment: This was not Sol; whether it was a contributor to this volume will remain undisclosed.

The Beginnings of Radar Astronomy

Richard Goldstein, now a fellow Caltech PhD student, had been keeping tabs on the development of the DSIF as its implementation progressed. During one of my Friday afternoons at JPL he said to me, "You know, we're getting close. All we need is another 10 db of signal-to-noise improvement." "For what?" I asked. "To bounce radar off of Venus." He replied.

I then asked the question that I now consider to be one of my most least insightful, "Why would you ever want to do a thing like that?" I would shortly come to realize the naivete of that question.

Eb Rechtin and Walt Victor had a policy that no new piece of equipment would be put into operational use until its theory had been developed, documented, and thoroughly tested. Some of the less extensive engineering model fabrications and proof-of-theory experiments were done in our labs, either by us or our technicians. But big theories required big equipment. Bouncing radar signals off of Venus was no lab experiment.

Besides the operational DSIF sites at Goldstone, California, there was another site there, known as Deep Space Station 13 (or DSS13), that was dedicated to demonstration of new equipment capabilities. When the station was completed, it had an 85-foot parabolic antenna, 4-kelvin receiver, and megawatt transmitter. The 10db that Goldstein had needed had become available.

By this time Goldstein, with Sol's help, was able to convince Caltech that the planetary radar experiment was an acceptable PhD thesis and that it was OK to work on it at JPL as his assigned task. Caltech at that time had the policy, "No pay for thesis work!" But since he was actually already a Caltech employee, and because there was a mutual interest in JPL and Caltech collaborative projects, he was able to get dispensation.

Moreover, Eb Rechtin was able to convince NASA headquarters that radar astronomy would not only provide a means to test new equipment and ideas in a non-operational venue, but it would also serve to demonstrate NASA's ability to support interplanetary spacecraft missions. NASA agreed to fund the effort, so Walt Victor and Sol commissioned a team that would be led by Goldstein to complete the task.

And so it was that one night in early 1961, at Venus solar conjunction, that the team used the Goldstone DSS13 system to successfully detect the bounce of a radar signal from Venus. The uplink signal was modulated with a message recorded by Conrad Foster, Goldstein's technician. The taped message was recorded at 15 ips, but sent at the slowest rate available on the tape recorder they had, I think around 1.75 ips. Conrad was chosen because he had the deepest (i.e., least bandwidth) voice. The return signal was similarly recorded at low speed and then played back at the recorded speed. In this way the message could be confined to a very narrow band about the carrier for noise rejection purposes. The playback was still very noisy, but Conrad's message was understandable. I was on site at the time and I remember the jubilation exhibited by all.

Goldstein not only demonstrated the possibility of using the DSIF as a radio science instrument, but soon he also, in continued studies, became the co-discoverer of the retrograde motion of, and other scientific facts about, Venus He further conducted similar studies on Mercury, Mars, the asteroid Icarus, and a comet.

DSS13 was subsequently named the Venus site. In 1964, I conducted range-gated radar experiments with the goal of tracking the variations in distance to the face of Venus, with the goal of measuring surface elevation irregularities. What I actually found was that the ephemeris I had been given, and which was used to program the range gates, was in error by several kilometers. At first, I couldn't acquire the face at all, my gate were just too narrow. Finally, I widened the gate distances enough that I could distinguish the face, and then narrowed it gradually, over several tracks, until I was able to get a precision of about 10 meters in light-time measure.

Concurrently, I would give my data to JPL Navigation, who used it to improve the ephemerides of the orbits of the Earth and Venus and to refine their value for the Astronomic Unit (mean Earth–Sun distance) by several of orders-of-magnitude in accuracy.

Butler Building 167 and the JFK Academy

By 1961, office spaces for the Communications Research and Information Processing groups had moved into an isolated steel building that was located along the boundary fence of the Laboratory. It had one entry from the adjacent off-Lab parking lot, and one door leading into the Laboratory. The secretarial pool was located at the door, deputized to buzz people having the proper clearance in and out of the Lab proper. It had been fabricated by the Butler company and designated Building 167 on the JPL site map.

President Kennedy had just been elected and had expressed his desire to tackle the Nation's ill health. In the spirit of compliance, many of us walked the few miles between the Laboratory and restaurants in nearby La Canada-Flintridge each Friday for lunch.

There was a mobile unit with sanitary facilities adjacent to Building 167, on the parking lot side. In respect and good humor, Gus Solomon dubbed our sanitary trailer as "The JFK Academy," or just the Academy. The euphemism stuck, as it seemed much more polite to say "I'm going to the Academy," than to express the intent in alternate verbiage.

The Academy appellation was dropped after the assassination of the president in 1963.

Because it was technically "off-Lab," it was deemed proper to provide office space for foreign nationals not having the clearance for unescorted on-Lab access. There were two such individuals, Werner Fogy and Phillip Hartl, from the German Institute for Air and Space Research (Deutsche Versuchsanstalt fur Luft- und Raumfahrt, or DVLR) that were here for a year, as part of a NASA cooperative interchange with the newly forming European Space Agency. I shared a

Solomon Golomb in the Faye Zlok Days at JPL 317

desk in their office while finishing up my PhD at Caltech, then working 20 hours a week.

By this time, Stan Lorens had left the Lab and Andy Viterbi had become supervisor of the Communications Research Group. Sol had been promoted to be Walt Victor's Deputy Section Manager, and Edward Poser had taken over the leadership of the Information Processing Group.

Several groups and new faces of note were added to the Communications Research Section. Bill Lindsey, Tom Kailath, Bill Kendall, and Howard Rumsey were all fresh-out PhDs. Len Baumert by this time was employed full time, but had not yet finished his thesis. He, Sol, and Marshall Hall, Jr., at Caltech, were the co-discoverers of the 92×92 Hadamard Matrix in 1961.

Goldstein's effort had been cast into a Radio Research Group with Richard Sydnor, Robin Winklestein, Duane Muhleman, Paul Reichley, and Shalav Zohar.

It was during this time that Sol had a paper approved for presentation in Tokyo, Japan. We were all aware that he could speak all the Scandinavian languages as well as Hebrew, Russian, French, Spanish, and Italian. He claimed to be fluent in at least fourteen languages, in all. So when it became known that he was to present that paper, it was no surprise to us that he took the challenge to master enough Japanese that he could present the paper so that locals would understand. From time to time during the learning period he would give us short insights into the Japanese language, the Kanji alphabet, and a few words of interest.

A year later, he did the same for a Chinese conference.

A Unique Intellectual Environment

The Communications Research Section became staffed with extremely talented, intensely intelligent, and elite researchers being mentored by a congenial genius who led by inspiration and example, considerate persuasion, and cordial communication. They responded by attacking and solving the most difficult telecommunications problems facing the emerging field of deep space communications. Most of them were either PhDs or grad students working on PhDs.

Sol also engaged many academic consultants to work with us, among which were Marshall Hall, Joel Franklin, and Gary Lorden from Caltech; Bob Gagliardi, Chuck Weber, and Bob Scholtz from USC; Leonard Kleinrock from UCLA; and Elwyn Berlekamp from UC Berkeley.

Staff members were encouraged to publish their research. They became so successful in their endeavors that soon, almost every month, a new paper from someone in the Section was being published in a tech journal or presented at a research conference. Within a decade, almost all of the significant theoretical problems of deep-space communications had been solved. Research that followed was primarily tweaking the older theory to gain a db here or there.

Their achievements were ultimately rewarded in many ways. Many became Fellow Members of the IEEE, AAAS, and AMS; entrepreneurs in very successful companies; and noted contributors to the profession. Many achieved significant recognition in the form of NASA medals and industry prizes and awards, including:

- NASA Exceptional Service Medal
- NASA Exceptional Engineering Achievement Medal
- Marconi Prize
- Claude E. Shannon Award
- National Medal of Science
- George Polya Prize
- IEEE Edwin H. Armstrong Achievement Award
- IEEE Alexander Graham Bell Medal
- IEEE/RSE James Clerk Maxwell Medal
- IEEE Medal of Honor
- IEEE Milestone Award
- IEEE Prize Paper Award

The Accident

One morning in late February of 1961 I turned in my candidacy papers for my PhD degree and proceeded toward JPL for work. I had finished all my Caltech course work and was ready to embark on my thesis, topic as yet TBD. I was heading north, but had stopped for a red light behind a green Buick on Orange Grove Boulevard in the area used to stage floats for the Rose Parade. The light turned green and I began forward.

My next remembrance was waking up in nearby Huntington Hospital with a doctor probing my face for broken glass shards. I later learned that a Culligan Soft Water truck was also at the same intersection, going south. When the light turned green and the Buick cleared the intersection enough for him to turn, he did so, smashing into my small Austin-Healy Sprite in T-bone fashion. The collision pushed my car onto a nearby curb and resulted in a concussion that left me unconscious for an hour, with double vision for a month, and a broken mandible that required surgery to wire back together. Since there are no jaw-splints, my mouth was wired shut for a period of 8 weeks thereafter. I still retain the tantalum wire in my jaw.

Further, the mental nerve in my lower jaw had been severed, rendering the left side of my face numb and immovable. I was told that the nerve would regenerate over time, but that they grow very slowly. For several years I had a crooked smile. I have no memory of seeing the collision to this day.

I know that the accident had pushed my Sprite onto the curb because it had occurred right in front of the home of a Los Angeles County supervisor. He was at home at the time and came on scene in time for photographers to snap him assisting the EMT personnel removing the unconscious me from the car. The picture was published in the *Pasadena Star News* the next day.

Gus Solomon jokingly claimed jealousy over the publicity I had received, as he always was looking for media coverage. Ed Posner came to see me in the hospital and brought a bottle of sherry to cheer me up. The staff said no, so he proceeded to drink it all himself as we conversed.

My hospital stay extended for quite a while, it seemed. I remember watching Astronaut John Glenn in his Friendship 7 flight soon after arriving. Also, my son, Robert Drew, was born on March first while I was still a patient. When discharged, neither my wife nor I could drive or do much to support our newborn or handle his two-year old sister. Luckily my brother Bill, a Master Sergeant in the Air Force, was given special leave to come to our aid.

As one added note, my youngest grandson Zane was born while Senator John Glenn was on the space station in his October 29–November 7, 1998 return to space.

When I could finally drive again, I returned, jaws still wired shut, to JPL for work and to began looking for a PhD thesis topic.

Sol and My PhD Thesis

By this time, Sol had been granted a joint appointment to the Caltech EE department staff. Hardy Martel, the Caltech professor who would have otherwise been my PhD advisor, deemed himself not acquainted with the state of the art in communications and coding theory sufficiently to take on another PhD student in that field (he was Goldstein's advisor). Because Sol was an acknowledged expert in the field and now was a member of the EE department, and since Goldstein had won the battle of JPL employees having the right to theses on their JPL work, he was approved to be the PhD advisor to Jack Stiffler, also in our 331 group. As mentioned earlier, Jack became Sol's first PhD student, to graduate in 1962.

Sol was also assigned to be my PhD advisor. He was only a little more than two years older than I was.

On learning that Sol would be my advisor, I approached him one day to discuss a topic for my thesis. He said, "You know, if you could count the number of equivalence classes necklaces (i.e., cyclic codes) of m-colored beads under rotations and reflections that would make a good thesis."

I thought about it overnight and came back the next say to tell him I had solved the problem. I had merely applied Polya's formula to the group of necklace reflections and rotations and showed him the simple result. I had been able to apply a method similar to the one that Sol and Lloyd Welch had just published a year earlier on the enumeration of polygons to determine the m-sequence formula. He looked at it and said "That will make a good chapter for your thesis."

We then decided to look into optimizing Boolean combinations of shift-register codes having short co-prime lengths into very long cyclic codes that could be used for spacecraft ranging. The idea was to transmit the longer composite code and to decode the received spacecraft-retransmitted signal by cross-correlations against local versions of each component sequence to determine the phase delay of each.

I developed a Fourier-like transform theory of such codes using the Rademacher-Walsh functions as its orthogonal basis. I was able to determine the optimum Boolean function for combining the component pseudonoise sequences for uplink transmission and the maximum likelihood decoding method by which the range could be determined in least time for a given probability of success by a receiver with only a single cross-correlator, sequentially shared in acquiring the cyclic phase offsets of each of the locally generated component pseudonoise codes.

That part of the thesis required an additional six chapters, ranging from modulation, synthesis, optimization, detection, decoding, and analysis, before the thesis was deemed complete. My thesis committee, which Sol chaired, included Hardy Martel, Marshall Hall, Olga Tausky Todd, and Dave Bachman. I successfully defended my thesis one day in mid November 1962. My formal graduation took place in the ceremony of June 1963.

I convened a number of friends and colleagues at what I called my "PhD Phling" party soon after the graduation ceremony. Sol, Gus Solomon, Bill Lindsey, Jack Stiffler, and Scott VanSant (and maybe others) from JPL were in attendance. I also had a quite a number of musician friends there, so there was rife entertainment, libation, and hilarity. A photo from that evening appears in Fig. 2, showing Sol

Fig. 2. Photograph of some of the attendees of the PhD Fling of 1963. Note Sol in conversation on the right, with Gus Solomon and Jack Stiffler in the doorway, similarly engaged. All appear to be painfully awaiting the conclusion of the current entertainment.

The Snake in the Toilet

By June 1962 my thesis work was well underway. My wife and I had decided that it was time to move from our four-unit apartment very near the Caltech campus to a house in Altadena, to the north. We had been out hunting all one Sunday afternoon, found what we wanted, and had returned home. I had gone into the bathroom for what I had intended to be a short, stand-up visit, lifted the closed lid, and discovered the head of a very large snake peering up at me.

Not knowing what to do, I called my landlady, who told me to call the SPCA. I tried calling the SPCA, but they were closed on Sundays. So, in desperation, I called the police, who told me to call the SPCA. On realizing I couldn't, they said they had an emergency number they would call.

Shortly, a patrol car and animal control van, both with police insignia arrived. The policeman and animal control officer attempted to pull the snake from the toilet, but it decided it wanted none of that and slithered back down into the toilet and drainpipe.

They decided to check on my neighbors in the four-unit house, only to find that the two undergrad Caltech students next door were indeed the owners of the snake. It had escaped from them by slithering down their commode, just on the other side of mine, earlier that day. They had tried to retrieve it by pulling on the tail, but it had escaped them also.

The two officers decided to remove the commode from the floor in order to gain better access. But in order to do that, they needed a wrench. They didn't have a wrench. When a policeman needs a wrench, they call the Fire Department to bring them one. The Fire Department policy was to keep squads together when called upon, so their delivery came by fire engine and the supervisor's van.

By now, in front of my California Avenue apartment, had gathered the police car, the police animal control van, the fire engine and supervisor van, and another police car belonging to the police unit supervisor. Plus a photographer from the *Pasadena Star News*. Plus a crowd of some 25–50 curious neighbors.

On removing the commode from the floor one could see the snake, soon to be identified as a seven-foot anaconda, dangling from the floor-side opening. The firemen do carry sledge hammers, so the device was taken into the back yard, broken apart, and the snake removed. It was promptly removed to SPCA care.

Remember, I had gone into the bathroom for a specific purpose, as yet unfulfilled. But by now I had no toilet. So the firemen removed the one from the Caltech students' side and installed it in my bathroom. At last I was able to complete my quest.

When the photographs of the incident were published the next day in the *Star News*, Gus Solomon again feigned jealousy over the publicity.

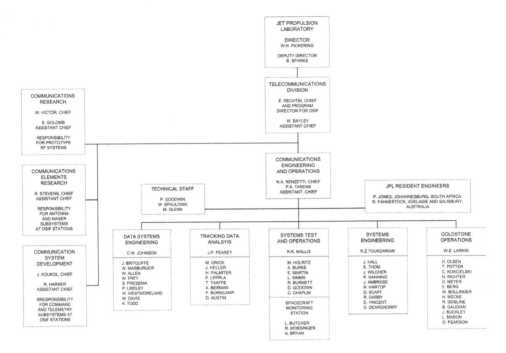

Fig. 3. A 1961 Organization of the JPL telecommunications division. Note at the top-left that Sol was Deputy Section Manager of the Communications Research Section.

Promotion to Group Supervisor

A 1961 organization chart of the Telecommunications Division, headed by Eb Rechtin appears in Fig. 3. It shows the Communications Research Section with Walt Victor as manager and Sol Golomb now promoted to Assistant Chief. His office, however, remained with us in Building 167. Ed Posner was given Sol's Information Processing Group.

Shortly after this chart was distributed, Walt Victor was promoted to be Rechtin's Deputy Division Manager. An organization chart of October 1962, seen in Fig. 4, shows Mahlon Easterling as being appointed to manage the newly named Communications Systems Research Section and Sol still on the Section staff, but now with the title Staff Scientist and Assistant Manager. It was only a part-time position for Sol, as he was by now preparing to join the USC staff.

Andy Viterbi had left for UCLA and was working only part-time at JPL as a consultant. I was made Group Supervisor of the combined remnants of Mahlon's and Andy's groups, now called the Digital Communications Group.

The 1962 organization chart lists me under my birth name. In December 1964 my last name was changed from Titsworth to Tausworthe. That chart of October shows the staff and grouping of the Section. Names that USCers will recognize in my group include Bill Lindsey, Gus Solomon, and Andy Viterbi.

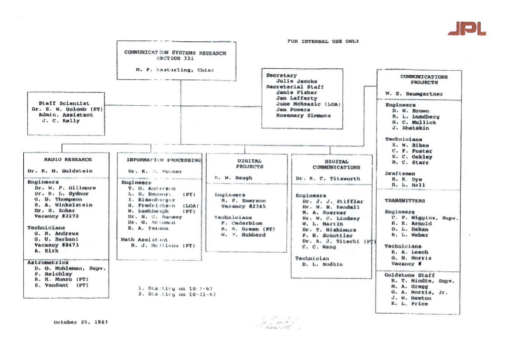

Fig. 4. A 1963 organization chart of the Communication Systems Research Section showing group breakdown.

The Twist Demonstration

Tom Kailath came to JPL in 1961 straight out of MIT for about a year before he departed for a very distinguished career in the Stanford University EE Department. He had probably been recruited by Andy Viterbi, by now a group supervisor and MIT graduate himself.

As a side note, Tom exclusively used public transportation while living in Boston and going to MIT. But on coming to California, he realized that he needed to learn to drive. He bought an old clunker of a car to learn on. His learner's permit allowed him to drive only when a licensed driver was present in the car in case of emergency. I lived fairly close to him and was on his route to JPL, so I agreed to be his chaperone during the learning period. It was sometimes a bit harrowing, but after a few weeks his license was granted. I gave silent thanks to the California DMV.

Gus Solomon, besides being a brilliant mathematician and co-inventor of the Reed-Solomon codes, was a Hollywood wannabee. I told him he should use the pseudonym Gilbert Sullivan in his theatrical activities because his initials would then remain the same. He regularly regaled us with titles and plots for his hoped-for theatrical and literary works, some of which were cast in humor. I mentioned his dubbing of the JFK Academy earlier. But also, on the publication of the 1962 Helen Gurley Brown book "Sex and the Single Girl," he proposed to publish a forthcoming paper under the title "Sex and the single errorcorrecting code."

Another, more serious, endeavor was entitled *The Master Singers of Newport*, which was seriously considered as a weekly television series, reaching the casting and costing stage by Hollywood studios. The producers, however, ultimately found that it would require both acting and musical talents of the cast, ingenuity and innovation of writers for a new score each week, and was just too costly to pursue further.

One day on coffee break, Gus was explaining to those gathered there, including Sol, Tom, and myself, that there was a new dance craze sweeping the country, called "the twist." Because of accentuated motion of the hips, it was the topic of many controversies among critics, who felt it was too provocative. Gus sang "Come on baby, let's do the twist" in demonstration of the dance movements. Shortly, coffee break was over and Gus left the building for a meeting, still humming and twisting.

Vey soon afterward, someone came looking for Gus and asked Tom, still nearby, if he knew where Gus was. Tom quite calmly, but with a rakish smile, said, "Yes, he just left in a fit of torque."

Sol's Departure from JPL and Afterward

Sol left JPL in the spring of 1963 for USC, despite its relatively inferior reputation at the time as compared to Caltech and UCLA. He was given a joint appointment as professor of Electrical Engineering and Mathematics. He remained there for the remainder of his life, 53 years.

Irwin Jacobs came to JPL in1964 for a year as a Resident Research Associate from MIT. His contribution to JPL research at this time was in garnering interest in convolutional codes. Andy Viterbi, still on staff as an academic consultant, had become interested in the codes and began researching their properties. Later, Irwin and Andy formed the Linkabit Corporation in 1968, and then Qualcom in 1985.

Eb Rechtin continued to expand the Telecommunications Division until his departure in 1967. He subsequently became the head of the Defense Advanced Research Projects Agency (DARPA), then the CEO of Aerospace Corporation, and finally, as a professor of engineering at the University of Southern California. Walt Victor then became the Division Manager.

I saw Sol from time to time and corresponded in holiday messages. Perhaps the last time that I saw him was at his 75th birthday celebration at USC in 2007, when a number of the JPL Section 331 members were also present. Figure 5 shows a photograph of the JPL alumni present. From the left, me, Bill Lindsey, Solomon Golomb, Herb Taylor, Andy Viterbi, Hal Fredericksen, Elwyn Berlekamp, Bill Hurd, Bob McEliece, and Laif Swanson.

Fig. 5. At Sol's 75 Birthday at USC. Right-to-left are Robert Tausworthe, William Lindsey, Sol, Herb Taylor, Andy Viterbi, Hal Fredericksen, Elwyn Berlekamp, Bill Hurd, Bob McEliece, and Laif Swanson.

The Random Number Generator

I met Joel Franklin, a mathematics consultant from Caltech at the time, in a hallway of Building 167 one day, clearly deep in thought. I greeted him and asked what seemed to be troubling him. He replied "You know, what the world needs is a really good random number generator."

Had he made that statement to any of Sol's other proteges, the result would have been the same, as we were all well versed in Sol's theory of pseudonoise shift-register sequences. Any one of us could have done what I did then, as it was already well known that all sequential bit-tuples of lengths up to the shift-register length taken from the binary sequence were uniformly distributed across its period.

I responded, "Joel, I've got just what you need." I wrote the paper, "Random Numbers Generated by Linear Recurrence modulo Two," which was published in the *Mathematics of Computation* journal in 1965. It merely extended Sol's results to show that m-tuples taken sequentially from a binary pseudorandom sequence with constraint length n and interpreted as binary numbers scaled to lie within the interval $[0, 1)$ would exhibit a two-level autocorrelation function, and that the numbers generated would be uniformly distributed over the cube of dimension n/m. and would have period $(2^n - 1)/m$.

The set of numbers generated from a 521-bit shift register thus will produce an uncorrelated sequence of 16-bit random-like numbers that will not repeat for 10^{139} years. Moreover, vectors of such numbers will appear uncorrelated up to dimension 32.

The article spawned a huge response from academic and industry researchers treating further theory, implementations, and testing of that which then became known as the "Tausworthe generator." It would have perhaps have been more accurately called the Golomb generator.

An internet search today by someone using my name alone will produce about 20,000 hits, almost all about the generator. Consequently, my other works are sometimes hard to find.

Planetary Ranging, the Tau Machine

About 1965, when it became recognized that a ranging data type was a useful thing to have in a deep space mission, two competing designs were identified. Robertson Stevens, by then my Division manager, was comparing them side by side one day on a blackboard in his office. As a sort of shorthand, he labeled the column bearing the advantages and disadvantages of the method that I had researched in my thesis by the Greek T(tau). The other contender was one using a sequentially-narrowed single-component period 2 code that had been proposed by Goldstein, but researched by Warren Martin, a member of my group. That column, after a pause, was labeled μ (mu). Thus the Tau and Mu ranging system nomenclature was coined.

The Tau ranging method was chosen over the 12 db more efficient Mu system for the Mariner 1967 mission to Venus (later renamed Mariner V) because it had been more thoroughly researched and because the radio-system engineers believed that a spread-spectrum signal was much less likely to have sidebands that could interfere with locking onto the carrier signal at the spacecraft. After two successful missions, the sideband issue was resolved, so the Tau machine approach was retired and replaced by the Mu system in future spacecraft operations.

But in the Mariner V mission, I was able to apply the method of my thesis as the first interplanetary ranging system, which successfully tracked Mariner V in its encounter of Venus and its data was later used as the spacecraft transited the Sun in further validation of Einstein's general relativity {ED: Sol's insights}".

The 12 db disparity in performance of the two systems was because there was only one cross-correlator each receiver, because of cost considerations. This correlator had to be shared sequentially in detecting each of the 75 possible phase combinations of the subsequences. The transmitted code sequence combined a period-2 clock with pseudonoise subsequences of lengths 7, 11, 15, 19, and 23, yielding a total disambiguation range of slightly over a million, but only 75 positions of subsequences to be determined.

Nevertheless, the Tau system was chosen over the Mu system at that time because the radio-system engineers believed that its spread-spectrum signal was much less likely to have sidebands that could interfere with locking onto the carrier signal at the spacecraft. After two missions, research and demonstrations allayed false-lock fears and The Tau machine approach was retired. The Mu machine was used on subsequent missions.

Today's GPS codes are formed and detected in a similar, but much simpler manner, than those used in the Tau system. Advanced technology and proximity relaxed many of the constraints that were placed on the 1967 Tau system.

In 1987, when it became possible to implement the 75 needed correlators digitally via VLSI, I was able to show that there was an acquisition procedure, not unlike the Viterbi algorithm, that gained back the 12 db loss and even produced performance slightly better than that of the Mu system. By this time, however, little interest was expressed within JPL to redo something that had worked for over a decade by another technique with so little to gain.

The Viterbi Algorithm

By 1967 my group had grown to around twelve engineers. So I was thus very busy running the group as well as keeping up with the Tau ranging missions. A number in my group were studying convolutional codes, and Andy, in particular, was considering methods of decoding them.

I had been too busy with other matters at the time to devote much time to the technology, so I asked Andy one day to explain to me how the decoding process worked. He began at the left-hand side of a blackboard with a lattice diagram, proceeding rightward as the stages of decoding developed, and explaining each step with the mathematics involved. When he got near the right-hand edge of the blackboard, he saw he needed more room to continue.

So he began to erase the leftmost part of the decoding lattice, saying "I'll erase this, I don't need it anymore." At that point, a member of my group entered my office announcing a problem that needed my immediate attention. When I turned my attention back to Andy, he was standing there gazing at the blackboard with his arms crossed and one hand at the chin, clearly in deep thought.

He said, "Let me get back to you on this. I just had a thought." He left and I proceeded to handle the issue that had arisen.

Not too long after that, Andy published his now-famous algorithm. It became the basis of his cofounding of Linkabit with Irwin Jacobs about a year later.

In a conversation after the publication, Andy said he didn't recall just how it all started, as he had thought about the decoding lattice a lot before coming up with the algorithm. But in retrospect, it is my belief that it was that moment in my office, when he erased the oldest lattice nodes, that he may have realized that keeping only a finite history of the process would produce a maximum likelihood convolutional decoding method.

Final Thoughts

I must admit to having some pangs of reverent nostalgia coming over me as I strove to remember and reconstruct the people, environment, and events in the initial decade of Sol's influence on JPL's Communications Systems Research effort. It is somewhat sad to note that none of the original stars whose names highlighted the state of the art of that era still remain. Some have passed away from us, and the rest have, by now, retired and can only look back in nostalgia, probably equally as poignant as mine.

It was the early reign of King Solomon, *ducis exemplum a magna*. It was technology's version of an earlier legendary king's Camelot.

© 2023 World Scientific Publishing Company
https://doi.org/10.1142/9789811234378_0034

Early Days with Sol Golomb

Lloyd Welch

I first met Sol in 1956. He was hired by JPL for his expertise n the area of linear shift sequences. At the time I was a graduate student at Cal Tech and worked at JPL part time.

The purpose of sequences was for anti-jamming purposes. The Army was developing radio guided missiles. Linear shift register sequence had the ideal statistical properties to reduce effects of jamming. However, there is a disadvantage to linear sequences, namely if you add a linear sequence to a delayed version, an advance version is obtained, so the jammer can overcome the distance problem. However for long sequences it is possible to find recursions for which the delay and add property would advance the sequence too far to be of any use to the jammer. So this led Sol to work on factoring trinomials.

He had a number of graduate students working on this problem including Al Hales. I may have worked on it too, I don't remember. But in general he did a lot of analysis and computation of shift register sequences which culminated in his 1967 book *Shift Register Sequences*.

These sequences had another advantage which was thought of later on. Dick Goldstein had access to the antennas at Goldstone in the Mojave desert, and he thought about bouncing signals off of Venus to measure the time of flight of the signal distance precisely. This gave precise measurement of the distance. However since the value was not well enough known, it would be a difficult task to search all the delays of the returning sequence to see which one was the right one.

So Sol came up with the idea for using three short sequences with relatively prime periods and using a majority logic to combine them. The resulting sequence had a reasonably good correlation with each of the short sequences. So you could find out what the relative timing was of each sequence and therefore accurately find the distance. I believe that Dick Goldstein in fact had the best estimate of the astronomical unit at the time from that data.

Other areas that Sol looked at were comma free codes. It was known that the number of nucleotides in a DNA sequence was 3 times as long as the amino acid sequence that they produce. So the thought was there was a dictionary mapping the 64 combinations of DNA to the 20 combinations of amino acids. To determine the right triplets to read ,the dictionary needed the property that when you strung words together the overlapped segments were not in the dictionary. That's called comma free codes.

It turned out that this had nothing to do with biology. But we did get some good theoretical results on comma free codes and they were useful in communications.

When Sol was recruited at the USC EE department in 1963, he recruited me to also join the staff in 1965. So we worked at USC together for 30 years and continued to work together on projects until his death.

We both lived in La Canada and during a fire in the early 80's, which did get near his house, I sent my daughters up there to help him and his wife Bo remove valuable artwork from their house and store it safely at my house.

© 2023 World Scientific Publishing Company
https://doi.org/10.1142/9789811234378_0035

The Consummate Scholar and Gentleman: Professor Solomon Wolf Golomb

Alan Eli Willner

Andrew and Erna Viterbi Professorial Chair and
Distinguished Professor of Electrical and Computer Engineering
Ming Hsieh Department of Electrical and Computer Engineering
Viterbi School of Engineering, University of Southern California
Los Angeles, CA 90089, USA
willner@usc.edu

Professor Solomon Golomb was a giant in science and engineering. However, he was an even more impressive human being. I spent ~25 years talking to, listening, and observing Sol. He was a person and mentor like no other, and I consider myself his disciple. Everything he did and said was with forethought, wisdom, and compassion. This chapter will highlight some examples of how Sol enriched my life.

1. Introduction

This book on Prof. Solomon Wolf Golomb contains many other chapters that deal with his extremely impactful scientific and engineering contributions to society. I'm sure that embedded in these other chapters are references to Sol as a wise and giving person with an encyclopedic memory. Indeed, this book is put together by his distinguished and devoted daughter, Dr. Beatrice Golomb, to whom I am indebted for inviting me to contribute a chapter.

My chapter, however, will take a different approach. When I met Sol in 1990, I had no idea who he was. Even today, although I know of some of his technical accomplishments and the impact they have had, I am simply not in the same research area as Sol. I fell in love with Sol the person long before I became in awe of Sol the technical luminary.

I knew Sol by having an office down the hall from his, interacting with him on numerous occasions at many levels within the University of Southern California, and learning from his extraordinary character. I also had the great fortune of having many wonderful discussions with Bo, Sol's wife of more than 60 years. They were almost always together, and she provided Sol with an added measure of her wisdom. Moreover, our common strong Jewish heritage and identity was a deep connection that bound us.

332   *The Wisdom of Solomon*

This chapter will deal with Sol as the "consummate scholar and gentleman", as mentioned in his Obituary published in the *Los Angeles Times*, May 20, 2016. It will give a picture of Sol through several examples and vignettes of my interactions with him. Although he passed away 4 years ago, I find myself constantly asking myself when confronted with a difficult decision, "What decision would Sol make?"

I was blessed and enriched to know him and learn from him, as were legions of other people. I hope you enjoy reading this chapter about a most special person, namely:

Solomon Wolf, the son of Rabbi Elhanan Tzvi, from the Jewish Priestly Kohen Group, may his name be remembered as a blessing

שלמה זאב בן הרב אלחנן צבי הכהן ז"ל

Shlomo Ze'ev ben Harav Elhanan Tzvi HaKohen, z"l

2. First Meeting

I first met Sol in the summer of 1990 when I interviewed for a faculty position in the Department of Electrical Engineering at the University of Southern California. Of course, Sol influenced my decision to come to USC.

Sol was on my interview schedule, although I had no idea who he was. After 5 minutes of his interviewing me on technical topics, the conversation changed dramatically and for the rest of the interview he described: (1) the great USC Hillel (i.e., Jewish Student Organization) — in which he was heavily involved, and (2) the wonderful Jewish community on campus — in which he was also heavily involved. He said that he would put me in touch with the Rabbi leading the Hillel if I joined USC.

Sol had an incredibly clear mind. He didn't need more than a few minutes to come to an opinion about someone or something. Once he made up his mind, he did not waiver and was an unfailing supporter.

Sure enough, soon after I joined USC, I received a call from the Rabbi. Sol had given him my name, and I was promptly invited to join the Hillel Faculty Board. Soon after, Sol became the Board Chair. Moreover, every year for decades Sol would attend High Holiday Prayer Services with the students and read the Torah portion with perfect cantillation.

For the first several years, I didn't know him as a giant, only as Sol.

3. Renaissance Man

Often I would see Sol in the hallway and we would start chatting. A quick hello would usually turn into a half hour discussion on topics that might include religion, politics, languages, and cultures. The next day, I would usually find a long handwritten note on several pages of his famous yellow paper pad detailing more insights into what we discussed. He would draw the outline of a letter, a word, or an artifact to illustrate his point. He would typically end with, "this is probably more than you wanted to know" — which was the one thing I disagreed with him about. Indeed, I always wanted to learn as much as possible from him and to be inspired by his incredible depth and breadth.

He was a true Renaissance man, seamlessly adept at almost any subject. A fun tidbit about "Renaissance Sol" — as might be mentioned in other chapters — is that he was able to count to 10 in nearly 70 languages. It is said that King Solomon of old could speak 70 languages. Sol Golomb was our King Solomon.

Furthermore, words were a passion and compulsion for him. As Mark Twain is quoted as saying, "The difference between the right word and the almost right word is really a large matter, it is the difference between a lightning bug and lightning." Sol was also our Mark Twain.

Moreover, Plato said, "Wise people speak when they have something to say, and fool's speak because they have to say something." Sol was no fool. Every word out of his mouth was measured, thought through, and wise. Even one word from Sol carried volumes of experience and wisdom. Indeed, Sol was our Plato.

With all that Sol had learnt and taught, Sol knew the value of time and hated to waste it. It was in this vein that, two days before he passed away, I asked Sol if he would consider writing an autobiography for the legions of people who cared, loved, respected, and admired him. He looked at me and said "I don't have any time."

4.   Jewish Identity

Sol used to say something interesting about his Judaism, which his daughter Beatrice repeated when she spoke at her father's funeral. She said that her father described himself as "a non-practicing Orthodox Jew." His heart was strongly Jewish, the Jewish traditions and practices held great meaning to him, his knowledge of the Hebrew Bible was astounding, and his bond to the Jewish people was unbreakable. Moreover, he was the son and grandson of Rabbis, and he was a member of the ancient Priestly Kohen group as a direct patrilineal descendant of Aaron. Sol's background combined with his unique intellect made him one of the few people who knew the entire Hebrew Bible literally by heart.

In Section 2, I described Sol's connection to USC's Hillel organization. Some other examples of his Jewish connection include the following:

(a) Sol's Second Bar Mitzvah

In 2015, when I knew Sol's 83rd Hebrew birthday was coming, I broached a subject with him. In King David's Psalms, it says that the normal lifespan is 70 years. Therefore, there are some Jewish people who have celebrated their 83rd birthday, 13 years after their 70th birthday, as a quasi-second Bar Mitzvah; a traditional "first" Bar Mitzvah being commonly celebrated on a boy's 13th birthday. I asked Sol if we could celebrate his second Bar Mitzvah with him and Bo. They came to our Synagogue, and Bo sat with my wife Michelle while Sol was called up to the

334 *The Wisdom of Solomon*

Torah, recited his Bar Mitzvah portion from the Biblical Prophets (i.e., HafTorah), and then gave the same speech he gave at his Bar Mitzvah 70 years earlier. There were around 20 teenage boys who then got up to dance and sing with Sol. His face was beaming underneath the prayer shawl that was draped over his head.

(2) Sol's Connection to Israel

Sol's roots to Israel — pre and post the establishment of the State — run deep. Branches of his family immigrated to Palestine in the early part of the 1900's, and they were considered pioneers. Sol's sister, Ms. Edna Sharoni, lives in Israel today with her children and their offspring.

Sol would have regular visits to Israel, and I would love seeing his photos and hearing his family stories. Furthermore, he was generous with his time and money to Israeli institutions.

One particular Israeli institution is the Jerusalem College of Technology (JCT), an Orthodox Jewish school in which secular and Torah Judaic studies flourish at a high level. Specifically, Sol and Bo, along with Sol's sister Edna, endowed at JCT an annual mathematics seminar in honor of their father and grandfather, Rabbi Elhanan Tzvi the son of Reb Yehuda Lev Ha-Cohen Golomb (officially titled the "Yehuda Leib Golomb Annual Lecture on Applied Mathematics"). Each year, Sol would travel to attend the annual seminar in which a distinguished scholar would deliver the lecture. Prof. Joseph Bodenheimer, a long-serving and impactful Past President of JCT, upon hearing that Sol passed away, wrote the following:

> *Prof. Sol Golomb was a close friend and supporter of JCT. He identified strongly with the unique aspect of the college that combines science and Judaic studies at the highest academic level. He founded an annual lecture at JCT in memory of his grandfather, ..., and made it a point to visit the college every year for the memorial lecture. Sol took a great interest in [our] growth, ... His special relationship with [us] is remembered with admiration and love for his unique personality.*

Another Israeli institution with which Sol had close contact was the Technion (Israel Institute of Technology) in Haifa. Sol was a member of their Board of Governors, and he would travel to Israel to attend their meetings.

At both the above institutions, he had numerous close friends and colleagues who loved seeing him.

Finally, Sol played an instrumental role in an important USC mission to Israel which included: (a) USC President Max and Niki Nikias placing a wreath at Yad Vashem Holocaust Museum, and (b) USC Viterbi School of Engineering Dean Yannis Yortsos signing a Memorandum of Understanding with the Technion on collaborative activities.

(3) Sol's Last Passover

In 2016, Sol came to our house for lunch on the seventh day of Passover (unfortunately without Bo since she was sick in the hospital), which turned out to be just

The Consummate Scholar and Gentleman

two days before he passed away. Michelle and I had our four sons as well as other guests. We were all singing, and Sol sat with such a lovely smile on his face. He also led the after-meal prayers completely out loud, punctuating each Hebrew word with deep meaning, as only he could do.

I mentioned to everyone that Dr. Golomb was the only non-family member who had attended all four of our sons' Circumcision ceremonies ("Brit Milah") and all four of the Bar Mitzvahs. Indeed, when I spoke at the naming our 4th son, Jacob Solomon, I looked at Sol and said, "May our new son be like Solomon the Wise of old and like our own Solomon the Wise."

I should mention that Sol left our house and went straight to USC to grade his students' final exams and turn in the semester grades on Friday afternoon. On Sunday, he didn't wake up.

5. The Champion of Both the Student and the Underdog

Sol and I were in the same academic department at USC, and we attended an untold number of meetings together. At these meetings, there are typically many discussions which are often followed by a vote of the faculty by a show of hands (unless, perhaps, if it is for a promotion case). In many instances, the votes had to do with passing or failing a student on a screening exam.

For the first few years, there were a few times I would vote and then notice that Sol would have a different vote. Each time, I would go over to Sol after the meeting and ask him why he voted the way he did. After a few words, I understood that he thought more deeply than I did about the issue, and I would regret my vote.

After several years, I would wait to vote at meetings until Sol would raise his hand. His opinion and his vote significantly influenced me — and many others. Compassionately, he would always take the side of the underdog, of the student in need. He was that gentle touch in the background, whispering impactful advice.

For example, we would vote every semester on the names of the Ph.D. students who should pass the department's screening exam. The faculty would look at the data, which included the screening exam and various other information about the students. Sol was the master of looking at the data in many different ways. Raw scores, how many sub-exams were passed or failed, the trajectory of each student's grades, etc. Typically, the faculty would agree on which students would pass and then discuss the border cases.

Sol's method of operation was masterful. He would say, "If we passed student X, we should also pass student Y since if you look at the data this way then student Y should be higher than student X and it wouldn't be fair to fail the student." Needless to say, student Y was almost always passed.

Whenever I give out grades to students at the end of the semester, I look at the students names and ask myself, "What grade would Sol give?" I look at the border cases and try to judge them using different criteria and metrics, like Sol would. Can I find a way to bump up this student's grade? Almost inevitably, I follow the "Sol Method."

6. Unique Dedication to USC

There was something unique about Sol's devotion to USC, and something unique about USC's devotion to Sol.

As may be mentioned in other chapters in this book, Sol was offered faculty positions at USC, UCLA, and Caltech. Although USC was the least "prestigious" at the time, he chose USC since he could have a greater impact on an up-and-coming institution. And what an impact he made! In nearly every metric, he moved USC engineering upward — and felt a personal responsibility to do so. When I was informed that I was elected into the US National Academy of Engineering, I felt that Sol was nearly as happy as I was. He was that gentle presence always caring and helping in the background at USC.

I would like to quote from two distinguished people about what Sol meant to USC, C.L. Max Nikias and Yannis Yortsos; when Sol passed away, Max was President of USC and Yannis was Dean of the USC Viterbi School of Engineering. Over the years, Sol described both of these people to me as true "Mensches," which is a Yiddish word used colloquially to describe people of extremely fine character. Both cared deeply about Sol.

(i) With Yannis' permission, I want to quote him from two eloquent speeches he gave about Sol:

  (a) 2007 Engineering Faculty and Staff Awards Luncheon: "In reviewing my records and my memory for [the faculty service award], a number of many worthy recipients crossed my mind. Among them was a person who I would have never thought has not won this award before. It gives me great pleasure to announce that this year's faculty service award goes to Professor Sol Golomb ... As far as I am concerned he is the soul of the School, its wisdom and compass. With loyalty, integrity and high standards, he has steered the school, in his own way, by mentoring innumerable faculty, students and staff, deans, provosts, and presidents. For a long overdue honor, I am very happy to present this year's faculty service award to Sol. Please join me in a standing ovation for Sol's outstanding and life-long service to the school."

  (b) Celebration of Life Ceremony held at USC a few months after Sol passed away: "[After becoming Dean], I discovered an amazing and kind man possessing the rarest combination of logic, humanity, and culture. And in February 2012, in Haifa, Tel Aviv and Jerusalem, when we visited Israel as part of the USC delegation. Sol guided me day after day in that trip through the history of the biblical lands, never forgetting to shrewdly connect everything (!) with Greek history- so as to make me feel included and connected. His brilliance was indeed kind. And his knowledge deep and illuminating. In that trip I realized that Sol was an archeologist of knowledge, uncovering layer upon layer of history, and relating it with a childlike elation and amazement, just as a child discovering a magical new world ... Today we shine a bright light on Sol's legacy because of the light he shined so brightly upon us throughout his professional and his personal journeys. So today is more than a celebration of a career. It is a celebration of an inspirational and aspirational life, one that transformed us, delighted us, helped us reach higher, and made us realize how fortunate we have been to be associated with a true genius, who solidified our faith in humanity."

Yannis memorably dedicated a relief of Sol's face in the lobby of USC's Electrical Engineering Building (EEB), and all are reminded of the "soul" of our department when entering.

(ii) With Max's permission, I want to quote him from his own chapter elsewhere in this book (which I encourage you to read) as well as from the heart-warming Eulogy he gave at Sol's funeral on May 4, 2016. (*I note that, in Jewish tradition, burial occurs soon after death. Stating that he would not miss paying his last respects to Sol at his funeral, Max changed his travel plans so he could attend.*):

> For more than fifty years, Sol Golomb cared for the University of Southern California as deeply as it is possible to care. And for more than fifty years, he served it as magnificently as it is possible to serve ... For the past quarter-century, Sol was one of my own most beloved and inspiring mentors and advisers, while serving together on the USC Viterbi School faculty. But today, as USC President, I salute him as not simply as a mentor or friend; not simply as one of our best faculty, but as one of the most distinguished and consequential scholars in the 136-year history of our university.
>
> ...
>
> Throughout his life, Sol proudly drew upon the values of his Jewish heritage, allowing him to care in the most profound ways for his academic community, for America, for Israel, and for the global community. That heritage and those values fueled his desire to promote tolerance, human dignity, diversity and a universal respect for people from all walks of life.
>
> ...
>
> He was always a sounding board and a staunch supporter. In fact, there was no other single faculty member whom I called more frequently for input and guidance than my trusted friend, Sol Golomb
>
> ...
>
> In 2012, Golomb accompanied a USC delegation I led to Israel, and it became clear that he was also a great mentor and inspiration to many there, too. He embraced the values of tolerance, human dignity, and diversity throughout his life, and this resonated deeply with so many around the globe.
>
> ...
>
> Poetically symbolizing his love for Israel—and the bridges he forever built between institutions there and his academic home at USC, a

meticulously restored book rests on the shelves of the USC Libraries' special collections. It dates to the 1550s and explores the philosophical meanings of the Torah. Inside, a plate bears a single name: Solomon W. Golomb.

7. Closing

I have included several pictures at the end of this chapter. Each figure gives a little snapshot of Sol, and I hope the reader enjoys looking at them.

This book helps write Sol's biography in the minds and hearts of all those he touched directly and indirectly. May Sol's memory be a blessing for all those who knew and will know him.

Bo and Sol Golomb. They were inseparable for over 60 years and frequently seen together.

Program of the Celebration of Life Ceremony held at USC's Town and Gown.

Sol's biography distributed at the Celebration of Life Ceremony held at USC's Town and Gown.

A plaque in Sol's memory located in the Study Hall of the Talmudical Academy in Baltimore, MD. Sol grew up in Baltimore and attended Johns Hopkins University. (*Note*: Sol's plaque was dedicated by the author and his wife, Michelle and Alan Willner, and Sol's plaque is in the same room as a plaque in memory of the author's late father, Mr. Gerald Sol Willner, z"l.)

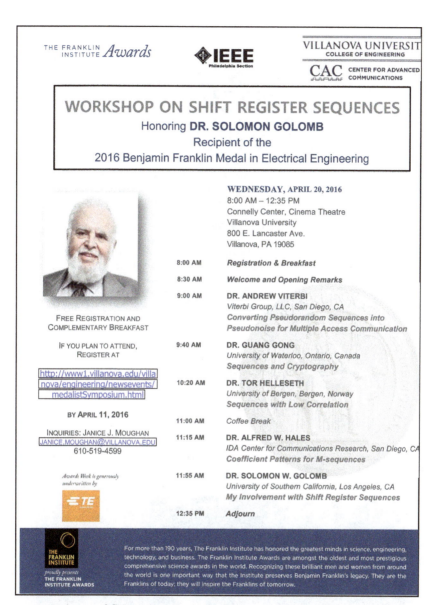

The program in honor of Sol winning the highly prestigious Franklin Medal, the last award he would receive.

שבת קודש אמֹר, ב' אייר ה'תשס"ג

Dear Moshe,
First, a hearty מזל טוב on this momentous day of your בר מצוה!

By reading your entire sedra (מקץ) from a ספר תורה, and not just the maphtir, you have distinguished yourself from many of your peers, and established a close personal bond with the sacred Torah, the foundation text of Judaism.

For many hundreds of years, many thousands of rabbis and scholars have found inspiration from the weekly Torah readings, where the opportunities for "drash" are endless.

This book of "Torah Commentary" contains essays or sermons by nearly 60 modern Jewish rabbis and thinkers on the 54 parashot into which the Torah is divided for reading throughout the year. The authors come from many backgrounds and all the main streams of Judaism.

You don't have to agree with everything they write. The Sages wisely taught us not to argue "back to a 'drash'", since what they offer are not facts but opinions and points of view; and as we read in Pirke Avot: "Which person is wise? The one who learns from every man; for as it is said (in the תנ"ך) 'From all my teachers I have become wise.'"

May you continue to grow in learning and wisdom, and have a long, happy, and prosperous life!

All the best —
Sol and Bo Golomb

A handwritten note — in English and Hebrew — from Sol and Bo to my son, Moshe, on the occasion of his Bar Mitzvah. Sol was famous for his handwritten notes that typically elucidated deep thoughts.

From left, Prof. Yoel Arieli and President Joseph Bodenheimer of the Jerusalem College of Technology, and Sol and Prof. Alan Willner from USC. This photo was taken at a fund-raiser in Los Angeles for JCT.

Postcard sent to the USC Trojan Family upon Sol receiving the National Medal of Science from the White House. For the photo, Sol chose to hold a copy of the Hebrew Chumash, the Five Books of Moses (Artscroll Publishers, Stone Edition).

A photo taken from the ceremony in which President Barack Obama conferred the National Medal of Science to Sol. President Obama whispered something into Sol's ear, and they both laughed.

A truly memorable marketing photo of two distinguished scholars at USC. Sol (pictured right) had a wonderful sense of humor, and he relished the opportunity to put on the USC Trojan Football uniform.

Obituary of Solomon W. Golomb in the *Los Angeles Times* on May 20, 2016. Indeed, he was the "consummate scholar and gentleman".

© 2023 World Scientific Publishing Company
https://doi.org/10.1142/9789811234378_0036

A Tribute to Sol Golomb, the Genius Who Transformed USC

Yannis Yortsos

Dean of Engineering, USC
Unveiling remarks for Relief of Solomon Golomb Jan 31, 2017

SOL GOLOMB RELIEF UNVEILING REMARKS

On behalf of everyone at USC Viterbi, the Ming Hsieh Department of Electrical Engineering, and our students, faculty and staff, I would like to welcome all of you to a historic occasion – the unveiling of Sol Golomb's relief – which is a small token of our respect and appreciation for our beloved colleague Sol Golomb.

Before we go on with the proceedings, please allow me to introduce some special guests:

[INTRODUCE FAMILY MEMBERS, FRIENDS AND SPECIAL GUESTS]

Thank you.

For more than five decades, Sol Golomb taught at USC Viterbi, touching the lives of thousands, students, faculty and administrators during his illustrious career. In the process, he helped elevate the school into one of the nation's best. In a most fundamental way, and seemingly effortlessly through his radiating brilliance, Sol Golomb built a field, attracted, mentored and promoted talented colleagues and created a culture of excellence that continues to this date. He was instrumental in attracting many talented colleagues, who helped built a great reputation for USC Engineering. It is unlikely that there would have been an Andy Viterbi at USC without Sol Golomb. Or a Bill Lindsey. Or a Lloyd Welch. The list is long.

But beyond engineering, Sol was also a man of many talents – a true Renaissance man. He could speak knowledgeably about nearly any subject, ranging from poetry to finance, from world politics to history. He had a great sense of humor, and he could speak multiple languages (although I don't think that he mastered Greek...).

The first-ever USC Engineering course was offered a century ago. A time when things we take for granted today, such as electricity, automobiles, refrigeration, were absent – the figment of a magical vision of the future. 100 years later, engineering and technology have truly transformed the world. In an exponential avalanche engineering has ushered in tremendous advances that have helped, among others, reduce extreme poverty, increase life expectancy, connect people globally and instantly by amazing communication technologies, many of them developed here in this place- and more importantly bring a global understanding of humanity's common destiny. This exponential pace carries the name of Moore's law, based on electronics, even though it is in fact widely more applicable. Electrical engineering has played a central role by empowering this transformation. One that carries the distinct fingerprints of Sol Golomb, Andrew Viterbi and their USC colleagues.

Indeed, in this revolution, which continues unabated today, USC Electrical Engineering has been a crucial contributor. In his book of the school's history since the mid-50s, *A Remarkable Trajectory*, Professor Emeritus George Bekey documents many of the seminal names and their impact on USC Viterbi's and EE's history. Key among them is the colossal figure of Sol Golomb. With unparalleled scholarly contributions and distinction to the field of engineering and mathematics, Sol's impact has been extraordinary, transformative, impossible to measure.

Very few people can say that they impacted the very nature of our lives. Sol could. (And so could his friend and colleague, Andrew Viterbi). His pioneering work in communications technology helped spark the digital communications revolution. I echo Andrew Viterbi, who recently said ... "we've lost a great mind, a great heart and a great sense of humor. The universe will miss him as much as we will." He will be an everlasting legend.

Last October, we celebrated the 10th anniversary of the naming of the Ming Hsieh Department of Electrical Engineering. To commemorate that occasion we unveiled a relief of Ming Hsieh. You can see it behind us. But today, we are going to mark another important milestone, by unveiling a relief to honor Sol Golomb. If I close my eyes I could vividly see Sol walking in this very place, with his characteristic determined pace on the way to his office. This place was his academic home for more than 30 years (ever since EEB was constructed).

I consider myself fortunate and honored to be able to recognize Sol in this modest, but permanent way. His relief joins those of his friend and colleague Andrew Viterbi and his wife Erna Viterbi, namesakes of our school, which are in nearby Ronald Tutor Hall. This is a symmetry that Sol would have appreciated, as it symbolizes a friendship that started more than half-a-century ago- and was further solidified earlier this year, when we established the Golomb-Viterbi Chair in Communication.

With these two reliefs in this building we honor Ming – who gave USC EE its name; and Sol, who gave USC EE its soul. I am sure that this beautiful symmetry would not have been gone unnoticed by Sol. And these two reliefs will make this location a special – a historic – place.

UNVEILING OF RELIEF
[[INVITE ANY FAMILY MEMBERS TO JOIN YOU @ PODIUM]] .
RETURN TO PODIUM

Thank you for coming today. I will look forward to seeing you at Sol's Celebration of Life which starts at 3 pm in Town and Gown.

Yannis Yortsos
Dean of Engineering, USC
USC Event to Honor Solomon Golomb, Jan 31, 2017

SOL GOLOMB CELEBRATION OF LIFE REMARKS

The last time I saw Sol was at the Franklin medal awards in Philadelphia, last April, the week before he passed away. The Benjamin Franklin Medal in Electrical Engineering was the last major award that had eluded him in his decorated career. We were very happy that he was fittingly recognized with that final distinction – thus joining the ranks of Andrew Viterbi, and Albert Einstein and Marie Curie, as a Franklin medalist. He was recognized for his pioneering work in space communications and for the design of signals in applications such as cryptography, missile guidance, defense, space and cellular communications, radar and GPS.

The official ceremony was held on Thursday evening. The School was an official sponsor – I had just made it in time taking a flight early that morning from Los Angeles, after the Viterbi awards, held the previous night in Beverly Hills – so that I could be there and celebrate with him. He was justifiably proud of this recognition – in a certain way it was the coronation of a brilliant career, the ultimate coming home for someone who grew up in nearby Maryland. At the ensuing dinner, also attended by his friend Andrew Viterbi and his son-in-law Terry Sejnowski, Sol was serene, kind and brilliant, as always. And peaceful. With the benefit of hindsight I think that he was anticipating the end.

Brilliance can be intimidating – but Sol's was uplifting; it was awe-inspiring. It flowed from his remarkable mind naturally, fluidly, effortlessly. *His* was the kindest, most magical brilliance. Being a chemical engineer, I did not know or interact with Sol as a faculty member, prior to becoming the dean of engineering. Of course I knew him from reputation – my EE friends referred to him as the "wise Solomon" – a perfect biblical analogy that went spectacularly well with the way he looked. I got to know him well, though, after I became the dean of the school. Now it is a common truism that Sol, with more than 50 years as a faculty member at USC, was always consulted in key appointments and/or strategic directions for USC (and definitely for engineering!) And I am certain with probability one that he had to approve and give his consent for my appointment. So, in my first meeting with him, 12 years ago, I was very apprehensive – worried that he will quickly uncover whatever flaws I had! Instead I discovered an amazing and kind man possessing the rarest combination of logic, humanity, and culture.

I certainly did not waste my time feeling inadequate (although I am sure he had sized me instantly) – because I was simply enthralled. Ever since that first meeting I had regular meetings with Sol, where I always, but always, marveled his weaving in his unique and magical way, numbers, puzzles, events and people, into shining objects of logic and connected narrative. With Sol, numbers came alive – and in words and events he discovered astonishing numerological significance.

350   *The Wisdom of Solomon*

I had the unique fortune and pleasure to engage this incredible mind several times: In June 2008, in Finland, when we attended the Millennium Prize event, where Andrew Viterbi was a finalist. In long walks in the town of Helsinki, in the midnight summer, he would revel on the linguistic connections of the Nordic languages (Finnish, Danish, Norwegian), uncovering the deep common roots that underpin a humane, global world. In May 2011, at Tsinghua University in Beijing, where he would enlighten us with a logic that only he would be able to impart on Chinese characters. And in February 2012, in Haifa, Tel Aviv and Jerusalem, when we visited Israel as part of the USC delegation. Sol guided me day after day in that trip through the history of the biblical lands, never forgetting to shrewdly connect everything (!) with Greek history – so as to make me feel included and connected. His brilliance was indeed kind. And his knowledge deep and illuminating. In that trip I realized that Sol was an archeologist of knowledge, uncovering layer upon layer of history, and relating it with a childlike elation and amazement, just as a child discovering a magical new world.

Sol spent every June in Israel, being on the board of Technion, the premier technological institute in the Middle East. His Judaism defined him as much as his academic achievements. The son and grandson of rabbis from the Jewish Theological Seminary in New York and from Vilna, Lithuania, he was a deeply spiritual man, full of humanity and of the wonder of life, deeply shaped by his faith.

Back on campus, Sol continued to teach a freshman seminar well into his 80s. It is really impressive that he felt compelled to continue teaching, but it speaks to the kind of person he was – in more ways than one, a quintessential teacher. And one of the most curious. His longtime friend, Professor Emeritus Bekey, remembers Sol as a renaissance man who could – and would – speak knowledgeably about any subject, from poetry and finance to history and world politics. He could do so, to varying degrees, in Hebrew, Danish, Norwegian, Swedish, French and German... But not in Greek (Smile).

Sol joined USC in 1963. He was recruited by all three prominent Southern California universities, USC, Caltech and UCLA, but he chose to become a Trojan, even though at the time USC was far from the academic powerhouse it would later become.

As he told "USC Viterbi Magazine" in 2012: "Some people were surprised I chose USC, but the question I asked myself was the one I ask my students: 'Where can you make the most difference?'"

Even before joining USC as a full-time faculty member in 1963, though, Sol had developed a relationship with the school. In addition to teaching part-time, he mentored an up-and-coming graduate student named Andrew Viterbi, who received his Ph.D. in 1962. Viterbi met Golomb his first day at JPL, where they both worked. "From that day," said Andrew, "he wasn't just my supervisor, he became my mentor."

In his book of the history of USC Viterbi since the 1950s, *A Remarkable Trajectory*, George Bekey documents many of the seminal names and their impact. But no other faculty member had more impact than Sol (I am just saying among the faculty – leaving administrators and various previous deans out of this comparison, so that I don't get in trouble. . .)

In a most fundamental way, and seemingly effortlessly through his radiating brilliance, Sol Golomb built a field, attracted, mentored and promoted talented colleagues and created a culture of excellence that continues to this date. It is unlikely that there would have been an Andy Viterbi at USC without Sol Golomb at USC. Or a Bill Lindsey. Or a Lloyd Welch. And I would even venture to say, or a Max Nikias. The list is long. For over 50 amazing years, Sol devoted his professional life to USC Engineering.

You know you have made a lasting impact when concepts are named after you. Sol was the inventor of Golomb coding. Golomb rulers, used in astronomy and in data encryption, are also named for him, as is one of the main generation techniques of arrays, the Lempel-Golomb generation method. For his remarkable contributions he was elected to the National Academy of Engineering and the National Academy of Sciences, and was awarded the prestigious Shannon Award among many others distinctions.

He was also the man who created the basic structure of Tetris. Sol just loved recreational mathematics, contributing many articles to "Scientific American". Information Theory Society members may be most familiar with his mathematics puzzles that appeared in the Society Newsletter.

In 2013, Sol went to the White House to receive the National Medal of Science from President Barack Obama – one only of 12 eminent scientists so honored that year. Just as with the Franklin Medal, he joined his friend and colleague Andrew Viterbi who had won the same in 2008 – awarded to Andy by President George W. Bush. I doubt that there is another such couple of mentor and mentee (Sol Golomb and Andrew Viterbi) in history who have won so many of the same prestigious awards, but each based on their own unique accomplishments. The school was an official sponsor of the National Medal ceremony, which I attended along with his beloved wife Bo and a table of other USC Viterbi representatives. I will never forget how happy and content Sol was. He radiated brilliance and contentment and shared his joy with all of us, when his name was announced and he marched into the hall among our enthusiastic cheers. His medal was well overdue – but in a testament to the wisdom, fairness and due process of the scientific community, it was justly awarded to Sol, a true wonder of nature, in time.

Today we shine a bright light on Sol's legacy because of the light he shined so brightly upon us throughout his professional and his personal journeys.

Earlier this afternoon, we paid tribute to Sol by unveiling a beautiful relief, which will be housed forever in the Electrical Engineering. His relief is in opposite walls to the relief of Ming Hsieh, USC Trustee and namesake of the EE department, which was installed last October, on the occasion of the Department's tenth naming anniversary. The symmetry would have been appreciated by Sol. One person gave the department its name. The other, its soul.

So today is more than a celebration of a career. It is a celebration of an inspirational and aspirational life, one that transformed us, delighted us, helped us reach higher, and made us realize how fortunate we have been to be associated with a true genius, who solidified our faith in humanity.

I will close with a very personal note. My father passed away on Greek Easter Sunday in 2002. Seven years later, in 2009, my mother passed away – on Greek Easter Sunday. And then, seven years later, in 2016, Sol Golomb passed away – on Greek Easter Sunday. I don't have the ability to explain this coincidence. But I am sure that Sol would have been delighted to connect these numbers, words and events, to provide an illuminating insight in his magical, kind brilliance.

© 2023 World Scientific Publishing Company
https://doi.org/10.1142/9789811234378_0037

Meeting Young Sol

Neal Zierler

Sol and I grew up in Baltimore. We went to high school at Baltimore City College, to Johns Hopkins University for the BA, and to Harvard for M.A. and Ph.D. degrees. We both were put in the advanced academic program at City College, which included four years each of French, German and Latin. This probably contributed to our shared love of language, our interest in learning other languages and enjoying the rich experience of visiting a foreign country while speaking the language reasonably well.

I am 5.6 years older than Sol and was several years ahead of him at high school and college, so our paths never crossed there. I was delayed a few years by service in the Navy during World War II which caused our times at Harvard to overlap.

It was 1955. I was working at the Lincoln Laboratory of M.I.T. and writing a thesis for the PhD degree at Harvard. I lived with my wife and three-year old son in an idyllic spot on the Assabet River in the Kalmia Woods section of Concord. A young man phoned one day and introduced himself as Sol Golomb. We spoke of our shared backgrounds, and agreed that he would visit for a few hours the next day.

He arrived in the early afternoon of a sunny day in summer (every day I lived there was a sunny day in summer). We reminisced about our times at City College and Hopkins. We didn't have much technical discussion, but I suspect we both had already worked out the remarkable and useful statistical and correlation properties of maximal length linear recurring sequences, shortened to m-sequences, as I called them. Sol called them "maximal length shift register sequences" to emphasize the implementation. Strictly speaking the qualification "with linear feedback" is necessary, and some nonlinear feedback shift registers are a bit longer. I developed them when investigating bit streams for use as (pseudo noise PN) sequences in the F9C (NOMAC) communications system designed by Wilbur Davenport, Robert Price and Paul Green. They had been using a linearly generated sequence that I showed was equivalent to a linear recurring sequence with a smaller number of bits in the register. I observed that certain linear recurring sequences with far superior properties, the m-sequences, were available. I also pointed out that linearly generated sequences could not be used unmodified for this application because they could be predicted from brief observation by an enemy who could then coherently jam the system, shutting it down. Sol probably developed the equivalent shift register sequences as part of his work at the Glenn L. Martin Company, a Baltimore aircraft manufacturer where he worked several summers. A few years later, under Sol's direction at the Jet Propulsion Laboratory, they were put to good use for space communication and radar ranging.

Our mathematical interests were otherwise somewhat disjoint, his in number theory and mine in analysis, so we talked of school days and a few people we both knew, but the details have faded away.

I took the year 1960 off from Lincoln Lab to work at JPL I think Sol was busy at USC then and the only time I saw him for more than a few minutes was when he invited me for dinner at his house. When I arrived he asked me to help him with a little project. We walked to the top of a hill on his property and connected a flat cable to a TV antenna. We strung the cable across about a hundred yards of desert scrub and up to his second floor bedroom window. Once the cable was safely secured we proceeded downstairs for some rest and relaxation. A little later I had a delightful dinner with Bo and Sol.

My wife met Sol for the first time during a dinner at the Princeton University Faculty Club on the occasion of my retirement from CCR in September 1996. She reported the following impression. "He had a warmth that drew you in. He enjoyed life and liked sharing his enjoyment with those around him." These qualities, this generosity of spirit enlivened by a sense of humor, were evident throughout his prodigiously productive life, as they were in the bright young man I met sixty-four years ago on a sunny day in Concord.

Part IV

© 2023 World Scientific Publishing Company
https://doi.org/10.1142/9789811234378_0038

Elhanan Golomb

Introduction by Michal Reuven (with discussion with her mother/Sol's sister Edna Sharoni)

Elhanan H. Golomb was born in 1887 in Lithuania. He received a Jewish education and emigrated to the USA at the age of 19. While working to support himself, he attended high school to learn English and then undertook academic studies first at Amherst College and afterwards at Dropsie College for Hebrew and Cognate Learning in Philadelphia. His graduate studies were in medieval philosophy and mathematics. His PhD thesis was entitled "Judah Ben Solomon Campanton and His Arba'a Kinyanim". Elhanan married Minna Nadel and they had two children: Edna Golomb Sharoni, living in Israel since 1947, and the late Professor Solomon W. Golomb. Elhanan taught Hebrew and Jewish studies at the Hebrew Union College in Baltimore for many years. After his retirement he came to Israel with his wife to be close to their daughter Edna. He passed away in 1956 at the age of 69. After his death, one of his Hebrew Union College colleagues described him as a "saintly man"; he was always ready to help everyone and was loved by students and fellow teachers.

Elhanan and Minna, Sol's Parents by Beatrice Golomb

According to those who grew up with Sol in their Baltimore neighborhood, Elhanan and Minna were both deeply respected for their learnedness and scholarship.

Elhanan was reportedly conversant with on the order of 80 languages and alphabets, although per Sol he had a strong accent in each of those he spoke; while Minna spoke "just" six, but spoke each like a native. Minna was the first of 11 children from a prominent family in Vilnius, which valued education for women as well as for men. Two of her sisters attended medical school, and at least one other had a science PhD. Sol's cousin Monique Stern (daughter of Minna's sister Riva, who received her PhD in biochemistry in Paris), noted that Jews were not permitted to receive graduate education in the sciences in Lithuania and had to travel to other countries to do so. Minna's parents (Solomon Wolf Nadel and Paula Nadel) had a

357

special place in their community, and indeed Minna's father was reportedly one of few Jews permitted to travel ~freely outside the designated Jewish area.

Monique notes that Elhanan was given great credit within the family for strongly urging others to leave "Lithuania" — at a time when it was incredibly difficult for others to grasp the threat to them posed by the Nazi regime. The decision to leave was especially difficult for those who left behind wealth and position in their community, often to face poverty. Much of the family did get out, and Elhanan assisted with this.

Sol's father had arranged a meeting with a high-ranking U.S. State Department official known for anti-Semitic propensities. (I do not recall the name, although it was mentioned by Sol.) Elhanan persuaded him to permit the entry into the U.S. of Hirsh Nadel, Minna's youngest sibling, for the couple to sponsor. The fact that Elhanan succeeded given the known proclivities of the party he met was a testament to the tremendous influence he exerted through his reasoning and palpable goodness — much the way that his son Sol could make a simple statement that could put an issue under discussion in a new light and change the opinion of an entire room.

Hirsh was a credit to the nation that took him in: after procuring a degree in civil engineering, he served with distinction in the U.S. Army in World War II. Injured twice in battle, he rose to the level of Captain and witnessed the surrender of Japan in the Philippines. He was awarded the Army's Distinguished Service Medal by General Douglas McArthur.

Minna, Sol and cousin Hirsh Nadel, Passover c. 1947.

World War II touched others in Sol's family, and some were lost in the Holocaust. Minna's sister Eva (forced to leave medical school by adverse circumstances) lost her husband and son in the Holocaust. When she was ultimately released from the Stutthof concentration camp (see picture) she went to live with Sol's family.

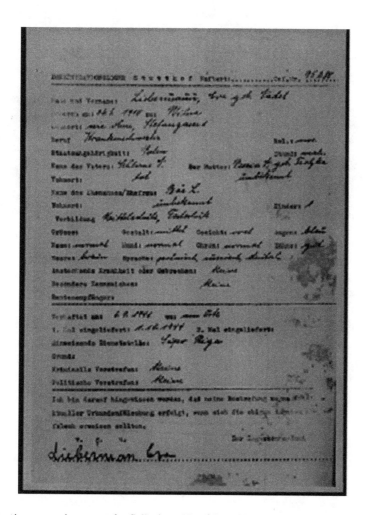

Concentration camp document for Sol's Aunt Eva (shared by Marshall Nadel, son of Hirsh).

Sol had great admiration for Eva and strongly supported women in academics before that was fashionable.

Sol's sister Edna shared that Elhanan had arranged for Minna to attend graduate school courses at Johns Hopkins in Russian Literature, and she loved this. Elhanan had attended Amherst at a time when that institution was not known for particular friendliness to Jews. Nonetheless, Elhanan was so patently learned and scholarly, and so beloved there, that they urged that when he secured his PhD (at the time presumed to be in mathematics) he return to become a member of their faculty.

Elhanan, Sol, Belle (Hirsh Nadel's wife), Minna, Aunt Eva, friend of Eva, c. 1949.

According to his obituary, Elhanan H. Golomb was a professor of the Bible and Talmud for 27 years. As reflected in the introduction by Michal Reuven, as is noted in a May 21, 1956, obituary of Elhanan H. Golomb in the *Baltimore Sun*, "'He was a beloved teacher, a saintly man and a scholar who was a veritable walking encyclopedia in the fields of Jewish literature and history. He was a mainstay of [the] faculty, a man who made a real contribution to the education of hundreds of young men and women.'"

DR. E. H. GOLOMB DIES IN ISRAEL

Was Professor For 27 Years In Baltimore Hebrew College

The death in Israel of Dr. Elhanan H. Golomb, a professor of Bible and Talmud at the Baltimore Hebrew College for 27 years, was announced yesterday by Dr. Louis L. Kaplan, dean of the local institution.

Dr. Golomb, who had lived in Israel since his retirement in 1953, died suddenly April 30 of a cerebral hemorrhage, according to information received by Dr. Kaplan.

In addition to his professional duties at the local college, Dr. Golomb also served as the librarian. He came to Baltimore in 1926.

"A Saintly Man"

Dr. Kaplan, in a tribute to his former colleague, said of Dr. Golomb: "He was a beloved teacher, a saintly man and a scholar who was a veritable walking encyclopedia in the fields of Jewish literature and history. He was a mainstay of our faculty, a man who made a real contribution to the education of hundreds of young men and women."

Dr. Golomb is survived by his wife and a daughter, Mrs. Abraham Sharoni, who are also living in Israel, and a son, Solomon, who is a graduate teaching fellow in mathematics at the University of Oslo, in Norway.

Obituary of Elhanan Hirsch Golomb. *Baltimore Sun.* May 21, 1956.

Neither Sol, nor his wife Bo (for whom the Nazi occupation of Denmark meant the actual occupation by Germans of her home, as I learned after her death from her brother, Flemming), ever spoke to me of their experiences related to World War II. I suspect for both the topic was simply too painful.

© 2023 World Scientific Publishing Company
https://doi.org/10.1142/9789811234378_0039

Sol's Historical Artifact: National Honor Society Letter

Introduction by Beatrice Golomb

Edna Sharoni: Sol's sister, born 1927, mailed this letter to me and it arrived on March 21, 2017.

A handwritten, 2-page letter from the National Honor Society dated May 6, 1949 addressed to "Mr. Golomb" states that he scored among the highest ranking students on the General Aptitude Test (given March 29), in the National Honor Society Scholarship contest, which selected top students from among hundreds of thousands of seniors from "all the states, the district of Columbia, Hawaii, Alaska, Canal Zone (I think that is what it says there — it is handwritten), and American Schools in Argentina and Japan", with his score placing him "in the 99.94 percentile or higher."

In addition to the Certificate, a letter of his achievement "will be sent to the admissions officer of the college you listed on the information report you recently sent to us" to aid in "admission or in obtaining aid as needed."

The letter was signed "Sincerely yours, for The Scholarship Board, Paul Elicker, Secretary-Director."

Since Sol was one year younger than his peers, and there is a powerful relation between age and scores on aptitude tests in that age range, this is an especial achievement.

National Honor Society
1201 16th Street, N.W. Washington 6, D.C
May 6, 1949

Dear Mr. Golomb:

You rated in the 243 highest ranking students in the General Aptitude Test given in your school on March 29, 1949, in the National Honor Society Scholarship contest. Participating in the scholarship contest were 5,915 students selected from 374,890 seniors in 1,492 schools. Based on a total of 374,890 seniors, you attained a rating in the 99.94 percentile or higher. Your exact score on the General Aptitude Test on March 29, 1949, was reported today to the Principal of your school. The Scholarship Board congratulates you on this high achievement. The formal announcement in the form of a Certificate of Merit (encased) was sent today to your high-school Principal who has been requested to award it to you at an early date.

2

To attain a place in the 243 highest ranking students is an outstanding achievement, and you merit high honor. All 5,915 contestants in the 1,492 schools in all the states, the district of Colombia, Hawaii, Alaska, Canal Zone, and American Schools in Argentine and Japan are considered outstanding students. In addition to the Certificate of Merit, a letter of your achievement on the General Aptitude Test taken on March 29, 1949, will be sent to the admissions officer of the college you listed on the information report you recently sent to us. It may aid you in admission or in obtaining aid if needed.

Sincerely yours,
for The Scholarship Board
Paul Elicker
Secretary - Director

Early Sol and Family

Sol's Father, Elhanan Golomb.

Sol's parents, Elhanan and Minna Golomb, Vilna, April 1922; wedding-week.

Early Sol and Family

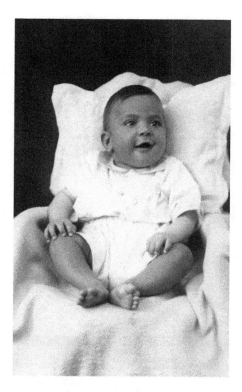

Sol c. 4-5 months old Oct 21, 1932 Postcard from Baltimore shared by Sol's sister, Edna Sharoni.

Sol and sister Edna. Photo courtesy of Edna Sharoni (Sol's sister).

Sol on pony. Photo courtesy of Edna Sharoni (Sol's sister).

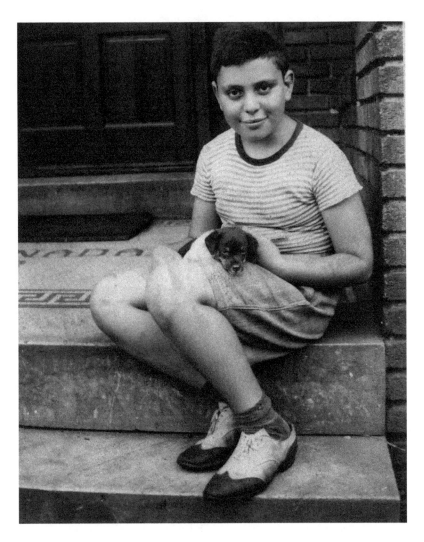

Sol with dog Victory. Photo courtesy of Edna Sharoni (Sol's sister).

Sol playing Chess Circa 1947 (from cousin Monique Stern).

Danish Newspaper Clipping reporting Marriage of Bo and Sol (top couple). (Shared by Bob Rosenstein.)

Sol and Bo (wedding day, Copenhagen).

Sol's wife, Bo.

Bo with Astrid and Beatrice.

© 2023 World Scientific Publishing Company
https://doi.org/10.1142/9789811234378_0041

JPL Section 331 — The Space Age Begins

In 1956, Sol Golomb returned to the US after his Fulbright year in Norway. He joined the staff of Caltech's Jet Propulsion Laboratory (JPL) in La Canada Flintridge, CA, as a Senior Research Engineer. He was assigned to Section 331 (Communications Systems Research), where he stayed until he moved to USC in 1963 — rising to the ranks of Assistant Section Chief and Staff Scientist. He was JPL's leader in digital communications and information theory research.

Below is an April 1957 picture of the full Section 331 staff, "Roots of Section 331", and an enlarged center part of the picture. Sol is in the middle, in the enlarged one, just to the left of center. Thanks to Eb Rechtin and Najmedin Meshkati for the picture.

Six months after this picture the USSR launched Sputnik. Soon JPL changed its main mission from missile guidance to space exploration/technology.

378 The Wisdom of Solomon

© 2023 World Scientific Publishing Company
https://doi.org/10.1142/9789811234378_0042

Life Magazine — **Explorer One**[*]

In early 1958 the US launched its first successful earth satellite "Explorer". It was designed and built at Caltech's Jet Propulsion Laboratory (JPL) in La Canada Flintridge, CA. Shortly thereafter Life Magazine featured an article about this titled "Explorer's News from Space", written by James Van Allen. After stressing the quality and quantity of valuable information which the satellite was providing, it goes on to say "... most of the data is being radioed back to scientists on the ground by an ingenious apparatus assembled in final form by the staff of the Jet Propulsion Laboratory of the California Institute of Technology".

At the end of the article is a picture captioned as follows:

"PLOTTING EXPLORER'S ORBIT, researchers at Caltech's Jet Propulsion Laboratory compute it by triangulating the signals from tracking stations on an aluminum surface representing parts of the earth. Lab assembled the satellite's instruments and built upper three stages of launching missile, Jupiter C". A young Sol Golomb can be seen in the middle back of this picture. See below.

---

[*]This chapter was originally published in *Life Magazine, "Explorer's News from Space,"* February 17, 1958, pp. 51–54.

379

© 2023 World Scientific Publishing Company
https://doi.org/10.1142/9789811234378_0043

Sol's Trip Report re Marshall Hall Visit

Introduction by Alfred W. Hales

In 1958 Marshall Hall, Jr., visited Caltech and Sol met with him there, afterwards writing a "trip report" about this and circulating it to some of the JPL staff (including me). As it turns out this was a recruiting trip, for a year later Hall left Ohio State and moved to Caltech. This had a big effect on many people — myself, Bob McEliece, and Len Baumert included. And it led to the Hadamard matrix result which is also included in the book.

Here we include a copy of the trip report.

The Wisdom of Solomon

JET PROPULSION LABORATORY

REPORT NO. ___8-246___

Page 1 of ___3___

SUBJECT ___Trip Report___

PROJECT _____ CONTRACTOR _____ CONTRACT or TGD NO. _____

ACTION REQUIRED BY _____

TELECON Initiated by _____	Report Prepared by ___S. W. Golomb___
CONFERENCE at ___Cal Tech___	Date Prepared ___7-12-58___
Date of Occurrence ___7-12-58___	

Participants	Distribution
Prof. Marshall Hall, Ohio State Univ. Solomon W. Golomb, JPL	E. Rechtin W. K. Victor C. S. Lorens L. R. Welch A. J. Viterbi P. Westlake R. Goldstein A. Hales S. Golomb

Professor Hall is not only the foremost American authority on combinatorial analysis, but also the original proposer of shift-register-type sequences for coding purposes. Having been in correspondence with Professor Hall for some time on a variety of problems, I was anxious to take advantage of his presence in the Los Angeles area during the month of July, 1958, to discuss problems of mutual interest. Professor Richard Dean of Cal Tech, who has been JPL's principle contact with NSA for several years, arranged the appointment, which was comparatively brief but highly fruitful. We hope to have a further discussion, probably at UCLA, during the next few weeks. The following areas were discussed:

1. Difference Sets (an algebraic viewpoint for certain pseudo-noise sequences)

   a. A new family of Hadamard designs (PN sequences with a balance of ONEs and ZEROs, and perfect two-level auto-correlation) has recently been discovered by Alfred Brauer:

Let p and q = p + 2 be a "prime pair", such as p = 3, q = 5, or p = 5, q = 7, or p = 11, q = 13, etc. Then a PN sequence of period $v = pq$ is constructed as follows:

Each of the numbers 0, 1, 2, 3,...., $v - 1$ is assigned a "zero" or a "one". The "zeros" go with the numbers a, $0 \leq a \leq v - 1$, for which $\left(\frac{a}{pq}\right) = + 1$,

where $\left(\frac{a}{v}\right)$ is the Jacobi-Legendre symbol of quadratic residue theory; and also with the numbers p, 2p, 3p,...., $(q-1)p$. For the remaining cases, namely b such that $\left(\frac{b}{pq}\right) = -1$ and 0, q, 2q,...., $(p-1)q$, the assignment of "one" is made.

Examples: i) $p = 3$, $q = 5$; $v = 15$.

Here the assignment is

```
 0  1  2  3  4  5  6  7  8  9 10 11 12 13 14
 0  1  1  1  1  0  1  0  1  1  0  0  1  0  0
```

The PN sequence thus obtained is coincidentally a maximum-length linear shift register sequence of degree 4, satisfying $a_k = a_{k-1} \oplus a_{k-4}$.

ii) $p = 5$, $q = 7$; $v = 35$.

Here the assignment is:

```
 0  1  2  3  4  5  6  7  8  9 10 11 12 13 14 15 16 17 18 19 20 21 22 23 24 25 26 27 28 29 30 31 32 33 34
 0  1  0  1  1  1  0  0  0  1  1  1  1  1  0  1  1  0  0  1  0  0  0  0  1  0  1  0  1  1  0  0  1  0
```

This case was not hitherto accounted for. Hall has a greatly simplified proof of the validity of this construction, based on his theory of multipliers of difference sets. The esthetic appeal of Brauer's theorem devolves from the relationship to the prime pairs, whose infinitude is a classical unsolved problem of number theory. From a practical standpoint, the primes pairs continue to appear as far as the tables reach (individual examples being known beyond 10^9), and even their statistical distribution is quite predictable. (The number of prime pairs up to x is believed asymptotic to $\dfrac{cx}{\ln^2 x}$, with $c = 1.320\ldots$.)

b. Generalizations of the Difference Set Approach

Ryser has generalized the incidence matrix technique to the study of matrices of ones and zeros with prescribed row and column sums. Ryser has discovered the elementary operations which change any matrix solving such constraints into any other solution.

Bruck has studied difference sets in groups. He has an example of $v = 16$, $k = 6$, $\lambda = 2$ in the sixteen element Abelian group A_2^4, with generators α, β, γ, δ, and difference set $D = \{\alpha, \beta, \gamma, \delta, \alpha + \gamma, \beta + \delta\}$. This is isomorphic to my example in the ring of 2 x 2 matrices over GF(2), with the six non-singular elements as the difference set, translated by additions of the matrix $(\begin{smallmatrix} 1 & 1 \\ 1 & 1 \end{smallmatrix})$. My example was new to Hall.

Hall agreed that my generalization of the (v, k, λ) problem to the (v, k, λ_i) problem could quite possibly shed new light on many areas. Hall was also interested in my theorem: "In a ring R with a difference set D, the following three conditions are equivalent: i) Left multipliers exist; ii) The left multipliers form a group, iii) The ring has a left identity.

2. Random Permutations

I mentioned to Hall the problem of cycle decomposition of N objects under random permutation, but not the context [as a statistical model for non-linear shift register sequences] in which it arose. After outlining my basic approach

-2-

and results, I posed the question of finding the expected length of the longest cycle. From the bounds I had obtained, Hall was firmly convinced that the limiting value would be $\lambda = \frac{\sqrt{5}-1}{2} = 61.8\%$, which I had suggested as the only interesting constant left in the range of possible answers. Time did not permit us to work out a proof of this conjecture.

3. <u>Other Subjects</u>

    a. When I asked Hall if he knew anything of the origin of the formula for counting equivalence classes, and specifically if Gleason had gotten it from him, he replied that he has always called "Gleason's formula."

    b. Hall is aware of the existence and difficulty of the comma-free coding problem. I gave him a reprint of the Golomb-Gordon-Welch paper which recently appeared in the Canadian Journal of Math.

    c. Hall mentioned that every consistent finite set of integers can be enlarged to a difference set for the integers, whereas consistent infinite sets frequently cannot be so enlarged. [Replacing each element of a difference set for the integers by its double produces a consistent set which cannot be enlarged to a difference set.]

SWO:bjg

-3-

© 2023 World Scientific Publishing Company
https://doi.org/10.1142/9789811234378_0044

Comma-Free Codes and Sol's Tribute to Max Delbrück*

Introduction by Al Hales

In 1958 Sol coauthored two papers on comma-free codes. Initially this work seemed to promise profound implications for the genetic code, though nature proved to have taken a different approach. Nevertheless the seriousness of the work is indicated by the fact that Max Delbruck (Nobel Prize winner) was one of the coauthors. Here we include three critical pages from a Danish biological journal paper, by Delbruck, Golomb and Welch, and the paper for the technical contents by Golomb, Gordon, and Welch, entitled "Comma-Free Codes". This was published by *Canadian Journal of Mathematics*, in 1958.

We also include the tribute Max Delbrück written by Sol himself which is the only such tribute he wrote. It contains numerous earlier developments and ideas of coding for biology. It was published by *The American Scholar* in 1982.

---

*The first chapter entry is 'Construction and Properties of Comma-Free Codes' by Solomon W. Golomb (with Lloyd R. Welch and Max Delbruck) in the Danish journal *Biol. Medd. Dan. Vid. Selsk.* Vol. 23 No. 9, 1958 pp. 3–34. Second chapter entry: "Comma-Free Codes" by Solomon W. Golomb (with Basil Gordon and Lloyd R. Welch), *Canadian Journal of Mathematics* Vol. 10, 1958, pp. 202–209. Third chapter entry: "Max Delbruck: An Appreciation" by Solomon W. Golomb, *The American Scholar* Vol. 51, No. 3 (Summer 1982), pp. 351–367.

Biologiske Meddelelser
udgivet af
Det Kongelige Danske Videnskabernes Selskab
Bind **23,** nr. 9

Biol. Medd. Dan. Vid. Selsk. **23,** no. 9 (1958)

CONSTRUCTION AND PROPERTIES OF COMMA-FREE CODES

BY

S. W. GOLOMB, L. R. WELCH, AND M. DELBRÜCK

København 1958
i kommission hos Ejnar Munksgaard

CONTENTS

Part I.

Pag.

Origin of the Problems, Summary and Discussion of Results by
    M. DELBRÜCK... 3

Part II.

Mathematical Developments by S. W. GOLOMB and L. R. WELCH. 14

  1. Definitions and General Theorems 14
  2. Classes of Dictionaries for $k = 3$ 16
  3. Reversible Portions of Dictionaries 20
  4. Characteristics of Messages................................... 24
  5. Mis-sense and Non-sense 26
  6. Extensions to Larger Values of k........................... 27
  7. Transposable Dictionaries................................... 29

Synopsis.

The sequence of bases in deoxyribonucleic acids is assumed to represent a coded message, embodying information concerning the sequence of amino acids in proteins. Crick *et al.* [7] suggested that the code might be a "comma-free triplet" code. This means that each amino acid is coded by a triplet of bases, and that the triplets are chosen such that no overlap between any pair of triplets codes for an amino acid. In such a code the triplets do not have to be separated from each other by some kind of comma; they can be run together without causing ambiguities in the message.

This paper concerns the following aspects of comma-free codes:

1) Procedures for the construction of all comma-free triplet codes involving the maximum number (20) of triplets. It is shown that there are five classes of such codes and a total of 408 codes.

2) It is shown that no message written with any of these codes ever contains a fourfold repeat of any base, and that in some of the codes certain threefold repeats are excluded.

3) Certain misprints in the coded message will produce nonsense (the resulting triplet does not code for any amino acid), other misprints will produce missense (the resulting triplet codes for a different amino acid). The codes were studied with respect to missense/nonsense ratio produced by various classes of misprints.

4) DNA has a directional symmetry. The basic structure is such that the message could be read in either direction. The question is posed whether codes could be devised such that if they are read in the wrong direction they give nonsense everywhere, i. e., no triplet or overlap between triplets read in reverse corresponds to any amino acid. Such codes are termed transposable codes. It turns out that a transposable triplet code can code for at most 10 amino acids, which is too few. Therefore quadruplet codes were taken into consideration. These are mathematically more difficult to handle and only a few fragmentary results have been obtained so far.

Printed in Denmark
Bianco Lunos Bogtrykkeri A-S

References.

1. WATSON, J. D., and CRICK, F. H. C.: "A Structure for Deoxyribose Nucleic Acid." Nature 171: 737 (1953).
2. HOAGLAND, M. B., STEPHENSON, M. L., SCOTT, J. F., HECHT, L. I., and ZAMECNIK, P. C.: "A Soluble Ribonucleic Acid Intermediate in Protein Synthesis." J. Biol. Chem. 231: 241–257 (1958).
3. BERG, P., and OFENGAND, E. J.: "An Enzymatic Mechanism for Linking Amino Acids to RNA." Proc. Natl. Ac. Scie. (U.S.A.) 44: 78–86 (1958).
4. SCHWEET, R. S., BOVARD, F. C., ALLEN, E., and GLASSMAN, E.: "The Incorporation of Amino Acids into Ribonucleic Acids." Proc. Natl. Ac. Scie. (U.S.A.) 44: 173–177 (1958).
5. GAMOW, G.: "Possible Mathematical Relation Between Deoxyribonucleic Acid and Proteins." Biol. Medd. Dan. Vid. Selsk. 22: no. 2 (1954).
6. BRENNER, S.: "On the Impossibility of All Overlapping Triplet Codes in Information Transfer from Nucleic Acids to Proteins." Proc. Natl. Ac. Scie. (U.S.A.) 43: 687–694 (1957).
7. CRICK, F. H. C., GRIFFITH, J. S., and ORGEL, L. E.: "Codes Without Commas." Proc. Natl. Ac. Scie. (U.S.A.) 43: 416–421 (1957).
8. GOLOMB, S. W., GORDON, B., and WELCH, L. R.: "Comma-free Codes." Can. J. Math. 10: 202–209 (1958).
9. MESELSON, M., and STAHL, F. W.: "The Replication of DNA in *Escherichia coli.*" Proc. Natl. Ac. Scie. (U.S.A.) 44: nr. 7 (1958).
10. SHAPIRO, H. S., and CHARGAFF, E.: "Studies on the Nucleotide Arrangement in Deoxyribonucleic Acids. II. Differential Analysis of Pyrimidine Nucleotide Distribution as a Method of Characterization." Biochem. Biophys. Acta 26: 609–623 (1957).
11. JACOBSON, N.: "Lectures in Abstract Algebra." D. van Nostrand Co., Inc., 1951, Vol. I, p. 207.

Indleveret til selskabet den 9 juni 1958.
Færdig fra trykkeriet den 7. november 1958.

COMMA-FREE CODES

S. W. GOLOMB, BASIL GORDON AND L. R. WELCH

1. A General Combinatorial Problem. Let n be a fixed positive integer, and consider an alphabet consisting of the numbers 1, 2, ... , n. With this alphabet form all possible k-letter words $(a_1 a_2 \ldots a_k)$, where k is also fixed. There are evidently n^k such words in all.

Definition: A set D of k-letter words is called a *comma-free dictionary* if whenever $(a_1 a_2 \ldots a_k)$ and $(b_1 b_2 \ldots b_k)$ are in D, the "overlaps" $(a_2 a_3 \ldots a_k b_1)$, $(a_3 \ldots a_k b_1 b_2)$, ... , $(a_k b_1 \ldots b_{k-1})$ are not in D.

The problem to be investigated here is that of determining the greatest number of words that a comma-free dictionary can possess. We denote this number by $W_k(n)$.

THEOREM 1.

$$W_k(n) \leqslant \frac{1}{k} \sum \mu(d) \, n^{k/d},$$

where the summation is extended over all divisors d of k, and $\mu(d)$ is the Möbius function, defined by

$$\mu(d) = \begin{cases} 1 & \text{if } d = 1 \\ 0 & \text{if } d \text{ has any square factor} \\ (-1)^r & \text{if } d = p_1 p_2 \ldots p_r, \text{ where } p_1, \ldots, p_r \text{ are distinct primes.} \end{cases}$$

Proof. Let d be a divisor of k. We say that a word $(a_1 a_2 \ldots a_k)$ has subperiod d if it is of the form $(a_1 a_2 \ldots a_d \, a_1 a_2 \ldots a_d \ldots a_1 a_2 \ldots a_d)$, and if d is the smallest number for which this is true. For example, if $k = 6$, then $(a \, a \, a \, a \, a \, a)$ has subperiod 1, $(a \, b \, a \, b \, a \, b)$ has subperiod 2 if $a \neq b$, $(a \, b \, c \, a \, b \, c)$ has subperiod 3 if $a \neq b$ or $b \neq c$, and all other words have subperiod 6. Any word w of a comma-free dictionary must have subperiod k because otherwise ww would contain an overlap of w. (Consider for example, $[a \, b \, c \, a \, b \, c] \, [a \, b \, c \, a \, b \, c]$.) We shall call words of subperiod k *primitive*.

For later purposes it is convenient to call two words *equivalent* if one is a cyclic permutation of the other, and to speak of $(a_1 a_2 \ldots a_k)$, $(a_2 \ldots a_k a_1)$, ... , $(a_k a_1 \ldots a_{k-1})$ as forming an equivalence class. If $(a_1 a_2 \ldots a_k)$ is primitive, then its equivalence class is also called primitive, and consists of k distinct words. At most one of these can be a word in D, for otherwise a contradiction

---

Received July 11, 1957.

390   *The Wisdom of Solomon*

COMMA-FREE CODES   203

would again arise upon considering the overlaps of ww. Hence, if $P_k(n)$ is the total number of primitive words, then

$$W_k(n) \leqslant \frac{1}{k} P_k(n).$$

But since each of the n^k words has some subperiod, $P_k(n)$ satisfies the equation

$$\sum_{d/k} P_d(n) = n^k,$$

from which we obtain

$$P_k(n) = \sum_{d/k} \mu(d)\, n^{k/d}$$

by Möbius inversion. For the Möbius inversion formula, cf. (**3**, p. 28).

2. Results for k odd. Theorem 1 gives a general upper bound for $W_k(n)$, of which a few examples are

$$W_1(n) \leqslant n, \quad W_2(n) \leqslant \tfrac{1}{2}(n^2 - n), \quad W_3(n) \leqslant \tfrac{1}{3}(n^3 - n), \quad W_4(n) \leqslant \tfrac{1}{4}(n^4 - n^2).$$

In many cases this upper bound is actually attained. We believe this to be true for all odd k, and have proved it for all odd $k \leqslant 15$. Note, from the proof of Theorem 1, that the upper bound will be attained if and only if a word can be chosen from each primitive equivalence class so as to form a comma-free dictionary.

THEOREM 2. *For arbitrary n,*

$$W_k(n) = \frac{1}{k} \sum_{d/k} \mu(d)\, n^{k/d}$$

if $k = $ 1, 3, 5, 7, 9, 11, 13, 15.

Proof. For $k = 1$, the proof that $W_1(n) = n$ is immediate. For the other values of k we shall show how to select a word from each primitive equivalence class in such a way that a comma-free dictionary is obtained.

(i) In the case $k = 3$, let D be the set of all words $(a\, b\, c)$ satisfying the inequalities $a < b \geqslant c$. It is immediately seen that D is comma free. In order to show that the number of words in D is $\tfrac{1}{3}(n^3 - n)$, one could, of course, count the number of solutions of the inequalities $a < b \geqslant c$, where a, b, c are integers between 1 and n. But it is simpler to observe that if $(a_1\, a_2\, a_3)$ is any primitive word (that is, one for which $a_1 = a_2 = a_3$ does not hold), then some cyclic permutation of it clearly satisfies $a < b \geqslant c$. In particular $W_3(4) = 20$, a fact which will be useful in section 5.

(ii) For $k = 5$, the procedure is similar but more complex. Let D consist of all words $(a\, b\, c\, d\, e)$ satisfying $a < b \geqslant c,\ d \geqslant e$, and also of all words satisfying $a < b < c < d \geqslant e$. It can be readily verified that D is comma-free. In order to show that the number of elements in D is the upper bound (in this

204 S. W. GOLOMB, BASIL GORDON AND L. R. WELCH

case $\frac{1}{5}(n^5 - n))$, we must prove that every primitive equivalence class contains a word of D. For this purpose let $+$ denote any number which is > 0, and $-$ any number which is $\leqslant 0$. Using this notation the elements of D can be characterized as those words $(a\ b\ c\ d\ e)$ for which the sequence of differences $b - a, c - b, d - c, e - d$ is of one of the forms $+ - - -, + - + -$, or $+ + + -$. These patterns are precisely those which begin with an odd number of $+$'s and end with an odd number of $-$'s, a property which we shall call property P. (Incidentally, in the case $k = 3$ our dictionary consisted of words $[a\ b\ c]$ for which the differences $b - a, c - b$ were of the form $+ -$, that is, possessed property P.) Given any primitive word $(p\ q\ r\ s\ t)$ we form the differences $q - p, r - q, s - r, t - s, p - t$ obtained by representing p, q, r, s, t as points on a circle. We call $p - t$ the improper difference. By performing a suitable cyclic permutation on $(p\ q\ r\ s\ t)$ we can arrange matters so that any one of the five differences becomes the improper one.

Now by primitivity, both $+$'s and $-$'s appear among the differences, and since the total number of signs is 5, there must occur someplace a run of $-$'s followed by a run of $+$'s, the lengths of these runs being of opposite parity (note that this result depends only on the fact that the total number of signs is *odd*). Permuting cyclically we can put the run of $+$'s at the beginning, the run of $-$'s at the end, and make the improper difference have the sign which occurred an even number of times. The proper differences will then satisfy property P, and hence, given any primitive word, some cyclic permutation of it is in D.

(iii) For $k = 7$ we use the same method. Every primitive word has some cyclic permutation with property P. Its proper differences will then have one of the following 8 patterns:

$$(+ + + + + -) \ (+ + + - + -) \ (+ + + - - -) \ (+ - + + + -)$$
$$(+ - + - - -) \ (+ - - + + -) \ (+ - - - + -) \ (+ - - - - -)$$

Letting D consist of all such words, we find that D is comma-free. (The verification begins to become tedious, but is straightforward. The first overlap of two words in D begins with an even number of $+$'s, hence is not in D, the second overlap ends with a $+$, etc.)

(iv) When $k = 9$, the difficulty arises that there may be more than one word in a primitive equivalence class with property P. This happens for words $(a_1\ a_2\ a_3\ a_4\ a_5\ a_6\ a_7\ a_8\ a_9)$ with $a_1 < a_2 \geqslant a_3, a_4 < a_5 \geqslant a_6, a_7 < a_8 \geqslant a_9$. Here the permutations $(a_4\ a_5\ a_6\ a_7\ a_8\ a_9\ a_1\ a_2\ a_3)$ and $(a_7\ a_8\ a_9\ a_1\ a_2\ a_3\ a_4\ a_5\ a_6)$ also have property P. But notice that these words consist of three blocks of three letters, each of the type used for $k = 3$. This suggests the idea of ordering the 3-letter words $(a\ b\ c)$ with $a < b \geqslant c$ in some fashion (say lexicographically), and choosing for the dictionary D that one of the three possibilities which is of the form $w_1 < w_2 \geqslant w_3$ in this ordering. For example, in the case of the word $(1\ 3\ 1\ 1\ 2\ 2\ 2\ 3\ 1)$, the permutation $(1\ 2\ 2\ 2\ 3\ 1\ 1\ 3\ 1)$ would be selected for D, because $(122) < (231) \geqslant (131)$ if lexicographic ordering is

392 *The Wisdom of Solomon*

COMMA-FREE CODES 205

employed. Adopting this convention, the dictionary which results is comma-free.

(v) For $k = 11$, 13, and 15 the same methods can be used, but the work becomes increasingly cumbersome. It is conceivable that all odd k can be treated in this manner, but we have stopped with the proof that $15W_{15}(n) = n^{15} - n^5 - n^3 + n$.

This case, the first where k has two distinct prime factors, is particularly powerful evidence for the validity of the general conjecture.

3. Results for Even k. When k is even, the results are much less complete, and we cannot even formulate a plausible conjecture as to the value of $W_k(n)$. We begin with

THEOREM 3. $W_2(n) = [\frac{1}{3}n^2]$, where $[x]$ denotes the integral part of x.

Proof. Let D be any comma-free dictionary, and define A to be the set of all integers which begin some word of D but never end a word of D. Similarly, let B be the set of integers which both begin and end words of D, and C the set of integers which only end words of D. For example, if $D = \{(43), (41), (35), (25), (15)\}$, then

$$A = \{4, 2\}, \quad B = \{3, 1\}, \quad C = \{5\}.$$

D must evidently consist of words of the forms $(a\ b)$, $(a\ c)$, $(b_1\ b_2)$, or $(b\ c)$, where $a \in A, b, b_1, b_2 \in B$, and $c \in C$. But $(b_1 b_2)$ cannot occur, for there is some word in D ending in b_1, and some word beginning with b_2, and the comma-free property therefore excludes $(b_1\ b_2)$. This leaves only words of the forms $(a\ b)$, $(a\ c)$, $(b\ c)$, and it is immediately seen that the set of all these is comma-free. If α is the number of elements of A, β of B, and γ of C, then the number of words in D is at most $\alpha\beta + \beta\gamma + \gamma\alpha$. Maximizing the quantity $\alpha\beta + \beta\gamma + \gamma\alpha$ subject to the constraint $\alpha + \beta + \gamma = n$, we see that α, β, γ should be chosen as nearly equal as possible, in which case

$$\alpha\beta + \beta\gamma + \gamma\alpha = [\tfrac{1}{3}n^2].$$

For example, if $n = 3$ we would take $A = \{1\}, B = \{2\}, C = \{3\}$, and obtain $D = \{(12), (13), (23)\}$. It is not difficult to see that for arbitrary n we may choose D to be the set of all words w congruent to one of these words (mod 3), where

$$(a_1 a_2 \ldots a_k) \equiv (b_1 b_2 \ldots b_k) \pmod{m}$$

means

$$a_j \equiv b_j \pmod{m}, \qquad j = 1, 2, \ldots, k.$$

Thus for $n = 5$, $D = \{(12), (15), (42), (45), (13), (43), (23), (53)\}$.

THEOREM 4. *If k is any even integer, then the upper bound given by Theorem 1 is not attained by $W_k(n)$ provided that $n > 3^{\frac{1}{2}k}$.*

206 S. W. GOLOMB, BASIL GORDON AND L. R. WELCH

Proof. Let $k = 2j$, and let L be a comma-free dictionary. We define S_1 to be the set of all j-tuples $(a_1 a_2 \ldots a_j)$ which form the first half of some word in L, and S_2 to be the set of k-tuples $(a_{j+1} a_{j+2} \ldots a_k)$ which form the second half of some word in L. Then we put

$$A = S_1 \cap S_2', \quad B = S_1 \cap S_2, \quad C = S_1' \cap S_2, \quad D = S_1' \cap S_2',$$

where the prime denotes complementation. The four sets A, B, C, D are mutually exclusive and mutually exhaustive, so that any j-tuple is in one and only one of them. Hence to every k-letter word we may associate a pair (AA), (AB), $\ldots$ or (DD) depending on which set its first half falls into and which set its second half falls into. As in the proof of Theorem 3, it is seen that for words in L the type (BB) cannot arise, and hence only (AB), (AC), and (BC) remain.

The upper bound of Theorem 1 was the number of primitive equivalence classes. To prove Theorem 4, we will show the existence of a primitive k-letter word, such that no cyclic permutation of it has any of the forms (AB), (AC), or (BC).

Consider the following particular blocks of length j:

$$(1, 1, 1, \ldots, 1, m) \qquad 1 \leqslant m \leqslant n.$$

Let T_i be the cyclic permutation which shifts each letter i units to the left. Define

$$F_m(i) = \begin{cases} 1 \text{ if } T_i (1, 1, \ldots, m) \in A \cup D \\ 2 \text{ if } T_i (1, 1, \ldots, m) \in B \\ 3 \text{ if } T_i (1, 1, \ldots, m) \in C \end{cases}$$

For each m, $F_m(i)$ is a function with a domain of j elements and a range of 3 elements. There can be at most 3^j such functions, and since $n > 3^{\frac{1}{2}k} = 3^j$, there exist two distinct integers p and m such that $F_p \equiv F_m$ for all i. We now claim that no cyclic permutation of the word

$$w = (1\,1 \ldots p\,1\,1 \ldots m)$$

is of the form (AB), (AC), or (BC). For any permutation of w consists of a cyclic permutation of $(1\,1 \ldots p)$ followed by the *same* cyclic permutation of $(1\,1 \ldots m)$ or vice versa. Since $F_p \equiv F_m$, we therefore get only the forms (AA), (AD), (DA), (DD), (BB), or (CC).

In particular, when $k = 4$, Theorem 4 proves that $4W_4(n) < n^4 - n^2$ for $n > 9$. By more delicate arguments it can be shown that this inequality is true for $n \geqslant 5$. On the other hand, if $n = 1, 2, 3$, then $4\,W_4(n) = n^4 - n^2$, as is seen by considering the dictionary D of words $(a\,b\,c\,d)$ satisfying $a < c$, $b \geqslant d$. The question of whether or not $W_4(4) = 60$ is still open. The best that can currently be proved is $W_4(4) \geqslant 56$.

4. Asymptotic Results. In this section we shall prove some theorems about the asymptotic behavior of $W_k(n)$ when k is fixed and $n \to \infty$.

COMMA-FREE CODES

THEOREM 5. *The limit*

$$\lim_{n \to \infty} \frac{W_k(n)}{n^k} = \alpha_k$$

exists.

Proof. We shall show that if n_0 is any fixed integer, then

$$\liminf_{n \to \infty} \frac{W_k(n)}{n^k} \geqslant \frac{W_k(n_0)}{n_0^k} .$$

This fact, coupled with the obvious boundedness of the ratio in question, proves the existence of the limit.

Consider then the integer n_0, and let D be a comma-free dictionary containing $W_k(n_0)$ words. For any arbitrary n, form the set S of all words w such that

$$w \equiv w_0 \quad (\text{mod } n_0),$$

where $w_0 \in D$. (The definition of congruence is given after Theorem 3 together with an example of the present procedure.) S is clearly comma-free, and so if it contains $S_k(n)$ elements, then

$$W_k(n) \geqslant S_k(n).$$

But it is easy to see that

$$\lim_{n \to \infty} \frac{S_k(n)}{n^k} = \frac{W_k(n_0)}{n_0^k} .$$

This completes the proof of the theorem.

THEOREM 6. *If k is odd, then $\alpha_k = 1/k$.*

Proof. By Theorem 1,

$$W_k(n) \leqslant \frac{1}{k} \sum_{d/k} \mu(d) \, n^{k/d}.$$

If k is fixed and $n \to \infty$, the right hand side is asymptotically n^k/k. Hence,

$$\lim_{n \to \infty} \frac{W_k(n)}{n^k} \leqslant \frac{1}{k} .$$

On the other hand, consider the dictionary D defined as follows: Put $k = 2j - 1$ and let D consist of all words $(a_1 \, a_2 \, \ldots \, a_k)$ such that a_j is greater than any of the other a_i's. D is comma-free, as is easily verified, and the number of elements in D is equal to

$$\sum_{m=1}^{n-1} m^{k-1} \sim \frac{n^k}{k}$$

This shows that

$$\lim_{n \to \infty} \frac{W_k(n)}{n^k} \geqslant \frac{1}{k} ,$$

and thus establishes Theorem 6.

208 S. W. GOLOMB, BASIL GORDON AND L. R. WELCH

THEOREM 7. *If k is even, then $1/ek < \alpha_k \leqslant 1/k$.*

Proof. The first part of Theorem 6 holds for any fixed k. Hence, $\alpha_k \leqslant 1/k$. To obtain a lower bound, we divide the integers from 1 to n into two disjoint classes U and V. Then let D be the set of all words $(a_1 a_2 \ldots a_k)$ such that $a_1 \in U$ and $a_2, \ldots, a_k \in V$. D is clearly comma-free, and if the number of elements in V is v, then D contains $(n - v)v^{k-1}$ words. If v could take on all real values, then the maximum of this expression would occur for

$$v = \frac{k - 1}{k} n \,,$$

and would have the value

$$\frac{n^k}{k} \left(1 - \frac{1}{k}\right)^{k-1} .$$

The fact that v must be an integer has no effect, since taking

$$v = \left[\frac{k - 1}{k} n\right]$$

gives a lower bound for $W_k(n)$ which is still asymptotically

$$\frac{n^k}{k} \left(1 - \frac{1}{k}\right)^{k-1} .$$

Hence

$$\alpha_k \geqslant \frac{1}{k} \left(1 - \frac{1}{k}\right)^{k-1} > \frac{1}{ek} .$$

For $k = 4$, Theorem 7 gives the bounds

$$\frac{27}{256} \leqslant \alpha_4 \leqslant \frac{1}{4} .$$

A better bound can be obtained from Theorem 5. As shown after Theorem 4, $W_4(3) = 18$, and hence

$$\alpha_4 \geqslant \frac{W_4(3)}{3^4} = \frac{18}{81} = \frac{2}{9} .$$

The exact value of α_k for even k is still an open question.

5. Applications. From their researches in the transfer of genetic information from parent to offspring, Crick, Griffith, and Orgel **(1)** advance the following hypothesis. Genetic information, they suggest, is encoded into a giant molecule (chromosome) by means of an affixed sequence of nucleotides, of which there are four types. Each such sequence is uniquely decodeable into a new protein molecule, consisting of a long sequence of amino acids, of which there are twenty types. They propose that each amino acid is specified by three consecutive nucleotides. However, only twenty of the sixty-four sequences of three nucleotides "make sense." Crick, Griffith, and Orgel

COMMA-FREE CODES 209

theorize that the twenty sequences of nucleotides actually corresponding to amino acids form a comma-free dictionary. As we have seen, $W_3(4) = 20$, which agrees with the number of amino acids. The reasonableness of this condition can be seen if we think of the sequence of nucleotides as an infinite message, written without punctuation, from which any finite portion must be decodeable into a sequence of amino acids by suitable insertion of commas. If the manner of inserting commas were not unique, genetic chaos could result.

In their search for optimum coding techniques, Shannon, McMillan, and others have studied codes which are uniquely decipherable *in the large*—that is, when the entire message is available. This is a larger class than the comma-free messages, which must be uniquely decipherable *in the small*. In communications applications where only disjointed portions of a message are likely to be received, comma-free codes may indeed be useful. An excellent discussion of codes uniquely decipherable in the large is presented in **(2)**.

REFERENCES

1. H. C. Crick, J. S. Griffith, and L. E. Orgel, *Codes Without Commas*, Proc. Nat. Acad. Sci., *43* (1957), 416–421.

2. B. McMillan, *Two Inequalities Implied by Unique Decipherability*, IRE Transactions on Information Theory, *2* (1956), 115–116.

3. T. Nagell, *Introduction to Number Theory* (Uppsala, 1951).

Max Delbrück
An Appreciation

SOLOMON W. GOLOMB

IT WAS MY PRIVILEGE TO KNOW MAX DELBRÜCK during the final third of his life. When I first met him in 1956, J. D. Watson's *Double Helix* had not yet appeared, DNA was not a household word, and Delbrück's name was widely known only within the several scientific disciplines to which he had contributed. Yet it was quickly apparent that this was not merely another competent Caltech scientist, but a major intellectual force, and a man who exercised a unique form of moral leadership over those with whom he came into contact. His role as final arbiter, authority, and judge of research developments and speculations during the seminal era of the formulation of molecular genetics ("What will Delbrück say?") is amply attested in *The Double Helix* and subsequently in *The Eighth Day of Creation* by H. F. Judson. When portions of *The Eighth Day* were being serialized in *The New Yorker* prior to its publication in book form, Max was amused to read that at a much earlier date he was already a legend in his own time. "If I was a legend then," he observed, "by now I must be merely a myth."

Inevitably, Max Ludwig Henning Delbrück will someday be the subject of a major biography. I will merely sketch some aspects and events of his life as I heard him describe them. His father, Hans Delbrück, was Germany's leading military historian, and more than half a century older than Max, the youngest of seven children. Until he came to the United States in his early thirties, Max was known principally as Hans Delbrück's son. To establish an independent identity for himself, and because he thought he had some talent in that direction, Max found himself attracted at an early age to science, and specifically to astronomy. His graduate studies in astronomy were unrewarded in that he never successfully completed a thesis. Eventually he switched to physics. In his own view, the undistinguished dissertation that finally qualified him for the Ph.D. in physics was unmemorable in the extreme. I envision the young Max Delbrück of this period as having a voracious intellectual appetite for facts and information in almost all fields (cer-

---

⊙ SOLOMON W. GOLOMB is professor of electrical engineering and mathematics at the University of Southern California.

398 *The Wisdom of Solomon*

THE AMERICAN SCHOLAR

tainly including philosophy, history, literature, and the arts) and developing very rigorous canons of scientific proof and validity, but not yet having learned to see the broad picture and to identify the characteristic features and unexpected patterns as required for the highest levels of scientific achievement.

In the late 1920s and early 1930s, Hans Delbrück's son spent his time at the world's most active centers of physics research: the University of Göttingen during the golden age; the Bristol of Dirac, Powell, and Blackett; the Zurich of Wolfgang Pauli; the Copenhagen of Niels Bohr; and the laboratory of Otto Hahn and Lise Meitner at the Kaiser Wilhelm Institute in Berlin. At Göttingen, Max also learned about Hilbert's "formalist" school of mathematical thought and the spectacular 1930 results of young Kurt Gödel (born the same year as Max) in the logical foundations of mathematics. At Bohr's Institute, Max met and befriended the polymath George Gamow, whose scientific interests and contributions encompassed physics, cosmology, and years later, molecular genetics. Max also became a convert to Bohr's "Copenhagen school" of the philosophy of science and was profoundly influenced by a lecture by Bohr in 1932 or 1933 on biology and physics, in which the question was raised (not for the first time) whether or not the laws of physics are sufficient to account for the phenomena of biology. Several years later, as the theoretician in Lise Meitner's lab, Max was asked to explain the presence of traces of unexpected elements after uranium (U^{238}) was bombarded with neutrons. Max proposed that these were new "transuranium" elements that had been formed, and he developed an elaborate physical theory of these elements. His colleagues found this highly plausible and, being physicists, delayed for some time the chemical testing that showed that the "new" elements were, in fact, barium and krypton. So Max missed the boat on that one, and it was Lise Meitner who finally came up with the fission explanation. (Years later, Max observed wryly that his theorizing probably set back nuclear physics in Germany by about three years. The unintended benefit to the world from this may have been enormous!) Thus at age thirty, Max Delbrück had all the earmarks of the mediocre son of a famous father.

In fact by this time, Max was already preoccupied and obsessed with a revolutionary idea, motivated in no small measure by Bohr's question about the physical basis of life. The central process of biology is genetics. Rather than studying the genetics of peas or fruit flies or hybrid corn, all of which are comparable in complexity to humans or elephants or oak trees, the emphasis should be on finding the simplest organisms that exhibit this property (that is, self-replication) and studying the genetics of these organisms in detail.

In 1937 Max Delbrück came to Caltech on a Rockefeller grant as a

MAX DELBRÜCK: AN APPRECIATION

biologist. The simplest organisms with any genetic properties at all, he reasoned, are viruses. In the 1920s, he learned, there had been a flurry of interest in trying to control various bacterial diseases by means of the viruses (called bacteriophages, or phages for short) parasitic on those bacteria. Delbrück subsequently selected the phages of the ubiquitous colon bacillus, *Escherichia coli* (found abundantly in all the world's sewage) as the appropriate experimental animals, and in particular the "T series" of phage, T1 to T7 as they were later designated. (At Caltech in the late 1930s, Delbrück was able to collaborate with Emory Ellis, a bacteriophage biologist. This collaboration was fruitful in developing many of the basic experimental techniques for the study of phage.) In the early 1940s Max persuaded the small coterie of viral geneticists to concentrate their efforts on these phages of E. coli, in what became known as the "phage treaty."

The insight that the way to study genetics is through viruses rather than elephants (or *Drosophila*, or *Neurospora*) was an idea whose time had come. Besides Delbrück, both Salvador Luria and Alfred Hershey had independently come up with the same notion. Indeed, all three were recognized in the award of the 1969 Nobel Prize for physiology and medicine. By about 1940, Delbrück and Luria had found each other and for a while worked together in this endeavor when Luria would visit Delbrück at Vanderbilt University in Nashville, Tennessee, and, during summers, at Cold Spring Harbor on Long Island. Hershey first met Delbrück when he visited Nashville at Max's invitation in January 1943. Delbrück, Luria, and Hershey formed the nucleus of what became known as "the phage group." But adequate funding for such research was an idea whose time had not yet come. Several of the people attracted to this area became discouraged and turned their attention elsewhere. Delbrück, however, resolutely persisted.

In 1941, stranded in the United States by World War II after his two-year fellowship at Caltech in Pasadena, Delbrück obtained a position as a physics instructor at Vanderbilt University (where he stayed until 1947), and where, on his own time, he continued to develop the concepts and experimental techniques that would be needed for phage genetics. Vanderbilt was not a high-pressure research university. As long as he met his physics classes, Max was free to pursue whatever other interests he wished. He even received a few hundred dollars a year to support his research work. In retrospect, Max observed that this was the ideal environment in which to develop a new branch of research that was not yet recognized as part of the mainstream. In 1947 Max was able to return to Caltech with a regular faculty appointment in biology. In August 1941 he married "Manny" (née Mary Bruce), and their home in Pasadena became a charmingly Americanized version of the European-style intel-

400 *The Wisdom of Solomon*

THE AMERICAN SCHOLAR

lectual salon, and remained so for more than thirty years.

Several members of the Delbrück family in Germany were persecuted and even imprisoned for failing to support the Nazi regime and its policies. Several perished as a result of the abortive 1944 plot on Hitler's life. Some of the survivors (including Max's brother) were released from German prisons only to die in Russian ones. Max and Manny maintained a lifelong commitment to helping others, especially scientists, obtain civil rights, emigration permits, and political asylum. Once, when asked by a Caltech undergraduate (possibly looking for support for resisting the draft) if he hadn't refused to work on the Manhattan Project as a matter of conscience, I heard Max answer that he was spared the necessity of facing that particular moral dilemma. "You forget that I was an enemy alien. I was German and not Jewish. There was no way they would have let me work on the atomic bomb." Max was often at his Socratic best when refusing to take credit for morally commendable behavior. Even his doing science, he would contend, stemmed from no lofty motives, but was an uncontrollable compulsion. He was merely fortunate that this vice of his was more socially acceptable than alcoholism or drug addiction.

After the war, the pace of molecular genetics quickened. Avery had demonstrated in 1943 that it was the nucleic acid (the DNA in a bacterium) that contained the genetic information, rather than amino acids (proteins, polypeptides) as had been previously supposed. Others originally trained in physics, including George Gamow and Francis Crick, began to take an interest in molecular genetics. Delbrück's doctoral and postdoctoral students began to spread the gospel according to Delbrück far and wide. One of Max's protégés, James Watson, went to Cambridge and collaborated with Crick to discover the double helix structure of DNA. Delbrück's students, associates, and disciples began to receive Nobel Prizes with monotonous regularity. It was almost an anticlimax when the Nobel committee finally acknowledged Max with the prize in 1969.

In fact, by the early 1950s, Max's interests had already begun to shift in another direction. The next frontier after the physical basis of *life*, Max reasoned, is the physical basis of *mind*, including perception, intelligence, and awareness. Applying the proven principle of looking for the simplest instance of the phenomenon in question as a basis for study and understanding, Max decided to focus on perception at the cellular level. As his "experimental animal" he selected *Phycomyces*, a one-celled fungus with a phototropic organelle, a "sense of sight" on a subcellular scale. Many summers thereafter, there was a research program on *Phycomyces* at Cold Spring Harbor, a biology research facility currently directed by Watson. But interesting as *Phycomyces* may be, it

354

MAX DELBRÜCK: AN APPRECIATION

has not been the touchstone to mind and intelligence that phage was to genetics. That would have been too much to expect.

After Max reached the normal retirement age of sixty-eight in 1974, the trustees of the California Institute of Technology named him a Board of Trustees Professor, exempting him from ever having to retire, a distinction rarely conferred even on Nobel Prize winners, in which Caltech abounds. In the spring of 1978, while preparing to undergo heart bypass surgery for angina pectoris, Max was examined and found to have an advanced case of multiple myeloma. He spent the rest of his life on a succession of chemotherapy regimens, and remained remarkably active, both physically and intellectually, until nearly the very end. He delivered the commencement address, a profound and moving essay, "The Arrow of Time," at Caltech in June 1978. Less than six weeks before his death on March 9, 1981, he appeared publicly at the midyear commencement of the University of Southern California, where he received an honorary degree. At the memorial service held in his honor at Caltech on Easter Sunday, April 19, 1981, Max provided much of the entertainment, in the form of audio-taped messages to those assembled, photos from throughout his career projected onto a screen, and recordings of songs that he sang at several student-produced Caltech skits in years gone by.

My own involvement with Max Delbrück began in 1956 as a result of the "genetic coding problem." It was Delbrück's old friend George Gamow who had started the excitement in the early 1950s with one of the most successful speculative leaps of the scientific imagination since Jonathan Swift—in the third voyage of *Gulliver's Travels*—announced that Mars had two small moons, some one hundred fifty years before they were actually discovered by astronomers.

Gamow observed that nucleic acid could be viewed as a sequence of four different bases, or nucleotides. It had long since been discovered by Oswald Avery that nucleic acid, rather than the protein in the chromosomes, somehow contained the genetic information; and Crick and Watson had just announced the double-helix structure of DNA. Obviously, said Gamow, the four nucleotides are the symbols of a code, the *genetic code*. And what is being coded for? Obviously the sequence of amino acids that form a protein. More specifically, he said, there are some twenty amino acids to be coded for, and since two symbols from the four-symbol alphabet allow only sixteen possible combinations, "clearly" the code words are triplets of nucleotides.

Now the problem arises that there are sixty-four different code triplets, but only twenty amino acids to be coded for. No matter, said Gamow, the code is "degenerate": There are, typically, several code words for each amino acid—as few as one code word, as many as six, per amino acid.

355

402   *The Wisdom of Solomon*

THE AMERICAN SCHOLAR

The incredible thing is that all of this pure, whole-cloth speculation, as far as I have gone, turned out to be entirely right. Nature does use a degenerate triplet code, for exactly twenty amino acids. This may be the greatest achievement of Aristotelian reasoning in the history of science—certainly better than anything Aristotle himself accomplished. The Aristotelian method is to ask: "How would Nature have done it, if she were as clever as I?" Unfortunately, Nature is rarely as clever as I. All too often, one of us has thought of something that the other overlooked. An information theorist, for example, might have proposed a variable-word-length code, like Morse code, with shorter codes for the more common amino acids, and longer codes for the less common. Fortunately, Gamow did not outsmart Nature, as many of us later did, on these basic aspects of the code. That is, not until he tackled the problem of how to synchronize the decoding.

Suppose our four symbols are A, E, H, T, and that in the coded message we see H, A, T, E, A, T. What are the code words? If they are HAT and EAT, what about the "overlap" words, ATE and TEA? Gamow visualized the amino acids swimming around in the soup, looking for their call letters. If the intended reading was HAT, EAT, but the amino acid corresponding to ATE showed up first, it would spoil everything. Gamow's solution was to propose an overlapping triplet code, where the successive amino acids correspond to the triplets HAT, ATE, TEA, EAT. A very efficient packing, but this scheme seriously limits the possibilities of what successors a given code word can have. Thus, HAT can be followed only by ATA, ATE, ATH, or ATT. It was Francis Crick, in Cambridge, around 1953–54, who showed that Nature could not be using Gamow's overlapping triplets, at least in the case of insulin, one of the few proteins whose amino acid sequence was known at that time. The entropy in the insulin sequence is too high for Gamow's overlapping hypothesis (that is, there is too much freedom as to which amino acid can follow which other amino acid). By 1956 Sydney Brenner, a colleague of Crick, using data from several protein sequences, was able to show that all overlapping triplet codes are impossible, provided that the code is "universal" (that is, that all earth organisms use the same code).

Gamow was also the first of many to be seduced by a clever scheme for reducing sixty-four triplets down to twenty amino acids. His hypothesis was that the order of the three symbols in the triplet didn't matter, so that AAT, ATA, and TAA would all correspond to the same amino acid. The amino acid corresponding to AAA would have only one code word, while the one corresponding to ATE would have six (AET, ATE, EAT, ETA, TAE, TEA). Remarkably, there are exactly twenty such classes of code words. This was so compelling to Gamow that I don't believe he ever reconciled himself to this part of his speculation being incorrect.

356

MAX DELBRÜCK: AN APPRECIATION

(As a matter of strict historical accuracy, the idea of triplet code words did not originate with Gamow. One earlier proposal was put forward in 1952 by Alexander Dounce, who had a scheme that allowed for code words for as many as forty amino acids. Dounce's exposition was heavy on biochemical details and did not attract the widespread attention it properly deserved. In retrospect, Gamow's principal contributions were to interest a great many people in the genetic coding problem and to propose a model for the code that was specific enough to be disproved.)

The next major synchronization hypothesis was "comma-free codes." In a paper called "Codes Without Commas," Crick, Griffith, and Orgel proposed, in effect, that if HAT and EAT are code words, then neither of the overlaps, ATE and TEA, can be code words at all. Again, miraculously, the largest comma-free dictionary using three-letter words written from a four-symbol alphabet has exactly *twenty* words, the magic number of amino acids.

It was at this stage of the problem that I entered the picture. In August 1956, having returned from a Fulbright year in Norway, I started work in the communications section of the Jet Propulsion Laboratory of the California Institute of Technology. My office-mate was Lloyd Welch, who was then finishing his Ph.D. in mathematics at Caltech and working part-time at JPL. Another young mathematician at Caltech was Basil Gordon, who had just gotten his doctorate there and who had been a good friend of mine since undergraduate days at Johns Hopkins.

Max Delbrück had asked Gordon about the problem of finding different maximum-sized comma-free codes. Gordon, Welch, and I collaborated on this problem. This was not part of my job at JPL, but I remember that during the fall of 1956 it absorbed many of my evenings and weekends. What we had done, in typical mathematical fashion, was not so much to solve the original problem as to generalize it. We created the theory of maximum-sized comma-free codes for word-length k over an n-symbol alphabet. Our paper, "Comma-Free Codes," appeared early in 1958 in the *Canadian Journal of Mathematics*, within a few days of the launching by JPL of *Explorer I*, the United States' first artificial satellite.

I frequently attended Delbrück's seminars at Caltech during 1957, and even before the "Comma-Free Codes" paper appeared, he got after me to get back to the "real" problem of what can happen in the very special and mathematically trivial case of writing three-letter words from a four-symbol alphabet. Gordon was drafted into the army in the summer of 1957, but Welch and I continued our collaboration and obtained a complete characterization of the maximum-sized, comma-free-triplet dictionaries for all alphabet sizes. With a four-symbol alphabet, there are basically five different dictionary structures, all having twenty code

404 *The Wisdom of Solomon*

THE AMERICAN SCHOLAR

words. (This fact was independently discovered, at about the same time, by Hans Freudenthal in the Netherlands.) One nontrivial result that we obtained was that in any message constructed from any one of these dictionaries, the same nucleotide symbol could not occur as many as four times in a row. Our results appeared later, in 1958, in the paper "Construction and Properties of Comma-Free Codes" by Golomb, Welch, and Delbrück.

In the very first series of experiments to determine the occurrence of repetitions of symbols in nucleic acid, it was found that the same symbol *did* occur four and, in fact, even five times in a row. The comma-free-triplet hypothesis was gravely wounded by this discovery and never recovered.

There is, however, an amusing postscript to all this. In the fall of 1958, the Nobel Prize in medicine and physiology was awarded to Dr. George Beadle, who was on leave from Caltech and staying in England. The Delbrück group at Caltech composed a telegram, using a set of comma-free triplets from the DNA alphabet of A, C, G, T (standing for *a*denine, *c*ytosine, *g*uanine, and *t*hymine) to represent the letters of the English alphabet, including a code word for the space symbol. The plaintext of this message was: "Congratulations. Break this code or give back Nobel Prize. —Sterling, Max, and Marco." The signatories' names, sent in the clear, included Sterling Emerson and Max Delbrück. We shall return later to the question of who is Marco.

A few weeks later a coded telegram was received in reply. It used a different triplet code, not even comma-free, with the same A, C, G, T alphabet. Delbrück called on me to decipher it. It said: "Thanks for your telegram. I'm sure it's a fine sentiment, but I haven't yet decoded it. —George." Soon after, Delbrück received an air letter, postmarked Oxford, November 14, 1958, as follows: "Dear Max, I'm a darned poor cryptographer, but so many friends and relatives have already helped spend the money that I can't give it back. What shall I do? —Beets." Obviously he had, finally, decoded the original telegram.

Copies of that original telegram, without explanation, had been sent off by Delbrück to friends and acquaintances far and wide. Some of them sent coded replies. A typical reply, written in the same code, decoded as follows: "MESSAGE WAS TOO SIMPLE SEND HARDER ONE."

Most baffling of the coded replies was a string of thirty-six beads, in four colors, from someone in Princeton. On the assumption that it was a triplet code, the message had the form of a cryptogram, with pattern: ABCDEFDGHIJC. The Caltech group had observed that it could not be in the original code, because the set of triplets was not comma-free. Moreover, such a short pattern could be fitted to English text in a great many ways. One such fit is PHYSICS·TODAY, with no space between

358

Comma-Free Codes and Sol's Tribute to Max Delbrück 405

MAX DELBRÜCK: AN APPRECIATION

words. When they decided the situation was hopeless, I was called in. As a mere cryptogram, I realized, it had literally thousands of solutions, including everything from CHERBORG·LINE to DISTANT·LOVES to WATSON'S·DIRT (obviously a reference to *The Double Helix*).

One might easily give up on the grounds that the intended decoding could not possibly be divined. But I persisted. Suppose, I reasoned, that it is in the original code, but some English letter not in the original message was needed, and a new triplet was created. The composer may not have realized this was a *comma-free* triplet code, and the new triplet may destroy the comma-free property. The original code had a *space* code word. Suppose that is the code word in positions four and seven. That hypothesis led immediately to the decoding "WHO·IS·MARCO." The *W* at the beginning was the letter that never appeared in the original telegram.

In December 1976 I talked to Delbrück on the phone regarding all of these coded messages. His recollections were rather hazy. When I told him about the first telegram, signed by Sterling, Max, and Marco, his exact words to me were "Who is Marco?" Incidentally, I didn't know either. (I later learned that Marco is Marco Zalokar, at Caltech in 1958 and subsequently in Paris.)

Max sent me the file from his archives from this period. At the Nobel ceremony for Beadle, a special double helix made of toothpicks in four colors was presented to him. Decoded from its triplet code, the message read: "I AM THE RIDDLE OF LIFE KNOW ME AND YOU WILL KNOW YOURSELF." Obviously Beadle had little difficulty decoding this one, because on January 7, 1959, a message was received in reply. It began: "Know myself but" in plaintext, and then switched over to code. Max said the coded portion was never decoded.

In July 1960 I was in England to present a paper on the theory of enumeration at the Fourth London Symposium on Information Theory. At the end of the conference, I took the train to Cambridge, where I met Francis Crick for the first time. He spent the entire day with me, showing me the adjacent lab where the original model of the hemoglobin molecule filled half the room, and took me to lunch at the Eagle, the pub where Crick and Watson worked out the structure of DNA on the back of a napkin—or was it the tablecloth? As I visualize it, Watson told Crick that it had been shown experimentally that in samples of DNA from many organisms, the symbols A and T had the same frequency, as did the symbols C and G. Now what did this suggest about the structure of DNA? To which Crick undoubtedly replied, "Elementary, my dear Watson."

We strolled back along the banks of the Cam, and continued our discussion of possible models for the genetic code. Some two years later,

359

406   *The Wisdom of Solomon*

THE AMERICAN SCHOLAR

in October 1962, I received a monograph called "The Recent Excitement in the Coding Problem," by Francis Crick. It arrived in the mail the same morning that the announcement came over the radio that Crick had won the Nobel Prize.

A few years later, when Francis Crick was visiting Pasadena, Max and Manny Delbrück invited him and the Golombs over to their house. Many stories were exchanged. Several were about a famous physicist who always introduced his wife as the sister of a well-known colleague, rather than as his own wife. Well, on one occasion, when the wife in question (or sister, if you prefer) was despairing of getting the family budget back into balance, she said, "Dear, why don't you go to Stockholm and get another one of those prizes they give out. We need the money."

Invitations to the Delbrück home were always a treat. When our children were small, there was an egg rolling contest on the Delbrück lawn on Easter Sunday. There were dinner parties, and *party* parties.

One evening I was invited to a supper where the other guests were Murray Gell-Mann and Roger Sperry. The main topics of the evening seemed to be whether behavioral biology was ready to become a science and whether Caltech should grant it institutional recognition. Gell-Mann at that time had been developing the theory of quarks and their "eightfold-way" group of symmetries, and Delbrück was suggesting he abandon all that—"since anybody competent could eventually work out all the details"—and undertake something truly challenging, like figuring out the physiological basis of thought and behavior. At another point in the conversation, Delbrück mentioned that I was his link with what was happening in prime number theory. I said that progress was painfully slow in that ancient field, and that we still could not prove mathematically something as simple as the existence of infinitely many "twin primes" (pairs of prime numbers differing by 2, such as 11 and 13, or 197 and 199), although the statistical evidence was overwhelming. Gell-Mann commented that this was because mathematicians were doing things all wrong. They were ignoring the result of Gödel that not everything that is true is provable. They could solve all of these intractable problems by adopting an appropriate set of more powerful axioms, incorporating what one could reasonably expect statistically. (Actually, such axioms were formulated decades ago by Cramér and by Schinzel, among others, and they do yield "proofs" of the infinitude of twin primes, and much more. To the physicist Gell-Mann, that may be enough, but to mathematicians, that leaves unanswered the question of whether these new and seemingly stronger axioms are consequences of, independent of, or inconsistent with the older and more intuitive ones.)

360

Comma-Free Codes and Sol's Tribute to Max Delbrück 407

MAX DELBRÜCK: AN APPRECIATION

The guests at a Delbrück party were not necessarily past, present, or future Nobel Prize winners (though indeed many of them were). There were large numbers of graduate students, undergraduate students, research associates, lab technicians, faculty members from innumerable institutions, visitors from Europe, publishers, neighbors, and relatives. I remember one party to which my wife and I were invited where it was important that the Delbrücks know in advance precisely who would be coming. It turned out that they had meticulously planned a treasure hunt, and each of the competing teams required the same composition: one team captain, one genius, one ingenious, one woman's intuitionist, and one linguist. Husbands and wives were invariably assigned to different teams. I was the "genius" on my team; my counterpart on one of the other teams was Richard Feynman. The cryptic clues had the various teams chasing on foot and by car all over Pasadena. Different teams were being run through the same sequence of clues, written on scraps of paper hidden in tree trunks or on window ledges, et cetera. A team finding a clue was to replace it carefully for the other teams. The winning team was the first to reach the final destination, which turned out to be a pizza parlor where we were all rewarded with food. I still have some shark vertebrae that I received as a prize for being on the winning team.

Max had a marvelous sense of humor and was fond of jokes of all kinds, including practical jokes (but always provided that no one would be really hurt, either physically or emotionally). I remember that his original reaction to Tom Lehrer's songs was one of disapproval—they were too cynical, he said. Twenty years later his objections seemed to have evaporated. In the context of the 1970s, Lehrer's humor had lost its sting, and was just good-natured fun.

In the fall of 1961, while Max was serving as guest director of the newly formed Institute for Genetics in Cologne, West Germany, he invited me to come and give a lecture on mathematical aspects of the genetic coding problem. He also arranged for me to give a similar lecture at the Genetics Institute in Copenhagen. Not very long thereafter, the experimentalists finally caught up with and passed the theoreticians, and the correspondence between codons (three-symbol code words in the nucleic acid alphabet) and amino acids became explicit and verifiable. My own role, such as it was, had come to an end.

I remember asking Max in the late 1950s if all this new genetic knowledge wasn't going to have a tremendous impact on the treatment and cure of disease, on the elimination of birth defects, and on the development of whole new strains of organisms, if not actual new life forms. I had suddenly been struck by these fantastic possibilities. "Yes," Delbrück said matter-of-factly, "this will all happen in due course. Some

361

408 *The Wisdom of Solomon*

THE AMERICAN SCHOLAR

of these things will occur fairly soon. Others, like the new life forms, may be a century or more away." "But aren't you interested in working on these problems?" I asked. "No," he explained, "what you've just described is biological engineering. Just as nuclear physics gave rise to nuclear engineering, genetic biology will give birth to genetic engineering. Some people are scientists and some are engineers. I am a scientist. When the time comes, there will be no shortage of bioengineers." At that time, I found all of this mind boggling. Now, of course, it has become commonplace.

By 1963 Max had returned to Caltech from Cologne, and I had left the Caltech-affiliated Jet Propulsion Laboratory for the University of Southern California. We still saw each other, though not as frequently. I remember Max acting astonished when I told him I was planning to write seven different books. His reaction seemed to say that serious scientists publish research papers, not books. Three of these books actually appeared in the mid-1960s. The failure of some of the others to materialize may have been related to a subconscious belief that "Max would not approve."

My spending academic year 1971–72 at Caltech as visiting professor of applied science had little effect one way or the other on the frequency of contact between the Golombs and the Delbrücks. Max at that time had developed interests involving the lives of great scientists. We attended two lectures he gave on 'Copernicus. Years later, at the time of the Einstein centennial, Max gave a fascinating lecture on Einstein to a packed auditorium at Caltech. He pointed out that on several occasions, both in Berlin and in America, his path and Einstein's had crossed. (Actually, their "world lines" had intersected, as he relativistically phrased it.)

In the later years, Max would occasionally call and suggest an evening of chess, sometimes at my home, sometimes at his. Early on I introduced him to the chess clock, which is really two clocks, each running only during a player's turn to move. Since I was the stronger player, I proposed giving him a time handicap. He would have thirty minutes for all his moves during the game, and I would have only five minutes for all of mine. He thought this would be an overwhelming advantage, but I won as easily as before. I suggested giving him an hour to my two minutes. (He was unwilling to accept the handicap of material, such as pawns or pieces, since in his view the game would no longer be chess, and even if he won, he couldn't claim it as a victory.) With that advantage, he said, he would surely win. But as we reached the end game, and he was far behind in material, the clock showed that he had only five minutes left to my one minute. "Now I know how you're doing it," he said in mock indignation. "You're thinking on my time."

MAX DELBRÜCK: AN APPRECIATION

Chess is an emotionally demanding game, and when his angina condition became severe, Max frequently could not stand the strain of more than one game at a session. He decided to teach me to play backgammon, at which he had considerable skill, and since I was a rank beginner, this imposed no emotional strain on him at all.

From his Göttingen days he remembered that David Hilbert, the high priest of mathematics at that sacred shrine, had pronounced the Riemann Hypothesis as the most important unsolved problem in mathematics. Unlike other famous unsolved problems (for instance, the twin-prime problem, or Fermat's Last Theorem), which any amateur can understand, and hundreds annually claim to have solved, the Riemann Hypothesis cannot even be stated without using the terminology of the theory of functions of a complex variable. In the middle of the last century, Bernhard Riemann conjectured that "all complex zeroes of the Zeta-function have real part one-half." In the mid-1970s, I was able to report to Max that Norman Levinson of MIT had proved, in a remarkable tour de force of mathematical analysis, that at least 70 percent of the zeroes of the Zeta-function have the required property. Meanwhile, Pierre Deligne, a Belgian mathematician living in France, had proved André Weil's generalization of the Riemann Hypothesis to function fields and would certainly receive a Field Medal (the nearest thing in mathematics to a Nobel Prize) for this discovery, but unfortunately Deligne's result did not contain the *original* Riemann Hypothesis as a special case.

As Max listened to all this, he seemed struck by a revolutionary thought. "You know," he said, "that's an entirely different form of reality. It has nothing to do with *physical* reality. If all the laws of physics were different, if gravity obeyed an inverse cube law, if quarks came in groups of five instead of three, that would have no effect on whether or not the Riemann Hypothesis is true. The zeroes of the Zeta-function would still be wherever they are."

I am sure the underlying observation here is very old, but it reveals something important about Max as a multidisciplinary scientist. Most mathematicians are realists (in the sense of Plato) not only about mathematics but about the physical universe as well. Not only were there infinitely many prime numbers before Euclid proved it, but the orbit of Mars was an ellipse before Kepler discovered it, and the neutrino existed before Fermi proposed it. Modern physicists tend to the opposite view. The meteor that strikes the surface of Mars makes no sound if there is none to hear it. Fundamental particles have existence only after they have been postulated. All the laws of our universe could be entirely different if random events had occurred just a little differently during the critical first nanosecond after the Big Bang. Max did not

410 *The Wisdom of Solomon*

THE AMERICAN SCHOLAR

find it necessary to choose between these two metaphysical views. It is not that one approach to epistemology is right and the other is wrong; rather, the mathematician and the physicist are investigating two distinct and independent realities. Each is correct about his own reality. It is only because mathematical reality so often provides such a good fit as a model of physical reality that confusion between the two arises.

I suggested that there may even be a hierarchy of realities. If we think of biology in terms of earth life, all of which involves the same genetic mechanism, with nucleic acid corresponding to protein, and even the same basic genetic code for all earth organisms, this may be a biological reality limited to one planet. There are even narrower cultural realities, which differ from one country or one language to the next. Max took a more absolutist view. The fundamental distinction, he said, is between mathematical reality and physical reality. The other distinctions don't involve reality.

Between the extremes of mathematical reality and cultural reality, I offered the following example. The fifth element in our alphabet sequence (that is, the letter *e*) is a *vowel*. This is a mere cultural reality. It is not true, for example, in the Russian alphabet. On the other hand, the fifth element in our numerical sequence (that is, the number 5) is a prime. Even a new set of laws for the physical universe would not change that reality.

From time to time Max would decide to teach an undergraduate course at Caltech—in elementary physics, in the philosophy of science, in mathematical logic. In the early spring of 1980, he called to ask if I would be interested in doing a joint book review with him of Douglas R. Hofstadter's *Gödel, Escher, Bach: An Eternal Golden Braid*. I was familiar with the book, but had not yet read it. "It may not be worth reviewing," Max cautioned. "Only if it has something very valuable in it, or possibly if it is so false and misleading that others must be warned; but if it is merely frivolous, that doesn't warrant a review."

He had a small class of Caltech seniors that was to meet at his house every other Wednesday during the spring quarter. The Hofstadter book was to be the text, and several chapters of reading would be required between meetings.

It was an interesting course. The discussion was almost as good as Manny Delbrück's pastries. There were some social and musical events, too, and a final banquet. But what to make of Hofstadter's book? I thought it was amusing, witty, at times a bit too precious, and certainly rather long. Max applied his usual critical standards. What was merely entertaining, though it did no harm, was essentially frivolous. The treatment of Gödel's Incompleteness Theorem was evidently rigorous but, despite all the attempts to entertain, no easier to follow than it ever

364

MAX DELBRÜCK: AN APPRECIATION

had been. The description of molecular biology and genetics tended to confuse and obscure what was already widely known and understood. But worst of all, in Max's view, was the treatment of how the brain works. "A disgrace," Max pronounced.

We each wrote a review, and the two reviews tended not to overlap. My approach was lighthearted, in what I thought was the spirit of the book itself, but pointed out errors of omission and commission, which were mostly minor, often amusing. Max took the book very seriously, and had serious criticisms.

Max made the first attempt to cut and paste our two critical documents together into a single review. As our class of students observed, they seemed to fit together far better than anyone should have expected. We went through a few rounds of revising, mostly to hide the seams even better, and in due course the review appeared in THE AMERICAN SCHOLAR (Autumn 1980).

After Max was diagnosed as suffering from multiple myeloma, he was put on chemotherapy. By the spring of 1980, none of the more traditional chemotherapeutic agents seemed to be working well, and it was suggested that Max would be an ideal candidate for interferon treatment. (The interferon mania in the press and on Wall Street was then at its peak.) After his first treatment with interferon, Max reported to me the symptoms that it had induced. These included headache and high fever lasting several hours.

I described to Max a speculation that I found very attractive. "You know," I said, "for centuries the human race has lived with the idea that the common cold is a disease that ought to be conquered, cured, eradicated. But maybe that's all wrong. If the cold were all bad, I suspect we would have evolved to not come down with it. So the fact that we still get colds suggests that the cold may have survival value and that it may be therapeutic. And here you've even suggested a mechanism. The cold is brought on by a virus. Every virus infection triggers the body's production of interferon. That prevents us from getting any *other* virus infection, one that might really be serious. And apparently it has anticancer properties. By getting colds every so often, the body may stimulate itself to make enough interferon to get rid of cancers when they are still small groups of cells, before they get too big to eliminate. Even the cold symptoms—the fever and the headache—may be brought on, not by the virus directly, but by the interferon whose production it triggers. And these, too, have therapeutic value. Hyperthermia—raising the body temperature—has proven cancer-fighting effectiveness. And headache helps persuade us to take it easy, to lie down and rest, while the body replenishes itself. Even the coughing and sneezing that occur in the later stages of a cold—why, that may be the mechanism for

412 *The Wisdom of Solomon*

THE AMERICAN SCHOLAR

expelling potentially harmful particles from the lungs before they get too well established and cause problems. What do you think of this hypothesis?" I asked, believing I had just presented an excellent, nearly irrefutable case.

"Oh, that's nonsense," said Max. "Utter nonsense."

I was startled. "How can you say that?" I asked. "What part do you disagree with?"

"The whole thing. Oh, I admit, it's one possible hypothesis. But there are hundreds of others, at least equally likely. What evidence do you have for picking that particular one?"

Of course. This was not a casual conversation between longtime friends. I had just formulated a tentative scientific hypothesis and presented it to Max Delbrück. And he had responded, as he had been responding for many decades under such circumstances: What are the alternative hypotheses? Why do you prefer that particular one? Where is your evidence? How can you go public with such half-baked ideas? Why don't you go back and do your homework, take it apart and examine it critically, then put it back together, and if you still think you have something, come back and we'll find more holes in it. But don't expect me to endorse some nonsensical idea as scientific truth just because we're old friends, and I'm not supposed to hurt your feelings.

That was the quintessential Delbrück role—to maintain the highest standards of scientific judgment and criticism, and to force others to do likewise.

(Incidentally, I still think the idea of the common cold as a therapeutic agent is "cute," and I still have no terrific suggestions about how to test it. One could try to analyze some vast medical data base to correlate frequency of colds against subsequent onset of various types of cancer. Even so, any conclusions arrived at by this method would be tentative at best.)

By September of 1980 it was obvious that interferon had done Max no good, but, true to its name, had interfered with his receiving effective chemotherapy. Max now spent most of his time lying down, and required frequent blood transfusions. He had two symmetrically situated wounds on the backs of his hands, a result, he said, of the low platelet count characteristic of his disease. Such were undoubtedly the stigmata of Saint Francis of Assisi, Max announced, and added that Saint Francis died when only about forty-four years old, looking very aged, some two years after his stigmata had first appeared. Max manifested an almost scientific detachment as he studied the progress of the disease that would inevitably take his life.

When he had been nominated in early 1980 for an honorary degree at USC, ritual required that the nominator solicit letters of support from

366

Comma-Free Codes and Sol's Tribute to Max Delbrück 413

MAX DELBRÜCK: AN APPRECIATION

Max's peers. Among those contacted were several of the Nobel Prize winners with whom Max had worked at various stages of his long career. There was one theme common to almost all the letters, a statement that also appears in *The Eighth Day of Creation*: "Max Delbrück is the most intelligent man in biology today."

Looking forward to receiving the honorary degree on January 27, 1981, may have extended Max's life by several weeks. When the occasion arose, he was very much up for it. He even attended the dinner the night before the ceremony and the luncheon that followed it. The following weekend there was a conference at Caltech, sponsored by the Leakey Foundation largely at Delbrück's instigation, on sociobiology, and Max, although quite ill, attended several of the talks. Characteristically, he was not overly impressed with any startling new truths. In fact, he had negative reactions to most of what he heard. Caltech did well to be out of that business, he so much as stated.

Two weeks later, on February 15, I saw Max for the last time. He had a visitor from Europe who wanted a Rubik's Cube. Yes, I could supply one. He also had a copy of the March 1981 *Scientific American* with a cover story by Douglas Hofstadter on Rubik's Cube, and which generously referenced my observation of an analogy between twists on the corners of the Cube and quarks in physics. Yes, I wanted to read it. My copy probably wouldn't arrive for another week.

Max was in good spirits, and his health seemed as good as at any time in the previous three months. He told me he had been having a new type of pain in his back, which he believed represented some healing of previous damage that was taking place, but which his doctor believed was quite ominous. "Well then," I replied, "I certainly believe your interpretation." "Yes," said Max, "but you see, the doctor and I have disagreed many times. And on all the previous occasions, his interpretation was the correct one." The cancer had reached his spinal column, and some three weeks later he was dead.

The family circumstances, the educational system, the entire cultural milieu that molded Max Delbrück belong to a long-vanished world. We will not see another like him, but we are fortunate indeed that he lived and worked among us for nearly three-quarters of a century.

© 2023 World Scientific Publishing Company
https://doi.org/10.1142/9789811234378_0045

U.D.C. 621.396.969.181.4: 523.42

Radar Measurements of the Planet Venus*

By

L. R. MALLING, B.Sc.†

AND

S. W. GOLOMB, Ph.D.†

Presented at the Convention on "Radio Techniques and Space Research" in Oxford on 5th–8th July 1961.

Summary: The Jet Propulsion Laboratory established contact with the planet Venus using a planetary radar system from 10th March to 10th May 1961. Using both velocity and range data, a new value of 149 598 500 km was determined for the astronomical unit which is presently accurate to 500 km. Further data reduction is expected to improve accuracy towards 150 km. The rotation of Venus has been determined by spectral analysis as ~ 225 days.

A super-sensitive receiving system was employed with a 10 kW transmitter operating at 2388 Mc/s. The receiver range capability is a 50 mW signal located on Venus at 30×10^6 miles. The data received included (1) received signal level, (2) power spectrum of the Venus-reflected signal, (3) Venus-Earth velocity and (4) Venus range. The first two used open-loop receivers; the second two used closed-loop automatic tracking receivers. All significant radio frequencies were derived from an atomic frequency standard.

1. The Experiment

At intervals of about 584 days, the planet Venus passes through inferior conjunction and approaches within 25 million miles of the earth. A maximum separation of 162 million miles occurs at superior conjunction. The inverse fourth-power law expressed by the radar range equation and the present state of the art make it most practicable to attempt radar contact with Venus during the few weeks just before and after each inferior conjunction. For the year 1961 this period is defined between 10th March and 10th May.

The decision was made to modify the Goldstone radar system, originally designed for tracking the *Echo* satellite, and reflect a continuous-wave signal off the surface of Venus. The necessary capability was to be obtained by using a super-sensitive receiving system rather than transmitting with extremely high power.

Some of the high performance characteristics specified and achieved for the Venus radar experiment were the following:

(i) A high-gain, low-noise receiver antenna feed which, in conjunction with the 85-ft parabolic reflectors, provided 54 dB gain but with a noise temperature of no more than 30° K.

(ii) A 2388 Mc/s maser amplifier, capable of two months of nearly continuous operation, with a noise temperature of 20 to 30° K.

(iii) A low-noise post-maser parametric amplifier, with a noise temperature of 300° K.

(iv) An ultra-stable transmitter crystal oscillator slaved to an atomic frequency standard, stable to 1 part in 10^9 over the 10 hours or so per day that Venus is visible, and stable to 1 part in 10^{10} over the 6-minute interval required for a radio signal to go from Earth to Venus and back.

(v) A receiver local oscillator, stable to 1 part in

Fig. 1. The receiving antenna.

† Jet Propulsion Laboratory, California Institute of Technology, Pasadena, California, U.S.A.

Journal Brit.I.R.E., October 1961
D

*This chapter was originally published in the *Journal of the British Institution of Radio Engineers*, Vol. 22, No. 4, October 1961, pp. 297–300.

L. R. MALLING and S. W. GOLOMB

10^9, which can be programmed in accordance with predicted data to follow the Doppler-shifted return signal. With these specifications, the range capability of the receiver was extended so that a 50 milliwatt omnidirectional signal located on Venus could be detected at a 30 million mile range. Figure 1 is a picture of the receiving antenna showing the mounting of the maser and parametric amplifier.

On 10th March 1961, the Goldstone transmitter was aimed at Venus for the first time. Several minutes later, the receiver was also pointed at Venus. During the 68 seconds of signal integrating time, one of seven stylus recordings was seen to deviate significantly and remained centered at a new mean signal level until the transmitter was deliberately allowed to drift off Venus, nearly an hour later. A typical recording for 29th

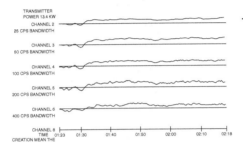

Fig. 2. Non-synchronous open-loop receiver signal level data. 29th March 1961 (P.S.T.).

March is shown in Fig. 2. Six and one half minutes later (the time of flight) the stylus recording reverted to its previous mean value. For positive confirmation, this experiment was promptly repeated, and with the same result. The first real-time detection of a radar signal returning from Venus had been accomplished.

2. Description of the Equipment

The 85 ft transmitting and receiving antennas are located 7 miles apart, with the receiving antenna shielded by intervening mountains. The desert location is shown in Fig. 3 for the transmitter site. The transmitter has a c.w. output of 12·6 kW at a frequency of 2388 Mc/s which, with an 0·35 deg conical beam, illuminates Venus with 10 watts. Table 1 gives the radar system parameters. About one watt of this is re-radiated into space and produces a signal of 10^{-20} watts, or -170 dBm, at the input terminals of the receiver. For a receiver bandwidth of 1 c/s and a system temperature of 60° K with the antenna aimed at Venus, the receiver threshold is -181 dBm, providing a typical signal/noise ratio

of 11 dB. The contribution of Venus to the system temperature was measured as being approximately 0·7° K.

Several different receiver configurations were utilized to determine received-signal levels, the power spectrum of the Venus reflected signal, Venus-Earth velocity and Venus-Earth range. Closed-loop, synchronous automatic-tracking receivers obtained both velocity and range data. Open-loop receivers, with predicted-Doppler local-oscillator tuning, provided signal strength and range data, spectrum analysis and reflectivity characteristics. A non-synchronous open-loop receiver block diagram is

Table 1
Venus Radar System Parameters

Unmodulated transmitter power (12·6 kW)	+ 71 dBm
Transmitter antenna gain	53·8 dB
Transmitter line loss	0·3 dB
$\left(\dfrac{\sigma}{4\pi R^2}\right)$ at 31 million miles	− 84 dB
Power intercepted by Venus	+ 40·5 dBm
$\left(\dfrac{\lambda}{4\pi R}\right)^2$ at 31 million miles	−255 dB
Receiving antenna gain	53·5 dB
Maximum received signal level	−161 dBm
Apparent reflection and propagation loss	9 dB
Typical received signal level	−170 dBm
Receiver threshold ($T = 60°$ K, $BW = 1$ c/s)	−181 dBm
Typical signal/noise ratio	11 dB

Fig. 3. The transmitter site.

RADAR MEASUREMENTS OF THE PLANET VENUS

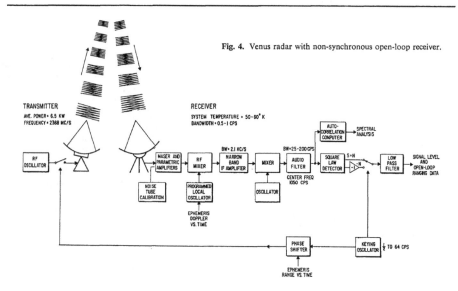

Fig. 4. Venus radar with non-synchronous open-loop receiver.

shown in Fig. 4. The transmitter is keyed with a low-frequency square-wave which also provides a phase-controllable reference signal for the receiver. The output of the low-pass filter provides both signal level and ranging data. An auto-correlation computer preceding the square-law detector permits spectral analysis of the reflected signal on a long term basis. Spectrum analysis was also accomplished in $1\frac{1}{2}$ minutes using a signal digitizer and recorder connected to the output of the i.f. amplifier.

The velocity of Venus with respect to Earth was measured using a c.w. transmitted signal and a synchronous closed-loop automatic tracking receiver with a tracking bandwidth of 5 c/s.† This configuration permits direct comparison of the Venus reflected signal with the transmitted frequency. Using a non-synchronous receiver and a closed-loop coded-ranging system accurate instantaneous ranging data could be accumulated.

3. Data Obtained

The principal result of the experiment was the more accurate determination of the Astronomical Unit by measurement of the range and velocity of Venus relative to the Earth. Three independent determina-

† This mode of operation has been discussed in the literature. See M. H. Brockman, H. R. Buchanan, R. L. Choate and L. R. Malling, "Extra-terrestrial radio tracking and communication", *Proc. Inst. Radio Engrs*, **48**, pp. 643–54, April 1960.

tions of the A.U. were obtained with a best number at the present time of 149 598 500 km, which is considered to be accurate to 3 parts in 10^6, or 500 km. Further data reduction, it is hoped, will reduce this to ± 150 km.

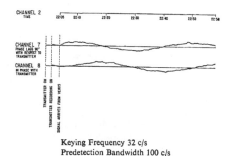

Keying Frequency 32 c/s
Predetection Bandwidth 100 c/s
Received Signal Level −168 dBm

Fig. 5. Non-synchronous open-loop receiver ranging data. 6th April 1961.

A typical recording of the Venus reflected signal obtained with a non-synchronous open-loop receiver for determining range is shown in Fig. 5. Relative to a perfectly conducting sphere, Venus was shown to have a reflectivity of 10% at 2388 Mc/s. A sample spectral analysis of the reflected signal is shown in Fig. 6, where the 3 dB bandwidth is 9 c/s. The spectrum appeared quite stable over the two-months' observation period and, assuming that the axis of rotation is not pointed directly at the earth, Venus appears to rotate very slowly, possibly with a rotation period of 225 Earth days.

The accuracy of the velocity measurement is 1 part in 10^5 without further data smoothing. The width of the closed-loop range gate was 8·2 ms, and it is believed that the accuracy of the time of flight measurement was about 1 ms for a round trip propagation time of 300 000 ms.

Date analysis and reduction began as soon as the first data became available. Complete recordings of all data obtained have been collected, and this raw data is being published in its entirety. It is hoped that numerous hypotheses concerning Venus can be confirmed or refuted on the basis of these recordings, over and above the conclusions discussed in this paper.

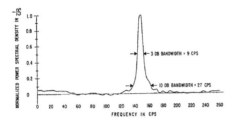

Fig. 6. Spectral analysis of Venus reflected signal using special auto-correlation computer with non-synchronous receiver. Integration time 1 hr 23 min. 21st April 1961.

Manuscript received by the Institution on 29th June 1961. (Contribution No. 38.)

© The British Institution of Radio Engineers, 1961.

DISCUSSION

Dr. S. W. Golomb: My first reaction to reading the Soviet press release on Venus radar in *Izvestia* on 16th May was that we should congratulate our Russian colleagues on the discovery of a new planet. It surely wasn't Venus! It was some 30 000 miles closer than Venus and rotating 25 times *faster*.

To-day Professor Kotelnikov has cleared up part of the mystery. To the figure 149 457 thousand kilometres for the astronomical unit which appeared in *Izvestia*, he has added the term ±130 *p* thousand km, for a suitable integer *p*. With *p* = 1, the figure 149 587 can be reconciled with the value obtained by J.P.L. as well as Lincoln Laboratory and Jodrell Bank. The decision we reached at J.P.L. two years ago, when first designing our ranging systems for deep space measurements, to leave no range ambiguities, however, gross, unresolved, definitely paid off here. As it turned out, 130 000 km was bigger than the uncertainty in the A.U., and in picking the most reasonable value on the basis of previous experiments, the Russian choice was originally 0·1% in error. So much for the A.U. The problem about the discrepancy in rotation rate requires more elaborate comment.

First, allow me to digress briefly on the history of the Goldstone system. When J.P.L. was building *Explorer I*, the Western World's first successful satellite (Jan. 1958) and the first of the highly successful series which discovered the Van Allen belts, we had a very sensitive receiver system called Microlock. During the summer of 1958, the receiver antenna was installed at Goldstone to track our intended escape shot, *Pioneer III* (Dec. 1958). In Mr. Pardoe's paper† this, the first *Juno II* launching, was listed as a failure, but it was a remarkably successful failure Although the *Jupiter* booster underperformed slightly, so that an escape orbit was not quite achieved, all the high speed stages as well as the payload electronics worked perfectly, and the five-day trip up to an altitude of 70 000 miles and down again, resulted in the best profile of the radiation belts that Van Allen has ever obtained to date. In March 1959, the *Pioneer IV* shot was successfully tracked on its escape orbit out to several Moon-orbit-diameters, when, as anticipated, its transmitter power was exhausted. The sensitivity of the Goldstone receiver at that time was about −120 dBm.

The Goldstone transmitter was built during the second half of 1959 in preparation for Project *Echo*, and the receiver sensitivity was improved to −150 dBm. During 1960, this two-antenna system was used to bounce signals and voice off *Echo*, the Moon and even a desk-sized *Courier* satellite. Voice via the Moon was communicated to Woomera, and since *Echo* travelled from West to East, it was the Goldstone antenna system that had to *acquire* the *Echo* balloon, while the Bell Laboratories antennas at Holmdel were slaved in angle to the output of the co-ordinate converter data from J.P.L.

For the Venus experiment, the transmitter power was up some 30%, the frequency was more than doubled, and an overall receiver sensitivity of better than −180 dBm was obtained—one of the most sensitive receivers at any frequency, for any purpose, anywhere in the world. The very first time the system was tried on Venus, on 10th March 1961, there was a clear-cut indication that success had been achieved. With only a 68 second time constant in the post-detection filter, when Venus was brought into the beam of the receivers, their output could be seen to move off its old centre value and climb to a new value,

---

† G. K. C. Pardoe, "The engineering aspects of satellites and their launching rockets", *J. Brit.I.R.E.*, **22**, pp. 145–160, August 1961.

DISCUSSION

within the duration of the time constant, and remain there until Venus was removed from the beam, when it dropped back again just as quickly. Almost every single day, many hours per day, for the next two months, contact with Venus was established, with literally hundreds of hours of data gathered, representing four major experiments and a few minor ones as well.

The most surprising and characteristic feature of our experiment was the narrowness in the bandwidth of the returning signal. Using the extra flexibility of a two-antenna system, for spectral analysis we had the transmitter turned on and off during alternate seconds, keeping the receiver always aimed at Venus. Processing the even and the odd seconds separately, we obtained spectra for pure noise and for signal-plus-noise. The difference of these two spectra gave us our very narrow signal spectrum, which by no stretch of the imagination contains Doppler spread in excess of 81 c/s at our 2388 Mc/s frequency, however far down in the noise. The 81 c/s spread would correspond to Doppler spread from the limbs of a "synchronous" Venus, rotating on its axis once every Venus year of 225 Earth days. While our frequency was much higher than the Russian frequency, the results of Lincoln Laboratory at a slightly lower frequency than the Russian one fully confirm our conclusion of the absence of Doppler spread. If one were to chop the signal-plus-noise spectrum into vertical strips, the resulting diagram would resemble Professor Kotelnikov's Fig. 5. The possibility that his experiment failed to remove all the noise power from the

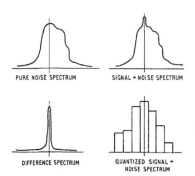

Fig. A.

"signal" spectrum should, therefore, be investigated (see Fig. A). Also, there are a variety of component malfunctions that have the apparent effect of broadening the received spectrum, but I am certainly not aware of any hardware problems that would *narrow* the spectrum. Another possibility is that when the wrong astronomical unit value is used (say $p = 0$ instead of $p = 1$), what is really a Doppler *shift* due to relative orbital motion between Earth and Venus may be misinterpreted as Doppler *spread*, suggesting a non-existent rotation.

One of our "minor" experiments involved polarization reversal to measure the surface roughness of Venus. The result, far from indicating a highly polished Venus, indicated a roughness very similar to that of the Moon, and I have never heard anyone call the surface of the Moon particularly smooth. If a Venus of that roughness were rotating as rapidly as the Russian experiment suggests, the non-coherent spectral return would certainly not be too far buried in the noise, and we are convinced we would have detected it.

A few comments about the precision we hope for in the A.U. may be of interest. We actually had several independent measurements, two of them computing the Earth-Venus round trip time very accurately in light-microseconds, one computing the instantaneous Earth-Venus relative velocity from the Doppler shift in the returned signal, and one computing the cumulative change in distance between Venus and Earth over a several hour period by tallying Doppler cycles cumulatively. Using the very accurate angle measurements compiled by optical astronomical observations, we converted each of these experiments into a value for the A.U., getting numbers in the range of 149 598 000 km to 149 599 200 km, each experiment consistent with itself to a few kilometres. However, by the time one examines sixth, seventh and eighth decimal places, such problems as the conversion from light seconds to kilometres and the propagation properties of the interplanetary medium become quite significant. Also, near inferior conjunction Venus has an apparent diameter of about 1 minute of arc, of which only a narrow crescent is visible. If the ephemeris angle to the "centre" of the Venus disc is in error by as much as 1 second of arc, differences of more than 1000 km can arise. A careful recheck and recomputation of the Venus ephemeris have enabled us to bring all our measurements together at the value 149 598 000±500 km, where the uncertainty of 3 parts in 10^6 is not experimental fluctuation, but uncertainty in other physical constants and conversion factors. We have as a goal for ourselves the ultimate resolution of the A.U. to one or two parts in 10^7, but the computations involved are quite prodigious.

Fig. B.
Venus: crescent and centre.

Finally, some brief comments on the outlook for the future. Radar with Venus is 65 to 70 dB harder than with the Moon. Illuminating Venus with 10 kW is comparable to illuminating the Moon with a milliwatt. Think of turning on your flashlight, covering it with your hand, pointing at the Moon, and trying to detect the difference in lunar brightness! An extra 65 or 70 dB beyond Venus would make it possible to establish radar contact with most of the smallest and remotest bodies in the solar system—Neptune, for example, and either of Mars' tiny moons.

With only 12 to 16 dB beyond our present system, we could follow Venus all the way around its orbit, track

420   *The Wisdom of Solomon*

DISCUSSION

Mercury and Jupiter all around their orbits too, and follow a sizeable portion of the orbit of Mars. However, it is only fair to warn you that Pluto presents a special problem. The round trip delay time is almost exactly twelve hours, so that if you aim your transmitter at Pluto now, by the time the signal returns you will have rotated clear around to the other side of the earth!

My concluding proposal is definitely "blue-sky". I suggest we rush out and turn our transmitter on the planetary system, if any, of the star Arcturus. It is true that successful detection of the returning reflection is some 280 dB harder than for Venus. However, the round trip delay is 66 years, and surely by that time our grandchildren will have no trouble at all with the incoming signal.

Dr. J. H. Thomson: As the Jodrell Bank Venus radar system has already been described[†] I will confine myself to a very short account. Unlike the two systems we have just been hearing about, it is a fairly conventional pulsed radar system at 408 Mc/s, with a peak power of 60 kW in 30 millisecond pulses at a repetition rate of 1 pulse per second. Unlike the Russian system, the Doppler shift is removed at the receiver rather than at the transmitter. As the signal/noise ratio expected was less than unity it was necessary to use an integrating device to examine the receiver output to detect the echo. The output of the final intermediate frequency amplifier which had a bandwidth of 67 c/s was applied to a linear detector; a variable frequency pulse output was then produced, linearly dependent in frequency on the detected voltage. The integrator sampled this pulse output at eight adjacent positions on the time-base, each sampling period being 30 ms. The outputs from corresponding sampling periods on successive sweeps were counted in eight separate counting channels. Thus if only noise is present the counts in the eight channels will show random fluctuations about a mean, but if an echo is present the count in one channel will show a significant rise from the mean.

During the observations the range of Venus was changing by as much as a sampling period in a few minutes so an automatic system was employed to alter the position of the sampling channels to keep the echo always in one channel. Echoes were first detected on 8th April as soon as the equipment was functioning correctly and continued to be detected till observations ceased on 25th April.

By 8th April, both J.P.L. and Lincoln Laboratory had been obtaining echoes for some weeks, and the Lincoln value of the parallax was available to us by private communication. The echo obtained agreed almost exactly with the Lincoln value, and also with the J.P.L. result when that became available. Nevertheless a complete search was made over the full range of possible values of parallax, covering unambiguously the full spread of the astronomical results. Our value for the A.U. is 149 600 000 ± 5000 km.

It is interesting to compare the four radar results now available. The Lincoln and J.P.L. results agree together

within their limits of error, and are well under the large umbrella of the Jodrell Bank limits of error—which at ± 5000 km are ten times as great as those on the admirable J.P.L. result we have just heard. Even allowing for the ambiguity now revealed in the Russian work, the corrected value of 149 587 000 km is still about 11 000 km from the mean of the three western results, well outside the limits of error, and it appears therefore that a real but small disagreement still exists.

As to the width of the spectrum of Venus, our pulsed equipment with its wide radiated spectrum is fundamentally rather unsuitable for such an investigation. Nevertheless a frequency analysis of the many tape recordings taken during the observations may shed some light on this matter.

Prof. V. A. Kotelnikov: I should like to reply to the remarks of Dr. Golomb.

The supposition that the enlarged spectrum of reflected signals in the experiments on radar contact with Venus carried out in the Soviet Union was due to an inaccurate value of the Astronomical Unit used in the preliminary calculation is not correct. Indeed, the Doppler shift of the frequency during the time of radar contact amounted to 20 000–30 000 c/s for the signals used. Therefore, a discrepancy in the Astronomical Unit of 1 part in 10^3 from its real value would result in an error in the Doppler shift of only 20–30 c/s. Moreover, this Doppler shift would be systematic. Evidently this cannot explain the enlargement of the spectrum by about ± 200 c/s.

The enlargement observed in the spectrum of the signals reflected from Venus also cannot be due to stochastic or systematic measurement errors, since all necessary measures were taken to exclude them.

Special measurements, in which signals imitating the attenuated transmitter signals were applied to the input of the receiver, did not result in an enlargement of the spectrum.

I believe that the discrepancy in the character of the reflected signal spectrum obtained in the U.S.A. and the U.S.S.R. can be explained as follows:

Let us assume that when sending radar signals to Venus a large part of the signal energy is reflected from the point nearest the radar set (just as would be the case for reflections from a shiny sphere), and moreover, that the dispersed reflection from all of the remaining surface is due to irregularities. Then, the reflections from the nearest point will give a narrow spectrum, while the reflections from the rest of the surface due to the Doppler shift of the frequency caused by the rotation of Venus will give a wide spectrum. The narrow spectrum in our measurements must fall in the sixth filter. From the energy distribution among the filters obtained in our experiments we can deduce that the energy in the sixth filter is about of the same order of magnitude as the total energy in all the other filters. Thus, for purposes of orientation we can assume that the energy in the narrow and wide parts of the spectrum are about equal.

Comparing the width of the narrow part of the spectrum (which according to the data obtained in the U.S.A. amounts to 9 c/s or in terms of our frequency 2·6 c/s) with the wide part (which according to our data amounts to

[†] J. H. Thomson, J. E. B. Ponsonby, G. N. Taylor and R. S. Roger, "A new determination of the solar parallax by means of radar echoes from Venus", *Nature* (Lond.), **190**, No. 4775, pp. 519–20, 6th May 1961.

400 c/s), we see that the spectral density of the energy in the wide part of the spectrum must be about $400/2\cdot6 \simeq 150$ times smaller than in the narrow part.

In the Soviet Union a preliminary analysis of the spectrum was carried out with filters having a bandwidth of 60 c/s. As a result, the narrow component of the spectrum could not be detected. Apparently most attention in the U.S.A. was given to the narrow component, while the wide component of the spectrum having a much smaller spectral density could have been undetected.

The spectral density curve given in the paper by Mr. Malling and Dr. Golomb does not refute what was said. Indeed, beyond the bandwidth of 30 c/s it contains irregularities, whose ordinates comprise 1/30th of the maximum value even if we assume that the abscissa should be taken to be slightly curved as Dr. Golomb indicated. Such a value of the spectral density is even larger than the approximate value of the spectral density of the wide part of the spectrum obtained above.

In the future we will attempt to reproduce the narrow part of the spectrum, also, using the magnetic tape recordings that we have. Maybe it would be possible to reproduce the wide part of the spectrum in the U.S.A. using the recordings that they have.

It would be also very interesting to attempt to obtain the form of the spectrum using the recordings of the signals reflected from Venus, which were obtained at Jodrell Bank.

I hope that these investigations will confirm the suppositions that I have made.

Dr. S. W. Golomb: I would very much like to thank Professor Kotelnikov for all the interesting information he has given us about the Soviet Venus-radar experiment. Apparently my suggestion of signal-plus-noise instead of signal spectrum does not underlie our discrepancy. Neither, however, does Professor Kotelnikov's suggestion about the total off-centre power in our experiment, since originally we used seven consecutive filters each 200 c/s wide, which when adjusted for wave-length is exactly parallel to the Russian's use of 60 c/s filter; and we detected *no* signal power except in the centre filter. I think Professor Kotelnikov will join me in urging our colleagues at Jodrell Bank to speed the reduction of their spectral data and thereby assist us in resolving our conflicting conclusions.

DETERMINATIONS OF THE ASTRONOMICAL UNIT

In an article in *The Times Science Review*, Autumn 1961, entitled "The Scale of the Solar System", Dr. J. H. Thomson of the Nuffield Radio Astronomy Laboratories, Jodrell Bank, discussed recent determinations of the astronomical unit by radar techniques and their relation to values obtained by other methods. By kind permission of the Editor of *The Times Science Review* and Dr. Thomson, the useful tabular presentation of the various results given in the article is reproduced here as an appendix to the discussions which took place at the Institution's Convention.

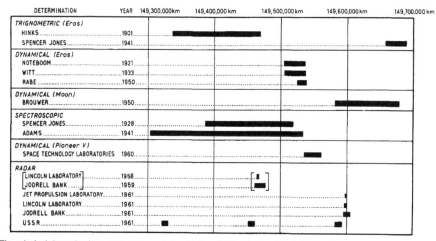

The principal determinations of the astronomical unit since 1900. The lengths of the blocks corresponding with each determination show the limit of error as determined by the workers concerned. The two radar results enclosed in square brackets have since been repudiated by their authors. The reason for ambiguity in the recent Soviet determination (here presented by three blocks) is explained by giving different values to the integer p in Professor Kotelnikov's final result.

© 2023 World Scientific Publishing Company
https://doi.org/10.1142/9789811234378_0046

Hadamard Matrix Discovery

Discovery of An Hadamard Matrix of Order 92

It has been conjectured that there exists a Hadamard matrix of order $n = 4k$ for every k. The smallest order that cannot be constructed by a combination of Sylvester's and Paley's construction is 92. Sol, together with Leonard Baumert and Marshall Hall, Jr give the first construction of a Hadamard matrix of order 92. Here, we include: (1) JPL's report on this discovery in 1962; (2) A blog from Cooper, Julie A. 2013. "Hadamard Matrix," Slice of History. NASA/JPL-Caltech: Pasadena, CA. Available from https://www.jpl.nasa.gov/blog/2013/8/hadamardmatrix; and (3) The original paper: Leonard Baumert, Solomon Golomb, and Marshall Hall, Jr. "Discovery of an Hadamard matrix of order 92", *Bull. Amer. Math. Soc.* **68** (1962), no. 3, 237–238.

424 *The Wisdom of Solomon*

OFFICE OF PUBLIC EDUCATION AND INFORMATION
CALIFORNIA INSTITUTE OF TECHNOLOGY JET PROPULSION LABORATORY
NATIONAL AERONAUTICS AND SPACE ADMINISTRATION
PASADENA, CALIFORNIA. TELEPHONE MURRAY 1-3661, EXTENSION 3351

FOR RELEASE: February 5, 1962

PASADENA, Calif.--Solution of a long-outstanding mathematical problem was announced today by a three-man team composed of CalTech and Jet Propulsion Laboratory mathematicians.

The importance of the discovery for divising codes for space communications was pointed out by Dr. Solomon W. Golomb, assistant chief of JPL's Communications Systems Research Section; Prof. Marshall Hall, Jr., CalTech Mathematics Department; and Leonard D. Baumert, predoctoral student at CalTech, who all cooperated in the solution.

Titled "A Hadamard Matrix of Order 92," the discovery involved construction of a pattern of variable combinations of positive and negative signs (a Hadamard Matrix) consisting of 92 rows and columns. Jacques Hadamard, the great French mathematician, first studied such combinations in 1893. Later scholars, among them R. E. Paley in 1933, described a variety of methods for constructing Hadamard matrices. Of the sizes less than 200 for which a Hadamard matrix could exist, there were only six cases where Paley did not give a solution.

In 1944, another mathematician, John Williamson, succeeded in constructing a Hadamard matrix of 172 columns by 172 rows, thus reducing this list of unsolved sizes to five. Since 1933, many mathematicians had attempted to prove or disprove the existence of a 92 by 92 combination.

"Due to the large number of possible internal combinations," Dr. Golomb stated, "even an electronic computer working at the rate of a million a second, would take billions of years to test all possible solutions

Hadamard Matrix Discovery

Hadamard Matrix:　　　　　　　　　-2-

of the problem." After useful suggestions by Prof. Hall, the JPL IBM 7090 computer was programmed by L. D. Baumert to search for a solution similar to Williamson's 172 by 172. An example of the long-awaited 92 by 92 Hadamard matrix was discovered after less than an hour of computation. In fact, according to Dr. Golomb, there turned out to be one and only one example of the Williamson type for size 92.

The large square in the photograph presents positive (dark) and negative (white) squares arranged into rows in such a way that between each pair of rows there are 46 positions in agreement (dark and dark or white and white) and 46 positions in disagreement (dark and white).

In the photo, from left to right are Sol, Leonard Baumert and Marshall Hall, Jr, who hold a picture of the Hadamard matrix of order 92 of the Williamson type.

BLOGS | SLICE OF HISTORY | AUGUST 1, 2013

Hadamard Matrix

By Julie Cooper

Hadamard Matrix -- Photograph Number 331-3717Ac
› View Full Image

In 1961, mathematicians from NASA's Jet Propulsion Laboratory and Caltech worked together to construct a Hadamard Matrix containing 92 rows and columns, with combinations of positive and negative signs. In a Hadamard Matrix, if you placed all the potential rows or columns next to each other, half of the adjacent cells would be the same sign, and half would be the opposite sign. This mathematical problem had been studied since about 1893, but the solution to the 92 by 92 matrix was unproven until 1961 because it required extensive computation.

From left to right, holding a framed representation of the matrix, are Solomon Golomb, assistant chief of the Communications Systems Research Section; Leonard Baumert, a postdoc student at Caltech; and Marshall Hall, Jr., a Caltech mathematics professor. In a JPL press release, Sol Golomb pointed out the possible significance of the discovery in creating codes for communicating with spacecraft.

The team used JPL's IBM 7090 computer, programmed by Baumert, to perform the computations.

428   *The Wisdom of Solomon*

DISCOVERY OF AN HADAMARD MATRIX OF ORDER 92[1]

BY LEONARD BAUMERT, S. W. GOLOMB AND MARSHALL HALL, JR.
Communicated by F. Bohnenblust, December 29, 1961

An Hadamard matrix H is an n by n matrix all of whose entries are $+1$ or -1 which satisfies $HH^T = nI$, H^T being the transpose of H. The order n is necessarily 1, 2 or $4t$, with t a positive integer. R. E. A. C. Paley [3] gave construction methods for various infinite classes of Hadamard matrices, chiefly using properties of quadratic residues in finite fields. These constructions cover all values of $4t \leq 200$, except $4t = 92, 116, 156, 172, 184, 188$. Further constructions have been given by J. Williamson [5; 6], A. Brauer [1], M. Hall [2] and R. Stanton and D. Sprott [4]. Williamson's first paper gave an Hadamard matrix of order 172, incorporating a special automorphism of order 3. The same method may be applied to 92, 116, 156, and 188, but Williamson did not do so, principally because of the amount of computation involved.

Williamson's method has been applied to $4t = 92$ using the IBM 7090 at the Jet Propulsion Laboratory. The matrix H has the form

$$
H = \begin{vmatrix}
A & B & C & D \\
-B & A & -D & C \\
-C & D & A & -B \\
-D & -C & B & A
\end{vmatrix}
$$

where each of A, B, C, D is a 23 by 23 symmetric circulant matrix. We give here the first row of each of A, B, C, D writing $+$ for $+1$ and $-$ for -1.

```
    1 2 3 4 5 6 7 8 9 10 11 12 13 14 15 16 17 18 19 20 21 22 23
A   + + - - - + - - - + - + + - + - - - + - - - +
B   + - + + - + + - - + + + + + + - - + + - + + -
C   + + + - - - + + - + - + + - + - + + - - - + +
D   + + + - + + + - + - - - - - - + - + + + - + +
```

REFERENCES

1. A. Brauer, *On a new class of Hadamard determinants*, Math. Z. **58** (1953), 219–225.

---

[1] The work reported in this paper was conducted at the Jet Propulsion Laboratory of the California Institute of Technology under a program sponsored by the National Aeronautics and Space Administration, Contract number NASw-6.

238   LEONARD BAUMERT, S. W. GOLOMB AND MARSHALL HALL, JR.

2. M. Hall, Jr., *A survey of difference sets*, Proc. Amer. Math. Soc. **7** (1956), 975–986.

3. R. E. A. C. Paley, *On orthogonal matrices*, J. Math. Phys. **12** (1933), 311–320.

4. R. G. Stanton and D. A. Sprott, *A family of difference sets*, Canad. J. Math. **10** (1958), 73–77.

5. J. Williamson, *Hadamard's determinant theorem and the sum of four squares*, Duke Math. J. **11** (1944), 65–81.

6. ———, *Note on Hadamard's determinant theorem*, Bull. Amer. Math. Soc. **53** (1947), 608–613.

CALIFORNIA INSTITUTE OF TECHNOLOGY

Eight JPL *Lab Oratory* Issues

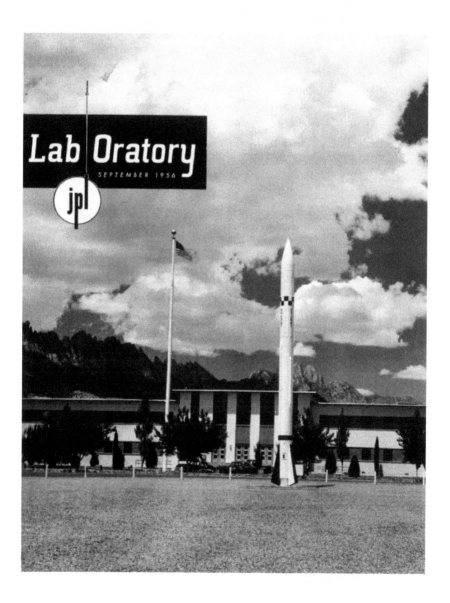

432   *The Wisdom of Solomon*

Welcome ...

New JPL Employees
August 1 to 31, 1956 Inclusive

Aaron. Clyde	Plumber
Adler, Sharon	Clerk-Typist
Aguirre, James	Engineering Designer "B"
Atchison, Dale	Clerk-Stenographer
Barnard, Constance	Clerk-Typist
Barnes, Rulon	Administrative Assistant
Batone, Clarence	Jr. Technician (Electronic)
Beck, William	Jr. Mathematician
Belanger, Joyce	General Clerk
Bell, Charles	Truck Driver (Medium Duty)
Bills, Ella Folder	Trimmer
Brayshaw, James	Research Engineer
Brubaker, E. Lee	Inspector (Experimental)
Burkhart, Milton	Truck Driver (Medium Duty)
Burkhart, Nancy Jo	Clerk-Typist
Button, Daniel	Research Engineer
Cass, Robert M.	Personnel Supervisor
Christensen, Elmer	Sr. Development Engineer
Cleminshaw, Marian	Mathematical Assistant
Colbert, Harry	Electrician (Maintenance)
Cortese, George	Drawing Checker
Courtney, Richard	Mail Man
Crowford, Gordon	Technician (Electronic)
Dakon, Sharon	Clerk-Stenographer
Dick, Dean	Engineering Assistant
Divone, Louis	Development Engineer
Eisner, Mary	Secretary
Estes, David	Precision Machinist
Fearing, Paul	Stock Clerk
Framan, Elliot	Development Engineer
Garrison, Frank	Storekeeper
Golomb, Solomon	Sr. Research Engineer
Hart, Thomas	Mechanic (Model)
Hendrickson, Burton	Tabulator Operator
Hildebrandt, Alvin	Sr. Research Engineer
Howick, John	Millwright
Johnson, Joan	Material Control Clerk
Kindt, Donald	Research Engineer
Kinkead, Maxwell	Truck Driver (Medium Duty)

By Gary Weber

Sol Golomb Wins Chess Tournament

A perfect score of six wins, no defeats, put Sol Golomb, Section 8, on top in the Chess Club's second tournament, which ended March 4. Sol's was the only perfect score in the six-round Swiss-style Tourney, which matches players with similar records.

The top eight finishers are listed, with their relative position determined by an examination of their respective opponent's scores.

Sol Golomb	6 - 0
Jo Reger	5 - 1
Jovan Moacanin	4 - 2
Gary Weber	4 - 2
Jim Creson	4 - 2
John Coy	4 - 2
Norm Jacobson	4 - 2
Anton Havlik	4 - 2

The popularity of the tournaments is increasing, with about thirty players entering. Plans are already underway for a third tournament to begin soon.

Club Holds Instructional Classes

The club is also holding classes for all people interested in either learning chess or improving their present game. Instruction is led by John Earnest, past Oklahoma State Champion and winner of JPL Chess Club's first tournament of the current series. The classes are held in Bldg. 122 Conference Room, Wednesday noon on the weeks the club doesn't have a regular meeting. Anybody interested in attending is very welcome.

A Bird in the Hand...

The badminton enthusiasts on the Lab, members of the Pasadena Badminton Club, have announced playing-meetings at the Pasadena City College Girl's Gym every Thursday night. Admission is 75¢ for the use of a choice of 6 courts, lights and showers. Birds are furnished also.

434   *The Wisdom of Solomon*

"SCIENTIFIC AMERICAN" ARTICLE BY SOL GOLOMB

or

ANYONE FOR POLYOMINOES?

I imagine nearly everyone is familiar with the game of dominoes, and the little domino itself. Well, Sol Golomb, Electronics Research, has taken the lowly domino, sophisticated it. and developed it into a whole family of little multi-squared creatures which he cells polyominoes. In addition to the two square domino, there is now the single square monomino, the three square tromino, the four squared tetromino, the pentomino, etc.

Then Sol applied this new family to his favorite battlefield. the 8 x 8 chess-checker board, and came up with a fascinating mathematical game. The May, 1957 issue of Scientific American magazine devotes a considerable section of its mathematical game section to the problems raised when one tries to fit certain polyominoes onto a checkerboard, the article being excerpts from an original article by Sol in the *American Mathematical Monthly* magazine.

The variations of problems arising seem infinite, and if you go for brain twisters, this should do the trick. A typical theorem: "A necessary and sufficient condition for the checker-board to be coverable with 21 straight trominoes and one monomino is that the monomino be placed on one of only four squares of the board". Try it.

Sol has quite a few copies of his article for anyone interested, and will even put on a demonstration using some prettily-colored wooden pentominoes, if you stop by his office in Building 125.

Jetmen Need More Fans

Our team isn't exactly getting off to a flashing start this season, posting a record of one win, two ties, and one defeat for the first four games. If I'm not mistaken, though, as the weather warms up, so will the Jetman. The team has looked good despite the record, and with a break or two in our direction, things might have been different. The new uniforms really look fine on the field, and certainly add a psychological boost that carries all the way to the fans.

The season opened with a tight 1-1 game which even went into extra innings against Ives & Warren, one of last year's top threats. The following week saw the Lab whip Solteros by a 4-0 shut-out, Jack Seymour getting credit for the whitewash.

Then the Independent Garage Owners turned in an upset, defeating us by a score of 2-1.

In the last game before press time, the Jetmen battled from behind to tie Bent & Son 2-2, but were unable to push across a winning run. Bent got off to a two run load in first inning, but from then on, their offensive made little headway, only one man advancing to third the rest of the game. Our Side started getting to the Bent pitcher in the third inning, and in the fourth combined a single by Ray McLain, Jerry Downs towering ground rule double, and a sacrifice fly to drive in two runs.

the smallness of these crowds. Whassamatter? No spirit? C'mon out end yell, and have a lot of fun. Bring your wife and have her yell, too. She might get laryngitis and not be able to talk for a week.

Truthfully, I have watched a lot of major league baseball back in Cleveland, but never have I enjoyed any games as much as our own JPL. Conversely, these games do a lot to make me enjoy JPL, too.

"Splinter" Weber

P.S. Thanx, Bill Merrick. Glad to know I have a reader.

Sol Golomb Wins Second Chess Tournament

Sol Golomb continued his mastery of the game of chess, winning his second straight JPL chess tournament and remaining undefeated in tournament play with 12 consecutive victories. In second and third place were Jim Christman and Gary Weber (who?) with scores of 5 wins and 1 defeat. Here is the preliminary results, not necessarily in final order as determined by a point basis:

Sol Golomb	6-0
Jim Christman	5-1
Gary Weber	5-1
Dan Button	4-2
John Coy	4-2
Anton Havlik	4-2
Ray Newburn	4-2
Marv Perlman	4-2
Jo Reger	4-2

The Chess Club now plans to concentrate on presenting a series of planned instruction lectures in a drive to acquaint more people with the game. Further notice will appear on the bulletin beards.

Summer Bowling

Summer bowling is well under way in its 1957 season, with most of the old teams and old faces still in there competing. Only the Glendale Tuesday League has decided to sit out the summer months.

In the Pasadena Thursday League, as of May 23, the Pin Wheels are off to a bit of a lead, closely followed by a whole slew of teams. For the week of the 23rd, high team series honors go to the Holy Bowlers with 2956. High Team Game - Inspection 5 - 1010; Men's High Individual Series - Ralph Bliven - 566; Men's High Game, a three-way tie with Jack Brimhall, Herb Grainger and Roger Barnett all at 202; Women's High Series - Orvo Bokarica - 490, Women's High Game - Martha Brimholl - 178.

SPORTS

By *Gary Weber*

Chess Club Elects Officers

Jim Creson was named President of the JPL Chess Club in a closely contested semi-yearly election of officers. Position of Vice President/Tournament Director was voted to Dorothy Sweitzer, while Secretary-Treasurer will be Vickie Pandrea. To complete the slate of officers, Bud Moulton will continue as membership chairman; Gary Weber, publicity director and Charles Goodell, librarian.

Among President Creston's objectives for the coming season are four tournaments, including Swiss end ladder type, increased membership through increased activity and continuation of the instructional program. He also would like to call attention to the club's excellent lending library and to the availability of chess men and boards for players who lack these necessary items.

The club is well underway with its weekly instructional class, which is drawing ever increasing interest as it progresses deeper into the game. This is probably the best way to become acquainted with chess, because it offers personal insight on each phase of the game by some pretty experienced players. Classes are held every Tuesday noon in the conference room. Bldg. 122.

BALANCE OF CLASSES SCHEDULED

Tuesday	Sept. 3	Offensive Tactics	Gary Weber
	Sept. 10	Art of Sacrifice	George Lindskog
	Sept. 17	Pitfalls	Anton Havlik
	Sept. 24	Basic Objectives	
		Middle Game Play	Jovan Moacanin
	Oct. 1	Pawn Advance	John Coy
	Oct. 8	Checkmates	Sol Golomb

Leisure Living

An interesting item by Bruce M. Jones, in a recent issue of THE SATURDAY EVENING POST supports our view that fishing is a beneficial and dignified way of doing nothing. The item, about as follows, is based upon a 1907 Georgia court decision:

Easygoing Joe was arrested by the town policeman and charged with vagrancy, defined by ordinance as "idleness and a failure to furnish oneself with employment." To prove the charge, the prosecutor sought to establish at the trial that Joe spent most of his time fishing.

THIS TAKES THE CAKE! The word got out that Marge Boyle (Sect. 14) would share her termination Cake if somebody in the Photo Lab would take a picture. So what happens — six cameramen and one Camerawoman showed up (only one really had film); but all got their cake. To insure our continuously fine cooperation from the Photo Lab we'd better identify them (left to right): Marge Ryon, Ray Buch, Bob Pace, Mrs. Boyle, Hal Sobotker, Norm Fryslie and Don Maxiener, Jim Rayle was present but someone had to take the picture. (Note to future mothers: Don't count on this cooperation just anytime as most of the time we're lucky to get just one cameraman.)

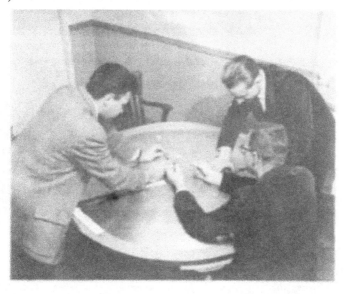

TRACKING INSTRUMENT DEVELOPED BY JPL MEN. This is the "sphere-dop" table from the message center in Bldg. 111. Busily engaged in using this highly specialized instrument are: (l to r) Dr. Sol Golomb (Sect, 8) Dr. William McDonald (Sect. 23) and in the foreground, Dr. Stan Lorens (Sect. 8).

Little Things In Life

"HEY! WATCH THAT YOU'RE DOIN' BACK THERE!" This is the birth picture of Rachelle Denise, born to CAROLE ANN JOHNSON (formerly Sect. 30) and husband, Richard. It all happened at the Huntington Memorial, May 13. The young lady weighed 6 lbs. 4 oz.

A FEW YEARS DIFFERENCE. Louis Wyant (Prec. Machinist) is so proud of his grandson Timothy Joseph Lett, born to his daughter Norma Jean, he came through with a rare shot of himself for comparison purposes. Timothy, about 5 months, is on the left and Louis, about 2 years of age, is on the right.

It's the first child for SOL GOLOMB (Sr. Research Engineer) and his wife BODIL. Girl child Astrid Elizabeth was born at the Arcadia Methodist Hospital on May 8. She weighed 6 lbs. 3½ oz.

IRV NOREN SCRATCH LEAGUE

The first round of bowling has ended with the following results: Kagey Keglers lead; Vagabonds in second place with the Bowleros in a very close third piece. All tied at 4th are the Odd Balls. Orbiters and Five Fabs.

The following bowlers have average of 165 or better: Walt Padgham - 181, Jack Brimhall - 180, George Cummings - 177, Norn Fryslie - 173, Ralph Bliven - 173, George Griggs - 170 and Chuck Thore - 166.

The Bowleros have the highest series with 2579 pins and the Orbiters have the highest game with 919. (Is it true that the Odd Balls are buying Norm Fryslie a book on how to keep score?)

SANTA ANITA LEAGUE

No report.

IRV NOREV TUESDAY L EAGLE

Standings for November: weekly team series - 9 & A Wiggle - 3162, weekly men's game - Carl Bieber - 235; weekly teem game - Cherry Pickers - 1041; weekly women's game - Val Dembrowski - 190. The Hits & Misses are two games ahead of the Cherry Pickers.

PASADENA BOWLING LEAGUE

Results for the month of November; men's high game - Roy Walker - 230; women's high game - Kit Barker - 193; men's high series - Eugene Otto - 585; women's high series - Flo Courtney – 456. The Bowl Weevils are in the lead with seven games to their advantage. Rocket Rollers are in second with Inspection 5 in a close third.

Latest Moves of JPL Chess Club

At the Wednesday meeting, December 3, Sol Golomb and Steve Szirmay, played a full 13 games in one half hour against other members of the chess club. They played three and four boards at a time, consecutively, and didn't lose one game.

Don't let the above "powerhouse" scare you. If you'd like to participate, they welcome all beginners who would like to learn. Steve is club prexy, Gary Weber, Publicity, reports that all tournaments are now under way.

JPL Baseball Team Historical Display

Jack Seymour (Sect. 26) retrieved all the post available photos of our baseball teams, organized them into a display which now hangs in the Main Cafeteria dining room.

The teams are from 1947 to 1958 except for 1950. No picture could be located for this particular year. Jack says the fellows who deserve the credit for most of the work were: Jr. Edwards (Sect. 55) George Chisman (Sect. 37) and Johnny Carter (Sect. 26).

This is an appropriate time to have this done since the Athletic Club has disbanded and is now reorganizing into the JPL Baseball Club.

April 1959

PICTORIAL DISPLAY OF PIONEER IV PROBE. If you might like to see this scale model based on the orbit of the Pioneer IV shot, it is in Dr. Sol Golomb's office. Bldg. 125 - 106. Sol is the one responsible for this excellent model which is the only true scale model of the Earth - Moon system at JPL. It is also the largest model. There is a duplicate located at Goldstone. Sol is Research Group Supervisor of Section 8.

ADVANCE NOTICE! At a recent meeting held at the campus with representatives from JPL and CWT, a date has been set for another Hobby Arts & Crafts Show for March 19, 20 and 21 in 1960. So if you haven't got a hobby — now is the time to start one. Your hobby will be displayed in (and outside of) Dabney Lounge, Caltech campus, as indicated by the picture above which was from the last show we had in March, 1956.

Beatrice Alexandra was born to DR. & MRS. SOLOMON W. GOLOMB on May 16 at 2:45 a.m. at the Methodist Hospital in Arcadia. She weighed 6 lbs. 11-1/4 oz. at birth. Her older sister is Astrid Elizabeth. Sol is in Sect. 8; Supervisor of Information Processing Group.

A most novel announcement came through from our ABMA Liaison at Huntsville, Alabama, heralding the arrival of Charles Hunter Richmond, first-born son of ANN and AL RICHMOND, putting a Confederate dollar bill to its first good use in some time.

© 2023 World Scientific Publishing Company
https://doi.org/10.1142/9789811234378_0048

SHIFT REGISTER CODERS — THE GOLOMB ALGORITHM

In 1959 Sol Golomb was working at the Jet Propulsion Laboratory on an Army Ordnance grant, focusing on missile guidance. He wrote two related papers. The first one, unclassified, introduced linear correlation invariants for Boolean functions and studied their properties and their relation to Rademacher-Walsh functions. This was published in 1959 in an international conference proceedings and then republished in 1967 in Sol's opus "Shift Register Sequences" which was widely circulated.

The second 1959 paper, "Principles of the Design of Shift Register Coders", was classified and dealt with the applications of these above invariants to secure missile guidance and, more generally, to the cryptanalysis of stream ciphers. It was not declassified and released to the outside world until April 2022. The declassification request was submitted by Al Hales in late 2021, after several years devoted to locating and accessing the paper.

In retrospect ("hindsight") it is hard to understand how these applications could have escaped the outside academic community for 29 years. But it was not until 1988 that Xiao and Massey (building on work of Siegenthaler and fully aware of Sol's 1967 book) published their IEEE paper "A spectral characterization of correlation-immune combining functions". As a result of this paper the technique was sometimes referred to as the Xiao-Massey theorem. A more reasonable description, in view of the above history, should be the Golomb algorithm!

Enclosed are the title pages, table of contents, abstract and introduction for Sol's classified paper, along with the cover letter (from the Army).

---

*This entry was provided for this volume by Al Hales.

DEPARTMENT OF THE ARMY
UNITED STATES ARMY INTELLIGENCE AND SECURITY COMMAND
FREEDOM OF INFORMATION/PRIVACY OFFICE
FORT GEORGE G. MEADE, MARYLAND 20755-5995

APR 2 0 2022

Freedom of Information/
Privacy Office

Mr. Alfred W. Hales
4320 Westerra Court
San Diego, California 92121
hales@ccr-lajolla.org.

Dear Mr. Hales:

This is in further response to your Freedom of Information Act (FOIA) request of December 19, 2021 and undated follow-up letter approximately one month later, and supplements our letter of March 4, 2022.

Coordination with other government agencies has been completed and a record was returned to this office. Please be advised, upon receipt of the record, several missing pages were identified within the record. We have no way of obtaining a copy of these missing pages as our office does not maintain the record. We have completed a mandatory declassification review in accordance with Executive Order (EO) 13526. As a result of our review the information no longer warrants security classification.

If you have any questions regarding this action, feel free to contact this office at 1-866-548-5651, or email the INSCOM FOIA office at: usarmy.meade.902-mi-grp.mbx.inscom-foia-service-center@mail.mil and refer to case #0134F-22. Please note that you now have the ability to check the status of your request online via the U.S. Army Records Management and Declassification Agency (RMDA) website: https://www.foia.army.mil/FACTS/CaseStatus.aspx. Please refer to FOIA Control Number: FP-22-011286. You may also contact the INSCOM FOIA Public Liaison, Mrs. Joanne Benear, for any further assistance and to discuss any aspect of your request at 301-677-7856. Additionally, you may contact the Office of Government Information Services (OGIS) at the National Archives and Records Administration to inquire about the FOIA mediation services they offer. The contact information for OGIS is as follows: Office of Government Information Services, National Archives and Records Administration, 8601 Adelphi Road-OGIS, College Park, Maryland 20740-6001, email at ogis@nara.gov, telephone at 202-741-5770; toll free at 1-877-684-6448; or facsimile at 202-741-5769.

Sincerely,

HEATON.MICHAEL.TODD.1160922075
Digitally signed by HEATON.MICHAEL.TODD.1160922075
Date: 2022.04.20 13:37:18 -04'00'

Michael T. Heaton
Director
Freedom of Information/Privacy Office
Investigative Records Repository

Enclosure

Best Copy Available

~~SECRET~~
UNCLASSIFIED

PROGRESS REPORT NO. 20-554

PRINCIPLES OF THE DESIGN OF SHIFT-REGISTER CODERS

SOLOMON W. GOLOMB

This document contains information affecting the national defense of the United States, within the meaning of the Espionage Laws, Title 18, U.S.C., Sections 793 and 794, the transmission or revelation of which in any manner to an unauthorized person is prohibited by law.

JET PROPULSION LABORATORY
CALIFORNIA INSTITUTE OF TECHNOLOGY
PASADENA, CALIFORNIA
JUNE 30, 1959

Regraded UNCLASSIFIED on
13 Apr 2022
by USAINSCOM FOI/PA
Auth Encl 5, para 1-d, DOD 5200-01-V1

~~SECRET~~
UNCLASSIFIED

UNCLASSIFIED
~~SECRET~~

ORDCIT Project
Contract No. DA-04-495 ORD 18
Department of the Army
ORDNANCE CORPS

PROGRESS REPORT No. 20-386

PRINCIPLES OF THE DESIGN OF
SHIFT-REGISTER CODERS

Solomon W. Golomb

Walter K. Victor

Walter K. Victor, Chief
Electronics Research Section

Copy No. F 32

This document consists of 1 page and 34 pages

GROUP 1
Excludes from automatic
downgrading and declassification

JET PROPULSION LABORATORY
California Institute of Technology
Pasadena, California
June 30, 1959

~~SECRET~~
UNCLASSIFIED

Regraded UNCLASSIFIED on
13 Apr 2022
by USAINSCOM FOI/PA
Auth Encl 5, para 1-d, DOD 5200-01-V1

Progress Report No. 20-386

UNCLASSIFIED

Jet Propulsion Laboratory

Copyright © 1959
Jet Propulsion Laboratory
California Institute of Technology

This document contains information affecting the national defense of
the United States, within the meaning of the Espionage Laws, Title 18,
U.S.C., Sections 793 and 794, the transmission or revelation of which
in any manner to an unauthorized person is prohibited by law.

Regraded UNCLASSIFIED on
13 Apr 2022
by USAINSCOM FOI/PA
Auth Encl 5, para 1-d, DOD 5200-01-V1

UNCLASSIFIED

448   *The Wisdom of Solomon*

UNCLASSIFIED   Progress Report No. 20-386

~~SECRET~~   *Jet Propulsion Laboratory*

CONTENTS

		Page
I.	Introduction	1
II.	The Versatility of Code Generators	2
	A. Considerations of Equipment Complexity	2
	B. Bottlenecks, Precomputation, and Forbidden Regions	2
	C. Screening Sequences—Versatility vs Yield	2
	D. Families of Feedback Logics	3
	E. External Logic Generators	4
	F. Summary	5
III.	Invariants of the Logic Families	6
	A. The Zero- and First-Order Invariants	6
	B. The Higher-Order Invariants	7
	C. Canonical Forms	8
	D. The Rademacher-Walsh Expansion Coefficients	8
	E. Practical Methods of Computing the Invariants	9
	F. Application to Finding Asymmetrical Logical Functions	10
	G. Applications to Deciphering by the Jammer	12
	H. Anticountermeasures	14
	I. Evaluation of the Communicator-Jammer Conflict	15

Nomenclature   17

References   17

Appendices

		Page
A.	The Six Classes of Balanced Boolean Functions of Three Variables	18
B.	The Symmetry Classes of Balanced Boolean Functions of Four Variables	19
C.	Balanced Boolean Functions of Five Variables with All First-Order Invariants Different	29
D.	Examples of Balanced Boolean Functions of Five Variables with All First-Order Invariants Equal to 8	67
E.	Some Instances of Balanced Boolean Functions of Six Variables with All First-Order Invariants Distinct and Unequal to ½ T	71
F.	Multi-dimensional Correlations	74

Regraded UNCLASSIFIED on
13 Apr 2022
by USAINSCOM FOI PA
Auth Encl 5, para 1-d, DOD 5200-01-V1

Page iii

~~SECRET~~
UNCLASSIFIED

Shift Register Coders — The Golomb Algorithm 449

Progress Report No. 20-386

UNCLASSIFIED

Jet Propulsion Laboratory

SECRET

TABLES

		Page
1	Occurrence and Versatility of the Nonlinear Logics	4
2	Proto-invariants for the Function $y \bigcirc z \bigcirc zy \bigcirc yz \bigcirc yz$	13
A-1.	Description of the Classes of Three Variables	18
A-2	Representative Karnaugh Charts for Classes of Three Variables	18
B-1.	Truth Tables for Representatives of the Classes of Four Variables	19
B-2	Invariants of the Classes of Four Variables	26
B-3	Analysis of the Classes of Four Variables	28
C-1	The Ten Cases of T 8, T 7, T 6, T' 5, T 3	29
C-2.	The 205 Cases of T 8, T 7, T 6, T' 5 T, 4	32
D-1.	Eleven Balanced Boolean Functions of Five Variables	67
E-1	Five Balanced Boolean Functions of Six Variables	71
F-1.	One-dimensional Correlation, Case 1	75
F-2	One-dimensional Correlation, Case 2	75
F-3.	One-dimensional Correlation, Case 3	76
F-4	One-dimensional Correlation, Case 4	76
F-5.	Two-dimensional Correlation, Case 1	77
F-6.	Two-dimensional Correlation, Case 2	79
F-7.	Two-dimensional Correlation, Case 3	81
F-8.	Two-dimensional Correlation, Case 4	83
F-9.	Two-dimensional Correlation, Case 5	85
F-10	Two-dimensional Correlation, Case 6	87
F-11	Two-dimensional Correlation, Case 7	89
F-12	Three-dimensional Correlation, Case 5	91

Regraded UNCLASSIFIED on
13 Apr 2022
by USAINSCOM FOI/PA
Auth Encl 5, para 1-d. DOD 5200-01-V1

Page iv

SECRET

UNCLASSIFIED

450   *The Wisdom of Solomon*

ABSTRACT

A variety of considerations which influence the design of shift-register coders are investigated. The relative roles of length of the shift register, complexity of the feedback logic, time required for feedback computations, and external re-encoding are discussed and evaluated. The classification of Boolean functions is treated in detail, with applications to shift-register problems. The security of shift-register codes is analyzed, including a discussion of the effectiveness of certain countermeasures and anticountermeasures techniques.

I. Introduction

Shift register sequences are an excellent source of code keys in a variety of secure communications applications. In this report, the basic factors which influence the security of a coding system based on shift-register sequences are analyzed, and recommendations are made for the optimum design of shift-register coders.

Security is seen to depend on the size of the ensemble of possible codes as well as on the properties of individual members of this ensemble. The total number of codes available for a secure communications application called the *versatility* of the coding system is contrasted with the smaller number of codes called the *yield* of the system having certain favorable properties. Once the relative importance of versatility and yield is understood it is possible to formulate rules which specify the proper length of the shift register and the necessary complexity of the feedback logic for a given application.

Considerations directly relating to the security of shift register coding systems are bound to depend on the classification of Boolean functions of n variables - the feedback functions of the shift registers - relative to the group of permutations and complementations of the inputs. Aspects of this classification include the number, the sizes and the invariants of the symmetry classes. The sizes of the classes directly influence the versatility of the coding system while the invariants suggest a possible countermeasures technique for an enemy jammer or interceptor. The relevant theory of Boolean functions is developed in detail and the applications to various aspects of secure coding are explored. In particular several methods are suggested for overcoming countermeasures techniques based on the Boolean invariants.

© 2023 World Scientific Publishing Company
https://doi.org/10.1142/9789811234378_0049

The PapaVerse

Introduction by Beatrice Golomb

Introduction

Sol, in young adulthood, not infrequently penned verse, often playful, for family and friends.

Through humorous verse, Sol passed his love of language to his children. (Indeed, through Sol's example, neither of his daughters are contemptible as versifiers. Among a couple quite good examples by Astrid, I recall one written about — and during a time of much news about — the financial crisis in Greece, for which I recall suggesting the title "Owed on what Grecians Earn" — but the verse was perhaps better without the distraction of such a title.)

Beyond inculcating appreciation for words, a further benison of Sol's verse was that through childhood and young adulthood I could readily recite the fifty United States — another of the manifold advantages accruing to having Sol as a sire — having come across a handwritten version of *The States* (by Solomon W. Golomb) as a young child. After decades sans recent repetition, the final couplets of The States had become indistinct in memory, but I believe I have resurrected them with fidelity.

Most of Sol's poems are long lost. The verse included here is not necessarily the best, but what I remembered (like *The States*), or had once encountered — often on a scrap of paper — that managed to be saved. Others, often greater favorites, have not survived the ravages of time, and changes in domicile.

Some examples reflect Sol's fondness for internal rhyme schemes, construction of which could be construed as rather like a word puzzle.

Sol once shared with me which were some of his own favorite poems, and a lolloping rhythm, logic play, and/or internal rhyme schemes were features ...

Some of Sol's favorite poems include:

- Etiquette — by W. S. Gilbert
- The Highwayman — by Alfred Noyes
- How the Helpmate of Blue-Beard made Free with a Door — by Guy Wetmore Carryl
- The Raven — by Edgar Allen Poe

452   *The Wisdom of Solomon*

- The Cremation of Sam McGee — by Robert W. Service
- How they Brought the Good News from Ghent to Aix — by Robert Browning

(One I liked to which Sol introduced me was The Twins, by Henry Sambrooke Leigh.)

    The verse by Sol were mostly for family, or for an occasion, and not intended for broad viewing — so should be judged accordingly. However, my favorite among these also reflects Sol's subtle wordsmithing and devotion to Bo. His "translation" of "The Milky Way" is indeed my favorite poem in the Universe, with its beauty and poignancy — and perhaps too because I know Sol translated it for Bo, at a rough patch early in their marriage... I had saved it and read it at my own 10th anniversary, and though without forewarning that it would be read, having likely not thought about it in decades, Sol recited the full poem in the original Swedish. Already believing it exquisite, I was struck anew, by how remarkably Sol's translation captured the sound quality, beauty and meaning.

The Milky Way

Translated from the original Swedish by Solomon W. Golomb,
and dedicated to his beloved Bo.
Swedish poem by Zacharias Topelius

And now the lamp grows dim and now the night is crisp and clear
And now old memories arise from many a vanished year.
Nostalgic fancies waft about, then fleetingly depart –
So wonderful and touching, as they warm the frozen heart.

The shining stars come piercing through the dark of winter night...
Such blessed joy, as though no death exists beneath their light.
You understand their silent speech? I know a tale of yore
That I have learned from these, the stars. I'll tell you of their lore.

Remote upon a star he dwelled, beyond the evening sky –
She lived upon another sun, unseen by human eye;
And Salami, as she was called, and Sulamit was he,
Loved one another with a love as strong as love can be.

They loved each other long before, when on the Earth they dwelled,
Until by force of Night and Death, and Sin and Grief, expelled –
Broad wings grew out upon them as they lay beneath Death's bars,
And they were doomed to dwell apart, to live on distant stars.

The Papa Verse 453

They thought but of each other in their dwellings up on high –
Between them lay a realm of suns, in vast expanse of sky –
Unnumbered worlds, the handiwork of God's all-wise desire
Spread out, twixt Salami and Sulamit in burning fire.

And then one evening Sulamit, when longing did incite,
Began to build, from world to world, a bridge of purest light,
And then did Salami as well, beginning from her place,
Commence to build a bridge across the vastnesses of space.

A thousand years they built, and more, with never-tiring strength,
And thus the Milky Way was formed, a stellar bridge, its length
Unmeasured, spanning Heaven's highest arch, its starry sweep
Uniting, linking every shore along the cosmic deep.

The cherubs saw and trembled, and they flew to God, afraid –
"O Master, see what Salami and Sulamit have made!" –
But God Almighty laughed and said, emitting light and joy,
"Whatever Love builds in my world, I never will destroy."

Then Salami and Sulamit, the bridge complete at last,
Did spring into each other's arms. At once a star was cast –
The brightest in the sky above; their path it did illume –
As after centuries of grief, a heart burst into bloom.

And all who on this darkened earth have loved with all their might...
But forced apart by Sin and Grief, by Pain and Death and Night –
Have they but faith to build a bridge from world to world, be sure
They'll reach their loved ones, and their souls will be at rest once more.

The States

By Solomon W. Golomb

Reconstructed by BG 2019-08-17

There's Maine, Vermont, Rhode Island, Tennessee and Indiana
And Maryland, Missouri, Massachusetts and Montana

There's Pennsylvania, Florida, Ohio, Minnesota
Plus Oregon and Iowa, and North and South Dakota

454 *The Wisdom of Solomon*

New Mexico, New Jersey and New Hampshire and Nebraska
New York, Kentucky, Kansas and Hawaii and Alaska

There's Delaware, Wyoming, Texas and Louisiana
California, Utah, Washington, Wisconsin, Alabama

There's Georgia and Connecticut and also Oklahoma
Virginia, West Virginia, Illinois and Arizona

There's Mississippi, Michigan and Idaho, Nevada
The Carolinas, Arkansas and lastly "Colorada"

These are the fifty states which unto memory I've committed
Though there may soon be others they have yet to be admitted.

Sonnet
By Solomon W. Golomb
Praha. Sept 7, 1961
(handwritten)

Bride of my youth and wife of my desire
O fairest paragon of earthly grace
Divine of countenance and fair of face
I crave communion with thine inward fire.
Love is the food thy spirit doth require
     Most cherished daughter of the human race.
     Yea, but the contact of thy warm embrace
Basaltic rock to music would inspire.
Enthralled, my soul in torment doth endure
Love's separation, and until the hour
Of fond reunion there can be no peace...
Vast eons my addiction would not cure.
Enriched, ennobled by thy gracious power
Devotion unto death shall never cease.

(And, he did remain devoted unto death...)

The Papa Verse 455

The Candidate
(~1976 or 1977)

Now my name is Jerry Bennett and I'm aiming for the Senate —
    I desire to inspire you to hire me this time.
My ambition is explainable: positions are attainable.
    I'd clearly love you dearly if you'll merely hear my rhyme.

I'm for dealing out big pensions and repealing old conventions,
    And for ending needless spending while defending friendly groups.
I'm for Motherhood, the Voters and What's-Good-for-General-Motors;
    Full employment; more enjoyment; and deployment of our troops.

I am weary of the dèbates, and I'm leary of the rebates —
    Just another of Big Brother's ways to smother us from birth.
Yes, I've said it: lower taxes, and a credit card at Saks's,
    Might just worsen every person on this mercenary earth.

Though my rival uses slander, <u>my</u> survival's based on candor —
    I am truthful, I am youthful, and I'm ruthful, I'll admit.
Don't be duped by my "capacity to stoop to base mendacity" —
    That fable I am able just to label counterfeit.

Let me mention I would wish you pay attention to the issue
    That will enter front and center as frequenter on the scene —
Yes, that critical reality, <u>political morality,</u>
    Lies leveled and disheveled and bedeviled and unclean.

For the White House seems to weaken, like a lighthouse with no beacon;
    All those capers in the papers; all those tapers with their bugs!
While our previously president, so devious a resident,
    Stands pardoned in his garden, though he's hardened as a thug.

As a charmer I'm unvarnished, and my armor is untarnished.
    I stand gleaming 'mid the screaming and blaspheming of the horde.
Come November and elections, just remember these directions;
    "Jerry Bennett for the Senate" is the tenet to record.

Note: The political opinions expressed in the foregoing, as well as the syntax and prosody, are solely those of the Candidate, and the Poet disclaims all responsibility therefor.

© 2023 World Scientific Publishing Company
https://doi.org/10.1142/9789811234378_0050

Palindrominologist Gets Him Coming, Going*

By Solomon W. Golomb

I went to interview Professor Otto R. Osseforp in November of 1970, shortly after his appointment to the Emor D. Nilap Chair in Palindrominology at Harvard. The sign on his door said "Professor Osseforp" and the monogram on his tieclasp read "O.R.O.," but he asked me to call him simply "Otto."

Golomb, a professor of applied science at Caltech. wrote this story originally for the Harvard Bulletin.

I knew him to be a great fan of Harvard athletic teams, so I asked him first for his favorite football cheer.

"HARVARD-RAV-RAH!" he replied.

"Yes. 1 should have guessed," said I "And what do you call the annual lvy League track meet?"'

"Yale Relay."

"Of course. And haven't you proposed a new version of the University motto?"

"Satire Veritas." He intoned.

"Profound." I muttered. "By the way, I see the Harvard Corporation has published the annual financial report in Latin again. It runs to some 30 pages. I wonder if you could summarize it for me."

"Satis Revinue Universitas," came the rejoinder.

"Marvelous! And what altout your new novel? Could you tell me the title?"

"Dennis Sinned."

"Intriguing. What is the plot?"

"Dennis and Edna sinned."

"I see. Is there more to it than that?"

"Dennis Krats and Edna Stark sinned."

"Now it all becomes clear," I agreed. "Tell me, with all this concern about the ecology, what kind of car are you driving nowadays?"

"A Toyota."

"Naturally. And how about your colleague, Professor Nustad?"

---

*Los Angeles Times (1923–1995); May 21, 1972; ProQuest.

468   *The Wisdom of Solomon*

"Nustad? A Datsun."

"Yes. And do you drive to the campus via Belmont?"

"Not so. By Boston."

"I understand you are also an expert on foreign affairs. Whom do you see as the next leader to be deposed in Southeast Asia?"

"Lon Nol."

"Incredible! Is that his real name, or did he merely adopt it as his *phnom penh?* But no matter. Do you see your position as support for Administration policy in Vietnam?"

"No, it is opposition."

"What was the turning point of the war?"

"Tet."

"Getting back to Harvard — who is the outgoing president? Did he do a good job? And which way is the University headed?"

"Pusey. Yes. Up," came the terse reply.

"Indeed! I see you are wearing a campaign button recommending your nominee to succeed Mr. Pusey. Could you tell us what it says?"

"For President — Ned Iser, Prof."

Oh yes, the dynamic new assistant professor of Dravidian languages who spent his sabbatical in a Tamil monastery. What is that language he's such an authority on?"

"Malayalam."

"How arcane! Not only is he under 30 and wears beads, but I don't think he's the *kind* of Brahmin that Harvard traditionally selects as its president. But enough of campus politics. I heard you recently spoke at the University of Ethiopia. Don't they refer to the capital city as Addis?"

"Ababa."

"Wasn't there some new product you were urging them to export?"

"Lion oil."

I cleared my throat audibly, and then said, "Tell me the first thing that comes to mind for each of these: Lady Macbeth."

"Dame mad."

"Hitler's autobiography."

"I, Zany Nazi."

"Dostoevsky's mailing address.

"To-Idiot."

"Great. Could you tell me who are some of the other leading experts on palin-drominology, and where are they located?"

Without hesitating, he launched into an alphabetical recitation that began with

"Akaso—Osaka"

and

"Amoroso—Roma"

and seemed to continue interminably. Somewhere in the vicinity of

"Olson—Oslo"

I made my apologies and left the room. I am not entirely sure he noticed.

Puzzles from *LA Times**

'Squarish' Numbers for Arithmagicians
Golomb, Solomon W
Los Angeles Times (1923-1995); May 15, 1989; ProQuest
pg. J3A

ENIGMA

'Squarish' Numbers for Arithmagicians

By SOLOMON W. GOLOMB

We call a whole number n "squarish" if it is the product of two whole numbers, l and w, where l is at least as big as w, but less than twice as big as w. (That is, n is the area of a rectangle whose length is at least equal to its width, but less than twice its width.)

An example of three consecutive squarish numbers is 168=14x12, 169=13x13, 170=17x10. Can you find an example of six consecutive numbers?

Answers appear on Page 5.

Reproduced with permission of the copyright owner. Further reproduction prohibited without permission.

ENIGMA: Three-Way Partition, All the Same in Size and Shape
Golomb, Solomon W
Los Angeles Times (1923-1995); Sep 18, 1989; ProQuest
pg. A3

ENIGMA

Three-Way Partition, All the Same in Size and Shape

By SOLOMON W. GOLOMB

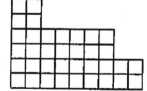

The region shown at the left is to be partitioned into three identical pieces—all the same in both size and shape (but mirror reflection is allowed). It can be done!

Answer appears on Page 6.

---

*This chapter was a collection of several pieces originally published in *Los Angeles Times* during the period from May 1989 to September 1991.

The Wisdom of Solomon

A Chance to Play Detective
Golomb, Solomon W
Los Angeles Times (1923-1995); Nov 4, 1991; ProQuest
pg. OCB9

A Chance to Play Detective

By SOLOMON W. GOLOMB

The police have rounded up six questionable individuals after a murder has been committed, and the Great Detective is brought into the room to confront the suspects. He scans their faces quickly and recognizes none of them. However, he then announces, "I am already sure of one thing. Either there are three of you who were previously mutually acquainted, or there are three of you who were previously mutually unacquainted." As usual, he was right, but how did he know?

Answer appears on B12

Reproduced with permission of the copyright owner. Further reproduction prohibited without permission.

ENIGMA: A Complicated Bar Bet
Golomb, Solomon W
Los Angeles Times (1923-1995); Sep 3, 1990; ProQuest
pg. OCB7

ENIGMA

A Complicated Bar Bet

By SOLOMON W. GOLOMB

There are several patrons watching television in the bar. The waitress demands to see the ID of each patron and notes the age of each. She then calculates the sum of every possible pair of ages, and gets the numbers 50, 53, 56, 58, 59, 62, 64, 65, 67 and 70.

She challenges the bartender to determine the ages of the individual patrons from these numbers. (All the ages are taken to be whole numbers.) How many patrons are there, and what are their ages?

Answer appears on B10

Reproduced with permission of the copyright owner. Further reproduction prohibited without permission.

A Matter of Weight and See
Golomb, Solomon W
Los Angeles Times (1923-1995); Aug 12, 1991; ProQuest
pg. VYB7

A Matter of Weight and See

By SOLOMON W. GOLOMB

Here is a two-part question. The answer will be a surprise to many people.

A. Which is heavier, an ounce of silver or an ounce of lead?

B. Which is heavier, a pound of feathers or a pound of gold?

Answer appears on B10.

Reproduced with permission of the copyright owner. Further reproduction prohibited without permission.

A Question for the Ages
Golomb, Solomon W
Los Angeles Times (1923-1995); Jan 7, 1991; ProQuest
pg. OCB9

A Question for the Ages

By SOLOMON W. GOLOMB

John is one-third as old as his mother was when she was half as old as his father will be 10 years from now, at which time John will be half as old as his mother is now. When John was born, his mother was three-fourths as old as his father was then. How old is John?

Answer appears on B12

Reproduced with permission of the copyright owner. Further reproduction prohibited without permission.

ENIGMA

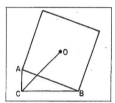

Knowing the Angles Squares With Puzzle

By SOLOMON W. GOLOMB

Inspired by Pythagoras, you have constructed a square on the hypotenuse of the right triangle ABC, as illustrated. Then you drew a straight line OC from the center of the square, O, to the vertex C at the right angle of the triangle.

Now, suppose that the angle ABC is smaller that the angle BAC. Then, is the angle OCB bigger than, equal to, or smaller than the angle OCA?

Answers appear on Page 8.

ENIGMA

Attention Tourists: VAT Refund Is Taxing

By SOLOMON W. GOLOMB

Each country in the European Common Market has a national sales tax, called a "value added tax," or VAT. The tax rate varies considerably from country to country. One of the highest VAT rates is in Denmark, where it is 22%.

An American tourist shopping in Europe can usually arrange to get the VAT refunded on large purchases taken back to the United States.

If an item purchased in Denmark cost X amount (including the VAT), and the VAT is refunded in full, what fraction of X should the tourist expect as a refund?

Answer appears on B10.

ENIGMA

Ask Not for Whom the Ax in Dyslexia Falls

By SOLOMON W. GOLOMB

In the land of Dyslexia, three suspects, whom we will call A, B and C, are arrested and imprisoned in connection with a crime. The legal system in Dyslexia requires that in a case with multiple suspects, the authorities will determine by lottery which one suspect to convict. That suspect will be executed, and the remaining suspects will be freed.

Suspect A has become friendly with the prison warden, and one day he asks if the warden can tell him anything about the progress of the case.

"The outcome has been decided, but the rules forbid me from telling you anything about your own fate. But I can tell you that Suspect B will go free," the Warden replies.

What is the probability that A will be executed?

Answer appears on B5.

ENIGMA

Beware in Forming Square

By SOLOMON W. GOLOMB

How is it possible to relocate only one of the four matchsticks in the configuration shown at right so as to form a perfect square?

(Obviously, there must be a trick.)

Answer appears on B6

ENIGMA

Birthdays Bunched by Happy Coincidence

By SOLOMON W. GOLOMB

Eight random, unrelated patients are in the dentist's waiting room. What is the probability that all eight of them have their birthdays within some consecutive half-year interval? (The half-year interval is allowed to start on any calendar date, and we assume that all dates are equally likely as birthdays.)

Answer appears on B14.

Can You Cut a Square Deal Out of This?

By SOLOMON W. GOLOMB

How is it possible to cut a 13 x 13 square into only four pieces that can be reassembled to form a 5 x 5 square and a 12 x 12 square? One of the new squares is allowed to be a single piece of the original square, in which case the other new square is to be formed from the remaining three pieces of the original square.

Answer appears on B6.

ENIGMA

A Common Denominator

By SOLOMON W. GOLOMB

Mr. Webster has been teaching his arithmetic class to reduce fractions to lowest terms.
This is fairly easy when it involves reducing 12/21 to 4/7, but one of the homework problems is to reduce 667/2047 to lowest terms.
How would you answer this one?

Answer appears on B6.

Diminishing Returns for the Halfback From Hades

By SOLOMON W. GOLOMB

A football player in Hades has been assigned the following torment: Starting at his own goal line, he runs 100 yards forward, but then he must reverse direction and run 50 yards back. He then runs forward 25 yards and then 12½ yards. He is condemned to reversing direction again and again, each time running only half as far as the previous run. Where will he ultimately end up?

Answer appears on B10.

ENIGMA

Dividing Lines for Piece Makers

By SOLOMON W. GOLOMB

Can you divide this figure into four identical pieces? Into three identical pieces? (Identical pieces must be the same in area and shape, but turning over is permitted.)

Answer appears on Page 10.

ENIGMA

Drawing Lines to Exceed Perimeters

By SOLOMON W. GOLOMB

A quadrilateral is drawn inside a triangle. How is it possible for the perimeter of the quadrilateral to exceed the perimeter of the triangle?

Answer appears on Page 14.

ENIGMA

Drawing to a Pointed Conclusion

By SOLOMON W. GOLOMB

A classic problem asks the solver to draw a continuous path, consisting of exactly four straight-line segments, that goes through all nine points in the diagram below, left. The solution to this one, which you have probably seen before is below, right.

Your task today is to draw a continuous path, consisting of exactly eight straight-line segments, that goes through all 25 points in a 5-by-5 array of dots, and returns to its starting point. The dots are to be regarded as ideal mathematical points, with no thickness.

Answer appears on Page B12

ENIGMA

Ask Not for Whom the Ax in Dyslexia Falls

By SOLOMON W. GOLOMB

In the land of Dyslexia, three suspects, whom we will call A, B and C, are arrested and imprisoned in connection with a crime. The legal system in Dyslexia requires that in a case with multiple suspects, the authorities will determine by lottery which one suspect to convict. That suspect will be executed, and the remaining suspects will be freed.

Suspect A has become friendly with the prison warden, and one day he asks if the warden can tell him anything about the progress of the case.

"The outcome has been decided, but the rules forbid me from telling you anything about your own fate. But I can tell you that Suspect B will go free," the Warden replies.

What is the probability that A will be executed?

Answer appears on B6.

Figure All the Angles to Come Up With a Square

By SOLOMON W. GOLOMB

How can you cut the triangular figure shown at the right into only three pieces that can then be reassembled to form a 6 x 6 square?

Answer appears on B12

ENIGMA

Heads or Tails: Computing the Odds

By SOLOMON W. GOLOMB

When a coin is tossed repeatedly, consecutive occurrences of heads form a run of heads, and consecutive occurrences of tails form a run of tails. For example, the sequence HTTHHHTHTT consists of six runs: H-TT-HHH-T-H-TT, three each of heads and tails.

A "perfect coin" is to be tossed 10 times in a row, and you have a chance to bet, at even odds, on any one of the following three situations. Which one gives you the best chance of winning?

A. The sequence will have at most five runs altogether.

B. The sequence will have at least three runs of heads.

C. The total number of runs will be either five or six.

Answer appears on Page 10.

ENIGMA

Here's Odds on Cashing In on California 6/49 Lottery

By SOLOMON W. GOLOMB

In the California Lotto 6/49, you have picked six of the numbers from 1 to 49, and then the Lottery selects six numbers from 1 to 49 as the "winning numbers" plus a seventh "bonus number."

How much more likely is it to have five of the winning numbers plus the bonus number than to have all six winning numbers?

How much more likely still is it to have exactly five of the winning numbers *without* the bonus number?

Answers appear on Page 8.

Inch by Inch: Try Designing a Ruler So It All Adds Up

By SOLOMON W. GOLOMB

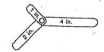

Our nearsighted carpenter, Mr. Wood, has a jointed ruler consisting of three segments, joined together as shown. The lengths of the segments are one inch, two inches and four inches, and he can fold it to measure any whole number of inches, from one to six inches, as the total length of one or more consecutive segments. (There are no inch marks on this ruler.) Can you show how to design a jointed ruler with five segments which can be used in this way to measure every whole number of inches from one inch to 15 inches? More than one joint is allowed.

Answer appears on B12.

ENIGMA

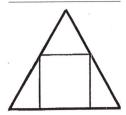

Ins and Outs of a Square in a Triangle

By SOLOMON W. GOLOMB

A square is inscribed in an equilateral triangle, as shown. Is more of the area of the triangle inside the square or outside the square, or is the area equal inside and out?

Answer appears on Page 12.

ENIGMA

It's All in the Bag

By SOLOMON W. GOLOMB

Three opaque bags contain two coins each. It is known that one bag contains two nickels, one bag contains two dimes, and the remaining bag contains a nickel and a dime. The bags are labeled "10 cents," "15 cents" and "20 cents," but it is known that no bag contains the amount listed on the label.

Without looking inside the bags, you are allowed to draw out one coin at a time. What is the smallest number of coins you need to extract to be able to determine the contents of each of the three bags?

Answer on B14

Puzzles from LA Times 465

ENIGMA

Knowing Prime Numbers Forward and Backward

By SOLOMON W. GOLOMB

A palindrome is a word that reads the same forward and backward, such as *noon* and *level*. A palindromic number, such as 88 or 737 or 14,641, is a number that reads the same in both directions. A prime number is a number (such as 2, 3, 5, 7, 11 and 13) that is divisible only by itself and 1. There are many "palindromic primes," such as 11, 101, 373 and 10,301. Are there any four-digit palindromic primes?

Answer appears on B12.

ENIGMA

Lotto—6 of 1, Half Dozen of Another?

By SOLOMON W. GOLOMB

The California Lotto game previously involved selecting six of 49 numbers correctly to win the top prize. Then the rules were changed to require picking six of 53 numbers correctly. What effect does the change have on the probability of winning (or at least sharing in) the top prize?

Answer appears on B10

Making Hexagon Sums Fall Into Line

By SOLOMON W. GOLOMB

Is it possible to assign the numbers 1 to 19 in circles forming a hexagon pattern in such a way that the 15 "line sums" (the sum of the numbers in each of the five rows, the sum of the numbers in each of the five left-sloping diagonals and the sum of the numbers in the five right-sloping diagonals) all come out the same? Can you find such a numbering?

Answer appears on B12.

Measured in Inches, Size of Square Meter Is a Gray Area

By SOLOMON W. GOLOMB

A rectangular sheet of cardboard measures exactly 31 inches by 50 inches. Is it larger than, equal to, or smaller than one square meter in area? Does the answer depend on whether the measurement was made in the United States or England or on whether it was made before or after 1959?

Answer appears on B8

466   *The Wisdom of Solomon*

squaring off_golomb

ENIGMA: Nickel-and-Dime Puzzle Is in the Bag
Golomb, Solomon W
Los Angeles Times (1923-1995); Sep 17, 1990; ProQuest
pg. B3

ENIGMA

Nickel-and-Dime Puzzle Is in the Bag

By SOLOMON W. GOLOMB

Three opaque bags contain two coins each. It is known that one bag contains two nickels, one bag contains two dimes and the remaining bag contains a nickel and a dime. The bags are labeled "10 cents," "15 cents" and "20 cents," but it is known that no bag contains the amount listed on its label. Without looking inside the bags, you are allowed to draw out one coin at a time. What is the smallest number of coins you need to extract to be able to determine the contents of each of the three bags?

Answer appears on B6.

Reproduced with permission of the copyright owner. Further reproduction prohibited without permission.

Numbers Squared, Cubed Show Power of Equations
Golomb, Solomon W
Los Angeles Times (1923-1995); Mar 4, 1991; ProQuest
pg. OCB7

Numbers Squared, Cubed Show Power of Equations

By SOLOMON W. GOLOMB

The numbers 25 and 27 furnish the example of a perfect square and perfect cube which differ by two. Can you find an example of a perfect cube and a perfect fourth power which differ (in either direction) by 28? No negative numbers, please.

Answer appears on B10.

Reproduced with permission of the copyright owner. Further reproduction prohibited without permission.

Playing All Triangle Angles Helps Find Rectangle Limits
Golomb, Solomon W
Los Angeles Times (1923-1995); Oct 2, 1989; ProQuest
pg. VY_A3

ENIGMA

Playing All Triangle Angles Helps Find Rectangle Limits

By SOLOMON W. GOLOMB

A rectangle sits inside a triangle, as illustrated. Show that the area of the rectangle cannot exceed half the area of the triangle.

Answer appears on Page 12.

Reproduced with permission of the copyright owner. Further reproduction prohibited without permission.

Scoring Decimal Points
Golomb, Solomon W
Los Angeles Times (1923-1995); Dec 2, 1991; ProQuest
pg. VYB7

Scoring Decimal Points

By SOLOMON W. GOLOMB

The decimal expansion of 1/7 is .142857142857 . . . . repeating with a "period" of six digits. Can you find a prime number p where the decimal expansion of $1/p$ has a period of *three* digits? How about a period of *five* digits?

Answer appears on B10.

Reproduced with permission of the copyright owner. Further reproduction prohibited without permission.

ENIGMA: The Barmaid's Challenge to Arrive at the Age of Reason
Golomb, Solomon W
Los Angeles Times (1923-1995); Dec 10, 1990; ProQuest
pg. OCB7

ENIGMA

The Barmaid's Challenge to Arrive at the Age of Reason

By SOLOMON W. GOLOMB

The barmaid has decided to check the identifications of the patrons at the bar, and she decides to take the sums of their ages two at a time in all possible ways. The sums, arranged in increasing order, are 44, 48, 50, 58, 60, 64, 66, 68, 72 and 82, all even numbers. She challenges the bartender to determine the individual ages from these two-at-a-time sums. Can you solve the puzzle?

Answer appears on B12.

Reproduced with permission of the copyright owner. Further reproduction prohibited without permission.

ENIGMA: Squaring Off
Golomb, Solomon W
Los Angeles Times (1923-1995); Jun 25, 1990; ProQuest
pg. VYB8

ENIGMA

Squaring Off

By SOLOMON W. GOLOMB

Suppose X and Y are two of the 10 digit symbols from 0 to 9. Is it possible for both XYY and XYYY to be perfect squares? (For example, if X = 1 and Y = 0), then XYY = 100 = 10^2 is a perfect square, but XYYY = 1000 is not.)

Answer appears on B12.

Reproduced with permission of the copyright owner. Further reproduction prohibited without permission.

Puzzles from LA Times

Thinking Person's Tile Job
Golomb, Solomon W
Los Angeles Times (1923-1995); Sep 23, 1991; ProQuest
pg. OCB9

Thinking Person's Tile Job

By SOLOMON W. GOLOMB

How can a rectangle be formed using an odd number of identical copies of the tile at left? Copies of the figure may be rotated and/or turned over as you wish in forming the rectangle. Maximum credit will be given for using the *smallest* odd number of copies in forming the rectangle. (There are several solutions with fewer than 20 copies of the figure.)

Answer appears on B12

Reproduced with permission of the copyright owner. Further reproduction prohibited without permission.

ENIGMA: Winning Money at Decco: Is Success in the Cards
Golomb, Solomon W
Los Angeles Times (1923-1995); Nov 12, 1990; ProQuest
pg. VYB7

ENIGMA

Winning Money at Decco: Is Success in the Cards?

By SOLOMON W. GOLOMB

The California Lottery's Decco game requires the player to select four cards, one from each suit of a bridge deck. For a $1 bet, the player receives $5,000 if all four of his selections match those chosen by the lottery. If three of the four match, he wins $50; if two of the four match, he wins $5. Finally, if there is one match, he gets a coupon allowing him to play a subsequent game of Decco for free. What is the expected return to the player for a $1 bet?

Answer appears on B10.

Reproduced with permission of the copyright owner. Further reproduction prohibited without permission.

ENIGMA: Working With the Odds
Golomb, Solomon W
Los Angeles Times (1923-1995); Jun 12, 1989; ProQuest
pg. OC_A9

ENIGMA

Working With the Odds

By SOLOMON W. GOLOMB

A lottery has been designed with n balls in a jar labeled from 1 to n, of which four will be selected at random as the winning numbers. The odds *against* your selection of four numbers turning out to be the winning ones are found to be 1,000 to 1. How many balls are in the jar?

Answer appears on Page 12.

Reproduced with permission of the copyright owner. Further reproduction prohibited without permission.

Zigzagging Toward a Square Deal
Golomb, Solomon W
Los Angeles Times (1923-1995); Jun 17, 1991; ProQuest
pg. SDB5

Zigzagging Toward a Square Deal

By SOLOMON W. GOLOMB

How can you cut the Z-shaped figure at the right into three pieces that can be reassembled to form a square?

Answer appears on B8

Reproduced with permission of the copyright owner. Further reproduction prohibited without permission.

Several Pieces from Golomb's Gambits*

---

*This chapter comprises five "Golomb's Gambits" columns, originally published in *Johns Hopkins Magazine* "Spelling the States" (Fall 2011), "More Megameanings" (Feb 1984), "Tiling Rectangles" (Feb 1986), "Word Games" (Dec 1988), and "*Probabilities and Executions*" (Oct. 1989).

Golomb's Answers
"Spelling the States" Solutions

Puzzle on page 17.

1. Letters adjoined: *b, t, o.*
State added: Ohio.

2. Letters adjoined: *a, w.*
States added: Iowa, Hawaii.

3. Letters adjoined: *d, n.*
States added: Idaho, Indiana.

4. Letters adjoined: *k, l, s.*
States added: Alaska, Illinois, Kansas.

5. Letters adjoined: *b, e, r.*
States added: Arkansas, Delaware, Nebraska, Rhode Island.

6. Letter adjoined: *m.*
States added: Alabama, Maine, Oklahoma.

7. Letter adjoined: *t.*
States added: Minnesota, Montana, North Dakota, Tennessee.

8. Letters adjoined: *c, g.*
States added: Colorado, Georgia, Michigan, North Carolina, Oregon, Washington, Wisconsin.

9. Letter adjoined: *u.*
States added: Connecticut, Louisiana, Massachusetts, Missouri, South Carolina, South Dakota, Utah.

10. Letter adjoined: *y.*
States added: Kentucky, Maryland, New York, Wyoming.

11. Letter adjoined: *v.*
States added: Nevada, Vermont, Virginia, West Virginia.

12. Unused letters: *f, j, p, x, z.*
States not yet spelled: Arizona, California, Florida, Mississippi, New Hampshire, New Jersey, New Mexico, Pennsylvania, Texas.

Extra credit:
The letter *j* appears only in New Jersey, and *z* only appears in Arizona. *F* can be found in California and Florida, and *x* in New Mexico and Texas. The letter *p* appears in Mississippi, New Hampshire, and Pennsylvania. Because the letter *b* occurs in only two states (Alabama and Nebraska), if we had omitted *b* from our list of 20 letters and adjoined *p* instead, we would have a total of 42 states using only 20 different letters.

Congratulations

To the recipients of the 2011 Johns Hopkins University Alumni Association Awards

Read more about these 41 notable alumni and faculty or nominate someone for next year's award at
alumni.jhu.edu/awards.

MORE MEGAMEANINGS!

Purportedly efforts toward a more complete conversion table for scientific units, the original Megameanings were devised by Solomon W. Golomb '51 and appeared in the August issue. Our thanks to the many readers who sent more, and to illustrator Robert Soulé for the drawings.

We open with Leo Meire's introduction of a term previously unknown to science:

1 Golomb = the amount of nonsense developed by a Solo Mon

Continued on next page

A table of the standardized prefixes corresponding to scale factor changes by powers of ten:

$10 = 10^1 =$ deca-
$100 = 10^2 =$ hecta-
$1{,}000 = 10^3 =$ kilo-
$1{,}000{,}000 = 10^6 =$ mega-
$1{,}000{,}000{,}000 = 10^9 =$ giga-
$1{,}000{,}000{,}000{,}000 = 10^{12} =$ tera-
$\frac{1}{10} = 10^{-1} =$ deci-
$\frac{1}{100} = 10^{-2} =$ centi-
$\frac{1}{1{,}000} = 10^{-3} =$ milli-
$\frac{1}{1{,}000{,}000} = 10^{-6} =$ micro-
$\frac{1}{1{,}000{,}000{,}000} = 10^{-9} =$ nano-
$\frac{1}{1{,}000{,}000{,}000{,}000} = 10^{-12} =$ pico-

In addition, there are well-established prefixes for the factors from 1 to 9, based on both Latin and Greek roots:

1 =	uni- OR mono-	OR holo-
2 =	bi- OR di-	OR diplo-
3 =	ter- OR tri-	
4 =	quadri- OR tetra-	
5 =	quinto- OR penta-	
6 =	sexa- OR hexa-	
7 =	septa- OR hepta-	
8 =	octa- OR okta-	
9 =	nona- OR ennea-	

Other factors for which there are agreed prefixes include:

$\frac{1}{2}$ = semi- OR hemi- OR demi-
$\frac{3}{2}$ = sesqui-
11 = undeca- OR hendeca-
12 = dodeca-
13 = triskaideca-
20 = icosa-

Femto- for 10^{-15} and *atto-* for 10^{-18} have been adopted by the scientific community (from Swedish *femton* = 15, *atton* = 18).

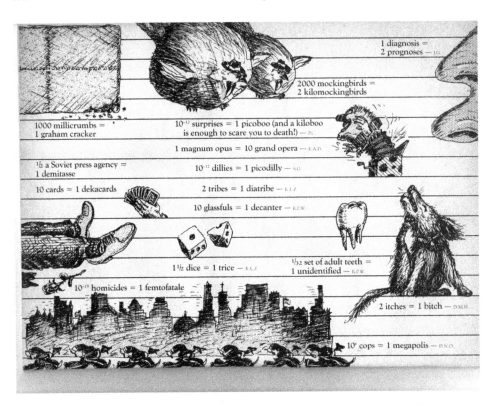

Several Selections from Golomb's Gambits

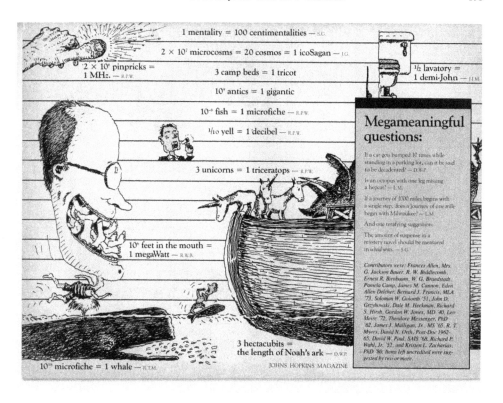

GOLOMB'S GAMBITS

© 1986 Solomon Golomb

Tiling Rectangles

There are many different shapes which can be used to tile a floor. The simplest tiles, of course, are squares or rectangles. Regular hexagons are popular as tiles, but if the floor to be covered is rectangular, the hexagons along the edges must be cut to fit:

More imaginative tilings are possible with fancier shapes. Here are some examples of shapes which can be used to tile squares or rectangles:

In each case, identical copies of the same shape are used in the tiling. For present purposes, we also allow tiles to be flipped over, so that

are considered to be the same tiling shape.

Each of the following shapes can be used to tile a rectangle. Your task is to find the *smallest* rectangle which can be formed from copies of the given tile shape. (It will be the smallest in area, and also will use the fewest copies of the basic tile.)

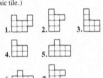

The shapes that *look* the most complicated are not necessarily the ones that require the most tiles to form a rectangle! In 1953, at a lecture to the Harvard Math Club, I coined the term "polyomino" (poly-OM-in-o) for any connected shape made up of unit squares. Two squares form the familiar *domino*: ⬚⬚ Three squares form a *tromino*, either the "straight tromino" ⬚⬚⬚ or the "right tromino"

There are five *tetromino* shapes:

and so on. (I readily admit that the "domino" was never really *two* of anything. Linguistically, "polyomino" bears the same relationship to "domino" that "cheeseburger" bears to "hamburger.")

It is fairly easy to show that certain polyominoes, such as ⬚⬚ and

⬚⬚ and ⬚⬚ , will not tile

any rectangle. (Again, we require the tiling to be accomplished using copies of a single shape.) For all other polyomino shapes listed earlier, tilings of rectangles *do* exist. It is reasonable to ask: Is there some systematic *test* that can be applied to any given polyomino which will tell us whether or not it tiles some (unspecified) rectangle? The surprising answer is "no." Some fifteen years ago I published a proof that the *general* problem of testing polyominoes to determine whether they do or don't tile any rectangle is "computationally undecidable." This can be decided for many individual polyominoes, but there is no systematic procedure which will give an answer in all cases!

This is not just a theoretical result. There are two simple examples for which we honestly don't know the answer.

These are the hexomino

and the heptomino.

For each of these shapes, no one has found a way to tile any rectangle, but neither has anyone discovered a mathematical proof that such a tiling is impossible. It is crucial that there is no limit, *a priori*, to how big the smallest rectangle might be. This defeats the strategy, even in principle, of testing all possible tilings of all possible rectangles up to the limiting size. I offer a prize of $500.00 to the first person (as determined by postmark) who can 1) tile a rectangle using only identical copies of   , 2)

tile a rectangle using only identical copies of

or 3) prove to my satisfaction that at least one of these two tilings is impossible. (Turning figures over is permitted in the tilings.)

(Since I routinely receive fallacious proofs of virtually all the famous unsolved problems in mathematics, and am often unable to convince their otherwise rational authors of the invalidity of their arguments even when I pinpoint the fallacy, the proof of impossibility must be "to my satisfaction" whether or not its author accepts the verdict.)

For those readers willing to tackle more tractable problems with no offer of financial reward, I propose the following:

8. Use copies of   to tile a rectangle except for a 2x2 hole somewhere inside it.
9. Use copies of   to tile a square except for a 1x1 hole in the center.
10. Use copies of   to tile a square except for a square hole in its center.

Solutions to "Tiling Rectangles"

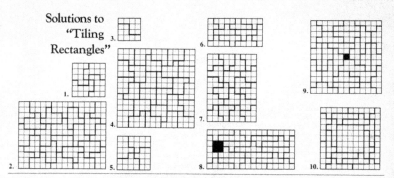

Long Answers to "Efficient Rulers"

Problems from the October issue

The two tables which follow, giving the best known candidates for "spanning rulers" and "covering rulers" when the number of marks is between 2 and 20 (inclusive), were compiled by Dr. Herbert Taylor of the University of Southern California from a number of published and unpublished sources, including his own research efforts.

Major references include:

1. John Leech, "On the Representation of 1,2, . . . ,n by Differences," *London Math. Soc. Journal*, vol. 31, April, 1956, pp. 160-169.
2. G.S. Bloom and S. Golomb, "Applications of Numbered Undirected Graphs," *Proceedings of the IEEE*, vol. 65, no. 4, April, 1977, pp. 562-570.
3. Martin Gardner, *Wheels, Life, and Other Mathematical Amusements*, W.H. Freeman and Co., 1983. Chapter 15, "Golomb's Graceful Graphs," pp. 152-165.
4. J.P. Robinson and A.J. Bernstein, "A Class of Binary Recurrent Codes with Limited Error Propagation," *IEEE Transactions on Information Theory*, vol. IT-13, no. 1, January, 1967.
5. M.B. Sverdlik and A.N. Meleshkevich, "Synthesis of Optimum Pulsed Sequences Having the Property of 'No More Than One Coincidence,' " *Radiotech. i Elektron.*, vol. 19, 1974, pp. 721-729.
6. J.P. Robinson, "Optimum Golomb Rulers," *IEEE Trans. Comput.*, vol. C-28, no. 12, December, 1979.
7. J.C.P. Miller, "Difference Bases, Three Problems in Additive Number Theory," in *Computers in Number Theory*, A.D.L. Atkin and B.J. Birch, Eds. Academic Press, 1971, pp. 299-322.

Best examples of spanning rulers come principally from References 3,4,5, and 6. The example for n=13 was recently discovered by Douglas S. Robertson (private communication) who conducted a month-long exhaustive search by computer. Robertson has verified that the examples of spanning rulers given here are best possible for all n up to 13. (This had been previously verified through n=12 by John P. Robinson.) The examples of covering rulers come primarily from References 1 and 7.

Number of Marks on the Ruler	Positions of the Marks on the Shortest Known Candidate(s) for "Spanning Ruler"
2	0,1
3	0,1,3
4	0,1,4,6
5	0,1,4,9,11
	0,2,7,8,11
6	0,1,4,10,12,17
	0,1,4,10,15,17
	0,1,8,11,13,17
	0,1,8,12,14,17
7	0,1,4,10,18,23,25
	0,1,7,11,20,23,25
	0,1,11,16,19,23,25
	0,2,3,10,16,21,25
	0,2,7,13,21,22,25
8	0,1,4,9,15,22,32,34
9	0,1,5,12,25,27,35,41,44
10	0,1,6,10,23,26,34,41,53,55
11	0,1,4,13,28,33,47,54,64,70,72
	0,1,9,19,24,31,52,56,58,69,72
12	0,2,6,24,29,40,43,55,68,75,76,85
13	0,2,5,25,37,43,59,70,85,89,98,99,106
14	0,5,28,38,41,49,50,68,75,92,107,121,123,127
15	0,4,5,15,33,57,59,78,105,117,125,139,142,148,155
16	0,6,19,40,58,67,78,83,109,132,133,162,165,169,177,179
17	0,8,31,34,40,61,77,99,118,119,132,143,147,182,192,194,199
18	0,2,10,22,53,56,82,83,89,98,130,148,153,167,188,192,205,216
19	0,1,6,25,32,72,100,108,120,130,153,169,187,190,204,231,233,242,246
20	0,24,30,43,55,71,75,89,104,125,127,162,167,189,206,215,272,275,282,283

Number of Marks on the Ruler	Number of Solutions	Positions of the Marks on the Longest Known Candidate for "Covering Ruler." (Only one listed in case of ties.)
2	1	0,1
3	1	0,1,3
4	1	0,1,4,6
5	2	0,1,2,6,9
6	3	0,1,2,6,10,13
7	6	0,1,2,6,10,14,17
8	2	0,1,2,11,15,18,21,23
9	3	0,1,3,6,13,20,24,28,29
10	1	0,1,3,6,13,20,27,31,35,36
11	1	0,1,3,6,13,20,27,34,38,42,43
12	2	0,1,3,6,13,20,27,34,41,45,49,50
13	6	0,1,3,6,17,24,27,38,45,49,53,57,58
14	2	0,1,2,5,10,15,26,37,48,54,60,66,67,68
15	1	0,1,2,5,10,15,26,37,48,59,65,71,77,78,79
16	1	0,1,2,5,10,15,26,37,48,59,65,71,77,78,79,82
17	1	0,1,2,4,11,18,30,42,54,66,71,76,81,86,87,89,91
18	1	0,1,2,3,4,5,6,40,49,57,64,71,78,84,90,96,101,106
19	1	0,1,2,3,7,14,21,28,43,58,73,88,96,104,112,120,121,122,123
20	1	0,1,2,6,12,18,24,37,50,63,76,89,102,109,116,123,130,131,132,133

GOLOMB'S GAMBITS

Word Games

A. In recent years, the suffix "-gate" has been detached from "Watergate" and made into a generic suffix for political scandals. This suggests several questions.

1. Since Watergate, in the early 1970s, what new **-gate** words have been coined in the media to refer to scandals, real or imagined, major or minor? *(4 examples?)*

2. List as many English words of at least two syllables, ending in **-gate**, as you can think of, *excluding* those in which -gate has the meaning of an entrance or doorway. (25 examples or more? More than 35?)

3. What other examples can you give of word endings that have been detached from their original settings and turned into generic suffixes to indicate some property, category, or characteristic?. For each example, give at least two instances of the use of the suffix. *(3 examples?)*

B. Written vertically in capital letters, some words will look the same in a mirror:

	M		H
e.g.	A	or	O
	T		T

1. Which letters of the alphabet have this mirror-symmetry?

2. List all words (including proper names) of more than four letters that you can think of that have this symmetry. (Give yourself one point for each letter in excess of 4 for each word on your list. *A score of 100 is fair, 140 is good, 180 is excellent.)*

3. How many two-word expressions of more than seven letters can you find with this symmetry?. (Examples of two-word expression—*without* the symmetry—include "Brussel sprouts," "handy-dandy," Dallas Texas," and "Isaac Newton.") *(Try for at least 6 examples.)*

C. Double letters occur often in English words, and most letters of the alphabet are frequently found doubled in common words. A few letters, however, are rarely doubled.

1. List all English words you can think of containing -hh-.

2. List all English words you can think of containing -uu-.

3. Those containing -vv-.

4. Give at least one example each of -jj-, -kk-, -ww-, and -yy-.

5. What common English word contains three double letters in a row?

6. Are there any English words in which a *triple* letter occurs?

Answers on page 52.

©Copyright 1988 Solomon Golomb

Several Selections from Golomb's Gambits   477

GOLOMB'S GAMBITS

Continued from page 4.

Answers to Word Games

A.1. a. Debategate
b. Koreagate
c. Irangate
d. Contragate

2. abjugate, abnegate, abrogate, aggregate, allegate, arrogate, castigate, centrifugate, congregate, conjugate, corrogate, corrugate, delegate, derogate, elongate, expurgate, frigate, fumigate, fustigate, instigate, interrogate, investigate, irrigate, legate, litigate, margate, mitigate, navigate, negate, obligate, profligate, prolongate, propagate, relegate, segregate, subirrigate, subjugate, subrogate, suffumigate, supererogate, surrogate, variegate, vulgate.

3. a. **-holic** (from *alcoholic*, to indicate a compulsion) as in *workaholic* and *chocoholic*.

b. **-burger** (from *Hamburger*, originally indicating "from the city of Hamburg," as also with Frankfurter—from Frankfurt, and Wiener—from Vienna; but now indicating food served between the halves of a sliced roll) as in *cheeseburger, beefburger, veggieburger,* etc.

c. **-omino** (from *domino*, as if the initial *d* indicated "two" and the suffix "omino" referred to squares) as in *tromino, tetromino,* and in general *polyomino,* a figure consisting of a number of connected squares of the same size.

d. **-ling,** a suffix indicating small size (as in *duckling, sapling, foundling, stripling, fingerling,* etc.) or pettiness in status (as in *princeling, lordling, hireling,* etc.), said by some etymologists to have resulted when the Saxon word *etheling* (from *ethel* = noble, + *ing,* meaning minor nobleman) was construed as if the ending was *-ling* indicating "minor."

B.1. A,H,I,M,O,T,U,V,W,X,Y, and perhaps Q if written "Q."

2. a. *(5-letter words)*: AMATI, AMATO, AMITY, ATAMI, AWAIT, AXIOM, HAITI, HAMMY, HYATT, MAMMA, MAMMY, MAXIM, MIAMI, MITTY, MOMMY, MOTHY, MOTTO, MOUTH, MUMMY, OMAHA, QUITO, QUOIT, QUOTA, QUOTH, TAMMY, THOTH, THYMY, TIMMY, TOMMY, TOOTH, TUMMY, TWIXT, VOMIT, WHITY, WITTY, WYATT, YAHOO, YOUTH, YUMMY.

b. *(6-letter words)*: AXIOMA, HAWAII, HOWATT, MAHOUT, MAXIMA,

MAUMAU, MOUTHY, MUUMUU, OTTAWA, OUTHIT, OUTWIT, QUOTHA, TAHITI, TATAMI, TATTOO, TOMATO, TOMIUM, TOMTOM, TOOTHY, TOYAMA, TOYOTA, TWO-WAY, VOMITO, VOMITY, WHAMMY, YAMAHA, YAMATO, YOUTHY.

c. *(7-letter words)*: AQUAVIT, ATHYMIA, AUTOMAT, HIIUMAA, IMAMOTO, MAHATMA, MAMMOTH, MAXIMUM, OTTAVIA, OTTAVIO, TAXIWAY, THUMMIM, TIMOTHY, TOWAWAY, WITHOUT.

d. *(8-letter words)*: AUTOMATA, AUTOMATH, AXIOMATA, HATHAWAY, HATOYAMA, HIAWATHA, MOTIVITY, THATAWAY, WAYMOUTH, YAMAMOTO.

e. Longer words arise if we assume that any adjective ending in *-tive* can be transformed into an abstract noun ending in *-tivity,* as: IMITATIVITY, AUTOMOTIVITY, MOTIVATIVITY, and perhaps MUTATIVITY and IMMUTATIVITY.

3. Some examples are HAMA-HAMA, HOITY-TOITY, MAHI-MAHI, MAMMA MIA, MAUI HAWAII, MIAMI OHIO, OAHU HAWAII, and TIMOTHY HAY. Surely there are other good ones. I have found the name "TIMOTHY HATHAWAY" in telephone books in several U.S. cities, but I doubt that any have the middle name of "HIAWATHA."

C.1. **-hh-** occurs in *bathhouse, beachhead, bushhammer, fishhook, highhanded, roughhouse, touchhole,* and *withhold (withheld, withholding)* etc., but **not** in *threshold.*

2. **-uu-** occurs in *continuum, duumvir* (and *duumvirate*). *Equus* (the horse genus), *Equuleus* (a constellation), *menstruum, residuum, vacuum,* etc.

3 **-vv-** occurs in *civvies, divvy, flivver, navvy, revved* (or *revving*), *savvy,* etc.

4. **-jj-** is found in *hajj* and *hajji;*
-kk- is found in *chukker* (a division of a polo match, also spelled *chukkar*), in *pukka,* in *Sikkim,* and in numerous other words of foreign (especially Indian) origin;
-ww- is found in *bowwow, glowworm, powwow,* etc.;
-yy- is found in *Iyyar* (a Hebrew month), and many obsolete words and spellings (e.g. *wyyfe*) listed in the Oxford English Dictionary.
Words with **-aa-** and **-ii-** are too numerous to list, but **-xx-** seems limited to proper nouns such as *Exxon, Fixx,* and *Foxx.* Only **-qq-** seems totally unrepresented, though I

suspect there may be place names in the Middle East which have **-qq-** in one of their variant spellings.

5. bookkeeper

6. I have not found any in a dictionary, but words ending in **-less, -like,** and **-ling,** preceded by **ll,** seem the most likely candidates, e.g. a *shelless* crab, *falllike* weather, and surely a baby troll is a *trollling.* Other plausible attempts include *stifffaced* (or *stifffisted*), *Hawaiiite,* and *blisssome.* Please write in if you find any word with a triple letter (no hyphens, please) actually sanctioned by an English dictionary.

Reach the
Educated Elite

Advertise in

Johns Hopkins
MAGAZINE

212 Whitehead Hall
Johns Hopkins University
Baltimore, MD 21218
301-338-7645

403(b)
Colleague

INVESTOR SERVICES

● Monthly Investment Newsletter for Johns Hopkins University 403(b) participants.

● Buy and Sell recommendations.

● Risk Management Computer Model.

● Model Portfolio and Performance Ratings.

● 24 hr. Telephone Hotline.

● Designed for active and retired faculty and staff.

● Developed by a Johns Hopkins staff Ph.D.

For information and free Investment White Paper write or call: The 403(b) Colleague, Box 5735, Baltimore, MD 21208. ☐ (301) 484-9420

GOLOMB'S GAMBITS

Probabilities and Executions
What are your chances?

©1989 Solomon Golomb™

1. Three identical opaque envelopes are known to contain paper money, as follows: one contains two $5 bills, one contains two $10 bills, and one contains a $5 bill and a $10 bill. You select one of the three envelopes at random, and extract a bill from it at random, which you then observe. It is a $10 bill. What is the probability that the remaining bill in that envelope is also a $10 bill?

2. Joe Sharpie is a master of the shell game. He has placed your valuable coin under one of three identical cups, and proceeded to scramble them so thoroughly that although he still knows which cup hides the coin, you haven't a clue. He explains the rules: You will point to a cup, which you allege hides your coin. He will then remove one of the other two cups, which he guarantees will not be the one hiding the coin. You may now either stick with your original choice, or switch to the other remaining cup. What gives you the best chance of getting your coin back?
a) Stick with your original designation.
b) Switch.
c) It makes no difference.

3. In the land of Dyslexia, three suspects, whom we will call A, B, and C, have been charged with a serious crime and imprisoned. In such a situation, with multiple suspects, the legal system of Dyslexia requires that one of the three will be chosen by lot and executed while the other two will be set free.

One day Suspect A, who has become friendly with the warden, asks if there have been any developments in the case. "Yes," the warden replies, "the decision has been made, but I am forbidden to tell you anything relating to your own fate. I can tell you, however, that Suspect B will be set free."

What is the probability that A will be executed?

4. At the Petersburg Casino, you are offered the opportunity to play the following

game, for which you must pay an entry fee of $100.

A perfect coin is tossed repeatedly until the first occurrence of tails. If tails comes up on the first toss, you receive $1. If there is one head before the first occurrence of tails, you receive $2. If there are two heads before the first occurrence of tails, you receive $4; and so on. That is, each additional *head* before the first *tail* doubles your payoff.

Is your expected payoff from this game bigger, smaller, or equal to the $100 entry fee?

5. Mr. Gottbucks, an eccentric but generous billionaire, has decided to give you some money. To make it interesting, he selects an amount of money at random, and writes it on a card. On a second card, he writes down twice this amount, changing the last digit (the "cents" digit) at random, so that it is not necessarily an even digit.

The two cards are placed face down on a table, and you are allowed to select one and look at it. The rules say that you can claim the amount on the card you selected, or ask for the amount written on the second, unseen card.

Suppose the amount on the card you selected is X. Then the amount on the other card is equally likely to be $1/2 X$ or $2X$, and the average of these is $1 1/4 X$. So your expected reward is 25% greater if you switch to the unseen card. Or is it?

6. A Latin American general, having staged an unsuccessful coup, has been arrested and imprisoned. The Presidente, who has survived the attempted putsch, comes to visit him in jail.

"Jorge," the Presidente says, "you know what the law requires in this case. You will be taken out at sunrise one day next week and executed by hanging. But since you are an old comrade of mine from our days at the military academy, I don't want you to worry needlessly. I promise that you won't know the night before that the execution will take place the next morning."

The Presidente leaves, and the general begins to reflect on his situation. "One day next week," he reasons, "limits the possibilities to only seven days. The execution must take place on Sunday, or Monday, or Tuesday, or Wednesday, or Thursday, or Friday, or Saturday. That means, if I am still alive Friday night, that the execution will take place Saturday morning, and I will know this the night before. That would violate the Presidente's pledge, so they can't possibly schedule the execution for Saturday. But *that* means that if I am still alive *Thursday* night, I will know that the execution must take place Friday morning, and that too would violate the Presidente's pledge. So they can't possibly schedule the execution for Friday! And *that* means that if I am still alive Wednesday night...."

Continuing to reason in this way, Jorge convinces himself that they cannot execute him at all, since the Presidente is known far and wide to be a man of his word. Much relieved, Jorge is able to sleep soundly at night.

Tuesday morning at sunrise, Jorge is led from his cell to the gallows, and sure enough, he is completely surprised! Was there a flaw in his reasoning?

Several Selections from Golomb's Gambits 479

GOLOMB'S GAMBITS

Solutions

1. Most people say that the probability is *one-half*, because there are two envelopes that the $10 bill could have come from, and in one case the second bill is also a $10 bill, but in the other case the second bill is a $5 bill. But most people are wrong.

The correct answer is *two-thirds*, because initially there are *six* equally likely *bills* that might have been extracted. However, when the extracted bill is now observed and seen to be a $10 bill, it is then one of *three* equally likely cases. In two of these cases, the second bill is also a 10, and in only one of the three cases is it a five.

There is another way to see this. Suppose we ask, "What is the probability that the envelope we select will contain two bills of the same amount?" The answer to this question is *obviously* two-thirds. If we then extract one bill from the selected envelope, its mate has a two-thirds chance of matching it, whether it is a five or a 10.

2. It surprises most people to learn that the best strategy is b) Switch. This actually *doubles* the probability of guessing right!

When you initially pointed to a cup, all three cups were equally likely—each with probability one-third. Since his removal of one of the other cups is independent of whether you actually guessed right or wrong, the probability of your original cup being correct remains at one-third. However, by removing an empty cup, he has increased the probability of the other cup still in contention to two-thirds.

To make this reasoning more explicit, suppose we call the cups A, B, and C, and that you originally pointed to A. There are three equally likely possibilities: the coin is either under A, B, or C, and you have no idea which it is. We will calculate the effect of the recommended strategy, which is to switch after he has removed a cup.

If the coin is actually under cup A, and you switch, you lose. This happens one-third of the time.

If the coin is actually under cup B, he must remove cup C. You will switch to B, and win. This happens one-third of the time.

If the coin is actually under cup C, he must remove cup B. You will switch to C, and win. This happens one-third of the time.

Thus, the recommended strategy wins two-thirds of the time.

3. There is insufficient information about the way in which the warden decided how to reply; so several answers are plausible. The answer "one-half" is based on the assumption that since it is now between A and C, each of them has probability one-half of execution. The answer "one-third" is based on the assumption that the warden truly gave no information that would affect A's knowledge of his own fate. (This assumption corresponds to the rules of the shell game in the previous problem.) The answer "zero" is based on the idea that the warden is trying to reassure A by mentioning something positive, and is really indicating that C is the loser. The answer "one" is based on the idea that the warden knows that A has been selected for execution, but is trying to spare his feelings. Perhaps you can invent a scenario that leads to still another answer.

4. When this problem was first popularized, in the 18th century, both mathematicians and gamblers were puzzled by the answer they got using the standard formula for "expected value," where you multiply each possible payoff by the probability of receiving that payoff, and then sum over all cases. This gives the "infinite series":

$$1/2 \times \$1 + 1/4 \times \$2 + 1/8 \times \$4 + 1/16 \times \$8 + \ldots = \$.50 + \$.50 + \$.50 + \$.50 + \ldots$$

which has an *infinite sum*! This became known as the "St. Petersburg Paradox," after the Russian city now called Leningrad.

The resolution of this paradox is that any actual casino has a finite limit to how much money it can actually pay out. Suppose we make the very generous assumption that our casino can pay out up to *$1 trillion*, but no more. This deflates the expected payoff dramatically, to a mere $21. (This is based on the approximation $1 trillion = 2^{40}. Since one trillion is actually a little less, the expected payoff should be reduced by a cent or two.)

Under the circumstances, I regard the $100 entry fee as exorbitant.

5. There is something deeply disturbing about the notion that whichever card you picked, you'd be better off with the other one! If switching once gains 25%, why not switch again and gain *another* 25%?

The fallacy underlying this paradox is rather subtle. One resolution to it is an economic argument, similar to the way out of the St. Petersburg Paradox. Specifically, there is some upper limit to how much money even Mr. Gottbucks can dispense. Hence, he cannot select an amount of money "at random," if this includes arbitrarily large numbers beyond his ability to pay off.

There is an even more basic difficulty, which is mathematical rather than economic. Even if Mr. Gottbucks had no economic limit on how much money he could give away, he could not select an amount of money "at random" in a way that all choices would be equally likely. In mathematical terms, there is no "uniform distribution" on the positive real numbers.

As a very simple illustration, suppose you know that Mr. Gottbucks always selects the amount of money he writes on the first card from the range of $800 to $1500. Then, if you see such an amount on the first card you look at, you ask to have the amount on the second, unseen card. However, if you see an amount in the range of $1600 to $3000, you should *not* switch to the other card.

6. This famous paradox, called "the unexpected hanging" by Martin Gardner, has been the subject of many articles, giving as many different explanations of where the difficulty lies. My own explanation, however, is different from any of the ones I have seen published.

There is an important distinction between "truth" and "knowledge." "Truth," which mathematics strives to attain, is independent of time. "Knowledge," however, is time-dependent. The promise to the general concerns only his *knowledge* of the timing of the execution. In reasoning backward in time, he is relying on knowledge that he would have were he still alive at some later time, but knowledge that he does not yet have. Now, truth is truth whether you've already discovered it or not, but knowledge is *not* knowledge until you have it. Reasoning based on a state of knowledge not yet attained is inherently invalid.

If you think the general may be right in concluding that he can't be executed without violating the assurances he was given, suppose he had been told that the execution would be "one day next year" instead of "one day next week." Just how far do you believe his argument can be stretched?

Marketing in USA

- 1980 Toy Fair New York — Ideal Toy Co.
- Hungarian connection was important

Zsazsa Gabor   *Prof. Solomon W. Golomb*

Taken from marketing material for Rubik's cube from Ideal Toy Company: URL: https://indico.cern.ch/event/contributions/attachments/Rubiks Cube and Marketing as Educational Tool.ppt

© 2023 World Scientific Publishing Company
https://doi.org/10.1142/9789811234378_0054

Word Ways Selections*

Jeremiah and Karen Farrell at Butler University are the masterminds behind *Word Ways*, "The Journal of Recreational Linguistics." They kindly mentioned that Sol's contributions could be made use of, and a selection are included here. All are by Sol with one exception ("Collector's Corner, Round 5") which has a paragraph about Sol that I thought would be of potential interest. These give a small taste of one of Sol's wide-ranging interests, involving words and language. Encompassed are a touch of physics and the erudition that invested so much of what Sol did. Please note that in the "Amalgamate, Chemist!" article, Sol did the anagramming all himself, without access to computers, which were not widely available at that time and he never made material use of – having far less need than the rest of us since he had so much information and calculation ability in that most remarkable computer, the one in his head. Regarding "A Letter to Martin" (Gardner) who for many years ran *Scientific American's* column entitled "Mathematical Games," many of us who knew Sol have received letters of the ilk illustrated, in which Sol draws from his wide-ranging knowledge to address a topic of casual interest of the moment. Regarding the column entitled "Hebrew or Japanese?" I note that the day after reading this to select items for inclusion in this section, I received an email from someone who stated that, based on my surname, I must be Japanese! (Not that the name "Golomb" is Hebrew.) In reading Sol's semihumorous contribution "Extraterrestrial Linguistics," I was struck by the quality and clarity of Sol's prose – noting that Sol was still in his 20s. The last tripartite entry (final three pages) is from the May 2016 issue of *Word Ways* in memory of Sol (Sol died on May 1, 2016). The items were selected by the editors, Jeremiah and Karen Farrell. The first piece is an introduction of "G.S. BLOOM" as the anagram of "S. GOLOMB," which Sol described in the second (bulleted) item "Amalgamate, Chemist!" in this collection. (See chapter by G.S. Bloom.) We separated the Farrells' selections from their own writing about Sol in that issue (included earlier in this book), to better preserve the materials written by Sol in one locus.

The *Word Ways* selections are:

- Extraterrestrial Linguistics
- Amalgamate, Chemist!
- A Letter to Martin (*Word Ways*: August 2010. Letter to Martin Gardener from: July 8, 2002)
- Collectors Corner, Round 5
- The Antiquity of the Dreidle
- The Meaning of Time
- Anagramming Co-authors
- Hebrew or Japanese?
- The Periodic Table of the Alphabet
- Three pieces from *Word Ways* May Issue 2016

---

*This introduction was contributed for this book by Beatrice Golomb.

Volume 49, Number 2
May 2016

The Journal of Recreational Linguistics

Word Ways

Word Ways Selections

WORD WAYS® The Journal of Recreational Linguistics

1968 Editor:	Dmitri A. Borgmann, Dayton, Washington
1969 Editor:	Howard W. Bergerson, Sweet Home, Oregon
1970-2006 Editor:	A. Ross Eckler, Morristown, New Jersey

EDITOR and
PUBLISHER

Jeremiah Farrell
9144 Aintree Drive
Indianapolis, Indiana 46250
wordways@butler.edu

BUSINESS and
SUBSCRIPTIONS

Karen Farrell
9144 Aintree Dr.
Indianapolis, Indiana 46250
wordways@butler.edu

ALPHAMETICS
EDITOR

Steve Kahan
78-51 220th St.
Hollis Hills, New York 11364

KICKSHAWS
EDITOR

David Morice
1304 Bloomington St.
Iowa City, Iowa 52240
drabc1946@gmail.com

US ISBN 0043-7980

Published Quarterly in February, May, August, November
Annual Subscription Rate: $33 (domestic), $35 (Canada and Mexico),
 $45 (Europe), $60 elsewhere, $60 libraries and institutions, $20 online

Back issues available (write for prices)

Website: www.wordways.com (Web specialist, David S. Dillon)

Editorial Board:
    Lacey Echols
    A. Ross Eckler
    Kirstin L. Ellsworth
    Katie Mohr
    David D. Wright

Printed at Campus Impressions, Butler University

Copyright © 2016 Jeremiah Farrell

486   *The Wisdom of Solomon*

11/1968

202

Extraterrestrial Linguistics *

DR. SOLOMON W. GOLOMB
Los Angeles, California

There are two questions involved in communication with Extraterrestrials. One is the mechanical issue of discovering a mutually acceptable channel. The other is the more philosophical problem (semantic, ethic, and metaphysical) of the proper subject matter for discourse. In simpler terms, we first require a common language, and then we must think of something clever to say.

As far as the channel is concerned, there would seem to be many different possibilities. On earth, we can communicate by speech (using the ear as the receptor), by writing or semaphore and pictographs (using the eye as the receptor), by tactile means (e.g., Braille), and, as demonstrated recently, by modulating an olfactory channel (aromarama). Electromagnetic relaying being involved or not, any message we receive must ultimately activate one of our many sense perceptions—sight, hearing, touch, smell, taste, temperature, equilibrium, pressure, acceleration, etc. There is also conceivably telepathy, or at least alpha-rhythm.

If we met a strange creature on an alien planet who seemed to be capable of intelligence, it could be quite difficult to decide which of the many sounds (whistles, clicks, snaps) or smells (of which we have no theory to speak of) or radiations at many frequencies (possibly, but not necessarily, including optical frequencies) would be information-bearing. It could turn out that none of these are significant, but that behavior patterns such as fluttering of appendages and agitation of membranes tell the story, as in human discourse, where gestures and glances can easily replace words. On our own planet and within our own species, there have developed such diverse systems as sign, whistling, and gesturing languages (not only in Southern Europe, but in such constructs as the deaf-mute language), not to mention Morse, Semaphore, Braille, and spoken languages using quite dissimilar phonemes and intonation patterns. How to recognize an attempt to communicate something, when you first encounter it, might prove quite a difficult matter.

Of course, there is the approach of Project Ozma. If we start with as many

---

* Reprinted, by permission, from the May, 1961, issue of *Astronautics*, copyrighted by the American Rocket Society, Inc.

WORD WAYS

assumptions about the reasonableness of our friends, the Extraterrestrials, as UFO enthusiasts do, we might end up with English-speaking, anti-Communist, white-Protestant Centaurians. Relaxing the constraints slightly, we find humanoids who build the same kind of radio systems we do, and who think and act much as we, even to the point of recognizing the 21-cm hydrogen line as the best of all possible frequencies. However, they are more interested in us than we are in them, which is why they are transmitting, whereas we are merely receiving.

It would be wonderful, indeed, if this approach would lead to success, but 100-to-1 would be good odds that no results will be obtained by Ozma in time to affect my intention to retire from active space probing in the year 2000. After all, if they are as similar to us as all that, they have a budget-minded Congress which has set their peak transmitter power level just below our minimum detectability threshold, if it hasn't cancelled the project altogether for having failed to produce results for lo! these many millennia.

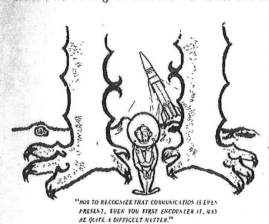

"HOW TO RECOGNIZE THAT COMMUNICATION IS EVEN PRESENT, WHEN YOU FIRST ENCOUNTER IT, MAY BE QUITE A DIFFICULT MATTER."

It would be particularly ironic if our portion of the universe is densely populated with gigantic receiving antennas, but with no one willing to undertake the additional expense and round-trip delay time required for transmitting. The conservative assumption is that, even if there is other intelligent life in our neck of the galaxy, we will not find it until we have brought a spacecraft fairly close to it.

Who's Intelligent?

Suppose that we have landed on or near some congenial planet, and we find there a species living in elaborate cities, and hence prime suspects in our quest for new intelligence. The mechanical problem of finding a mutually acceptable channel for the commencement of negotiations cannot be divorced from the semantic problem of convincing these creatures that *we* are intelligent. The solution to this problem depends on whether we are truly face to face with the creatures, or merely have a narrowband link in operation. The narrowband link, being more constrained, is easier to analyze. For it, we want a pattern too regular to be random noise, but too irregular to be a naturally produced pulsation phenomenon. Standard suggestions include the binary expansion of the number *pi*, the sequence of the first few prime numbers, or simple arithmetic, such as $3 + 4 = 7$.

To begin with, it is probably rank terrestrial provincialism to expect others to attach the same importance to *pi* that we do. Even in our own mathematics, such constants as *e* and *log 2* are considered important, and the exaggerated role of *pi* stems largely from the Greeks' undue efforts attempting to square the circle. Doing arithmetic has the drawback that such concepts as "plus" and "equals" must be brought into the picture.

PARDON ME, MA'AM, DO YOU SPEAK ENGLISH? I GOT SEPARATED FROM MY PARTY, AND.. ....

My own recommendation is the prime sequence 2, 3, 5, 7, 11, 13, 17, 19, 23, . . ., with a long period to prove the non-accidental nature of the signal. It isn't so much that I'm sure these Extra-terrestrials would *recognize* the primes; but if they don't, they must be dull fellows, indeed, and I would just as soon not get acquainted. Of course, we use the base 1, thus: 1l—111—11111—1111111— . . . We could equally well send portions of arithmetic progressions—for example, 1, 2, 3, . . .; 1, 3, 5, . . .; 1, 4, 7, . . . This makes fewest demands on pattern-recognition capability, and hence is a low-threshold intelligence detector.

The eminent Netherlands mathematician, Hans Freudenthal, is currently at work on a book to be called *Lincos* (for "lingua cosmica"), in which he will attempt to describe an ideal language for cosmic intercourse. This is all well and good, except that the Arcturists may be less interested in learning Lincos than in teaching us some equally ingenious invention of their own.

At closer range we can demonstrate gadgets, especially munitions, at which our species excels. Other than publicly killing one another, we can perhaps demonstrate our intelligence by collecting biological specimens to raise in captivity. Such scientific curiosity is an unmistakable sign of intelligence, although with a notable drawback: if our specimen belongs to the dominant species, his capture and imprisonment may be regarded as an act of war.

Now we come to the really significant question. Suppose we have passed the I.Q. test, resolved all the semantic questions, and have an effective communications link going. What do we talk about? Bell Telephone and Western Union assume their subscribers already have something in mind to say, so that classical information theory turns out to be of no help. Ruling out such commonplaces as baseball scores, the weather, and back-fence gossip, I have compiled the following list of likely topics for discussion with our extraterrestrial neighbors: (1) Help! (2) Buy! (3) Convert! (4) Vacate! (5) Negotiate! (6) Work! (7) Discuss!

Each of these topics merits a brief elucidation:

Extraterrestrial Linguistics

(1) HELP! Assumes we have encountered a superior civilization and want their help in solving our earthbound problems or fighting our internecine battles.

(2) BUY! Presupposes that some basis for mutually profitable trade exists and should be acted on.

(3) CONVERT! Suggests that as missionaries from the Chosen Planet we have undertaken to spread the Good Word that the Galaxy is coming to an End.

(4) VACATE! Means that we like the planet, and figure that we can defeat the inhabitants.

(5) NEGOTIATE! Means that we are looking for new members in OPTO (the Occidental Planetary Treaty Organization).

(6) WORK! Supposes that we've uncovered a good source of cheap labor.

(7) DISCUSS! Presumes that there is no common environment which we and they can share. Only in such a case does the history of our species offer encouragement for the prospect of free mutual interchange of ideas, experiences, and scientific theories.

Naturally, we must not risk telling too much until we know whether the Extraterrestrials' intentions toward us are honorable. The Government will undoubtedly set up a Cosmic Intelligence Agency (CIA) to monitor Extraterrestrial Intelligence. Extreme security precautions will be strictly observed. As H. G. Wells once pointed out, even if the Aliens tell us in all truthfulness that their only intention is "to serve mankind," we must endeavor to ascertain whether they wish to serve us baked or fried.

* * *

THE JOURNAL OF RECREATIONAL LINGUISTICS

490 *The Wisdom of Solomon*

2/1984

20

AMALGAMATE, CHEMIST!

SOLOMON W. GOLOMB
Los Angeles, California

When Douglas Hofstadter took over Martin Gardner's column in Scientific American, he changed the name from MATHEMATICAL GAMES to METAMAGICAL THEMAS. The new name is an anagram of the old one, that is, a permutation or reordering of the same set of letters. This naturally suggests the question of how easy is it to find an anagram on these letters which "makes sense".

Some sets of letters (such as XJRMQ) will form no English words or phrases at all, though the die-hard may find the call letters of a Mexican radio station! Some sets of letters (such as those in the word FLANK) will form only one word; some will form exactly two (NIGHT, THING); some will form several (POST, POTS, SPOT, STOP, TOPS, OPTS). How many "meaningful" expressions can in fact be formed from the 17-letter set AAAACEEGHILMMMSTT, the letters in MATHEMATICAL GAMES?

First, the purely mathematical question: how many distinct permutations of these 17 letters are there, whether they make any sense or not? If all 17 letters were different (which they clearly are not), the number of permutations would be 17!, pronounced "seventeen factorial," and defined to be $17! = 17 \times 16 \times 15 \times 14 \times 13 \times 12 \times 11 \times 10 \times 9 \times 8 \times 7 \times 6 \times 5 \times 4 \times 3 \times 2 \times 1 = 355,687,428,096,000$. Because there are repeats among the letters, we must divide this number by the number of ways of permuting identical letters among themselves (since interchanging identical letters produces no visible change in the resulting sequence). The four A's can be permuted in $4! = 4 \times 3 \times 2 \times 1$ or 24 ways, the three M's in $3 \times 2 \times 1$ or 6 ways, the two E's in 2 ways, and the two T's in 2 ways. Thus the total number of distinct sequences of the 17 letters in MATHEMATICAL GAMES becomes $17!/(4! \times 3! \times 2! \times 2!) = 617,512,896,000$. This number is increased if we have the option of introducing spaces between words and punctuation marks wherever we wish in the sequence.

How many of these mathematically distinct sequences are likely to make any sense as English? This is a question for Information Theory, which reformulates it as follows: What is the entropy of written English, regarded as a source of letters of the alphabet? (Entropy is a measure of the permitted degree of randomness.) If written English had an entropy as large as $\log_2 26 = 4.7..$ bits per letter, that would mean that essentially every sequence of letters is "meaningful". We know this is far from the case, and the best estimates of the entropy of written English are in the range of one to two bits per letter. This would suggest that there are between 2^N and 2^{2N} meaningful English messages which are N letters long, and in particular, between $2^{17} = 131,072$ and $2^{34} =$

Word Ways Selections 491

21

17,179,869,184 English "messages" 17 letters in length. I suspect, for sequences of 17 letters, the truth is closer to the lower than the upper estimate. In any case, this would only be approximated for large values of N, and would not correspond to only the permutations of a single set of letters. However, most of these messages, for large values of N, would be permutations of sets of letters with the typical letter frequencies of English.

The letters in MATHEMATICAL GAMES are not very typical: three M's but no N's, R's, or D's; four A's but only two E's and no O's, etc. My own estimate is that there are perhaps 3000 "meaningful" sequences of these letters (where punctuation and spaces can be inserted at will, and only fragments of ideas or phrases need be expressed), give or take a factor of 5. And even this estimate is sensitive to how strict or liberal we are with our notion of what constitutes "English". Thus, METAMAGICAL THEMAS consists of two "words" which are not in any standard English dictionary of which I am aware. METAMAGICAL is a Hofstadter coinage whose meaning we are to deduce by analogy with METAMATHEMATICAL and META-PHYSICAL; and THEMAS is an improper plural of THEMA. (Theme and scheme lead to themes and schemes; but from thema and schema come themata and schemata.) As we allow more and more foreign words, variant spellings, abbreviations, initials, contractions and proper names, the entropy of "English" increases, and we get more and more "meaningful" anagrams.

To illustrate what is possible, I have listed 100 "meaningful" anagrams of MATHEMATICAL GAMES. I have created a new literary form for this purpose, which I will call serve verse. An example of serve verse can be called a mope poem. A brief, self-referential mope poem is:

Revés: Sever verse, / Veers; serve.

My serve verse creation on MATHEMATICAL GAMES is divided into sections on the general themes of metamathematics, geology, zoology, sororities, athletics, culinary arts, and medicine. The metamathematics section is concerned with such things as THEMATIC LEMMAS; the geology section (or is it mineralogy?) with AMALGAMS, MAGMA, MALACHITE and STALAGMITE. The zoology section is obsessed with CAMELS and MAMMALS. The sorority section has GAMMA, ETA, THETA, SIGMA and CHI at its disposal, and seems to contain a vicious ad feminem attack against one EMMA GALE SMITH. The athletics section refers to TEAMMATES, GAMES, SETS, MATCHES and to ATHLETICS itself. The culinary section mentions STALE MEAT, STEAM HEAT, HAMS and CLAMS, and the medical section alludes to TEETH and MEASLEs. Readers are encouraged to invent even more imaginative anagrams of their own.

For several centuries, one of the forms of cryptography actually employed to conceal the content of diplomatic messages was the use of permutation ciphers, where the letters of the message were scrambled according to a procedure known only (it was hoped) to the sender and the intended receiver. Without the correct rule,

492   *The Wisdom of Solomon*

22

as we have seen, the same letters might be unscrambled to reveal any "secret" message from MAGMA HEATS CLIMATE to SAM, THE MAGIC TAMALE.

Mathematical Games

Thematic lemma saga
Meet math as magical
Magic sheet: lama mat
Image chats at lemma

Metamagical Themas

I get the lemma – a scam
Elastic math game, Ma!
The same magical mat
G.l. lemmata schemata

Amalgamatic Themes

Amalgamate, chemist!
Chemist ate amalgam
Magma heats climate
Magi select mahatma
Malachite set magma
Stalagmite came, ham
Almagest: teach imam

The Magmatical Seam

Misamalgamate, Chet!
Alchemist ate magma
Mahatma's climate, e.g.
Michael, taste magma
Malaesthetic magma
Ah, magmatic Maltese
Megalith caste, Mama

This Camel Ate Magma

Get this, Ma'am: a camel
Tight as a camel, Emma
Camel meat? I'm aghast!
As a camel might mate
I get mahatma's camel
Castigate mammal, eh?
Cage atheist mammal
Tia cages the mammal
That's ice-age mammal
The camel's gait, Ma'am
Age maims that camel
Camel stigmata? Ahem!

Magma? Camels Hate It

Get this, ace: a mammal!
It aches? Get a mammal!
Aghast, Mac? Mealtime!
I get a chaste mammal
Che agitates mammal
Acetate mammal? Sigh!
Hattie cages mammal
I teach stage mammal
Mammal at "eight aces"
The ice mammal's gate
Ice ages that mammal
Legit schemata, Ma'am

Hag Claims Teammate

Chi Gamma's late mate
Asthmatic Emma Gale
Met me, claims Agatha
Gamma Chi stalemate

"Hate Mate," Claims Mag

Theta Gamma's malice
(Emma Gale Smith: a cat?)
Me? Clammiest Agatha!
Eta Chi Gamma metals

Mag's Mealtime Chat

Masticate ham leg, Ma
Chi Gamma: taste meal!
Stage claim: ham meat
Came as tamale might

Mag's Calm Teatime? Ha!

Magic ham? Stale meat!
Teatime: ham, clam, gas
Malt, sage ethic, Ma'am
Sam, the magic tamale

Mama Mia! The Last Egg!

Aim, match, tag measle
Mica teeth amalgams

Ethical, Sam? Get Mama!

I calm the stage-mama
Asthma: age claim met

Word Ways Selections   493

23

Athletic Games, Ma'am	Thalamic Team Games
Michaelmas team tag	1 gash calm teammate
A late game mismatch	A lame gate mismatch
Claim gate at Thames	Time Malaga matches
Aim, Al: game, set, match	Lit matches age Mama
Sham cage title, Ma'am	Claim tag? Shame, team!
Met Lhasa magic team	Act the lamaism game
Mime act: Hamlet saga	Ham act: get me salami
1 match eaglets, Ma'am	1 tag the camel's mama
Team game: Cal/Smith	1 get Tech's alma mama
Hit game castle, Ma'am	Agile mama cast them

Editor's Note: One can ask what is the likelihood that any of the sentences given above have somewhere appeared in English writings in a meaningful (not wordplay) context. Obviously, this is a very stringent (and unverifiable!) measure of the plausibility of an anagrammed phrase; it is quite possible that none of these phrases (save the already-noted METAMAGICAL THEMAS) would pass. Put in other terms, it is not difficult for the reader to pick out the original phrase on which the anagrams were based. Cryptographers using permutation ciphers would be well-advised to select a message using convoluted grammar, with an anagram that sounds more plausible than the intended message.

OMNI GAMES

This paperback anthology of the best brain-teasers from the last five years of Scot Morris's Omni magazine Games column is devoted to an amazing variety of subjects: besides the usual mathematics and logic puzzles, it contains physics problems, geographical oddities, electronic calculator tricks, bar bets, juggling, and a tribute to Martin Gardner. The emphasis is on (1) problems that look difficult but can be solved easily with the proper insight (aha!), (2) problems that look easy but contain hidden traps, and (3) problems that look difficult and in fact are, but which lead to curious facts. Though many problems are golden oldies, Scot Morris writes about them in a lively and entertaining way reminiscent of Martin Gardner, and often succeeds in bringing in new angles. The book is enriched by reader comment and emendations to the original Omni material. A substantial part of the book is devoted to problems related to logology: four unsolved Enigma-style conundrums by Bishop Wilberforce, the Beale cipher, two tough spelling quizzes, the "world's hardest" word quiz (containing much Word Ways material), mnemonics, rebuses, anagrams, and the first crossword puzzle. The book is available from Holt, Rinehart and Winston for $11.95. Let's hope that this is the start of a long series.

494   *The Wisdom of Solomon*

8/ 2010

A LETTER TO MARTIN

SOLOMON W. GOLOMB
Los Angeles, California   July 8, 2002

Mr. Martin Gardner
3001 Chestnut Road
Hendersonville, NC 28792

Dear Martin,

In London recently, I bought a copy of the *Definitive Edition* of the *Annotated Alice*, at the bookshop in the British Library (where the books from the British Museum were relocated a few years ago). It was a wonderful reading experience. I had read your original *Annotated Alice* ages ago, but this was like reading the Alice books for the first time.

I made a few notes as I was reading.

1. On p. 171, when the Red Queen says: "When you say 'hill', the Queen interrupted, "I could show you hills, in comparison with which you'd call that a valley"; and Alice objects: "a hill can't be a valley, you know. That would be nonsense —"; I suspect that Dodgson was reacting to something in Hans Christian Andersen's story "Elverhøj" (the *Elf Hill*, which is very famous and was even made into a ballet). The Troll King (the Mountain King, or Dovregubben, in Ibsen's Peer Gynt, written later) from Norway, is visiting the Elf King in Denmark; and the Troll King's ill-mannered son says, regarding the "Elf Hill" of the title, "You call this a *hill*? In Norway, we would call it a *hole*!" (Denmark is very flat and Norway is very mountainous.) Alice expresses Dodgson's mathematical view that what is convex cannot be concave. (We would need to know when the English translation of "Elverhøj" reached Oxford, and if Dodgson is likely to have read it.)

2. It is interesting to compare Dodgson and Andersen. Both were men who liked children and never married, or even seem to have come very close. Both wrote children's stories with at least one eye on the adult audience, and both consciously dispensed with the need to end children's stories with a moral. Both were translated into dozens of languages, and are still popular today. (Denmark is planning a major celebration of the 200th anniversary of H.-C. Andersen's birth, in 2005.)

3. On p. 192, you quote Carroll as writing: "In composing the Walrus and the Carpenter, I had no particular poem in mind". It is certainly true that in its entirety, this poem is not a specific parody of any other. But the first three

Word Ways Selections   495

stanzas, and the first one in particular, were surely "inspired" by Coleridge's *Rime of the Ancient Mariner*, which has several stanzas about the way the sun was shining, e.g.,

> "The sun now rose upon the right,
> Out of the sea came he,
> And he shone bright, and on the left
> Went down into the sea."

(Carroll couldn't resist parodying the least poetic outputs of the Lake Poets.) But it also seemed to me that Carroll was also parodying the explicit *moral* of the Ancient Mariner: "He prayeth best who loveth best
All creatures great and small..."

with the treatment of the oysters by the Walrus and the Carpenter.

4. On p. 212 you have a note explaining what a "teetotum" is. Many of your readers would be surprised to learn that this is precisely the kind of spinning four-sided top, called a "dreidle", with which Jewish children play on Chanukah. (The four Hebrew letters *nun, gimmel, heh,* and *shin,* are on the four sides, instructing the player, respectively, to take a) nothing, b) everything, c) half, or d) a negative amount, from the pot.)

5. On your notes on page 235, you indicate that the inspiration for Carroll reintroducing the Hare and the Hatter as Anglo-Saxons named Haigha and Hatta is obscure. My vague recollection from high school English history (more than fifty years ago) is that Anglo-Saxon history in England begins with two warriors named Hengist and Horsa. Quite likely the Liddell sisters had learned something like this from one of their tutors, or seen it in the play that you mentioned, and Carroll couldn't resist wordplay on these names.

6. There are many Britishisms that are still in current use but unfamiliar to Americans (e.g., that a *pudding* is any sort of sweet or dessert, or even a different food entirely as in Yorkshire pudding); and things that have become less familiar even in England during our own lifetimes (like the old pound-shilling-pence currency; and "English" chess notation, which even in the UK has been giving way to "algebraic"). If there is ever a post-definitive edition (the "pluperfect Alice"?), it may be useful to include some glossaries, and two chessboards side by side to exhibit the two notations. (Even in English notation, **N** replaced **Kt.** for "knight" several decades ago.)

7. Another mathematician who had some things in common with Dodgson was the late Paul Erdös, who also never married and was very fond of children. But the most striking similarity was their common preoccupation with aging and death, starting at an earlier age than is customary.

496   *The Wisdom of Solomon*

8. I recall sending you (or at least mentioning) a clipping from a British paper, after a previous trip to England about ten years ago, with the information that you summarize in your note on p. 64 about the grin being all that Mr. Joel Birenbaum could see of the cat's head after kneeling down in a church in Croft-on-Tees where Dodgson's father had once been the rector. It is the custom in Anglican churches for the congregants to kneel (usually on a *prie-dieu*, several inches above the floor) during the service (a fact which Mr. Birenbaum may not have known?), so that seeing the cat's face disappear except for the grin would have been experienced by the entire congregation. In the clipping I saw, Mr. Birenbaum further speculated that Dodgson had perhaps misremembered the locale, and transferred it from Croft-on-Tees to Cheshire, where his father had earlier preached. But you make it clear that "to grin like a Cheshire cat" was an established simile in England long before the Alice books were written. The Carroll innovation (as far as I'm aware) was to have the cat disappear except for the grin.

9. In your notes on p. 23, referring to Carroll's parodies of once well known poems, you promised "...all the originals will be reprinted in this edition." But you never included Wordsworth's *Resolution and Independence*, a.k.a. *The Leech-Gatherer*, the original for the White Knight's *The Aged Aged Man* (a.k.a. *Ways and Means*, etc., etc.).

This is all I can think of for now.

Best regards,

Solomon W. Golomb

SWG:mat

Word Ways Selections 497

8/2011

COLLECTOR'S CORNER, Round 5
An Occasional Item of Interest to Gardner Collectors

Scientific American Magazine Revisited
Suggested by Jeremiah Farrell

Martin Gardner's 300th article for *Scientific American* appeared in their August, 1998 issue (Cover: New Thinking about Back Pain). It's title "A Quarter-Century of Recreational Mathematics," highlights Gardner's 297 "Mathematical Games" columns and his 1956 article on "Hexaflexagons" (he does not mention the 1952 article on "Logic – see Collector's Corner – Round 4").

Gardner's good friend Solomon W. Golomb of the University of Southern California was one of the first to supply grist for "Mathematical Games". The May, 1957 issue (Cover: Birds In The Museum) introduced Golomb's studies of the immensely popular polyominoes, i.e. shapes formed by joining identical squares along their edges. The 2-square domino can take only one shape but the tromino, tetromino, and pentomino can assume a variety of forms. "The study of polyominoes soon evolved into a flourishing branch of recreational mathematics," said Gardner. "Arthur C. Clarke, the science-fiction author, confessed he had become a 'pentomino addict' after he started playing with the deceptively simple figures." Golomb would contribute many more times, over the years.

A simple tetromino problem for the reader. Can you prove that the five different tetrominos cannot cover exactly a 4x5 checker board?

Gardner mentions several other multiple contributors including Denmark's Piet Hein and John H. Conway of Princeton. Conway's game of Life and Hein's game Hex are both still popular.

One of Gardner's favorite columns was April 1975 (Cover: Dinosaur Renaissance) "Six sensational discoveries that somehow or another have escaped public attention." The article was an April Fool's Day hoax! It told readers that Leonardo da Vinci had invented the flush toilet, that opening pawn to King's rook 4 was a certain chess winner, and supplied a complicated map purporting to require five colors to ensure that no two neighboring regions were colored the same. "Hundreds of readers sent me copies of the map colored with only four colors," reported Gardner. "Many said the task had taken days".

The four-color-map theorem is exploded

Every year or so Gardner would devote a column to the numerologist Dr. Irving Joshua Matrix (note the "666" provided by the number of letters in the first, middle and last names). There were 22 in all, starting in January, 1960 (Cover: *not seen*) and ending with Matrix' untimely "death" in the September, 1980 issue (Cover: Economic Development). However, Gardner, in *Penrose Tiles to Trapdoor Ciphers*, 1989, Freeman, discovered that the reported "death" of Matrix was premature when he found the mysterious doctor in Casablanca.

For a flavor of Matrix and his numerology we offer the following from the 1980 issue. An emirp is a prime that yields a different prime when it's digits are reversed (like 13 and 31). Leslie E. Card discovered these order 4 and order 5 emirp squares:

```
9 1 3 3              1 3 9 3 3
1 5 8 3              1 3 4 5 7
2 5 2 9      and     7 6 4 0 3
3 9 1 1              7 4 8 9 7
                     7 1 3 9 9
```

The order 4 square is unique save for rotations and reflections.

Surprisingly often so-called recreational mathematics can lead to serious, applied mathematics. Gardner was first to report, in his August, 1977 (Cover: Kangaroos) column, "A new kind of cipher that would take millions of years to break," about an "unbreakable" code discovered by Ronald L. Rivest, Adi Shamir, and Leonard Adleman, computer scientists at the Massachusetts Institute of Technology. Gardner notes that "It was the first of a series of ciphers that revolutionized the field of cryptology."

We would be remiss if we did not mention the work of the remarkable Scott Kim that Gardner so much admired. Kim was first featured in the June, 1981 issue (Cover: Accretion of Planets) where his "Inversions" were displayed.

Gardner cites Kim's "magical ability to take just about any word or short phrase and letter it in such a way that it exhibits some kind of striking geometrical symmetry." A recent example:

Word Ways Selections 499

11/2011

THE ANTIQUITY OF THE DREIDLE

SOLOMON W. GOLOMB
Los Angeles, California

Dear Rabbi Dov,
As I mentioned to you a few days ago, the dreidle is actually an ancient toy, but I may be one of the very few people who knows this, and I discovered it by accident.
A few years ago, I was reading the "ultimate" edition of "The Annotated Alice" about Lewis Carroll's Alice in Wonderland books, by the late Martin Gardner. There is a scene in "Through the Looking Glass" where Alice is on a railroad train in the same compartment with an old woman whose large handbag is open, and there is an inventory of the items it contains. One of these items is a "teetotum", and Gardner explains that this was a toy in 19th century England, and from his brief description I realized that it was basically a dreidle. So I did a little research. Here is what my Cassell's "Concise English Dictionary", UK edition, 1997 says:
tee-to'-tum, 1. a toy, originally four-sided, turning like a top, used in a game of chance.
　　　　　　　　　2. any toy that is spun using the fingers.
　　　　　　　　　(for T-totum, take all: T, + Lat. totum, the whole, marked on one of the sides) So the T in teetotum corresponds to the gimel on our dreidle.
Then I looked in my fairly ancient copy of the Encyclopedia Britannica, and found "teetotum" at the end of the article on toys; where it said "known to the ancient Greeks and Romans; used in more recent times where dice were forbidden (sometimes with six sides, to mimic dice)".
So the actual item goes back to "the ancient Greeks and Romans", and is thus not of recent origin. How and when the Jewish version originated is not yet clear to me, nor how it got associated with Chanukah...but it was probably known in ancient israel during the Hellenistic period, which was also the age of the Hasmoneans...and the borrowing could have gone in either direction. (Many Greek words entered Hebrew at that time, but there were also Hebrew words that entered Greek and from there to Latin, like "pita" for bread became the Greek word for "pie' to this day, and from there became the Latin word that in Italian became "pizza". Likewise, from "k'tonet", a Hebrew garment, the Greeks formed the word "kiton" which then became the Latin word "tunic".) In talking recently to an Iraqi Jew, they don't have the Yiddish associations (e.g. gantz, halb, nisht,...) for the four letters on the dreidle--he said that for them "nun" is the letter that means to take the entire pot!
The name "dreidle" is almost certainly Yiddish from the German verb dreien, to turn or spin. (Of course, in Israel, it is called "svivon", a "spinner".) Best wishes, Sol Golomb

I found another dictionary (my Webster's New World Dictionary of the American Language, ca. 1955) that actually listed the four Roman letters on the classical teetotum, and the Latin words they stood for. In addition to T for totum (all), there was N for nihil (nothing), D for depone (put), and A for aufer (take), closely corresponding to the Hebrew letter meanings associated with gimel, nun, shin, and heh. The definition given was "A kind of top spun with the fingers, especially one with four lettered sides used in a game of chance."

500 *The Wisdom of Solomon*

2/2013

THE MEANING OF TIME

SOLOMON W. GOLOMB
Los Angeles, California

The English word TIME has multiple meanings which would require at least three distinct words when translated into other languages. Consider the following three answers to the simple-sounding question "What time was it?"
a. It was 9 a.m. b. It was the third time. c. It was during the tenth century.

In German, to elicit these three answers, TIME would be: a. UHR, b. MAL, c. ZEIT. In French, it would be: a. HEURE, b. FOIS, c. TEMPS. In Danish, it would be: a. KLOKKEN, b. GANG, c. TID. In Hebrew, it would be: a. SHA'AH, b. PA'AM, c. Z'MAN. For meaning "a.", other languages ask "How much is the clock?" or "What hour is it?", or even "What is the clock?"

These questions are themselves subject to misinterpretation. To the question (e.g. in Danish) "How much is the clock?" the answer might be "The clock costs 300 kroner." To the question (e.g. in Hebrew) "What is the hour?" The answer might be "The hour is one twenty-fourth of a day." To the question (e.g. in Swedish) "What is the clock?" the answer might be (as once given mischievously by Pippi Longstocking) "The clock is a machine that goes tick-tock."

For meaning "b." (as also in "the next time" or "the last time") languages use the TIME of multiplication, as in "Two times three is six." For the concept of TIME in Physics, they use meaning "c."

Languages may use a single word, or their own idiom, for "spare time," as in "What did you do in your *spare time*?" (English can use the single word "leisure.") Two different expressions or idioms may be used to translate "good time" in the following two contexts: i) "Did everyone have a good time?" ii) "It will happen in good time."

A large English dictionary may list 30 or more meanings for TIME as a noun, but most languages content themselves with the three words corresponding to a., b., and c. above. (Many languages including English also use TEMPO for *musical time*.)

For "At the present time" most languages would simply use a word for "now." For "At this time" they would use their TIME-word for meaning "c.", with their word(s) for "at this"; but for "This time we'll do it right" the TIME-word would have meaning "b."

For "Now is the time..." some languages would use their TIME-word with meaning "a.", while some would use the one with meaning "c." (For meaning "a.", "Now is the hour..." would usually have to make sense, though in Danish "The clock has become many" can be used to mean "The time has gotten late.")

If TIME is used in compounds like *daytime, springtime, mealtime, standard time*, etc., the TIME-word with meaning "c." would be appropriate.

That's all till next time!

ANAGRAMMING CO-AUTHORS

SOLOMON W. GOLOMB
Los Angeles, California

A British mathematician, I. J. Good, who was a veteran of the World War II codebreaking group at Bletchley Park, once published a paper listing "K. Caj Doog" as a co-author (an obvious rearrangement of the letters of Jack Good). When asked subsequently about the status of his collaborator, Good replied that he lost contact with Doog after the Chinese Communists overran Tibet.

Famous mathematician Norbert Wiener once published a novel under the nom de plume of W. Norbert. It is unlikely that he really wanted to conceal his identity.
I was pleasantly surprised to see a mention of my former Ph.D. student Gary S. Bloom, in connection with the concept of "eodermdromes" in A. Ross Eckler's article (p. 265) in the November, 2013, issue of WORD WAYS. Bloom, later Chairman of the Computer Science Department at the City University of New York (CUNY), told me that often when he met people for the first time they were surprised to discover that he really existed, having thought that G. BLOOM was merely an anagram I had invented of GOLOMB, or that G. S. BLOOM was an anagram of S. GOLOMB. When I visited him at CUNY in 1981, he mentioned another logology project: identifying English words that sounded like they were Yiddish. The best one I remember from his collection was "farfetched". (Bloom passed away a few years ago of an apparent heart attack while hiking in New England.)

At the Caltech Jet Propulsion Laboratory (JPL) in the late 1950's, six of us published an article under the collective pseudonym "B. H. JIGGS", using the initials of our six surnames, plus the imaginary "I" for pronounceability. When asked to, I agreed to review it for Mathematical Reviews. When someone later asked me what had become of Jiggs, I replied that since he had published only one article, it seems he never got tenure. Another member of my JPL group was the late Gustave Solomon, best remembered as the co-inventor of "Reed-Solomon Codes", widely used to protect the integrity of both stored and communicated digital information. Gus told me he could never co-author a paper with me, since he wouldn't get proper credit for something referred to as, say, "Golomb-Solomon Codes". He was not mollified when I told him I would be happy to give him first billing!

502   *The Wisdom of Solomon*

HEBREW OR JAPANESE?

SOLOMON W. GOLOMB
Los Angeles, California

1. INTRODUCTION

It is quite common that, for any two languages, there will be words that look or sound alike, but with different meanings. French words that look like English words but with different meanings in the two languages are called *faux amis* ("false friends"). In some cases they are from unrelated roots (e.g. French *pain* means "bread" and *rue* means "street". In other cases, words that began the same evolved different meanings. From its Latin roots, *concurrence* basically means "running along side of", which in English came to mean "agreement", but in French it means "competition."

Except for recent borrowings from English, there should be no cognates between such unrelated languages as Hebrew and Japanese. And since neither of these languages is normally written in the Roman alphabet, they won't have words spelled the same except in transliteration. However, there will be quite a few homophones, words that sound the same, but with very different meanings.

In Japanese, the word pronounced *tora* (as in the movie title "Tora Tora Tora," the code name for the December, 1941, attack on Pearl Harbor) means "tiger." The same pronunciation in Hebrew, usually Romanized as *torah*, has the basic meaning of "Law", and is often used to mean the Pentateuch, the first five books of the Bible (often referred to as "The Five Books of Moses").

Another example is Hebrew *Shoah* (literally a "catastrophe", the name given to the Holocaust), which sounds like Japanese *Showa,* the "reign name" of Emperor Hirohito (1926-1988) which included Japan's wars from 1933 to 1945.

But figuratively, this is only the "tip of the iceberg". Here we will consider many more of these linguistic coincidences, where a Hebrew word sounds like an unrelated Japanese word.

2. SOME EXAMPLES

A Japanese restaurant called *SAKANA* recently opened in my neighborhood. In Japanese this means "fish", but in Hebrew it means "danger". Could sighting a shark elicit the same exclamation in both languages?

The words *ISHA* and *HA-ISHA* in Hebrew mean "woman" and "the woman". These words are also related in Japanese, where they mean "doctor" and "dentist".

The well-known Japanese brand *MATSU-SHITA* literally means "under the pine tree". The same two-word phrase in Hebrew means "they found an acacia tree." But don't think these two phrases involving trees have any relation. The word that means "pine tree" in Japanese (*MATSU)* means "they found" in Hebrew, and the word that means "acacia tree" in Hebrew (*SHITA*) means "under" in Japanese.

Hebrew has three pronouns, pronounced *MI, HU,* and *HI* (or *me, who,* and *he,* if you prefer), where, from Hebrew to Engish, *me* means "who", and *who* means "he", and *he* means "she"! (Confusing?) These three words are also closely related in Japanese, where (in the native Japanese names for the numbers) *MI* means "three", *HU* means "two", and *HI* means "one". For good measure, in Japanese (from their adaptation of the Chinese number system) *SHI* means "four".

93

Word Ways Selections 503

In both Israel and Japan, you can find men named *AKIBA, YANNAI,* and *ZAKAI* (from completely different roots).

In Hebrew, *SHISHI* means "sixth", but in Japanese it means "lion" (or more narrowly, the lion-looking dogs in Chinese–style statues).

In Japanese, *KAZE* means "the wind" (as in *KAMIKAZE,* literally the "divine wind" credited with a major Japanese naval victory over the Chinese, off the coast of what is now Indonesia, during the reign of Chinese Emperor Kublai Khan). In Hebrew, *KAZE* (or *ka-zeh*) means "like this".

In Japanese, *IMA* means "now". In Hebrew, *IMA* is the informal word for *mother* that everyone in Israel uses.

NISAN, the first Spring month in the Hebrew calendar, is pronounced like the Japanese car maker *NISSAN.*

URU, the verb "to sell" in Japanese, is the command "shine" in Hebrew.

ISHI, "my man" in Hebrew, is "a stone" in Japanese.

YATSU, "they went out" in Hebrew, is the (native) Japanese word for the number *eight.*

SHIRO, "his song" in Hebrew, means *white* in Japanese.

KANA, the Japanese name for syllabic writing, means "he acquired" in Hebrew.

KARA, in Japanese means "empty" (as in KARATE, "empty hand," and KARAOKE, "empty orchestra"). As a preposition, KARA in Japanese means "from". One meaning of KARA in Hebrew is "happened." With a Qoph instead of a Kaph for the K-sound, KARA in Hebrew means "(he) called" or "(he) read". Most Israelis don't distinguish these two K-sounds; so MAH KARA means "what happened?" while MI KARA means "who called?"

ATARASHI, "new" in Japanese, means "You are Rashi" in Hebrew. (From the acronym name RASHI of the most famous medieval Jewish commentator; this is now a common boy's name in Israel.)

In Japanese, *UTSUKUSHI* means "beautiful". In Hebrew, *UTSU KUSHI* means "they counseled an Ethiopian."

In Japanese, *ASA* means "morning". In Hebrew, *ASA* means *"(he) made"* or *"(he) did".*

Finally, in Japanese *DAI* is "big", but in Hebrew, it's "enough".

3. CONCLUSION

Many more examples could be given. There are literally dozens of one-syllable words that mean one thing in Hebrew and another (or several others, since Japanese has many one-syllable homophones) in Japanese. Each language has a number of sounds that the other lacks, which limits the number of potential homophones between the two. Even so, it is impressive that a list like the one presented here should exist.

504   *The Wisdom of Solomon*

8/1968

134

The Periodic Table of the Alphabet

DR. SOLOMON W. GOLOMB
Los Angeles, California

If that bearded nineteenth-century Russian, Dmitri Mendeleev, had turned his attention to the Science of Linguistics, instead of to the pseudo-scientific cult of Alchemy, then rather than merely discovering the Periodic Table of the Elements, he might have been led to invent the following "Periodic Table of the Alphabet."

Cols: Rows	V_b	IV_b	I	II	III	IV_a	V_a
1			^{1}A	^{2}B	^{3}C		
2		^{4}D	^{5}E	^{6}F	^{7}G	^{8}H	
3		^{9}	^{10}I	^{11}J	^{12}K	^{13}L	
4	^{14}M	^{15}N	^{16}O	^{17}P	^{18}Q	^{19}R	^{20}S
5		^{21}T	^{22}U	$^{23-24}$V*	^{25}X	^{26}Y	^{27}Z

* The rare—labial series ^{23}V | ^{24}W

FIGURE 1.—The Periodic Table of the Alphabet, Preliminary Form.

Note that Column I consists of the 5 vowels: A, E, I, O, U. No less striking is the group of 5 velar consonants (the K-family) in Column III: C, G, K, Q, X. Column IV_a contains the semi-vowels (H, L, R, Y) and Column IV_b the dental consonants (D, N, T). Note that Column V_a consists of the sibilants, S and Z. Finally, Column II features the labials or lip consonants (B, F, P) and the "rare-labial series" of V and W (called "double vee" in most languages). The fact that J *is* in Column II, whereas M is *not*, is clearly a case of experimental error, which is readily rectified by the simple expedient of interchanging the positions of J and M!

The Periodic Table of the Alphabet

We are now ready for the acid test (to borrow a chemical term) of our new theory. We shall attempt to plug the gap, to "fill hole in the dental column" as it were (no pun intended), in Position 9. There are other dental-consonant sounds in English, such as the one rendered by "th," but they lack a special symbol of their own. However, if we look back to the Greek alphabet, we find that between H and I (called "eta" and "iota" in Greek, and actually written H and I) there occurs the letter θ, "theta," the missing "th"! Our new theory is a success. (The old Semitic alphabet, from which both the Greek and the Roman were adapted, had as its first ten letters the ancestors of positions 1 to 10 in our revised chart, including as #9 the dental letter "teth," between #8 "heth" and "10 "yod"). Our revised table appears as Figure 2.

Cols: Rows	VI	I	II	III	IV	V	
1		¹ A	² B	³ C			I = vowels
2	⁴ D	⁵ E	⁶ F	⁷ G	⁸ H		II = labials
3	⁹	¹⁰ I	¹¹ M	¹² K	¹³ L	¹⁴ J	III = velars
4	¹⁵ N	¹⁶ O	¹⁷ P	¹⁸ Q	¹⁹ R	²⁰ S	IV = semi-vowels
5	²¹ T	²² U	²³ V ²⁴ W	²⁵ X	²⁶ Y	²⁷ Z	V = sibilants
							VI = dentals

Figure 2.—The Periodic Table of the Alphabet, Revised Form.

Even the format of Figure 2 is tentative. Perhaps there is another sibilant to be placed above "J." Perhaps W is not truly a separate letter, but merely an isotope of V. In French, for example, they sound exactly alike. (One wag has suggested that the entry in row 5, column II indicates that we have not yet gotten all the *bugs* out of our system. Bah!) But these are relatively minor details, which will surely be resolved during the course of our next Federal Research Grant. However, there is a bold new direction for our future research, which cannot help but excite the imagination—the next triumph for our new theory will be that we shall use it to predict the properties of the letters in Row 6! Once they have been predicted, we are confident that these new, heavier, and probably unstable letters will be discovered individually, by diligent experimental search.

From May 2016 *Word Ways*

Solomon Golomb, the author of "Amalgamate, Chemist!" in this issue, notes that he is an anagrammatic twin of Gary S. Bloom, author of "Ensnaring the Elusive Eodermdrome" in August 1980: S.GOLOMB has the same letters as G.S.BLOOM. Bloom at one time was a graduate student of Professor Golomb, and they coauthored several papers. Bloom, however, had trouble convincing people that he really existed, for readers assumed that G.S.BLOOM was merely a Golombian pseudonym!

Word Ways Selections 507

Johns Hopkins Magazine – Golomb's Gambits: Word Expansion

Golomb's Gambits: Word Expansion

DECEMBER 2, 2009 | BY SOLOMON GOLOMB

A. It is often easy to add a letter inside a short word to form a very different new word. Thus CAT can be lengthened to CHAT or CART or CAST, etc. There are fewer choices, and often none at all, for lengthening a long word. In the following, you get a clue for the shorter word, the letter to be inserted, and a clue for the longer word. The shorter words gradually increase in length from six to 10 letters as you proceed. See how many you can discover.

1. Add C to "defective" to get a "mental capability."

2. Add R to "on a current subject" to get "like a hot region."

3. Add L to "looking intently" to get a type of "bird."

4. Add L to an "outsider" to get a type of "murderer."

5. Add R to a "male gland" to get "lying horizontally."

6. Add C to a "supplement" to get "causing dependence."

7. Add L to a "runs aground" to get a type of "bottom fish."

8. Add E to a "military rank" to get "having a physical body."

9. Add T to an "attempt to reach a settlement" to get "deep thought."

10. Add R to "leaving no will" to get a "major highway."

11. Add R to "not merited" to get "inadequately provided for."

12. Add T to "imitation" to get "arousal."

B. The same word can be tacked on to the end of each word in each of the following lists to form very different new words. The added word is different for each list. Find the words to be adjoined.

1. HE, IMP, RAMP, REST.

2. CAR, DAM, DO, STAG.

3. COOPER, DESIGN, LITER,

PRIM.

4. CAST, GENE, MODE, PI.

508   *The Wisdom of Solomon*

The following piece appears in *Word Ways*: Vol. 41 : Iss. 2 , Article 17, 2009.

LOMOROCK

SOLOMON WOLF GOLOMB
Los Angeles, California

Guess what Sol's favorite vowel is! He claims this is "very poetic. Translation: It doesn't make much sense!" Can readers supply the sense?

VOODOO MOODS

Don't go to Tomorrow's Door,
Nor look down on London's poor- -
Frodo knows Toronto's flood- -
Otto's grotto- -pools of blood;
Fog rolls on to Sorrow's Moor.

Solomon Wolf Golomb

© 2023 World Scientific Publishing Company
https://doi.org/10.1142/9789811234378_0055

SOLOMON W. GOLOMB

Mathematics After Forty Years of the Space Age[*]

hen I was a graduate student at Harvard, in the early 1950s, the question of whether anything that was taught or studied in the Mathematics Department had any practical applications could not even be asked, let alone discussed. This was not unique to Harvard. Good mathe-

matics had to be *pure* mathematics, and by definition it was not permissible to talk about possible applications of pure mathematics.

This view was not invented by G.H. Hardy (1877–1947), but he was certainly one of its most eloquent and influential exponents. In *A Mathematician's Apology* (Cambridge U. Press, 1940) he wrote (p. 29), "Very little of mathematics is useful practically, and . . . that little is comparatively dull"; and (p. 59), "The 'real' mathematics of the 'real' mathematicians, the mathematics of Fermat and Euler and Gauss and Riemann, is almost wholly 'useless' "; and (p. 79), "We have concluded that the trivial mathematics is, on the whole, useful, and that the real mathematics, on the whole, is not." In order to force external reality into his rhetorical model, Hardy decided to include leading theoretical physicists in his canon of "real" mathematicians, but to justify this by saying that their work had no real utility anyway. Thus, he wrote (p. 71), "I count Maxwell and Einstein, Eddington and Dirac among 'real' mathemati-

cians. The great modern achievements of applied mathematics have been in relativity and quantum mechanics, and these subjects are, at present at any rate, almost as 'useless' as the theory of numbers."

Remember that this was in 1940; and Hardy also wrote (p. 80), "There is one comforting conclusion which is easy for a real mathematician. Real mathematics has no effects on war. No one has yet discovered any warlike purpose to be served by the theory of numbers or relativity, and it seems very unlikely that anyone will do so for many years." He also asserted (pp. 41–42), "Only stellar astronomy and atomic physics deal with 'large' numbers, and they have very little more practical importance, as yet, than the most abstract pure mathematics." Today, 50 years after Hardy's death, it seems incredible that a book so at odds with reality was so influential for so many years.

It is ironic that Hardy's *Apology* was in fact not directed to mathematicians at all. After the dreadful carnage of World War I, and the realization that "the War to end Wars"

An earlier version, titled *Mathematics Forty Years after Sputnik*, appeared in *American Scholar*, Spring 1998. Permission to reprint portions of that article is gratefully acknowledged.

38 THE MATHEMATICAL INTELLIGENCER © 1999 SPRINGER-VERLAG NEW YORK

---

[*]This chapter was originally published in *The Mathematical Intelligencer*, Vol. 21, No. 4, 1999, pp. 38–44.

510   *The Wisdom of Solomon*

hadn't really changed the world, pacifism was very widespread in England, and was effectively the established religion at Oxbridge between the Wars. The extreme attempts by Stanley Baldwin and Neville Chamberlain—who between them occupied 10 Downing Street from 1935 to 1940—to avoid antagonizing Hitler can only be understood in this context. It was primarily to the non-scientists at Oxford and Cambridge that Hardy wanted to proclaim the *harmlessness* of mathematics. Hardy indicates (p. 14) that the *Apology*, in 1940, was an elaboration of his inaugural lecture at Oxford, in 1920, when the revulsion at the horrors of war would have been particularly vehement; and that he was reasserting his position that "mathematics [is] harmless, in the sense in which, for example, chemistry plainly is not" (p. 15).

Chemistry, responsible for the poison gases and disfiguring explosives of the Great War, is Hardy's chief example of a "useful" science, closely followed by *Engineering*, which does helpful things like building bridges, but destructive things as well, like designing warplanes and other munitions. In *A Mathematician's Apology*, Hardy is anxious to persuade his readership that "real" mathematics (especially the kind done by Hardy himself) is a noble esthetic endeavor, akin to poetry, painting, and music, and has nothing in common with merely "useful" subjects like chemistry and engineering, which are also destructive in the service of warfare. A mere two years later, after the "blitz" bombing of London, Hardy's pacifist audience in England would have almost completely disappeared; but as a Mathematician's Manifesto, his *Apology* remained influential in mathematical circles for decades.

> "If nature were not beautiful, it would not be worth knowing, and if nature were not worth knowing, life would not be worth living."

David Hilbert (1862–1943), regarded by many as the leading mathematician of the first four decades of the twentieth century, and who largely defined the agenda for twentieth-century mathematics with his famous list of twenty-three outstanding unsolved problems, presented at the International Congress of Mathematicians in Paris in 1900, largely shared and advocated the view advanced in Hardy's *Apology*. However, Hilbert's list had several problems motivated by numerical analysis, and one asking for a proper, rigorous mathematical formulation of the laws of physics. Coming just ahead of the discovery of relativity and quantum mechanics, this problem led to interesting mathematical work in directions Hilbert could not have anticipated, but in which he actively participated.

Another famous professor at Göttingen during the Hilbert epoch was Felix Klein, who had a much broader appreciation of applications. According to a famous story, a reporter once asked Klein if it was true that there was a conflict between "pure" and "applied" mathematics. Klein replied that it was wrong to think of it as a conflict, that it was really a complementarity. Each contributed to the other. The reporter then went to Hilbert, and told him,

"Klein says there's no conflict between pure and applied mathematics." "Yes," said Hilbert, "of course he's right. How could there possibly be a conflict? The two have absolutely nothing in common." Since Hilbert, unlike Hardy, did work in areas of mathematics with obvious applications, and if the quote is authentic rather than apocryphal, the fundamental distinction he may have seen between pure and applied mathematics would likely have involved motivation—do we study it because it is beautiful or because it is useful?

Hilbert's illustrious contemporary and leading rival for the title of "greatest mathematician of the age" was Jules Henri Poincaré (1854–1912), a cousin of Raymond Poincaré (thrice Premier of France between 1912 and 1929, and President of France for seven years that included World War I). There is no question that Henri Poincaré worked in some of the most obviously applicable areas of mathematics. Yet even Poincaré asserted: "The scientist does not study nature because it is useful; he studies it because he delights in it, and he delights in it because it is beautiful. If nature were not beautiful, it would not be worth knowing, and if nature were not worth knowing, life would not be worth living." Many leading scientists who have made major practical discoveries would share this view, but it is significantly different from Hardy's message. Nowhere does Poincaré suggest that *applicable* science, or *useful* mathematics, is in any way inferior, but rather that the systematic study of nature turns out to be inherently beautiful.

Through the ages, the very greatest mathematicians have always been interested in applications. That was certainly true of E.T. Bell's "three greatest mathematicians of all time": Archimedes, Newton, and Gauss. It was equally true of Euler, Lagrange, Laplace, and Fourier. Even in the first half of the twentieth century, it was true of Hermann Weyl, Norbert Wiener, and John von Neumann. As we come to the end of the twentieth century, the earlier insistence on the desired inapplicability of pure mathematics seems almost quaint, though one lingering legacy is that the label "applied mathematics" retains a pejorative taint and an aura of non-respectability in certain circles.

I want to examine the questions of when and how the concept of inviolable purity became entrenched in many departments of mathematics by the end of the nineteenth century, and what has happened in the past 40 years to weaken this presumption.

In the United States, the beginning of the modern research university dates back only to 1876, with the founding of Johns Hopkins, which was based on a German model that was only a few decades older. Prior to this period, the modern division of knowledge into departments and disciplines was much less rigid. In Newton's day, the term "natural philosophy" covered all of the natural sciences. The chair which Gauss (1777–1855) held at Göttingen was in

Mathematics After Forty Years of the Space Age

Astronomy. Only when there were separate, clearly defined departments of mathematics was it necessary to invent a rationale to support their independence from either established or newly emerging fields which sought to *apply* mathematics. On the other hand, it was *not* necessary to justify the notion that every university *needed* a Department of Mathematics.

From the time of Plato's Academy, all through the Middle Ages, and into the rise of post-medieval universities, mathematics had always been central. The traditional "scholastic curriculum" consisted of two parts: the more elementary *trivium*, with its three language-related subjects—*logic, grammar*, and *rhetoric*; and the more advanced *quadrivium*, with its four mathematics-related subjects—*geometry, astronomy, arithmetic* (i.e., number theory), and *music* (i.e., harmonic relationships). At a time when Latin and Greek were indispensable parts of a university education, no one would have remotely considered eliminating mathematics as "impractical." Those students seeking a liberal university education, whether at Oxbridge in the U.K. or in the Ivy League in the U.S., were not thought to be concerned with learning a trade and earning a living. That came much later. And high-budget research, with the concomitant requirement to set funding priorities, was not yet a part of the university scene.

So, in the late nineteenth century, university mathematics departments had a firm franchise to exist, and considerable latitude to define themselves. Much was happening in mathematics at that time (as well as ever since). The abstract approach was being applied, especially to algebra. The algebraic approach was being applied, especially to geometry and topology; analytic function theory was in full bloom; and a new standard of rigor had emerged. In many areas, mathematics was running so far ahead of applications that it was widely assumed that most of these fields would never have any. This was also true of certain classical areas, like number theory, which was developing rapidly as the beneficiary of new techniques from function theory and modern algebra, and had no foreseeable applications. Rather than be apologetic about the lack of applications for many areas, leading mathematicians and mathematics departments decided to turn this possible defect into a virtue. (In this, they anticipated a basic tenet of Madison Avenue: "If you can't fix it, feature it.")

In fact, the best mathematics consistently found very important applications, but often not until many decades later. Riemann's "clearly inapplicable" non-Euclidean differential geometry (since everyone was certain that we live in a Euclidean universe), from the 1850's, became the mathematical basis for Einstein's General Theory of Relativity some 60 years later. Purely abstract concepts in group theory from around 1900 became central to the quantum mechanics of the 1930's and 1940's, and to the particle physics of the 1950's and onward. Finite fields, invented by Evariste Galois, who died in 1832, were considered the purest of pure mathematics, but since 1950 they have become the basis for the design of error-correcting codes, which are now used indispensably in everything from computer data storage systems to deep space communications to preserving the fidelity of music recorded on compact disks. George Boole's nineteenth-century invention of formal mathematical logic became the basis for electronic switching theory, from 1940 onward, and in turn for digital computer design.

Hardy's most precious area of inapplicable pure mathematics was prime number theory. Edmund Landau, in his *Vorlesungen über Zahlentheorie* ("Lectures on Number Theory", Leipzig, 1927), quotes one of his teachers, Gordan, as frequently remarking, "Die Zahlentheorie ist nutzlich, weil man nämlich mit ihr promovieren kann." ("Number Theory is useful because you can get a Ph.D. with it.") Today, the most widely used technique for "public key cryptography" is the so-called RSA (Rivest, Shamir, and Adleman) algorithm, which depends on several theorems in prime number theory, and the observation that factoring a very large number into primes (especially if it is a product of only two big ones) is much harder than testing an individual large number for primality.

I think it is fair to say that in a very special sense, Number Theory has become a type of applied mathematics, and I'm not referring to number theory's applications in communication signal design and cryptography. Rather, I refer to the fact that Number Theory, which has rather limited methods of its own, has been the beneficiary of powerful applications to it from analytic function theory, from modern algebra, and most recently from algebraic geometry, as with Andrew Wiles's proof of "Fermat's Last Theorem."

In 1940, *topology* would have been high on most people's lists of inapplicable mathematics. Within topology, *knot theory* would have seemed particularly useless. Yet today, knot theory has extremely important applications in physics (to both quantum mechanics and superstring theory) and in molecular biology (to the knotted structures of both nucleic acids and proteins). The topology of surfaces is also much involved in superstring theory, including the structures which superstrings may take in multidimensional spaces. Even graph theory, the "trivial" one-dimensional case of topology, has blossomed into a major discipline where the boundary between "pure" and "applied" is virtually invisible. Until recently, tiling problems were largely relegated to the domain of "recreational mathematics" (an obvious oxymoron to most non-mathematicians, but a pleonasm to true believers). Then, a decade or

> Rather than be apologetic about the lack of applications for many years, leading mathematicians and mathematics departments decided to turn this possible defect into a virtue.

The Wisdom of Solomon

so ago, Roger Penrose's work on small sets of tiling shapes which can be used to tile the entire plane, but only non-periodically, was found to underlie the entire vast field of "quasicrystals."

I could give many, many more examples of how topics and results from the "purest" areas of mathematics have found very important applications, but I believe I have made my point. It may still be necessary for some Mathematics Departments to defend themselves from being turned into short-term providers of assistance to other disciplines which are consumers rather than producers of mathematics; but the basic principle that good "pure" mathematics is almost certain to have very important applications *eventually* is now widely recognized. For most mathematicians today, the distinction that matters is between *good* mathematics and *bad* mathematics, not between *pure* mathematics and *applied* mathematics. To be fair to Hardy, this was the distinction he was trying to make in *A Mathematician's Apology* between "real" mathematics and "trivial" mathematics. Where he went off the deep end was in trying to insist that real mathematics is useless, and that useful mathematics is trivial.

Like many of my generation, I was attracted to mathematics not by Hardy's *Apology*, but by E.T. Bell's *Men of Mathematics*, and my early interest in number theory was partly motivated by the accessibility of the subject. The first mathematics book I ever bought, with my own almost non-existent disposable income while I was still in high school, was Carmichael's thin volume *Theory of Numbers*, in hard cover. Two years later, I was systematically reading Landau's *Vorlesungen über Zahlentheorie*, still on my own. When I came to Harvard as a graduate student, I already assumed that I would do a thesis in prime number theory. This was a respectable branch of mathematics at Harvard, although none of the faculty there specialized in it. David Widder, who had recruited me to be his student my very first day of classes at Harvard, and who included an analytic proof of the Prime Number Theorem in his book *The Laplace Transform*, was happy to sponsor my efforts. He had spent time as a post-doc of Hardy and Littlewood in Cambridge, where the highlight of his sojourn was attending a cricket match seated between these two famous gentlemen.

For help and inspiration, I drove to the Institute for Advanced Study in Princeton one morning in October, 1953, quite unannounced, and went to see Atle Selberg. I had discovered an identity involving von Mangoldt's lambda function, which I thought could be useful in analytic number theory. My identity looked slightly like Selberg's Lemma in his famous Elementary Proof of the Prime Number Theorem. His first reaction was to think that my identity was false. Trying to disprove it, he convinced himself in the next ten minutes that it was true. He then spent the rest of the day with me, exploring ways I proposed to use this identity, and making many helpful suggestions. I learned only later that Selberg had a reputation for being totally reserved and unapproachable.

The other number-theorist who was very helpful was Paul Erdős, who was always totally approachable if you wanted to talk about mathematics. I learned only years later of the supposed feud between Selberg and Erdős. Years later still, in November, 1963, I saw them both in the same room at a Number Theory Conference at Caltech. It was the day John Kennedy was assassinated. I remember the session chairman announcing the news flash that JFK was dead. After no more than half a minute, the meeting resumed as before. I suspect it was the only activity in the whole country that Friday afternoon that wasn't shut down. I mentioned this long afterward to an insightful mathematician friend, and I commented that these mathematicians hadn't reacted to the news of JFK's death. "No, you don't understand," he told me. "They're mathematicians. That *was* their reaction."

I could have finished up at Harvard in the spring of 1955, in time for my twenty-third birthday, but having been awarded a Fulbright fellowship for study in Norway, I decided to finish my thesis writeup there. I wasn't even sure who was still active in Norway when I applied for the fellowship. There were many famous Norwegian mathematicians, but I knew that Niels Henrik Abel, Sophus Lie, and Axel Thue were long dead, and that several others, including Osvald Veblen, Einar Hille, Øystein Ore, and of course Atle Selberg had resettled in the United States. From Landau's book I had learned about Viggo Brun's sieve method in prime number theory, but for all I knew Brun was also long dead. Fortunately this was not the case. Brun turned 70 the month I arrived, in June, 1955, but he did not retire until a year later, and he lived well beyond age 90. From Landau's austere *Satz, Beweis* approach, I could prove Brun's Theorem, that the series consisting of the reciprocals of the twin primes is either finite or convergent, but I had no understanding of what motivated it, or why it worked. It was only when Brun explained his method to me that it made sense. I included a sieve-derived result in my thesis, which I also published in *Mathematica Scandinavica*. A few years later, Erdős got an improvement on my result, which he published in the *Australian Journal of Mathematics*, in a paper titled "On a Problem of S. Golomb." Dozens of people have published papers titled "On a Problem of Erdős," but since Erdős published this one, "On a Problem of S. Golomb," I claim that my "Erdős number" is *minus one.*

For four summers while a graduate student, I worked at the Martin Company, now part of Lockheed-Martin, and became quite interested in mathematical communication theory, including Shannon's Information Theory, and especially "shift register sequences," which were of interest for a variety of communications applications, but which I discovered were modeled by polynomials over finite fields. Shannon's epic paper, "A Mathematical Theory of Communication," was published in 1948, the year after Hardy died, but had Hardy read and understood it, he would have called it "real" mathematics, secure in the be-

> I claim that my "Erdős number" is *minus one.*

Mathematics After Forty Years of the Space Age

lief that it was not really "useful." After all, Shannon gave existence proofs that codes could be constructed arbitrarily close to certain bounds, with no hint of how to find such codes. But what has happened in the past fifty years is that mathematicians have worked closely with communications engineers to develop Information Theory and Coding in a way that is simultaneously first-rate "real" mathematics and eminently practical and useful engineering. In several prominent cases, the same individual has spanned the entire range from developing the theoretical mathematics to designing the practical hardware.

I eventually discovered that an important early paper on linear recurrences over finite fields was published in 1934 by Øystein Ore, but if not for the wide range of applications to several areas of technology, including communications, I don't think we would have a subject classification today in *Mathematical Reviews* called "Shift Register Sequences," corresponding to literally hundreds of published research papers. Perhaps Hardy would have been disturbed to learn how practical the properties of finite fields have become—but then, none of his sacred cows has remained untainted.

When I returned from Norway in the summer of 1956, I came to Southern California to work in the Communication Research Group at the Jet Propulsion Laboratory, in Pasadena. This job, which grew out of the interest I had developed in mathematical communications during my summer jobs at the Martin Company, enabled me to continue my search for applications of "useless" mathematics to practical communications problems. Over the next seven years I formed and headed a group of outstanding young researchers who developed the systems that made it possible to communicate with space vehicles as far away as Neptune (three billion miles from Earth) and beyond.

In 1956, NASA did not yet exist, and JPL was funded by the Ordnance Command of the U.S. Army. They also supported Wernher von Braun's group at Redstone Arsenal in Huntsville, Alabama. The U.S. Army was prepared to launch a small artificial satellite in September, 1956, thirteen months before Sputnik, using a Redstone missile as the launch vehicle, and a small JPL-built payload with a radio transmitter, but General John Bruce Medaris, head of the Army Ballistic Missile Agency, was unable to get the permission of the Eisenhower administration to proceed. Unaware that we were in any kind of race with the Soviet Union, the Eisenhower administration had decided that the U.S. space program should be peaceful, and therefore should not use an Army missile as the launch vehicle. Instead, we had something called Project Vanguard under development, for which the launch vehicle would be a *Navy* missile!

The Soviet Union's launch of Sputnik 1, on October 4, 1957, took the world by surprise. It was visible to the naked eye in the night sky, and a fairly simple radio receiver could pick up its "beep-beep" signal. Around November 12, 1957, General Medaris not only had permission to launch a satellite using Army vehicles, but he had orders to proceed as quickly as possible. Meanwhile, on December 6, 1957, the first launch of a Vanguard satellite was attempted. With the entire world press corps watching, there was a spectacular explosion on the launch pad. The vehicle was consumed in flames from the bottom upward.

Eighty days after the authorization to proceed, the Army satellite was ready for launch. The first stage was a liquid-fuelled elongated Redstone rocket, from Huntsville. The second stage was a cluster of eleven solid-fuel Sergeant missiles from JPL. The third stage used three Sergeant missiles from JPL. And the fourth stage, built at JPL, was a cylinder about 5 feet long and 8 inches in diameter, packed with electronic equipment, and sitting atop a final Sergeant missile. This configuration had never been tested, but the launch of Explorer I, on January 31, 1958, was a success on the very first try. I had a lab at JPL at that time, where I studied the properties of shift register sequences experimentally. During the weeks leading up to Explorer I, a graduate student of James Van Allen was assembling a radiation detector in my lab, with the assistance of my technicians. It was this detector, flying on Explorer I, that "discovered" what came to be known as the Van Allen Radiation Belts around the earth. My own special assignment was to participate in the "early orbit determination" of the satellite.

After Explorer I was launched from Cape Canaveral, no signal was picked up by the down-range station on the Caribbean island of Antigua. We did not know that the signal had been successfully detected and recorded at our stations in Nigeria and Singapore until a few days later, when we were notified by air mail! Amateur radio groups in Australia and Hawaii reported nothing. We were understandably worried that we had lost our satellite. We had three tracking stations widely spaced in Southern California, connected to JPL only by ordinary telephone lines. The nominal time for the satellite to come into radio contact over California came and went, with no detection by any of our stations. Three minutes passed, then another three minutes, and still no detection. There were many long faces in our orbit determination room at JPL. Then, about eight minutes late, all three of our tracking stations called in almost simultaneously. Explorer I was alive and well. One of the upper stages of the launch rocketry of Explorer I had over-performed, slightly enlarging the orbit and lengthening the period, and incidentally increasing its lifetime in orbit. That was an exciting time to be at JPL.

While I was at JPL, I learned the real distinction between "pure research" and "applied research." In 1959, the Laboratory director, Dr. William H. Pickering, decided to form an *ad hoc* committee, with representatives from all over JPL, to report on the "research environment" and what could be done to improve it; and he appointed me to chair it. I discovered that every member had strong opinions about what was *pure* research and what was *applied* research, and surprisingly, it had absolutely nothing to do with the subject matter. It all boiled down to this. What *you* want to work on is *pure* research. What *your boss* wants you to work on is *applied* research.

Sputnik shocked the American public. The notion that

42 THE MATHEMATICAL INTELLIGENCER

514 *The Wisdom of Solomon*

the Soviets were ahead of us in rocketry contradicted a basic tenet in Vannevar Bush's influential book *Modern Arms and Free Men*, which argued that a closed non-democratic society like the Soviet Union couldn't possibly develop armaments and advanced weapons as well as we could in the U.S. Bush had developed a very early analog computer, called the "Bush differential analyzer," prior to World War II. During the war, he was Roosevelt's chief advisor on scientific and technological matters. He delivered the letter to FDR, drafted by Szilard and Wigner, and signed by Einstein, which led to the Manhattan Project and the Atomic Bomb. Beyond that, most of the things Vannevar Bush recommended—at least, the ones I am aware of—were ill-advised. For example, because of his bias in favor of analog computing, he delayed research on digital computers until after the war. In *Modern Arms and Free Men*, he not only asserted that the Soviets couldn't possibly come up with first-rate weapons, but also that neither we nor they could ever develop intercontinental ballistic missiles (ICBMs). In 1957, I got a letter from Bush on his letterhead stationery as Chairman of the Board of MIT. Martin Gardner had run an article in *Scientific American* about my "polyominoes," and included my proof that a particular covering of the checker-board with pieces of a certain shape was impossible, which was shown by coloring the board in a particular way. Bush wanted to know why the result would still hold if you *didn't* color the board in that particular way! Naturally I wrote a very polite and patient reply. (By the way, "Polyominoes" is also a subject classification in MR.) But perhaps I am too harsh on Dr. Bush. A newly published Bush biography credits him with creating the post-World-War-II structure of government funding for university research, which puts many of us in his debt.

The extent to which we were behind the Soviets in the development of large missiles was mostly a political matter, and secondarily an engineering issue. Basic science was not really involved at all. Nonetheless, both the public and the politicians were convinced that a much greater commitment to the support and funding of research, especially university research, in all the basic sciences, *including mathematics*, was an urgent national priority. I believe it is very fortunate that this happened, and that it contributed significantly to the U.S. winning the Cold War some 30 years later, but it had no relationship to the issue of whether the U.S. was behind in rocketry.

As you will remember, in *A Mathematician's Apology*, Hardy had contended that "real" mathematics is much more similar to poetry and painting that it is to chemistry or engineering. That is something that many of us, as mathematicians, might still like to believe; but the new government funding didn't extend to poetry and painting. The rationale for including mathematics in the new governmental largess required a comnitment to the principle that basic research in mathematics, like basic research in chemistry or engineering, will *ultimately* have practical, beneficial consequences. Suddenly, there was a reason for trying to show that one's mathematics had practical uses and implications. This was not the only reason for the change in attitude about whether good mathematics could be useful, but it certainly played a part.

Another major influence has been the development of digital technology, which has placed new emphasis on areas of discrete mathematics that were previously considered inapplicable—like finite fields, which I've already mentioned. Then there is Computer Science itself, which asks questions in pure mathematics like finding the computational complexity of various procedures, which turns out to be extremely practical. Another development has been Shannon's mathematical theory of communication, which asks questions motivated by applications, but which are more abstract mathematically than anything in physics. An atom, an electron, a photon, or a quark—these are all entities in the physical world whose behavior the physicist attempts to model. But Shannon's "bit of information" is a purely mathematical concept. It has no mass, no spin, no charge, no momentum—and yet the issues involved in measuring information in bits, in storing information, in moving information from one place to another are so important that we are told that we live in the "Age of Information."

It is also true that science and engineering have changed dramatically in the fifty years since Hardy's death. Semiconductors and lasers make nontrivial use of quantum mechanics, and the people who study them are not restricted to using what Hardy derisively referred to as "school mathematics." Biology at the University level has progressed from butterfly collecting to genome sequencing. It is not at all clear-cut whether "control theory" is a topic in mathematics or a branch of engineering.

I've never called myself an "applied mathematician." When I'm doing mathematics as mathematics I am a mathematician. When I'm focusing on applications to communications, I'm a communications engineer. For the first several years that I worked on mathematical communications problems, I didn't even realize that there were good journals in which new results in these areas could and should be published. That was a lingering aftereffect of my Hardy-style brainwashing. I hope I've finally outgrown it. In fact, I hope we've all outgrown it. Mathematics isn't "good" just because it's inapplicable, and it isn't "bad" just because it is.

In fairness to Hardy, there are many things in *A Mathematician's Apology* with which most mathematicians will agree or identify. Hardy asserts that mathematicians are attracted to the subject by its inner beauty, rather than by any overwhelming desire to benefit humanity. Most mathematicians I know would agree with that. Even more im-

> When I'm doing mathematics as mathematics I am a mathematician. When I'm focusing on applications to communications, I'm a communications engineer.

portant, Hardy identifies himself (p. 63) as a *Realist* (as the term is used in Philosophy) about mathematics. "I will state my own position dogmatically. . . . I believe that mathematical reality lies outside us, that our function is to discover or *observe* it, and that the theorems which we prove, and which we describe grandiloquently as our 'creations', are simply our notes of our observations. This view has been held, in one form or another, by many philosophers of high reputation from Plato onwards. . . . " The great majority of mathematicians share this view *about mathematics*. Plato went overboard, trying to extend mathematical reality to physical reality. Immanuel Kant explicitly distinguished between the "transcendental reality" of mathematics and the (ordinary) reality of the physical universe. My own version of this distinction is that if the Big Bang had gone slightly differently, or if we were able to spy on an entirely different universe, the laws of physics could be different from the ones we know, but 17 would still be a prime number. I recently found a very similar view attributed to the late great Julia Robinson (1919–1985) in the biography *Julia, a Life in Mathematics*, by her sister, Constance Reid. "I think that I have always had a basic liking for the natural numbers. To me they are the one real thing. We can conceive of a chemistry that is different from ours, or a biology, but we cannot conceive of a different mathematics of numbers. What is proved about numbers will be a fact in any universe." This is also reminiscent of the famous dictum of Leopold Kronecker (1823–1891): "Die ganzen Zahlen hat Gott gemacht; alles anderes ist Menschenwerk." ("God made the whole numbers; everything else is the work of man.")

Platonism (i.e., "Realism") about mathematics has dissenters. Some who, in my view, are overly influenced by quantum mechanics, would argue that $2^P - 1$, where P is some very large prime number, is neither prime nor composite, but in some intermediate "quantum state," until it is actually tested. Of course, the Realist view is that it is already one or the other (either prime or composite), and we find out which when we test it. Even less palatable to most mathematicians is the "post-modern" criticism of all of "science," that it is just another cultural activity of humans, and that its results are no more absolute or inevitable than works of poetry, music, or literature. The extreme form of this viewpoint would assert that "$4 + 7 = 11$" is merely a cultural prejudice. I will readily concede the obvious: it requires a reasoning device like the human brain (or a digital computer) to perform the sequences of steps that we call "mathematics". Also, culture can play an important role in determining which mathematical questions are asked, and which mathematical topics are studied. (Our widespread use of the decimal system is undoubtedly related to humans having ten fingers.) What I will *not* concede is that, if the same mathematical questions are asked, the answers would come out inconsistently in another culture, on another planet, in another galaxy, or even in a different universe. For example, the Greeks were interested in "perfect numbers," numbers like 6 (=1 + 2 + 3) and 28 (=1 + 2 + 4 + 7 + 14) which equal the sum of their exact

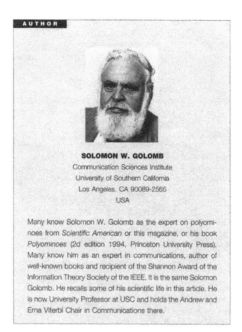

SOLOMON W. GOLOMB
Communication Sciences Institute
University of Southern California
Los Angeles, CA 90089-2565
USA

Many know Solomon W. Golomb as the expert on polyominoes from *Scientific American* or this magazine, or his book *Polyominoes* (2d edition 1994, Princeton University Press). Many know him as an expert in communications, author of well-known books and recipient of the Shannon Award of the Information Theory Society of the IEEE. It is the same Solomon Golomb. He recalls some of his scientific life in this article. He is now University Professor at USC and holds the Andrew and Erna Viterbi Chair in Communications there.

divisors (less than the number itself). I can readily imagine a "civilization" with advanced mathematics in which the notion of "perfect numbers" was never formulated. What I cannot imagine is a civilization in which perfect numbers were defined the same way as we do, but where 28 was no longer perfect.

It was Isaac Newton who wrote, "I do not know what I may appear to the world; but to myself I seem to have been only like a boy playing on the seashore, and diverting myself in now and then finding a smoother pebble or a prettier shell than ordinary, whilst the great Ocean of Truth lay all undiscovered before me." It is hard to give a better formulation of the Realist view of mathematical truth adhered to by most mathematicians.

Newton is firmly entrenched in the pantheons of both mathematics and physics. He certainly erected no artificial boundaries between theory and applications. In the three centuries since he published the *Philosophiae naturalis principia mathematica* (the *Principia*, for short), we have, with electromagnetism and relativity and quantum mechanics, waded deeper into that Ocean of Truth; but Newton's laws of motion and gravitation were actually sufficient for the launching of *Sputnik* and *Explorer*. Today, 40 years after those satellites first circled the earth, most mathematicians—myself included—have moved farther from Hardy's outlook and closer to Newton's.

© 2023 World Scientific Publishing Company
https://doi.org/10.1142/9789811234378_0056

GOLOMB'S REMINISCENCES *

Solomon W. Golomb
University of Southern California, Los Angeles, CA

Today, June 1, 2002, is the first day of my eighth decade. I have always tried hard to come up with clever names like *polyominoes* or *rep-tiles* or *graceful graphs* for the topics that I've introduced. So it's really a failure of my ability as an inventor of names if anything actually gets named after me. There have been a few failures. The things that I called minimum spanning rulers, Martin Gardner without my permission called *Golomb Rulers*. Other people have picked up on that, and I'm sort of stuck with it. I did a lot of work when I was at JPL looking at non-linear shift register sequences. I found a statistical model, and when I looked at cycle lengths of random permutations, one thing I showed was that the expected length of the longest cycle relative to the total number of elements had a limit, and I evaluated that limit. Lloyd Welch and several other people participated in this effort. But it was Donald Knuth who named this particular number *Golomb's constant*. If a constant had to be named for me, I would prefer it to be something like "3", rather than some strange transcendental number, but I guess the better constants were already taken. In a couple of papers around 1967 I asked the question: Suppose you had an infinite number of source messages that were geometrically distributed; you couldn't use Huffman's compression algorithm, because that requires you to start with the least probable message, and if there are an infinite number of possible messages you can't quite do that; but I came up with the equivalent of Huffman codes for that case. This result was subsequently rediscovered by various people. Irwin Jacobs told me he had rediscovered it many years later, but it was only when he tried to get a patent on this family of codes that he learned he had been anticipated. I had proved that these codes were optimal for certain special values of probability, p and $1 - p$. In a later paper by Robert Gallager and D.C. Van Voorhis, they showed that for every probability p between 0 and 1, one of my codes is in fact the optimum code to use. I think in was Gadiel Seroussi who decided to call these *Golomb's codes*.

There was a sequence that I described in 1963 that had the bizarre property that the histogram of the sequence (how often each value occurred) was the same as the original sequence. Then other people picked up on getting asymptotic

---

*This chapter was originally published in *Mathematical Properties of Sequences and Other Combinatorial Structures*, Kluwer Academic Publishers, 2003, pp. 235–245.

SEQUENCES AND COMBINATORIAL STRUCTURES

estimates on the growth rate of that sequence, and in the literature it became known as *Golomb's sequence,* or *Golomb's self-describing sequence.* And finally, two days ago, for the first time I heard what I have called "the lambda method" in analytic number theory, called *Golomb's Lemma* by Keith Conrad in his talk here. That's my current list of my failures in getting things not named after me.

A colleague of mine here at USC, Robert Biller, made an interesting observation, though he probably wasn't the first, namely, that careers exist mostly in retrospect. When you look back, you can figure out what it is that you were doing all of those years. So I'll try to summarize the major areas of research that I've been involved in. These are not in any particular order, but I've certainly done a lot of work on *polyominoes* and *rep-tiles* and similar things that involve tiling in two or more dimensions, a kind of discrete or combinatorial geometry. I've done several things involving numbers on graphs, the most famous probably being *graceful graphs.* My first love was prime number theory; and I'll come back to each of these areas in a bit more detail.

There are various families of specialized codes that I've worked on. *Comma-free codes* were mentioned by Prof. Levenshtein, along with other kinds of codes. Some new things that I thought were fairly interesting at one time included the theory of enumeration, with applications of finite groups. I think it was Gauss who claimed that everything he ever worked on was based on ideas that occurred to him by the time he was 20. I don't think I was nearly that precocious, but it is certainly true that most of the things I've worked on were areas I had gotten interested in by the time I was 25. One exception, something I've gotten interested in mostly in the last 20 years, has been designs based on square patterns like Costas Arrays, Tuscan Squares, and some others. Finally, the longest list comes under the heading of binary sequences. One thing I hope to have contributed to the Information Theory Society (violating one of my basic principles of something that I should never get involved in) began four years ago when I was persuaded by Alex Vardy to become the first "Editor for Sequences" for the *Information Theory Transactions.* The reason I agreed was that I always felt that sequences were short-changed by the IT Group. I don't think the best paper award by the IT Group has ever gone to a paper on sequences, even though for example Welch's paper on the minimum correlation of a set of sequences is certainly one of the more important results of that period.

I thought I'd mention a bit more on the discrete geometry of tiling with polyominoes. One of the strange experiences I have when I travel is people will know me for work in one particular area and be very surprised that I've done work in seemingly totally unrelated areas. Because I published a very early book on *Polyominoes,* many people associate that with me. I never published a book (I still think I ought to) on *rep-tiles,* featuring geometric figures that can be divided into smaller pieces that are similar to the original figure; and even in a book like Grünbaum and Shephard's *Tilings and Patterns*, they have quite a bit on rep-tiles but they never mention my name in connection with it. I decided that's because they read about it in Martin Gardner's column where Gardner attributes it to me, but that's a secondary attribution. I warn people that if you are only referenced by somebody else, that other person may end up being cited by subsequent people. Andy Liu and his colleagues very recently have been the first to take up the paper that I published in the early 1980's called "Normed Division Domains". I had been thinking for decades about the fact that if you think of a tiling figure as *dividing* a larger figure, you can develop a whole theory of division that never involves addition, subtraction and multiplication. This is a theory of division independent of the other operations of arithmetic. If you go back to look at how Euclid developed basic number theory, he has a very geometric viewpoint corresponding to one-dimensional tiling. And this one-dimensional view-point is really the best way to explain to students things like Euclid's algorithm for the greatest common divisor. It's totally unmotivated algebraically, but it's very directly motivated geometrically. You lay off one length against the other length and see what's left over. If there's nothing left over, you have the common divisor. And earlier than Euclid, to the Greeks, in looking at the theory of commensurable and incommensurable magnitudes, it was a great shock when the Pythagoreans discovered that the diagonal of a square and the side of that square are incommensurable. If you approach that with the Euclidean algorithm, it never ends. You never find the "greatest common divisor" of those two lengths.

There are lots of other examples like subgroups of a group, or normal subgroups of a group, where you have the notion of one thing dividing another. The notion of a factor group was well established a long time ago in mathematics, so I thought you could axiomatize this. It turns out you can always talk about greatest common divisors and least common multiples, and Andy Liu and his colleagues actually applied that back to polyominoes, which was one of the things I had been thinking about when I invented this notion.

With numbered graphs, *graceful graphs* have become so well known that it is rarely mentioned that I was the one who coined that name. The work with rulers is closely related to it. Gary Bloom and I have looked at "homometric rulers". Cyclic difference sets that correspond to finite projective planes also correspond to "circular rulers". I presented a paper last year in Bergen with a very clever construction by Herbert Taylor for getting simultaneous sets of rulers where all the measured differences are distinct.

In prime number theory I published a number of papers that I think may have some lasting value. With the benefit of hindsight you can see that analytic number theory as it was developed in the nineteenth century was really trying to get around the fact that there is no legitimate uniform probability distribution

238   SEQUENCES AND COMBINATORIAL STRUCTURES

on the positive integers, because if you want all the positive integers to have
the same probability, either the probabilities sum to infinity if it's greater than
0, or they sum to zero if it's equal to 0. You can't get them to sum to 1. And
probability theory wasn't yet a respectable branch of mathematics in the 19th
century. This is discussed in my papers on probability distributions on the
integers. I also found a number of "formulas for the next prime" and I was
really amazed to discover my name in the index of the fifth and last edition
of Hardy and Wright's *Introduction to Number Theory* (the first time there's
an edition of that book with an Index). I'm certainly not in the same league
with Euclid and Fermat and all those other people that should be referenced in
a basic book on number theory. The fifth edition added a section on formulas
for primes, which the first edition said could never exist. By the time of the
fifth edition many people had found such formulas, and my approach was one
that was cited. What I thought was my best contribution to Prime Number
Theory, but which no one picked up on till very recently, was what I called the
"lambda method". That's what Keith Conrad referred to as *Golomb's Lemma*.
So I was very happy that forty-six years after I'd finished my Ph.D. thesis,
someone finally noticed what I was doing. I also had one early paper proving
the infinite number of primes by a topology argument, and another paper in
which I coined the term "powerful numbers", numbers where every prime that
divides the number divides it to a power higher than the first power. This is
another case where terminology that I introduced has caught on.

In the area of specialized codes, there's the work I've done with Lloyd Welch,
Basil Gordon and others on *comma-free codes*, and *bounded synchronization
delay codes*. I mentioned the things that Seroussi called *Golomb codes*, and
other people have invoked my name in some generalizations of these that I never
thought of. Then there's the work I've done on generalized Barker sequences.
The first paper was with Robert Scholtz. My doctoral student Ning Zhang
has written several very good papers on these and other polyphase codes. And
finally there are Lee Metric codes where I've published a couple of papers, and
Tuvi Etzion mentioned that. Of course my interest in Lee Metric codes was
not unrelated to the work on polyominoes, because the Lee spheres are a type
of polyominoes, and perfect codes are ways of packing and tiling with those
polyominoes.

On what was once referred to as Polya-Burnside enumeration, people keep
finding not only Polya, and Burnside, but even Frobenius, and Cauchy still
earlier, in the 19th century, had all found this formula for counting the number
of orbits. My contribution was if you only wanted to count the number of
maximum sized orbits, there are some nice formulas for that; and it was Gian-
Carlo Rota who told me that my paper was actually the inspiration for his
work on Generalized Möbius Functions. One thing I discovered a few years
ago when teaching an elementary course, and I couldn't believe that no one

Golomb'S Reminiscences

had ever observed this before, was: How can you tell when two elements in a finite group are conjugate elements? If you write out the Cayley table, the multiplication table of the group, two elements in the group are conjugate if and only if they appear in symmetric positions relative to the main diagonal of the group table. It's trivial to prove; if you have the group table, it's trivial to check; and yet for some reason this doesn't seem to have been observed earlier.

In square-array designs, Costas arrays have already been mentioned here, as well as Tuscan Squares and other special cases. One thing that wasn't discussed is my work with Lloyd Welch and Jósef Dénes involving applying a special class of Latin squares for use as cipher tables in cryptography. We filed a patent application a little over two years ago on this idea. We are still waiting for a first office action, the first acknowledgment from the patent office that they've looked at it.

Here's my list of things on binary sequences. I started working with linear shift register sequences in 1953. I found their relationship to polynomials over finite fields. I'm guilty of coining the term *cyclotomic cosets*, because strictly speaking they are not cosets in the group theory sense; and I got interested in two-level correlation binary sequences and their equivalence to cyclic difference sets. I may have been the first person to point out the precise correspondence between the combinatorial cyclic (v, k, λ)-designs and binary sequences with two-level autocorrelation. My interaction with Nadav Levanon about twelve and half years ago when I visited Tel Aviv University enabled me to see that there are things the radar community didn't know in the literature familiar to the Information Theory Group, so I was able to show how to use two-level correlation sequences to design zero-sidelobe radar signals.

It has been gratifying to find my early work on shift register sequences being applied in radar, in cryptography, and in CDMA systems for cellular and wireless communications. I had done a lot of early work on non-linear shift register sequences: How to generate them, what properties they have, etc. I haven't done much further work on that since I was at JPL. I also did closely related work on Boolean functions, and I gave a paper at the 1959 International Symposium on Circuit and Information Theory (held at UCLA) that was published both in the *Information* and *Circuit Theory Transactions*, and which became Chapter 8 of my *Shift Register Sequences* book. That was rediscovered, I'm not sure how independently, 25 years later by Xiao and Massey, and has become known in cryptography as the Xiao-Massey algorithm. It's just that in 1959, one didn't dare talk about the fact that things had applications to cryptography, and I have been trying for quite a number of years, unsuccessfully, to get JPL to declassify my old reports. The work on non-linear sequences also got me interested in the general problem of cycles of random permutations. In cryptographic applications, if you want to combine sequences, you want to combine them in such a way that it would be as difficult as possible to pull them apart and

240 SEQUENCES AND COMBINATORIAL STRUCTURES

recover the pieces; but for the JPL ranging system the objective was exactly the opposite: Finding a way to combine short sequences of relatively prime period, so that the individual components would be as highly correlated as possible with the combined sequence. I gave a talk last week (at the McEliece 60th birthday symposium) where I finally got around to proving something that I never thought to doubt, which is that the majority decision function on an odd number of components is the function that is the most highly simultaneously correlated with these components.

Today we heard talks about properties of the de Bruijn graph. The only thing I'll add is that the first proof of de Bruijn's theorem appeared in the 1890's in a French journal called *L'Intermédiaire des Mathématiciens*, which was a journal for undergraduate math teachers, if not something even more elementary; and in one monthly issue in the year 1892, someone had posed the problem of how many periodic binary sequences are there, where each subsequence of length n occurs exactly once. And later in the same year the problem was solved. de Bruijn felt very strongly that these sequences shouldn't be named after him, but he discovered, as I've discovered, that once people name something for you, it has a life of its own. You can't disown it; you're stuck with it.

As I said, I wasn't going to tell you any new results. I was only going to try to organize this rather chaotic collection of things I've been working on all these years. And I really wanted to thank all of you who have come today. It's a marvellous experience for me to see people from many different aspects of my technical life, and I specially wanted to thank all the people involved in organizing this event, especially Mayumi Thrasher, Milly Montenegro and P. Vijay Kumar. Thank you very much.

1. A Selective Bibliography of S.W. Golomb

1.1 Doctoral Dissertation

1) "Some Problems in the Distribution of the Prime Numbers", May, 1956, Harvard University.

1.2 Journal Publications

1) "Sets of Primes with Intermediate Density", *Mathematica Scandinavica*, Vol. 3, 1955.

2) "Comma-Free Codes" (with Gordon, B., and Welch, L.R.), *The Canadian Journal of Mathematics*, Vol. 10, 1958.

3) "A Connected Topology for the Integers", *The American Mathematical Monthly*, Vol. 66, No. 8, October, 1959.

Golomb'S Reminiscences

4) "On the Classification of Boolean Functions", *IEEE Transactions on Information Theory*, June, 1959.

5) "Influence of Data Processing on the Design and Communication of Experiments", *Radio Science*, September, 1964.

6) "Codes with Bounded Synchronization Delay" (with Gordon, B.), *Information and Control*, August, 1965.

7) "Generalized Barker Sequences" (with Scholtz, R.A.), *IEEE Transactions on Information Theory*, October, 1965.

8) "Tiling with Polyominoes", *The Journal of Combinatorial Theory*, September, 1966.

9) "Run-Length Encoding", *IEEE Transactions on Information Theory*, July, 1966.

10) "Theory of Transformation Groups of Polynomials Over GF (2) with Applications to Linear Shift Register Sequences", *Information Sciences*, December, 1968.

11) "A Class of Probability Distributions on the Integers", *Journal of Number Theory*, Vol. 2, No. 2, May, 1970.

12) "The Lambda Method in Prime Number Theory", *Journal of Number Theory*, Vol. 2, No. 2, May, 1970.

13) "Powerful Numbers", *The American Mathematical Monthly*, Vol. 77, No. 8, October, 1970.

14) "Tiling with Sets of Polyominoes", *The Journal of Combinatorial Theory*, July, 1970.

15) "Perfect Codes in the Lee Metric, and the Packing of Polyominoes" (with Welch, L.R.), *SIAM Journal on Applied Mathematics*, Vol. 18, No. 2, January, 1970.

16) "A Direct Interpretation of Gandhi's Formula", *The American Mathematical Monthly*, August-September, 1974.

17) "Formulas for the Next Prime", *Pacific Journal of Mathematics*, Vol. 63, No. 2, pp. 401-404, 1976.

18) "On the Classification of Balanced Binary Sequences of Period $2^n - 1$", *IEEE Transactions on Information Theory*, Vol. IT-26, No. 6, 730-732, November, 1980.

242   *SEQUENCES AND COMBINATORIAL STRUCTURES*

19) "Obtaining Specified Irreducible Polynomials Over Finite Fields", *SIAM Journal on Algebra and Discrete Methods*, Vol. 1, No. 4, December, 1980.

20) "Normed Division Domains", *The American Mathematical Monthly*, Vol 88, No. 9, pp. 680-686, November, 1981.

21) "On the Characterization of PN Sequences" (with Cheng, U.), *IEEE Transactions on Information Theory*, Vol. IT-29, No. 4, July, 1983.

22) "Algebraic Constructions for Costas Arrays", *The Journal of Combinatorial Theory, Series A*, No. 37, pp. 13-21, July, 1984.

23) "Constructions and Properties of Costas Arrays" (with Taylor, H.), *Proceedings of the IEEE*, Vol 72, No. 9, pp. 1143-1163, September, 1984.

24) "Tuscan Squares - A New Family of Combinatorial Designs" (with Taylor, H.), *Ars Combinatoria*, Vol. 20-B, December, 1985.

25) "A New Result on Comma-Free Codes of Even Wordlength" (with Tang, B., and Graham, R.L.) *Canadian Journal of Mathematics*, Vol. 39, No. 3, pp. 513-526, December, 1987.

26) "Tuscan-K Squares" (with Taylor, H., and Etzion, T.), *Advances in Applied Mathematics*, No. 10, pp. 164-174, 1989.

27) "Polyominoes Which Tile Rectangles", *Journal of Combinatorial Theory, Series A*, Vol. 51, No. 1, pp. 117-124, May 1989.

28) "Sixty-Phase Generalized Barker Sequences" (with Zhang, N.), *IEEE Transactions on Information Theory*, Vol. 35, No. 4, pp. 911-912, July-August 1989.

29) "'Periods' of de Bruijn Sequences," *Advances in Applied Mathematics.*, vol. 13, no. 2, pp. 1181-1183, May, 1992.

30) "Two-Valued Sequences with Perfect Periodic Autocorrelation," *IEEE Transactions on Aerospace and Electronics Systems*, vol. 28, no, 2, pp. 383-386, April, 1992.

31) "Probablity, Information Theory, and Prime Number Theory," *Discrete Mathematics*, vol. 106/107, pp. 219-229, September, 1992.

32) "Polyphase Sequences with Low Autocorrelations" (with Zhang, N.), *IEEE Transactions on Information Theory*, vol. IT-39, no. 3, pp. 1085-1089, May, 1993.

Golomb'S Reminiscences 243

33) "On N-Phase Barker Sequences" (with Chang, N.), *IEEE Transactions on Information Theory*, vol. 40, no.4, pp. 1251-1253, July, 1994.

34) "A Symmetry Criterion for Conjugacy in Finite Groups", *Mathematics Magazine*, vol. 69, no. 5, pp. 373-375, December, 1996.

35) "The polynomial model in the study of counterexamples to S. Piccard's 'Theorem'" (with G. Yovanof), *Ars Combinatoria,* vol. 48, April, 1998.

36) "Binary Sequences with Two-Level Autocorrelation", (with G. Gong) *IEEE Trans. on Information Theory*, vol. 45, no. 2, March, 1999, 692-693.

37) "Exhaustive Determination of (1023, 511, 255)-Cyclic Difference Sets", (with P. Gaal) *Mathematics of Computation*, vol. 70, no. 233, pp. 357-366, March 1, 2000.

1.3   Convention Records

1) "On the Classification of Boolean Functions", *Proceedings of the 1959 Symposium on Circuit and Information Theory*, Los Angeles.

2) "Mathematical Theory of Discrete Classification", in *Information Theory, Proceedings of the Fourth (1960) London Symposium*, Colin Cherry, Editor, Butterworths, London, 1961.

3) "Arithmetica Topologica", *Proceedings of the (1961) Prague Symposium on General Topology and Its Applications*, Academia, Prague, 1962.

4) "The Information Generating Function of a Probability Distribution", *Transactions of the Fourth Prague Conference (1965) On Information Theory*, Academia, Prague, 1967.

5) "Algebraic Coding and the Lee Metric" (with Welch, L.R.), *Error Correcting Codes*, edited by H.B. Mann, John Wiley & Sons, New York, 1968.

6) "Numbering the Nodes of a Graph", *Symposium on Combinatorial Analysis and Computing*, Kingston, Jamaica, 1969.

7) "Irreducible Polynomials, Synchronization Codes, Primitive Necklaces, and the Cyclotomic Algebra", *Combinatorial Mathematics and Its Applications*, edited by R.C. Bose and T.A. Dowling, University of North Carolina Press, Chapel Hill, 1969.

8) "Shift Register Sequences – Solved and Unsolved Problems", *Shannon Lecture, International Symposium on Information Theory*, Brighton, England, June, 1985.

244   *SEQUENCES AND COMBINATORIAL STRUCTURES*

9) "Probability Distributions on the Integers and Formulas for Primes", *London Symposium on Analytic Number Theory*, July, 1985.

10) "Constructions and Properties of Tuscan Squares," invited paper, *Eleventh British Combinatorial Conference*, Goldsmith's College, London, July, 1987.

11) "Applications of Probability to Number Theory," *Proceedings of the Third Petrozavodsk Conference on Applications of Probability to Discrete Mathematics*, TVP Science Publishers, Moscow, Russia, 1992.

12) "Shift-Register Sequences and Spread-Spectrum Communications", Keynote Address, *IEEE Third International Symposium on Spread Spectrum Techniques & Applications*, Oulu, Finland, July 4-6, 1994. (Extended Abstract published in Conference Program.)

13) "The Use of Combinatorial Structures in Communication Signal Design", Proceedings of the *IMA Conference on the Applications of Combinatorial Mathematics*, C. Mitchell, Editor, 1996.

14) "Discovery of New Families of Cyclic Hadamard Difference Sets," Institute for Informatics, University of Bergen, Norway, July 14, 1997.

15) "Cyclic projective planes, perfect circular rulers, and good spanning rulers" (with H. Taylor), *Conference on Sequences and Their Applications*, Bergen, Norway, May, 2001.

1.4   Books

1) **(a)** DIGITAL COMMUNICATIONS WITH SPACE APPLICATIONS, Prentice-Hall, Inc., Englewood Cliffs, N.J., 1964. (Portions also authored by L. Baumert, M. Easterling, J. Stiffler, and A. Viterbi.)

    (b) DIGITAL COMMUNICATIONS WITH SPACE APPLICATIONS, Reprint Edition, Peninsula Publishing Co., January, 1982.

2) **(a)** POLYOMINOES, Charles Scribner's Sons, New York, 1965.

    (b) POLYOMINOES, George Allen and Unwin, Ltd., London, 1966.

    (c) POLYOMINOES, Russian Translation, Moscow, 1975.

    (d) POLYOMINOES, Revised and Expanded Second Edition, Princeton University Press, 1994.

    (e) POLYOMINOES, Princeton University Press, Paperback Edition of (d). 1996.

Golomb'S Reminiscences 245

3) **(a)** SHIFT REGISTER SEQUENCES, Holden-Day, Inc., San Francisco, 1967. (Portions co-authored by L. Welch, R. Goldstein, and A. Hales.)

   (b) SHIFT REGISTER SEQUENCES, Revised Edition, Aegean Park Press, May, 1982.

4) BASIC CONCEPTS IN INFORMATION THEORY AND CODING, Plenum Publishers, 1994. (Co-authored with R.E. Peile and R.A. Scholtz.)

2.   Other References
2.1   Books

1) *An Introduction to the Theory of Numbers, Fifth Edition*, G.H. Hardy and E.M. Wright, Oxford University Press, 1979.

2) *Tilings and Patterns*, B. Grünbaum and G.C. Shephard, W.H. Freeman and Co., 1987.

2.2   Journal Article

1) "Optimal source codes for geometrically distributed alphabets", R.G. Gallager and D.C. Van Voorhis, *IEEE Trans. on Information Theory*, vol. IT-21, pp. 228-230, March, 1975.

USC Engineer From JPL to USC and Beyond*

---

*This chapter was originally published as the cover story in the Fall/Winter 2004 issue of *USC Engineer* magazine.

COVER STORY

I n the smoggy 1950s at Pasadena's Jet Propulsion Laboratory, there were days when you could not see the mountain from the junior scientists' trailers at the base of the Arroyo Seco.

But inside those trailers, a rare clarity of thought emerged creating an unlikely alliance of mathematicians and engineers. Bridging disciplines that were as easily mixed as oil and water, a small progressive group laid the foundations of the digital age.

They were brilliant minds in their early twenties, fresh out of school and culled from every corner of the country. Later, most would join USC's engineering school, turning it into a leading presence in digital communications and information theory. One would put his name on that school. At the time, however, academic honors were far from their minds. The race for space was on, and the Soviets were in first place.

In 1957, William Lindsey, now a professor at the USC Viterbi School of Engineering, was an electrical engineering junior at Purdue University when the launch of *Sputnik* shocked America. Every 96 minutes the small aluminum can flew over the United States, and with every pass the country's reaction grew more dire. What was next: nukes in space?

More intrigued than alarmed, Lindsey hooked up a radio to a speaker to see if he could pick up any signals. There it was — a steady beep in the ham frequency range, put there by the Soviets to make sure every American got the message. Lindsey knew the signal came from *Sputnik*, because it was Doppler-shifted: that is, the beep rose in pitch as the satellite approached and dropped down as it receded, every 96 minutes, exactly.

"When I hooked the speaker up to this audio signal and we could hear the Doppler shift, the whole university came running. It spread like wildfire through campus. The whole student body came to listen to this tone," Lindsey remembers, still wide-eyed.

He realized two things: anything involving space was going to be a hot area for career-minded scientists; and, if you could pick up a signal from space, you could also modulate that signal and use it to carry information. It was at that moment Lindsey decided to become a communications engineer.

"I knew there was something there," he says.

Soon Lindsey joined the trailer group, formally known as Communications Section 331 in JPL's Division 33. It was home to the theorists, and had been their home for some time. Lindsey was one of its youngest members.

There was Eberhardt Rechtin, now an emeritus professor at USC, but then the long-serving head of Division 33 and in his words "chief architect of the whole shebang." He was in charge of the systems for space communications, networks and devices.

There was mathematician Solomon Golomb, chief of Section 331

MAN'S STEP ON THE FIRST RUNG OF SPACE TRAVEL, THE SOVIET UNION LAUNCHED AN 184-POUND *SPUTNIK* VEHICLE LIKE THIS MODEL INTO EARTH'S ORBIT, OCTOBER 4, 1957. LEFT, PERIOD POSTER FROM JPL CELEBRATING EXPLORER I.

and father of the shift-register sequences used universally in message coding. Golomb now holds the Andrew and Erna Viterbi Chair in Communications at the USC Viterbi School. His contribution goes beyond mathematics, however. Both at JPL and later at USC, Golomb demonstrated a keen eye for talent and team-building.

"He had collected a very striking group of mathematicians and mathematically oriented engineers," says Thomas Kailath, who spent a year at JPL and is now professor emeritus of engineering at Stanford University. "Section 331 was actually a powerhouse."

Kailath credits Golomb for having the vision to erase some traditional boundaries.

"In those days mathematics and engineering were still regarded as somewhat separate entities," Kailath says. "This was one of the groups that helped bridge that gap and showed the power of mathematical reasoning and mathematical modeling. That was the main contribution of the place."

Besides Lindsey, Golomb recruited Lloyd Welch, who with Leonard Baum developed the Baum Welch Algorithm, an important tool used in

COVER STORY

THE RACE IS ON...

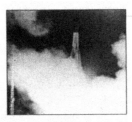

1957
Soviet Union launches first two earth satellites, Sputnik I and II, using R-7 rockets.

1958
Explorer I, first successful U.S. spacecraft, carries instruments to study cosmic rays & micrometeorites.

speech recognition and other areas. He is also now professor emeritus at USC. Golomb also recruited a 22-year-old Andrew Viterbi (Ph.D. EE '62), who would later become the co-founder of the cell phone giant Qualcomm. Today, Viterbi holds the Presidential Chair in Engineering at USC, and he and his wife Erna are donors of an endowed chair as well as the most generous naming gift to any engineering school in the U.S. Modern wireless technology is based in part on the Viterbi Algorithm for error decoding, the mathematics of which Viterbi says he learned from Golomb.

"I landed at JPL in a den of mathematicians," Viterbi says. What a contrast from his previous employer, Raytheon, whose engineers Viterbi remembers as being "rather dismissive" of mathematical theory. Viterbi himself was right in the middle: a prolific thinker with an innate nose for the most fruitful intersections of theory and practice. Lindsey, who remembers himself as a naïve theoretician when he began working at JPL, credits Viterbi for redirecting him to problems that were not only intriguing, but important. And just in time: Viterbi arrived in June of 1957, three months before the launch of *Sputnik* turned the place upside down.

While *Sputnik* is credited with giving the U.S. space program a much-needed kick in the pants, few realize that JPL was ready to launch a satellite a full year before the Soviets. Just like *Sputnik*, the American version would have been small and rudimentary and, most importantly, first in space.

In September 1956, Golomb recalls, the directors of JPL and a sister group in Alabama brought the satellite proposal to the White House. It was discussed by scientists on President Eisenhower's committee for basic research. That committee was supporting a rival project, called *Vanguard*. In addition, a majority on the committee felt that the exploration of space should be a non-military activity. Since JPL was run by the Army in the pre-NASA era, its proposal was dead on arrival.

"The Eisenhower administration was not aware that we were in any kind of a race at that point," Golomb says. The launch of *Sputnik* raised

BILL LINDSEY, CENTER, WITH JPL SUPERIORS, RECEIVING A DIGITAL DATA TRANS-TRACKING LOOP AWARD PATENT.

their awareness level substantially. Within three months the government was ready with its counterpunch. In December of 1957, the U.S. invited the world's press and television cameras to witness *Vanguard's* debut at Cape Canaveral, Florida. Embarrassingly, *Vanguard* blew up on the launch pad.

"And that was our competition," says Golomb.

A month later the scientists at JPL had their own answer to *Sputnik*, called *Explorer I*. Much to the Eisenhower administration's relief, this one made it off the ground in one piece on January 31, 1958. There were clenched fists and cold sweats in the Pasadena control room when several tracking stations around the globe failed to detect the satellite at the expected times. The communications team later realized that the thrust of the *Explorer I* boosters had been greater than planned, making its orbit longer and therefore delaying its progress by about 15 minutes.

The ultimate barometer of national enthusiasm — *Life* magazine — featured photographs of Golomb and Viterbi in the control room. More importantly, the satellite's success ensured the work of Section 331 would have far-reaching impact. By helping the satellite and mission control to communicate reliably, the scientists in the trailer had solved much more than the problem at hand. They had been drafting the blueprint for modern wireless communications.

The basic achievement had been the transport of information in a noisy environment with low-power transmitters. Out of that came satellite communications solutions, coding technology, the ability to track vehicles in space and time, the development of masers and rubidium, cesium standards for frequency and timing in all types of communications. It was a time of wide-open frontiers.

"What was our mission? We were defining it as much as we were being told what it was," Golomb recalls.

The science behind the group's achievements boiled down to two essential concepts. One was the idea of coding with shift-register sequences, developed by Golomb and further refined by himself, Viterbi

USC ENGINEER

COVER STORY

1962

NASA's Mariner 2 carries solar-powered instrument package past Venus. First successful planetary flyby.

1966

Soviet Luna 9 achieves soft landing on the moon.

and Welch. A shift register of n slots (n can be any whole number) is a device that takes an incoming stream of binary bits — ones and zeros — and, as each bit progresses from one slot to the next in the register, alters the message according to a predetermined formula. As each bit comes out of the register, it is fed back into the other side a certain number of times. The end result is a much bigger stream of bits with two advantages: it is coded to prevent jamming, and because it is redundant, with every original bit expressed multiple times, it is highly resistant to interference. Even if noise wipes out many bits in the stream, enough copies will be left to reconstruct the message. At the receiving end, a decoder inverts the coding formula and returns the original binary stream.

Another key technological building block was that of phase-lock loops, developed by Rechtin, Viterbi and Lindsey. The idea of a phase-lock loop is no different than every child's favorite bath hobby: making the water slosh higher and higher until it splashes out of the tub. An oscillator on the receiver tunes itself to resonate at the same frequency as the incoming signal. Resonance is a natural amplifier, helping to pull the signal out of deep noise. "Phase-lock" refers to the need for the oscillator and the signal to be in phase and to stay that way.

"Phase-coherent communications and tracking is key to modern digital communications," says Lindsey, who should know. While at JPL he worked secretly on the side for the U.S. intelligence community, monitoring Soviet communications. To his astonishment, he discovered that the other side was using phase-incoherent transmission — cheap, but very inefficient and not secure.

All these advances, though developed for communications in space, turned out to be vital for wireless transmission in a commercial airspace saturated with conflicting signals.

"I can say, without boasting at all, that we were the foundation of the global communications that we now have," Rechtin says. Lincoln Laboratories, run by the Massachusetts Institute of Technology, also contributed important advances, as did Purdue and Stanford. The legendary Bell Laboratories was a bastion of research but, according to Golomb, focused heavily on the land-based telephone system.

For years, no one in the group grasped the significance of their research. "It was just interesting work," says Welch. "I'm amazed, you see. We were basically in the digital communications era when nobody else was. Then as time goes on it's found a use in commercial applications. They were using all the theory that we had developed back in the fifties and sixties."

Years later, while consulting on the side, Lindsey was often asked how the group came together.

"I think God made this group," he replied.

After God came Zohrab Kaprielian, who did his best to live up to his predecessor. Revered, scorned, loved, feared (he was all four at once); Kaprielian re-assembled the JPL group at USC. Nominally Kaprielian's title was chair of the engineering department, but everyone knew he was much more.

EBERHARDT RECHTIN   THOMAS KAILATH

Golomb recalls that at one point around 1970, there were five levels of administration between himself and the president of the university, "and all of them were Zohrab Kaprielian."

ANDREW VITERBI   IRVING REED

"At that time we operated on the principle of one man, one vote," Golomb says, "and Kaprielian was the one man who had the one vote."

In matters of science Kaprielian used his vote brilliantly. Instead of trying to compete head to head with bigger universities, he focused on

continued on next page

"I CAN SAY, WITHOUT BOASTING AT ALL, THAT WE WERE THE FOUNDATION OF THE GLOBAL COMMUNICATIONS THAT WE NOW HAVE," RECHTIN SAYS.

The Wisdom of Solomon

COVER STORY

1973
Pioneer 10 reaches Jupiter, and then continues on outside the solar system.

1976
Viking lander sets down on Mars.

hot new fields where he knew USC could move faster than its stodgie rivals. Digital communications was one of those fields.

In 1963, on Viterbi's advice, Kaprielian recruited Golomb from JPL. Golomb's presence attracted others in the group: Welch and Lindsey, and later Rechtin and Viterbi (who had done his Ph.D. at USC because it was the only institution that would allow him to study while working at JPL). But Golomb's eye for talent went beyond JPL. Also in 1963, he persuaded Kaprielian to hire Irving Reed, a gifted computer scientist then at RAND Corporation. Reed was best known for having built the first computer on the west coast, a desk-sized machine that humbled eastern rivals ten times its size.

What was more intriguing to Golomb was Reed's research on error correction codes. Working with a collaborator, the late Gustave Solomon, Reed had shown that his algorithm for error correction was optimal, that is, unbeatable. At the time their finding was only of theoretical interest.

"Error correction coding was brand new," Reed remembers.

Reed-Solomon codes became considerably less theoretical on the Voyager spacecraft. JPL launched Voyager in 1977 to explore the outer solar system. As historian Peter Westwick explains in his forthcoming book, *Into the Black: A History of the Jet Propulsion Lab, 1976-2004* (Yale University Press, 2005), older error correction codes began to fail as the spacecraft sailed towards Uranus and Neptune. The Voyager project team then switched to Reed-Solomon codes, in combination with Viterbi decoding. The results were stunning: crystal clear photographs of the outer planets invaluable to scientists and inspirational to the public. Norm Haynes, the Voyager project manager, called this telecommunications success "the finest technological achievement of Voyager: being able to get images back from three billion miles away."

Reed-Solomon/Viterbi decoding — and the underlying Golomb codes — have been standard on every spacecraft since Voyager, including the recent Mars Rovers and Cassini. And for billions of people, Reed-Solomon codes are part of everyday life. They are inscribed into every single compact disc and DVD sold in the world. (A real-world tip from Reed's graduate student Gregory Dubney: when you clean your CDs, don't wipe in a circle, as that will erase the Reed-Solomon codes over

SOL GOLOMB AND ANDREW VITERBI AT THE SCHOOL NAMING CEREMONY LAST SPRING.

time and actually make the skipping worse. Clean the CDs by wiping towards the center.)

Together, these six pioneers — Golomb, Lindsey, Rechtin, Reed, Viterbi, and Welch — put USC on the map for information science and contributed to the USC Viterbi School's ascent into the top tier of engineering schools. Golomb, Reed and Welch won the prestigious Shannon Award, named for Claude Shannon, the first formulator of information theory. All six belong to the country's most select engineering society, the National Academy of Engineering. Golomb, Viterbi and Stanford's Kailath are also members of the National Academy of Sciences.

Despite their many achievements over the last four decades, members of the old JPL group look back on those days with a special fondness.

"It was the finest adventure we ever had, that exceeded anything I've done since," says Rechtin. "I don't think we could have done it singly, none of us could."

Says Lindsey: "So much work, so many things, so many areas started right in that Division 33, Section 331. We didn't know where we were headed. A cluster of the key guys in digital communications came together and worked together closely as friends, as colleagues.

"We sorted out the problems and solved them and someone started using the results. Almost anything we solved we could write this paper and it would be published, because it was that new."

That heady time may have come and gone in digital communications, but it repeats itself in emerging fields such as biotechnology, nanotechnology, bioinformatics and quantum information theory. Under Dean Nikias, the USC Viterbi School of Engineering has continued Kaprielian's legacy by seeking out the brightest minds in the newest areas. In less than three years, the number of tenure-track faculty has increased from 140 to 170, while the school as a whole now ranks sixth, tied with Caltech, according to *U.S. News & World Report*.

Many scientists over the ages have built new industries. The men of Section 331 can also say they helped to build an institution, the USC Viterbi School of Engineering.

© 2023 World Scientific Publishing Company
https://doi.org/10.1142/9789811234378_0058

Periodic Binary Sequences: Solved and Unsolved Problems*

Solomon W. Golomb

University of Southern California

Abstract. The binary linear feedback shift register sequences of degree n and maximum period $p = 2^n - 1$ (the *m-sequences*) are useful in numerous applications because, although deterministic, they satisfy a number of interesting "randomness properties".

An important open question is whether a binary sequence of period $p = 2^n - 1$ with both the span-n property and the two-level correlation property must be an m-sequence.

There is a direct correspondence between m-sequences of degree n and primitive polynomials of degree n over $GF(2)$. Several conjectures are presented about primitive polynomials with a bounded number of terms.

1 Introduction

Feedback shift register sequences have been widely used as synchronization codes, masking or scrambling codes, and for white noise signals in communication systems, signal sets in CDMA (code division multiple access) communications, key stream generators in stream cipher cryptosystems, random number generators in many cryptographic primitive algorithms, and as testing vectors in hardware design.

Notation:

- $\mathbb{F} = GF(2) = \{0, 1\}$
- $\mathbb{F}_2^n = \{(a_0, a_1, \cdots, a_{n-1}) | a_i \in \mathbb{F}_2\}$, a vector space over $\mathbb{F}_2$ of dimension n.
- A *boolean function of n variables*, i.e., $f : \mathbb{F}_2^n \to \mathbb{F}_2$, which can be represented as follows:

$$f(x_0, x_1, \cdots, x_{n-1}) = \sum c_{i_1 i_2 \cdots i_t} x_{i_1} x_{i_2} \cdots x_{i_t}, c_{i_1 i_2 \cdots i_t} \in \mathbb{F} \quad (1)$$

where the sum runs through all subsets $\{i_1, \cdots, i_t\}$ of $\{0, 1, \cdots, n-1\}$. This shows that there are 2^{2^n} different boolean functions of n variables.

An n-**stage shift register** is a circuit consisting of n consecutive 2-state storage units (flips-flops) regulated by a single clock. At each clock pulse, the state (1 or 0) of each memory stage is shifted to the next stage in line. A shift

S.W. Golomb et al. (Eds.): SSC 2007, LNCS 4893, pp. 1–8, 2007.
© Springer-Verlag Berlin Heidelberg 2007

---

*This chapter was originally published in Sequences, Subsequences and Consequences, International Workshop SSC 2007, Springer-Verlag LNCS Vol. 4893, p. 18.

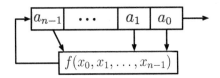

Fig. 1. A Diagram of FSR of Degree n

register is converted into a code generator by including a feedback loop, which computes a new term for the left-most stage, based on the n previous terms. In Figure 1, we see a diagram of a feedback shift register (FSR).

Each of the squares is a 2-state storage unit. The n binary storage elements are called the *stages* of the shift register, and their contents (regarded as either a binary number or a binary vector, n bits in length) is called a *state* of the shift register. $(a_0, a_1, \cdots, a_{n-1}) \in F^n$ is called an *initial state of the shift register*. The feedback function $f(x_0, x_1, \cdots, x_{n-1})$ is a boolean function of n variables, defined in (1). At every clock pulse, there is a transition from one state to the next. To obtain a new value for stage n, we compute $f(x_0, x_1, \cdots, x_{n-1})$ of all the present terms in the shift register and use this in stage n. For example, the next state of the shift register in Figure 1 becomes $(a_1, a_2, \cdots, a_n)$ where

$$a_n = f(a_0, a_1, \cdots, a_{n-1}).$$

After the consecutive clock pulses, a feedback shift register outputs a sequence:

$$a_0, a_1, \cdots, a_n, \cdots. \qquad (2)$$

The sequence satisfies the following recursion relation

$$a_{k+n} = f(a_k, a_{k+1}, \cdots, a_{k+n-1}), k = 0, 1, \cdots. \qquad (3)$$

Any n consecutive terms of the sequence in (2),

$$a_k, a_{k+1}, \cdots, a_{k+n-1}$$

represents a state of the shift register in Figure 1. A *state (or vector) diagram* is a diagram that is drawn based on the successors of each of the states. The output sequence is called a *feedback shift register sequence*.

If the feedback function $f(x_0, x_1, \cdots, x_{n-1})$ is a linear function, then the output sequence is called a *linear feedback shift register (LFSR) sequence*. Otherwise, it is called a *nonlinear feedback shift register (NLFSR) sequence*.

Examples. In Figure 2, we see a 3-stage shift register with a linear feedback function $f(x_0, x_1, x_2) = x_0 + x_1$.

Fig. 2. An LFSR of Degree 3

In Figure 3, we see an n-stage shift register with a linear feedback function.

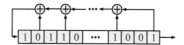

Fig. 3. An LFSR of Degree n

When f is linear, i.e.,
$$f(x_0, x_1, \cdots, x_{n-1}) = \sum_{i=0}^{n-1} c_i x_i, c_i \in \mathbb{F}_2$$
the elements of a satisfy the following recursion relation:
$$a_{k+n} = \sum_{i=0}^{n-1} c_i a_{k+1}, k = 0, 1, \cdots. \qquad (4)$$

m-**Sequences:** $f(x) = \sum_{i=0}^{n-1} c_i x_i$ is primitive over $\mathbb{F}_2$.

2 Polynomial Conjectures

The shift register

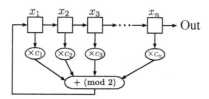

corresponds to the polynomial $f(x) = 1 + \sum_{i=1}^{n} c_i x^i$, and produces an m-sequence (starting from any initial state except $x_1 = x_2 = \cdots = x_n = 0$) if and only

538   *The Wisdom of Solomon*

4   S.W. Golomb

if $f(x)$ is **primitive** over $GF(2)$. This requires that $f(x)$ is **irreducible** over $GF(2)$, and is equivalent to: $f(x)$ divides $x^t + 1$ for $t = p = 2^n - 1$ and for no smaller positive integer value of t.

The number of such primitive polynomials of degree n over $GF(2)$ is known to be $\phi(2^n - 1)/n$, where ϕ is Euler's phi-function. This is the number of cyclically distinct m-sequences of degree n.

Many unsolved problems concern the existence of such primitive polynomials with a restricted number of terms. The two strongest conjectures are:

Conjecture 1. For infinitely many values of n there are **primitive trinomials**, $f(x) = x^n + x^a + 1, 0 < a < n$.

Conjecture 2. For all degrees $n \geq 5$, there are **primitive pentanomials**, $x^n + x^{a_3} + x^{a_2} + x^{a_1} + 1, 0 < a_1 < a_2 < a_3 < n$.

Perhaps the weakest conjecture of this type is:

Conjecture 3. There is a finite positive integer m such that, for infinitely many degrees n, there is a primitive polynomial $f(x)$ of degree n having no more than m terms.

Conjecture 1 is the special case of Conjecture 3 with $m = 3$. Conjecture 2 would imply Conjecture 3 with $m = 5$. I would hope that someone will prove Conjecture 3, perhaps with a very large value of m. Subsequent effort would then be devoted to reducing this value of m.

Even stronger than Conjecture 1 would be:

Conjecture 4. The trinomial $x^n + x + 1$ is primitive for infinitely many values of n.

For the foreseeable future I do not expect this to be proved.

An empirical observation that I made some 50 years ago, that there are no primitive trinomials when n is a **multiple of 8**, was proved by Richard Swan, who showed that $x^n + x^a + 1, 0 < a < n$ and n a multiple of 8, has an **even number** of irreducible factors.

I can easily prove: If there is a primitive trinomial of odd degree $n \geq 5$, then there is also a primitive pentanomial of this degree n. I conjecture that this is also true for even degree $n > 5$, and I expect that someone can prove this, as:

Conjecture 5. For all degrees $n \geq 5$, if there is a primitive trinomial of degree n, then there is a primitive pentanomial of degree n.

Over 75 years ago, Øystein Ore proved:

Theorem1. If $f(x) = 1 + \sum_{i=1}^{n} c_i x^i$ is primitive, then $f(x) = 1 + \sum_{i=1}^{n} c_i x^{2^i - 1}$ is irreducible.

3 "Randomness Properties" of m-Sequences

It is the remarkable "randomness properties" of the m-sequences that makes them so useful in many applications: **cryptography, radar, GPS, "Monte Carlo" random number generation, CDMA, etc.**

Here are the "randomness properties" of these m-sequences (the ones with maximum period $2^n - 1$.

P-1. In each period of $2^n - 1$ bits, there are 2^{n-1} 1's and $2^{n-1} - 1$ 0's. (The **balance property**).

P-2. (The **"Run Property"**.) In each period there are 2^{n-1} "runs". Since runs of 1's alternate with runs of 0's, half the runs (2^{n-2}) are runs of 1's, and half (2^{n-2}) are runs of 0's. Half the runs of each type have length 1; $\frac{1}{4}$ of the runs of each type have length 2; $\frac{1}{8}$ of the runs of each type have length 3; $\cdots$; there is **one** run of each type of length $n - 2$; finally, there is **one** run of 0's of length $n - 1$, and **one** run of 1's of length n. (This is the expected distribution of run lengths when tossing a **random perfect coin.**)

P-3. (**The "Span n" Property**.) As we slide a window of length n around one cycle of the sequence, we see every n-bit binary number, except $00 \cdots 0$, exactly once.

P-4. (**The Multiplier Property**.) If we take every second term of the sequence (repeating around the cycle to get a full $2^n - 1$ terms), we get back the same sequence (possibly rotated cyclically).

P-5. (**The 2-level autocorrelation property**.) If we compare the sequence of period $p = 2^n - 1$ with each of its (non-zero) cyclic shifts, we see 2^{n-1} **disagreements** and $2^{n-1} - 1$ **agreements**.

We define the **"autocorrelation function"** $C(\tau)$ of the sequence, at a shift of τ, to be $C(\tau) = \frac{A_\tau - D_\tau}{A_\tau + D_\tau}$, where $A_\tau = \#$ of agreements at a shift of τ, and $D_\tau = \#$ of disagreements at a shift of τ. We find

$$C(\tau) = \begin{cases} 1 & \text{at } \tau = 0 \\ -\frac{1}{p} & \text{for } 0 < \tau < p. \end{cases}$$

Example

τ	shifted sequence							A_τ	D_τ	$C(\tau)$
0	1 1 1 0 1 0 0							7	0	1
1	0 1 1 1 0 1 0							3	4	$-1/7$
2	0 0 1 1 1 0 1							3	4	$-1/7$
3	1 0 0 1 1 1 0							3	4	$-1/7$
4	0 1 0 0 1 1 1							3	4	$-1/7$
5	1 0 1 0 0 1 1							3	4	$-1/7$
6	1 1 0 1 0 0 1							3	4	$-1/7$

It is this **autocorrelation property** that makes these sequences ideal for use in **radar**.

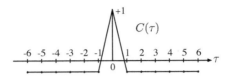

P-6. The **Cycle-and-Add Property**: Every m-sequence has the property that, when added, term-by-term "modulo 2", to any cyclic shift of itself, what results is merely a new cyclic shift.

Example

$$\begin{array}{cccc}
1110100 & 1110100 & 1110100 & 1110100 \\
+\ 0111010 & +\ 0011101 & +\ 1001110 & +\ 0100111 \\
\hline
1001110 & 1101001 & 0111010 & 1010011
\end{array}$$

3.1 Relationships of These "Randomness Properties" (Among All Binary Sequences with Period $2^n - 1$)

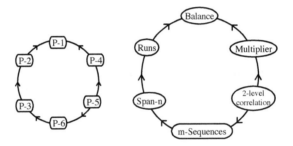

- There are $\binom{2^n - 1}{2^{n-1}} = \frac{(2^n-1)!}{2^{n-1}!(2^{n-1}-1)!}$ **P-1** sequences
- There are $2^{2^{n-1}-n} = \frac{2^{2^{n-1}}}{2^n}$ **P-3** sequences.
- There are only $\frac{\phi(2^n-1)}{n} \leq \frac{2^n-2}{n}$ **P-6** "m-sequences".

3.2 My Main Conjecture

If a sequence satisfies both **P-3** and **P-5**, it **must** be an m-sequence (**P-6**)!

Conjecture 6. A sequence of period $2^n - 1$ which has both span n and two-level autocorrelation must be an m-sequence.

Periodic Binary Sequences: Solved and Unsolved Problems   541

An example of a sequence with both **P-2** (the run property) and **P-4** (the multiplier property) which has neither **P-3** nor **P-5** already occurs at period $p = 2^5 - 1 = 31$.

Unsolved Problem: For what larger degrees n do such examples occur?

An example of a sequence with both **P-3** (the span n property) and **P-4** (the multiplier property), but without **P-5** (the 2-level correlation property) occurs at period $p = 2^7 - 1 = 127$.

Such an example will occur whenever $x^n + x + 1$ is primitive, with odd $n > 3$ (as here with $n = 7$).

Unsolved Problem: Are these the only examples of sequences with both **P-3** and **P-4** but without **P-5** (and therefore not m-sequences)?

An example of a sequence with both **P-2** (the run property) and **P-5** (2-level autocorrelation) but without **P-3** (the span n property) occurs at period $p = 2^7 - 1 = 127$. (This example was found in an example of a sequence in one of the six families of **P-5** sequences in Baumert's book.)

Unsolved Problem: What other examples are there of sequences with **P-2** and **P-5** but without **P-3** (and therefore not m-sequences)?

As early as 1954, I asked the question: **What are all the balanced binary sequences with 2-level autocorrelation?**

At that time, I knew only of the m-sequences and the Quadratic Residue (Legendre) sequences. This question can be formulated as: **"What are all the cyclic Hadamard difference sets?"**

These are the cyclic (v, k, λ) difference sets with $v = 4t - 1, k = 2t - 1$, and $\lambda = t - 1$.

Conjecture 7. For such a difference set to exist, v **must have one of three forms:**

i) $v = 4t - 1$ prime;
ii) $v = 4t - 1 = p(p + 2)$, a product of the twin primes p and $q = p + 2$; and
iii) $v = 4t - 1 = 2^n - 1$.

A stronger conjecture is:

Conjecture 8. The only cyclic (v, k, λ) Hadamard difference sets when $\mathbf{v \neq 2^n - 1}$ come from three constructions:

i) $v = 4t - 1$ prime with the **Quadratic Residue construction**;
ii) $v = 4t - 1 = 4a^2 + 27$ prime, with **Hall's sextic residue sequence construction**; and
iii) $v = 4t - 1 = p(p + 2)$, p and $p + 2$ are both primes, with the **twin prime construction** (using the Jacobi symbol) of Stanton and Sprott.

Extensive computer searches have failed to find any other examples, with $v \neq 2^n - 1$.

8   S.W. Golomb

3.3 Cyclic Hadamard Difference Sets when $v = 2^n - 1$

Finally, we have the question of cyclic Hadamard difference sets when $v = 2^n - 1$. The **known** examples include:

i) the m-**sequences**, for all $n \geq 2$ (Singer difference sets);

ii) the **Quadratic Residue sequences** when $v = 2^n - 1$ is a Mersenne prime;

iii) the **Hall's sextic residue sequences** when $v = 2^n - 1 = 4a^2 + 27 =$ prime, where it has been shown that there are only three such cases ($v = 2^5 - 1 = 31$, $v = 2^7 - 1 = 127$, and $v = 2^{17} - 1 = 131,071$);

iv) the **GMW construction and all its generalization**, when n (in $v = 2^n - 1$) is composite, $n > 4$;

v) **3-term and 5-term** sums of decimations of m-sequences, and the **Kasami power function constructions**;

vi) **the Welch-Gong transforms** of 5-term sequences;

vii) the three families of **"hyper-oval sequences"**: Segré sequences, and Glynn type 1 and type 2 sequences.

Details of all these constructions, including variant constructions that lead to the same sequences, can be found in the 2005 book **Signal Design for Good Correlation**, by S. Golomb and G. Gong.

Exhaustive computer searches have been completed for all cyclic Hadamard difference sets of period $2^n - 1$ for all **n $\leq$ 10**. (Starting at $n = 5$ in the early 1950's, one new value of n has been successfully searched in each succeeding decade.)

The aforementioned constructions account for all of the individual examples which have been found. Hence:

Conjecture 9. All the constructions which yield **cyclic Hadamard difference sets** are now known.

I am not sure that I believe this one.

References

1. Golomb, S.: Shift Register Sequences. Holden-Day Inc., San Francisco (1967); revised edition Aegean Park Press, Laguna Hills, CA (1982)
2. Golomb, S., Gong, G.: Signal Design for Good Correlation – for Wireless Communication, Cryptography, and Radar. Cambridge University Press, Cambridge (2005)

# Sol — Digital Pioneer, Sol Golomb Celebrates 50 Years at USC*

---

*This chapter was originally published in the Fall 2012 issue of the magazine *USC Viterbi*, pp. 38–41.

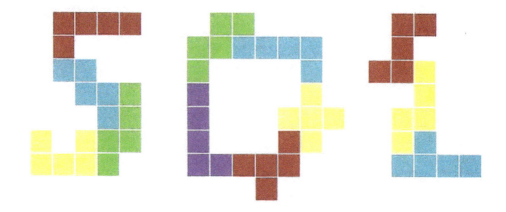

Digital pioneer **Sol Golomb** celebrates 50 years at USC

by Eric Mankin, photography by Noe Montes

More than six decades ago, a Baltimore-born prodigy graduated from Johns Hopkins University with a B.S. in mathematics before his 19th birthday. That individual remains brilliant in 2012, celebrating his 80th birthday and his 50th year on the faculty of the USC Viterbi School of Engineering – a school named after the young graduate student he mentored many years ago.

Polylingual, a USC leader, an international authority in communications, a celebrity in the world of mathematical games, winner of almost every honor in his multiple fields, Sol Golomb continues to teach (including freshmen seminars), to write (including deeply informed essays on history) and to research as he enters his ninth decade.

And his is the story of a remarkable individual embedded in a remarkable generation. Golomb was a leader in an extraordinary cohort of American thinkers who revolutionized understanding of the longstanding mysteries of mind and thought: breakthroughs as fundamental, or even more so, as the understanding of gravity, or the atom.

It all began in 1948 – three breakthrough IT discoveries at Bell Labs. That year, as Golomb recently wrote, saw the publication of "A Mathematical Theory of Information" by Claude Shannon, a figure Golomb compares to Einstein and Newton; the Allied invention of error correcting codes by Robert Hamming; and the invention of the transistor by John Bardeen, Walter Brattain and William Shockley.

Golomb recalls that as a young scientist working for his Ph.D. at Harvard, he was initially expecting to travel the abstract route of pure mathematics, but a summer job in engineering at Lockheed Martin, along with a Fulbright scholarship that took him to Oslo, Norway, brought his attention to the explosion of consequences of the 1948 revelations.

As a result, in 1956, after completing his Ph.D., he did not proceed directly to academia, but instead went to the Jet Propulsion Laboratory, where a wildly talented and diverse set of minds were working on projects including – four months after the Russian launch of Sputnik – the first U.S. satellite, Explorer.

"Our average age at the time was about 25," Golomb recalled, and the group included Andrew Viterbi (the future naming donor for the USC School), Lloyd Welch, William Lindsey, Thomas Kailath, Robert McEliece and others whose list of honors would fill this entire magazine.

And it was not just space that these minds

attacked. At the same time, the information theory implications of the discovery that DNA carried the coding of living things were sinking in. Golomb and one of his JPL colleagues were on that case as well.

As he recently wrote: "It was through Max Delbrück, the acknowledged father of molecular genetics, whom I met in 1956, that I met the others in this field. In 1960, on my first visit to London, I took a side-trip to Cambridge, where I spent the day with Francis Crick. He took me to lunch at the Eagle, the pub where Watson and Crick worked out the double helix structure of DNA, according to legend, on the back of a napkin."

Golomb insights enabled the Mars rovers to send back clear video images of what they saw on the surface of the red planet 100 million miles away.

Perhaps, Golomb's most outstanding contribution to the legacy of this outstanding group of colleagues was his analysis of shift register sequences, random-seeming series of bits that concealed a deep, unmistakable and highly conspicuous order to listeners who knew what to listen for.

Golomb created the mathematics behind them while still a graduate student at Harvard. A digital electronic message, he realized, could be modulated into a shift register sequence produced at a transmitting device. He saw that if the same sequence were built into the receiver, it would be possible to receive much fainter signals than would otherwise be detectable.

The shift register sequence, in effect, repeats the signal over time, so that more total signal energy can be concentrated at the receiver. Over time, the true signal builds while the random noise largely cancels itself out.

When Golomb subsequently came to JPL, he worked with colleagues in turning his insight into a critical tool for space communications, and in 1961, scored a spectacular success. A powerful radio transmitter aimed a specially designed signal at Venus, while a giant radar receiver that had the same sequence programmed in, listened — and detected the echo: clear, despite pervasive noise along its multimillion mile interplanetary journey.

The radar bounce off Venus provided another remarkable scientific benefit. Radar means "radio detection and ranging." Using the received signal for ranging gave the most accurate measurement yet of Venus's exact distance. It also had the unexpected effect of measuring the true value of an "astronomical unit" – the basic distance unit for space travel, the average distance of the earth to the sun. The accepted value, as calculations made with the new tool soon established, was off by 1/1000 – an enormous error for these purposes.

The same radio system, mounted on a space vehicle (Mariner 9, on its way to Mars), provided a test of Einstein's theory of relativity when it sent a signal back to JPL that traveled close to

---

THE MANY FACETS OF SOL

IN THE BEGINNING | THE WUNDERKIND | THE GREAT COMMUNICATOR | THE LEGEND | THE GAME-MASTER | THE POLYGLOT

POLYOMINOES

the sun. While the sun's gravity would bend the path of a physical object, general relativity predicted that gravity would instead alter the radar pulse's frequency in specific ways. Analyzing the Mariner signal confirmed Einstein's prediction more completely than any previous test.

The method has become a basic tool for NASA. Decades later, other Golomb insights enabled the Mars rovers to send back clear video images of what they saw on the surface of the red planet 100 million miles away. The system on the Rovers is known as "Low Complexity Lossless Compression for Images," or LOCO-I, and was developed by a three-man team of engineers working at Hewlett-Packard Laboratories in Palo Alto. Golomb's name occurs more than 34 times in their original 2000 paper.

Golomb's close JPL colleague Andrew Viterbi developed a related mathematical technique for encoding messages that worked for terrestrial purposes called Code Division Multiple Access (CDMA), sending out multiple signals simultaneously over a very wide frequency band at low power. The Viterbi algorithm, which allows users to efficiently recognize the signals coming for one receiver, is now at the heart of the systems that permit a cell phone to pick up messages sent to it - and to no other phone - was the foundation of the Qualcomm company.

And USC Engineering plays a key role in the Golomb-Viterbi relationship. After fruitful years at JPL, Golomb decided to pursue a new role as a professor in Southern California. While at JPL, "I taught part time at Caltech, at UCLA and at USC, and I got offers from all three. Some people were surprised I chose USC, but the question I asked myself was the one I ask my students: 'where can you make the most difference?'"

The answer was USC, where his JPL colleague Andrew Viterbi had already been studying as a graduate student, ultimately receiving his Ph.D. in engineering in 1962. Golomb not only mentored him for his degree, Viterbi recollected, but assisted in such things as helping find a place for his parents to live. And it was in Golomb's car, on a drive north to the Bay Area, that he had made his decision to propose to his wife, Erna.

Golomb's own wife is Bo, whom he met in Scandinavia while on his fellowship. From Bo he learned Danish, which he speaks along with varying levels of Norwegian, Swedish, Hebrew, German, French, Russian and Chinese.

At USC, Golomb's pursuits, and those of the school, progressed together-inextricably bound and in tandem. Not long after his arrival at the school, Golomb settled into teaching and researching, followed by a drumbeat of honors, awards and occupation of key USC positions. In 1976, Golomb became the first USC faculty member to achieve the distinction of joining the National Academy of Engineering (NAE) whilst still on staff. In 1985, he received the Shannon Award, the highest honor in IT; in 2000, he was awarded the Richard W. Hamming Gold Medal of the Institute for Electrical and Electronics Engineers (IEEE), followed by his election to the National Academy of Sciences (NAS) in 2003; he is a fellow of: American Academy of Arts and Sciences (AMCAD), the American Association for the Advancement of Science (AAAS) and IEEE.

His colleague, George Bekey was at USC Engineering a year before Golomb's arrival. "I have known Sol Golomb for almost 50 years, since I arrived at USC in 1962, and my admiration for his knowledge, wisdom, and kindness has continued to grow. I have known several geniuses over the years, and Sol outshines them all, because he is so broad in his interests and abilities. He is a true polymathic scholar, with a deep knowledge of mathematics, communication theory, the classics, European history, architecture of Scandinavian churches, the Book of Genesis, multiple languages, wine growing, and puzzle solving."

Golomb has long been thinking broadly about the enterprise of education, both about engineering and in a broader context. One area of interest: the growing dominance of non-U.S. students in engineering, particularly at the postgraduate level, a situation that has changed markedly from when he was a 19 year old.

Demographics are one factor. For a long time in the United States, engineers and scientists were second-generation immigrants, whose parents sacrificed "so their children could have a spectacular education." Golomb's parents and Viterbi's – "Andy himself was born in Italy and came to the U.S. at age 4" – are examples. For the third and fourth generation, there are easier routes.

And the problem, Golomb says, is intensified because now American would-be engineers are faced with students from China and India, where the competition for few slots is extreme, so the top one percent of Americans are facing off against the top one tenth or one-hundredth percent from other places: "It is very hard for even the brightest American students to compete and be standouts in that group."

But beyond this: in the United States, Golomb said, despite the fame and prominence of Steve Jobs, Bill Gates and Mark Zuckerberg, "We have not conveyed the notion that engineering is a glamorous profession. It's sad that for a significant portion of the population, an engineer is someone who drives a train."

"There have been a few attempts to have TV series that would appeal to high school age kids that would glamorize specific incidents in general area of science and technology – these have to be subsidized – and how do you get people to watch - there are no obvious or easy answers."

Which brings up, perhaps, the story of Sol Golomb: not obvious, not easy but certainly an inspiration. "He is also a kind and thoughtful person; truly, a giant of a man. I'm proud to call him a friend and colleague," said Bekey. "USC is indeed fortunate to have him on its faculty."

THE ORIGINALIST

THE HONOREE

© 2023 World Scientific Publishing Company
https://doi.org/10.1142/9789811234378_0060

BUSY
SIGNALS

The National Medal of Science recognizes Solomon Golomb's many contributions to communications technology.

THE TEENAGE SOLOMON GOLOMB, A mathematics prodigy at Johns Hopkins University, probably never considered that the nation's commander in chief would someday honor him for his intellectual prowess.

Yet there stood Golomb next to President Barack Obama at the White House in February, more than six decades after getting his first college diploma, his mind as curious at age 80 as it was at 18. In recognition of Golomb's profound contributions to technology, Obama placed a red, white and blue ribbon with a golden medallion around his neck: the National Medal of Science.

The award is America's highest honor for invention and discovery. Golomb was one of only a dozen researchers to receive it for 2011, the latest year awarded.

But Golomb's motivation originates from a different prize: the successful, thrilling hunt for answers to difficult questions.

"My research has always been directed by working on problems that I found interesting and challenging, and that I believed I had a chance to solve," he says. "I have never thought about receiving awards for my work, but it is always a pleasant surprise when they occur."

Golomb has often been pleasantly surprised in his career.

As Erna and Andrew Viterbi Professor of Communications and University and Distinguished Professor of Electrical Engineering and Mathematics, Golomb holds appointments at the USC Viterbi School of Engineering and USC Dornsife College of Letters, Arts and Sciences. But his ascent began thousands of miles from USC.

Born in Baltimore, he pursued his doctorate in mathematics at Harvard University and worked summers at Baltimore aerospace firm Glenn L. Martin Co. (now part of Lockheed Martin).

Out of this combination of rigorous academia and the working world grew Golomb's ideas. Among them: illuminating the mathematics behind "shift register sequences," seemingly haphazard series of 0s and 1s that conceal order behind their randomness. The feat would prove important.

While his Harvard professors took pride in promoting the purity of mathematics, Golomb sought its purpose—mathematics' practical side ultimately drew him in. So in 1956, he moved to the Jet Propulsion Laboratory in Pasadena, Calif., where he was swept up in the scientific fervor of the age. He and his fellow youthful, brilliant minds drove the race into space. JPL friends included Andrew Viterbi, whose naming gift later endowed USC's engineering school.

Golomb's work with shift register sequences bore fruit at JPL. Experts had said the mathematics behind them had no application, but he defied the purists. His insights enabled JPL engineers to build a communication system that bounced signals off Venus and accurately detected their echoes. The evolving technology eventually influenced everything from Mars exploration to cellular phones.

Then Golomb eased into his career's next phase: mentorship and performing research. After teaching part time at several Los Angeles-area universities, he made USC his academic home in 1963.

He has since amassed honors too numerous to list, including information technology's highest honor—the Shannon Award—and election to the National Academy of Engineering and the National Academy of Sciences.

Even after 50 years, Golomb continues to teach first-year students. He also is an expert on an array of topics including the classics, European history, puzzles, viniculture and several languages.

Citing Golomb's many contributions, USC President C. L. Max Nikias summed up the sentiments of the Trojan Family: "USC is so proud to have been Professor Golomb's academic home all these years."

Pamela J. Johnson, Eric Mankin, Robert Perkins and Alicia Di Rado contributed to this story.

Vision for the Future

USC chemistry alumnus Rangaswamy Srinivasan PhD '56, whose work formed the basis for LASIK, received the National Medal of Technology and Innovation in February.

Srinivasan was honored as a member of an IBM Corporation team at the same White House ceremony as USC's Solomon Golomb and other decorated scientists and engineers.

Called ablative photodecomposition, Srinivasan's technology uses pulses of ultraviolet light to erode layers of organic matter such as living tissue. His 1980 invention was inspired by work he performed 25 years earlier in pursuit of his chemistry doctorate at USC.

Researchers eventually applied the technology to laser eye surgery. About 700,000 LASIK surgeries are performed in the U.S. each year, according to the American Academy of Ophthalmology.

---

*This chapter was originally published in the Summer 2013 issue of the *USC Trojan Family Magazine*, pp. 32–33.

© 2023 World Scientific Publishing Company
https://doi.org/10.1142/9789811234378_0061

The "Golomb-Dickman Constant" —
E-mail re Lagarias Article

Introduction by Al Hales

In 2013, Jeff Lagarias wrote a long expository article for the *AMS Bulletin*, titled "Euler's Constant: Euler's work and modern developments". One section of this article deals with the "Golomb-Dickman" constant, which Sol introduced in connection with the expected length of the longest cycle in a random permutation. When I read the article I e-mailed Sol about it, and he replied as follows:

———————————— Original Message ————————————
Subject: Lagarias article
From: "Solomon Golomb" <sgolomb@usc.edu>
Date: Tue, October 1, 2013 7:18 pm
To: "Alfred W. Hales" <hales@ccr-lajolla.org.>

———

Dear Al,

I finally got my copy of the latest BAMS, and because you had called my attention to it, I read the article by Jeff Lagarias. The "Golomb-Dickman constant" (so named by Don Knuth) is now known to occur in some half-dozen connections, all having something to do with randomness. I did my work on this at JPL, when I decided to use cycles of random permutations as a possible model for the cycle behavior of cycles from nonlinear shift registers. I discovered the early paper of Goncharov, which I translated from Russian for myself (it was later translated as part of a project by the AMS), but he didn't consider the question of "the expected relative length of the longest cycle of a random permutation on n symbols". In a JPL report that subsequently became a chapter in my book on Shift Register Sequences, I proved that this number had a limit as n goes to infinity, and evaluated it to a few places. I also published a "Research problem" to find an "expression" for this constant, which led several people to find (different) expressions for it as definite integrals of non-elementary functions–as reported by Lagarias. It's interesting that there are problems where the solution involves this constant together with other famous constants, including pi, e, and Euler's gamma. (This expected relative

549

length is approximately, and asymptotically, the same as the probability that the "first" cycle is longest (where the "first" cycle is the one that contains one of the n objects that you have chosen to call "first").) There doesn't seem to be any theory of transcendentals powerful enough to show that this constant is transcendental, or even that it is irrational. It shares this distinction with Euler's gamma, and the Riemann Zeta function evaluated at odd integers > 3. Dickman's context involved the expected number of digits in the largest prime factor of a random n-digit number, and again one takes the limit as n goes to infinity. The constant also occurs in connection with the expected length of the largest limit-cycle loop in a random INTO mapping from n objects to the same set of n objects; and in the expected highest degree irreducible factor of a random degree-n polynomial over $GF(q)$, I believe as n (and not q, as I think Lagarias states) goes to infinity.

"My" constant inhabits the region of the real line that also includes the golden mean, five-eights, pi/5, and infinitely many other real values—but I'll have to settle for that; all the better constants seem to have already been taken.

Best wishes,
Sol G.

© 2023 World Scientific Publishing Company
https://doi.org/10.1142/9789811234378_0062

Letter to Granville; Erdos Number "Minus One"

Solomon W. Golomb
Distinguished University Professor
sgolomb@usc.edu

December 17, 2013

Professor Andrew J. Granville

University of Montreal

Department of Mathematics

6128 succ Centre-Ville,

Montreal, QC H3C 3J7 CANADA

Dear Professor Granville,

I recently read your paper "Bounded Gaps Between Primes". It is an excellent exposition of the dramatic results obtained during the past decade about the distribution of primes, up to and including Zhang's infinitely-often bounded differences (though not yet including Maynard's further improvement using sieve ideas).

My own interest in the twin prime problem began when I was in high school. My first "publication", submitted when I was 17, appeared as a problem in the *American Math Monthly* in May 1951: "Prove that there are infinitely many twin primes if there are infinitely many positive integers n not expressible in the form $6ab\pm a\pm b$ (where a and b are positive integers, and the two $\pm$ signs are independent of each other). There was only one solver other than the proposer, but this result has been 'rediscovered' and republished a number of times, often in the mistaken belief that this simple reformulation could somehow lead to a proof of the conjecture".

My PhD thesis (in Mathematics, at Harvard, submitted in May, 1956, with the degree awarded in 1957) was titled "Problems in the Distribution of Prime Numbers." Motivated by my interest in twin primes, I discovered the formula for the product of von Mangoldt lambda functions:

$$\prod_{i=1}^{k} \Lambda(a_i) = \frac{(-1)^k}{k!} \sum_{d|A} \mu(d)\log^k d, \ A = \prod_{i=1}^{k} a_i,$$

where all $a_i > 1$, and $(a_i, a_j) = 1$ for all $i \neq j$. This result has been used by GPY and the subsequent work by Zhang *et al.*, and appears in an equivalent form in your article.

I attempted to use this to prove

$$T(x) \sim Cx/\ln^2 x \quad \text{as } x \to \infty$$

by considering both

$$\sum_{n=1}^{\infty} \frac{\Lambda(2n-1)\Lambda(2n+1)}{n^s} \text{ as } s \to 1^+ \tag{1}$$

and

$$(1-z)\sum_{n=1}^{\infty} \Lambda(2n-1)\Lambda(2n+1)z^n, \quad \text{as } z \to 1^-. \tag{2}$$

In each case, what is missing from a rigorous proof is "only" the justification of taking a term-by-term limit inside a summation sign.

No one on the Harvard faculty at that time (early 1950's) knew as much analytic number theory as I had taught myself from books; so, when I discovered my identity for $\Lambda(a)\Lambda(b)$ and its potential application, I drove to Princeton (it was October 1953, when I was 21) and presented myself at the office of Atle Selberg. (I had seen Selberg's Lemma in his elementary proof of PNT.)

I wrote on his blackboard:

$$2\Lambda(a)\Lambda(b) = \sum_{d|ab} \mu(d)\log^2 d, \quad \text{if } a > 1, b > 1, (a,b) = 1.$$

His first reaction was "I don't believe it's true." (That's what convinces me it wasn't previously known.) He spent a few minutes at the blackboard, conceded that it was true, and then spent the rest of the day discussing my ideas for how to use this (and its generalization to $\prod \Lambda(a_i)$) to get results on twin primes and k-tuples for small k. (That day was the only time I had meaningful input for this work from an analytic number theorist.)

I spent academic 1955–1956 on a Fullbright Fellowship at the University of Oslo, where I interacted with Viggo Brun, Thoralf Skolem, and Ernst Selmer. (I had learned "Brun's Theorem" from Vol. I of Landau's *Vorlesungen*, but it was better motivated when Brun told me how he viewed it. I obtained (and published, *in Math. Scand.*) a sieve result ("Sets of Primes with Intermediate Density") that I also included in my thesis, and which was the basis for "On A Problem of S. Golomb" by Erdös in an Australian journal. (That's the basis for my claim that my Erdös number is "−1".) But I also used that year in Norway to put the finishing touches on my thesis. From Hardy's *Divergent Series*, I learned the three "Schur-Toeplitz conditions" for the "regularity" of a summation method. The Abelian theorems that would justify the term-by-term limits for my two analytic approaches to the twin prime distribution would follow easily from "regularity". But I *proved* that for the Dirichlet series approach, only one of the conditions were satisfied, while for the power series approach, the analogue to the regularity of Lambert summability satisfied two, but failed the third, of the Schur-Toeplitz conditions.

I gave a brief summary of all this in a paper, "The Lambda Method in Prime Number Theory", in *Journal of Number Theory*, vol. 2, no. 2, May 1970. A dozen or so years ago, a young number theorist, Keith Conrad, a student of Bateman, came across that article, got a copy of my thesis from me, and wrote a couple if papers showing (more generally) how "Golomb's Lemma" (his term for my product formula for $\Lambda's$) could be used to get the "Hardy-Littlewood Constants" for asymptotic formulas for many "sets-of-primes" problems more easily than their "circle method" approach.

Your probabilistic derivation of the "twin prime constant" which I write as:

$$C = 2 \prod_{\text{all } p > 2} \left(1 - \left(\frac{1}{(p-1)}\right)^2\right) = 1.32032\ldots,$$

first appeared in my article "The Twin Prime Constant", *American Math. Monthly*, vol. 67, no. 8, October 1960.

My approach to

$$\lim_{z \to 1^-} (1 - z) \sum_{n=1}^{\infty} \Lambda(2n - 1)\Lambda(2n + 1)z^n,$$

is similar to N. Weiner's proof of PNT using

$$\lim(1 - z) \sum_{n=1}^{\infty} \Lambda(n)z^n,$$

which "works" because "Lambert summability is *regular*"; and Hardy, in *Divergent Series*, comments that this regularity is "equivalent" to the PNT. In his book *The Fourier Integral and Some of Its Application*, Wiener uses his proof of PNT to showcase "Wiener's Tauberian Theorem", and I think he was angry that Hardy took a different view. My experience with applying Wiener's general approach to twin primes leads me to agree with Hardy. In both my approaches, the necessary Tauberian theorem was readily available, but the Abelian theorem was lacking.

Since my PhD, my career has consisted mostly of applying all the topics my Harvard professors took great pride in proclaiming their inapplicability, to problems in signal design for digital communication (which has gotten me into the NAE, the NAS, and most recently receiving the National Medal of Science), but my first love has remained prime number theory. I'm listed in the index of Hardy-Wright's *Introduction to the Theory of Numbers* from the fifth edition on, and I've taught both undergraduate and graduate courses in Number Theory.

I feel privileged to have lived long enough to witness the recent progress: GPY, primes in arbitrarily long arithmetic progress, and the existence of two (or more) primes infinitely often in bounded intervals. But the original twin prime conjecture still seems to remain just out of reach.

Sincerely yours,

Solomon W. Golomb
Distinguished University Professor
Ming Hsieh Department of Electrical Engineering

Sol — Too Famous for a Stamp*

An effort to make personalized stamps for the Sol Memorial at USC was rejected on grounds that "Design incorporates a celebrity or other famous person's name or likeness".

---

*This chapter reproduces an email sent to Beatrice Golomb on Janurary 20, 2017 from Zazzle Content Review Team, in response to an order for stamps with Solomon's photograph (ordered for use in relation to the USC event in honor of Solomon).

555

© 2023 World Scientific Publishing Company
https://doi.org/10.1142/9789811234378_0064

Shizef Raphaeli Obituary Translation*

Edna Sharoni

Translation by Edna Sharoni (Sister of Solomon Golomb) of article
on Solomon Golomb
From the daily Hebrew financial paper, "Calcalist".
By Shizef Raphaeli
June 2, 2016

Who is responsible for WIFI and Tetris? And what transformed the algorithms based on his research into the most userful / exploitable of all time?

Few people are acquainted with the work of the extraordinary mathematician Solomon Golomb, who passed away last month. Even fewer people are not influenced by his work. If you have a Smartphone, or have tried to break a record in TETRIS, you are numbered among this group.

Golomb, born in 1932, was already recognized as a mathematical genius in his youth. Unlike many of his colleagues, he was not involved only in the "pure" side of this profession, but employed his skills to turn in other directions as well: e.g. games based on two-dimensional moves. At age 16, he thought up the game "Cheskers" (a combination of "Chess" and "Checkers"). Here the ordinary pieces move like the pieces in checkers, and the kings like the Queens in Checkers, but there are also pawns and "Camels" whose moves resemble those of the knights in chess. For thinking children, and for parents who wish to present them with new challenges, this game can provide many hours of pleasure.

And this was just the beginning. At age 21, Golomb described the characteristic of "Polyominoes," forms made up of the various possible combinations of squares of equal size. He focused chiefly on "Pentominoes," forms composed of five squares; but those which really became famous were their cousins comprising four squares, the TETROMINOES. These starred in the computerized game of TETRIS which appeared in 1985, completely based on Golomb's theories, and which almost immediately became a giant, addictive hit.

But with all due respect to games, Golomb's most outstanding contribution to humanity is related to a very complex branch of Number Theory, called "Shift Registers." His work in this field made it possible to create efficient rapid algorithms to

---

*This article was originally published in *Calcalist* (in Hebrew) on June 2, 2016.

558   *The Wisdom of Solomon*

mimic number sequences that resemble random numbers. This may perhaps sound completely theoretical, but without this mathematical basis/foundation, Steve Jobs would never have been able to present us with the I-phones, the WIFI, the cellular Internet, GPS, or Blue Tooth – all technologies based on the principles pioneered by Golomb. Thus his complex mathematical ideas were transformed, incontrovertibly, into the most useful widespread devices of our times.

Another important researcher, Stephen Wolfram, who eulogized Golomb on his Blog, wrote that by a conservative estimate, the methods of the late mathematician have been used an octillion times so far, that is, a billion, billion, billion times!

Other mathematical innovations by Golomb have been eagerly/warmly adopted in the realm of astronomy. For example he developed methods for compressing and confirming data, employed by NASA space-craft vehicles launched to Mars, enabling the vehicles to transmit data back to Earth. Add to this his work in the realm of encryption, about which he remarked that this work serves the intelligence communities worldwide for purposes that have never been revealed to him.

Golomb, who up to the time of his sudden death at 84, had been serving as Professor at the University of Southern California (in Electrical Engineering), is credited with far more than ten thousand patents, only one of which is registered in his name. As befits an idealistic intellectual, he has consistently made his work available to the public in an open and free manner.

This tale cannot conclude without his connection to Israel. As a warm Jew, Golomb applied his talents to deciphering parts of the Dead Sea Scrolls and to Bible teaching. Among his many other awards, he recently received an Honorary Doctorate from the Haifa Technion. Along the way, he also wrote newspaper columns as well as amusing himself with linguistic puzzles and compiling crosswords.

How did he manage to do so much work? That is a riddle to be solved by some other mathematician.

                                        –Translated by Edna Sharoni.

(Transcribed from Edna's handwrilen translation by Beatrice Golomb – so, it is possible I may have introduced errors)

© 2023 World Scientific Publishing Company
https://doi.org/10.1142/9789811234378_0065

38

In Memoriam: Solomon W. Golomb

The Information Theory Society suffered a great loss earlier this year when Solomon Golomb passed away peacefully on May 1, 2016. Just ten days earlier he was awarded the prestigious Franklin Medal 2016 in Electrical Engineering for his revolutionary work on shift register sequences and their applications to space communications, satellite communications and cellular communications.

Sol was a giant of Information Theory and was responsible for several breakthroughs that changed the nature of the field and helped bring about the era we live in today. Much has been written over the past few months about his accomplishments (Indeed, we will address that ourselves below).

But Sol was more than a mathematical genius. He was a kind and generous man who loved his friends and family, designed math and wordplay puzzles, and spoke over 20 languages. While he will certainly be remembered for his contributions to information theory, his impact on the personal lives of those around him should be no less celebrated.

"Sol showed me what an academic advisor should be, taught me how beautiful fundamental research can be, how exciting application research can be, and changed my whole life from the time I was a PhD candidate until today." Said former student, Professor Hong-Yeop Song.

"During my time as a post-doc, Sol was the Vice Provost. Even still, he always made time for me and his PhD students." Said Tuvi Etzion. "Just three months after arriving at USC, he invited me to Thanksgiving Dinner – one of many learning experiences I shared with him over a meal."

Sol was as adventurous as he was academic. Andrew Viterbi, Sol's close friend of nearly 60 years, recalls a road trip the two of them took to San Francisco with their wives for an IEEE convention in 1957: "All went well on the way up, but returning on pre-Freeway Route 99, perhaps due partly to the heat, the oil cap would come loose and fall off every few miles. This happened half a dozen times, in each case resulting in a feverish but successful search over the last several yards just driven. We finally made it back to Pasadena in one piece."

In 1963 Sol was highly recruited by some of the best schools in the country. Having his pick of the litter, he chose USC – a school that at the time was not known as an engineering powerhouse. "The question I asked myself was the one I ask my students: 'Where can you make the most difference?'" Golomb told USC Viterbi magazine in fall 2012.

Over the next half century Sol helped turn USC into a leading center of communications research. An achievement this big does not come from groundbreaking research and genius alone. Sol understood the importance of community, mentorship, and service in building a lasting institution. It is for those qualities, as well as his unique mind, that we celebrate him today. Below, we summarize just a few of Sol's many contributions.

The start of Feedback Shift Register Sequences

As early as in June 1954, Solomon Golomb worked on a summer job with the Glen L. Martin Company in Baltimore (currently, Lockheed Martin). The leader of the Communications Group,

Thomas Wedge introduced him to a problem that was described as involving a *tapped delay line with feedback*. Sol called it a binary linear shift register with feedback and immediately called on his knowledge of pure mathematics to solve this problem.

An n-stage linear feedback shift register (LFSR) is a circuit consisting of n consecutive 2-state memory cells regulated by a single clock. At each time, the content of each memory cell is shifted to the next cell of line, and a new state is obtained by a feedback loop which computes a new term based on a linear Xor combination of the n previous terms. During that time, the experiments showed that for some combinations of taps of the memory cells, very long binary sequences would be produced, but for other combinations of taps, much shorter output sequences resulted. Sol's approach was to look at power series generating functions over finite fields, which yielded the correspondence between LFSR sequences and polynomials. The maximal length of those sequences is 2^n-1. These are called m-sequences and can be generated by a primitive polynomial over a finite field with order 2.

Sol proposed three criteria to evaluate randomness of those LFSR sequences, known as Golomb's Three Randomness Postulates: balanced 0–1 distribution, uniformly distributed consecutive 0's or 1's (i.e., the runs), and the two valued autocorrelation with all out-of-phase correlation values are equal to $-1/(2^n-1)$. The m-sequences possess all three randomness properties. Particularly, the autocorrelation of m sequences resembles white Gaussian noise with flat spectrum-like or impulse-like, so it is named pseudo noise (PN) sequences in many textbooks of digital communications.

Applications for Space and Satellite Communications and Radar

The first two applications of the 2-level autocorrelation property of m sequences, discovered by Sol, were in space communications when he worked at the NASA Jet Propulsion Laboratory (JPL). As the Leader of the Information Processing Group in JPL, his lab was tasked to provide a solution to early orbit determination of Explorer I, launched in 1958 after the launch of Sputnik (Russian) in 1957. The signal sent back from Explorer I was the binary phase shift keying (BPSK) pulse modulated by an m-sequence. Another amazing application of this property in 1958 shortly after launching Explorer I, was to prepare to launch a space probe to the vicinity of Venus. Sol was the leader on the Venus Radar detection project. He had designed an interplanetary ranging system at JPL, based on BPSK of an RF carrier using m sequences, that basically counted RF-cycles, and thus potentially provided extreme range accuracy. Venus was successfully detected by JPL in 1961, in a way which was more accurate than before. In their Venus experiment, which directly measured by radar, Sol showed that the distance between Earth and Venus reported before was wrong by far.

Applications in Cellular Communications

The m sequences are used as spreading codes in spread spectrum communications for anti-jamming and low probability interception. His revolutionary book, Shift Register Sequences [3], has long been a standard reading requirement for new recruits in many organizations, including the National Security Agency and a variety

IEEE Information Theory Society Newsletter

September 2016

---

*This chapter was originally published in the September 2016 issue of *IEEE Information Theory Society Newsletter*.

560   *The Wisdom of Solomon*

of companies that design spread spectrum communication systems for anti-jamming and low probability interception. These technologies, commercialized in cellular communications, are known today as CDMA (code-division multiple access) systems in 2G cellular systems as IS-95, and CDMA 2000 in 3G systems.

Exp-Golomb Code for Lossless Data Compression

Run-Length encoding, a paper Sol published in 1966 [2] became a widely used lossless data compression technique, adopted and termed as Golomb codes or Exponential Golomb (Exp-Golomb) codes. It was used to send back scientific data from the Mars Rover in 1960s. Currently, the Exp-Golomb code has been selected in the standards of multi-media communications such as MPEG-4 (or H.264).

Golomb Ruler

A pattern that Sol observed, originally for coded pulse radar, was a ruler of length L with n marks on it such that any distance d less than or equal to L can be measured in one and only one way as a distance between two of the n marks. This concept was popularized in Martin Gardner's column as *Golomb Rulers*. Since then these have been applied in fields of far ranging for Xray diffraction crystallography, radio antenna placement, and error correction. It is still open whether there exist infinite many shortest Golomb rulers. The latest exhaustive search results have been extended to the length $L = 27$.

Costas Arrays for Radar Detection

John Costas (1984) presented the following problem to Sol: Design an n by n frequency hop pattern for radar or sonar, using n consecutive time intervals and n consecutive frequencies where each frequency corresponds to one time interval and each pair of frequencies with their corresponding time intervals are different, i.e., correspond to an ideal ambiguity function for the frequency hop pattern. Sol discussed this problem with Lloyd Welch and Abraham Lempel and shortly they came up with three systematic constructions to this problem, known as the Welch, Lempel and Golomb constructions, respectively [5]. These constructions remain as general constructions till now. Recently, Sol tackled this problem with a slightly different angle [8]. The latest exhaustive search has been extended to $n = 29$.

Classification of Sequences

In 1958, when he was in JPL, Sol investigated how to generate a sequence with a long period given that the length of each LFSR is bounded. He experimented on using the periods of nonlinear shift register sequences in the radiation detector on Explorer I which discovered what is known as the *van Allen radiation belts*. (James van Allen was a graduate student who worked on this with Sol.) Currently, in cryptology, it is an important method to construct nonlinear generators using multiple LFSRs, called combinatorial generators.

In order to measure randomness of a random sequence with the same period as m-sequences, Sol defined a concept of span n sequences, i.e., any nonzero n-bit string occurs exactly once in a binary sequence with period 2^n-1. Then he conjectured that any sequence with 2-level autocorrelation and span n property must be an m-sequence (1980 [4]). Significance of the conjecture in cryptography is that a random sequence with large linear span (or linear complexity) has to compromise one of those two properties.

Golomb's Invariants and Nonlinearity in Cryptography

Most of the properties of nonlinear feedback shift register sequences were unknown until now and the known ones were collected in Sol's book, Shift Register Sequences. In cryptographic applications, in order to realize Shannon's concept of a one-time-pad, i.e., stream cipher encryption, m-sequences were used for their long periods in the 1950s and 1960s. However, m-sequences are linear which cannot be directly used in any cryptographic systems since 1968 and should be replaced by nonlinear sequences, generated by filtering single LFSR or multiple LFSRs. The question is then how to measure the cryptographic strength of those filtering functions. Sol (1959 [1]) investigated the invariants of a Boolean function, which measures the distances between the Boolean function and linear combinations of its inputs.

A good filtering function in cryptographic application should be far from or independent from input variables or linear combination of input variables. This can be measured by the invariants and was termed as nonlinearity in modern cryptography. Sol's results, presented in 1959, were rediscovered by Xiao and Massy in 1988. Currently, the nonlinearity of Boolean functions is one of the most important concepts for designing cryptographically strong Boolean functions.

Golomb's Invariants for Correlation Related Attacks

The immediate applications of Sol's invariants of Boolean functions in cryptography are the so-called correlation attacks. In other words, as long as a Boolean function is correlated with some of the input variables, the initial states, loaded as keys, those correlated LFSRs can be recovered individually by computing correlation between the output of the Boolean function and that input. This converts the overall time complexity from the multiplication of the number of states in each LFSR to the addition of the number of states of those correlated LFSRs plus the remaining part from exhaustive search. This is a significant reduction to the size of the exhaustive search. Although Sol did not explicitly mention this application in his 1959's paper, he was awarded a Medal for his contribution to cryptography by the National Security Agency in 1992.

Sol's concept of the invariants is to measure the correlation between a sequence and an m-sequence. However, there are many distinct LFSRs, corresponding to the number of primitive polynomials, which generate distinct m-sequences with the same period. By the end of 1990's, Sol, together with his collaborator, extended this measurement to any LFSRs, called the extended Hadamard transform. The use of the extended Hadamard transform as the measurement of the strength of a cryptographic function gave rise to new cryptographically strong functions, namely hyper-bent functions, which are widely investigated in cryptographic communities today.

Applications in Secure Communications

The field of shift register sequences (or equivalently, pseudorandom sequences), created by Sol is widely used in numerous cryptographic applications including stream cipher, block cipher, pseudorandom sequence generations, key deviation functions, pseudorandom functions, challenge number generations for authentication protocols, public-key cryptographic schemes, hardware test vectors, and

September 2016   *IEEE Information Theory Society Newsletter*

In Memoriam: Solomon W. Golomb

40

countermeasures for side-channel attacks. Especially due to the simple hardware structure of LFSR based pseudorandom sequence generators, they have been proposed to secure the Internet-of-Things (IoT) where constrained devices will be deployed in extremely large scales, especially, in radio frequency identification systems, for establishing trust among a variety of devices and ensuring confidential transmissions. Those embedded applications and systems have been used for protecting our daily digital world including on-line banking, shopping, health record transfer, medical care, and more future new applications in vision.

Polyominoes

The other branch that Sol investigated is recreational mathematics. Sol generalized a puzzle problem about putting dominoes on a checkerboard from which a pair of opposite corners had been removed, and created the subject of Polyominoes (1954). His book (1965, revised 1994) Polyominoes has a world-wide audience, and has lead to the invention of the computer game Tetris.

Inspiring Young Researchers

Sol loved to give talks in various occasions. Each of his talks were full of his intelligent and multi-faced knowledge, especially in the questions periods (e.g [6, 7]). His last talk was given at the Workshop on Shift Register Sequences for Honoring Dr. Solomon W. Golomb Recipient of the 2016 Benjamin Franklin Medal in Electrical Engineering, which was held in Villanova University, Philadelphia, on April 20, 2016.

Over his career, the impact of Sol's contributions were recognized with election into the National Academy of Engineering, the National Academy of Sciences, and a foreign membership in the Russian Academy of Natural Sciences. Within the University of Southern California, Sol was a University Professor and the recipient of the USC Presidential Medallion. He was awarded the National Medal of Science in a White House ceremony in 2011. Sol was a Distinguished Alumnus of Johns Hopkins University and received multiple honorary doctorates. He was a fellow of numerous societies including the IEEE, the American Mathematical Society, the Society for Industrial and Applied Mathematics, the American Association for the Advancement of Science, and the American Academy of Arts and Sciences.

In recent correspondence with a journalist from the New Yorker magazine, Sol recollected some of his experiences with Claude Shannon, "My Shannon Lecture was in Brighton, England, Shannon, who had delivered the first Shannon Lecture (in Ashkelon, Israel) quite a few years earlier, was in my audience – the only Shannon Lecture he attended since his own. I had spent most of the Fall semester of 1959 at MIT (on leave from my position at the Jet Propulsion Laboratory – JPL) where I had lunch three times a week with Claude Shannon and a few other members of the MIT communications faculty. I sat in on Shannon's lectures on his own course on Information Theory. On one occasion, he asked me about a problem in mathematical statistics, and the next day I presented him with a simple solution to it." Sol was also awarded the IEEE Richard W. Hamming Gold Medal.

Sol's passing has been deeply felt by the Information Theory community. Toby Berger wrote, "He is irreplaceable. A genius among our geniuses, he was a man who somehow knew almost

EVERYTHING in an era in which that seems to be impossible, but it was true." Further elaborating on Sol's many talents, "Perhaps less known but equally encyclopedic was Sol's philology. He read more than 120 languages and admitted to speaking over 20 of them; that admission meant that he could converse fluently with native speakers in each of those, which ranged from Ancient Greek, to Mandarin, to Norwegian; he knew three dialects of Norwegian but counted that as only one of his spoken languages."

Abraham Lempel reminisced about when his and Sol's paths first intertwined: "I first met him in 1967, at an IT workshop at the Technion, when I presented him with a proof of one of his conjectures re shift register sequences. At the time I was a fresh PhD at the Technion EE department and Sol invited me for my postdoc at USC which I was happy to accept. I arrived at USC in July 1968, shared an office with Lloyd Welch, and worked very closely with Sol for a year. Under Sol's influence and guidance I have traded my interest in network theory for the rich realm of digital sequences."

Sergio Verdu further adds, "What a towering figure he was. I vividly remember his Shannon lecture in Brighton."

Bob Gray did his PhD at USC under Bob Scholtz; his MS thesis advisor, Irwin Jacobs, strongly urged Bob to take as many courses as he could from Sol Golomb, "regardless of topic," noting that "Sol was a genius." After commenting on how much Bob learned from Sol, he reflected on Sol's human side: "Sol was the archetypal absent minded professor. He forgot to come to my PhD qualifying oral – he was a member of my committee – and wandered off to have tea. Zohrab Kaprielian grabbed Lloyd Welch in the hall and drafted him as a substitute. When Lloyd's turn came out of the blue he asked me prove the Möbius inversion formula, which thanks to Sol, I did easily."

Sol received his bachelor's degree in mathematics from Johns Hopkins in advance of his 19th birthday. He has an MS and PhD from Harvard University in Mathematics. Prior to joining the University of Southern California, he held senior positions at the Jet Propulsion Laboratory in Pasadena. Sol passed away the day after the centennial of Claude Shannon's birth, he was 83. He will be missed by family, friends and the information theory community.

Sol's technical contributions were summarized by Guang Gong and Tor Helleseth; additional memories were put together by Benjamin Paul, Urbashi Mitra and Vijay Kumar.

References

[1] Solomon W. Golomb. On the classification of Boolean functions. *IEEE Trans. on Inform. Theory*, 5:176–186, May 1959.

[2] Solomon W. Golomb. Run-length encodings. *IEEE Trans. on Inform. Theory*, 12(3):399- 401, 1966.

[3] Solomon W. Golomb. Shift Register Sequences. Holden-Day, Inc., San Francisco, 1967, Revised edition, Aegean Park Press, Laguna Hills, CA, (1982).

[4] Solomon W. Golomb. On the classification of balanced binary sequences of period 2^n-1, *IEEE Trans. on Inform. Theory*, 26(6):730-732, Nov. 1980.

[5] Solomon W. Golomb. Algebraic constructions for costas arrays. *Journal Comb. Theory (A)*, 37:13-21, 1983.

[6] Solomon W. Golomb. Costas arrays – solved and unsolved problems. Keynote Lecture presented at the Symposium on Costas Arrays, at *the CISS Conference*, Princeton, NJ, March 22–26, 2006.

[7] Solomon W. Golomb. A career in technology. Keynote Lecture presented at the *ECE Distinguish Seminar Series*, University of Waterloo, August 12, 2015.

[8] Solomon W. Golomb and Richard Hess. Seating arrangements and Tuscan squares. *Ars Combinatoria*, 2015, to appear.

GOLOMB'S PUZZLE COLUMN™ COLLECTION, Part 1

Beyond his extraordinary scholarly contributions, Sol Golomb was a long time newsletter contributor enlightening us all, young and old, with his beautiful puzzles. In honor of Sol's immense contribution to the newsletter, a collection of his earlier puzzles dated back to 2001 will appear in 4 complied parts over the next 4 issues. Part 1 is given below. He will be greatly missed.

Reprinted from Vol. 51, No. 2, June 2001 issue of Information Theory Newsletter

GOLOMB'S PUZZLE COLUMN ™

SUMS AND PRODUCTS OF DIGITS

Solomon W. Golomb

For every positive integer n, let $S(n)$ be the sum of the decimal digits of n, let $P(n)$ be the product of the decimal digits of n, and let $R(n) = n/S(n)$, the ratio of n to the sum of the digits of n.

1. The equation $S(n) \cdot P(n) = n$ can also be written $P(n) = R(n)$. One solution is $n = 1$, where $S(n) = P(n) = R(n) = n = 1$. There are larger solutions, but only finitely many. Which ones can you find?

2. The ratio $R(n) = n/S(n)$ is sometimes an integer (e.g. when $n = 12$, $R(n) = 12/(1+2) = 4$) and sometimes not (e.g. when $n = 15$, $R(n) = 15/(1+5) = 2.5$). Does every positive integer m occur as $R(n)$ for some positive integer n? If "yes", give a proof; if "no", find the smallest positive m which is never of the form $R(n)$.

3. For each positive integer k, determine which k-digit number n gives the minimum value (integer or not) of $R(n)$. (While this is a separate problem for each positive integer k, there is an interesting pattern to the solutions.)

4. For how many of the $9 \cdot 10^{k-1}$ k-digit integers is $R(n)$ an integer? (This value has been tabulated for $1 \le k \le 7$, but no closed form expression, or even a good asymptotic approximation, has yet been found.)

5. For each positive integer k, which k-digit number gives the smallest *integer* value of $R(n)$, and what are these values? (This behavior is far less regular than in Problem 3, and the explicit answer has only been found, by exhaustive search, for $1 \le k \le 7$.)

Note. Except for Problem 2, due to John H. Conway, the remaining problems are based on an unpublished paper of David Singmaster.

© 2023 World Scientific Publishing Company
https://doi.org/10.1142/9789811234378_0066

Puzzles in Memory of Solomon Golomb*

Joe Buhler ©, Paul Cuff, *Member, IEEE*, Al Hales©, and Richard Stong

(Invited Paper)

Dedicated to the memory of Solomon W. Golomb (1932–2016)

Abstract—We give 12 mathematical puzzles (and their solutions) that were presented at a special session in honor of Sol Golomb at the 2017 ITA meeting in San Diego. Some are "well known," and the very first one is a famous result due to Golomb.

Index Terms—Combinatorial mathematics, geometry, probability.

I. INTRODUCTION

PROFESSOR Solomon Golomb had an extraordinary breadth of interests and talents, including mathematics, electrical engineering, coding and information theory, discrete geometry, and languages. The Information Theory and its Applications Center (ITA), on the UC San Diego campus has an annual workshop every year. Its 2017 meeting included a special session devoted to the memory of Sol Golomb, and the organizers also chose to have a "puzzle session" in the evening to especially remember Sol's keen interest in constructing and solving puzzles of many kinds. His conundrum "in how many ways can the following fragment be parsed?"

a pretty light brown doll house

is a cute blend of his linguistic and mathematical interests. His invention of pentominoes traveled far and wide and was surely one of the key inspirations for Tetris. Moreover, Sol was a long-time editor for the IEEE newsletter puzzle column, and also wrote puzzles for several other publications, including the *Los Angeles Times*.

The four of us were asked to present puzzles to the audience in the evening session, and naturally focused on the sorts of mathematical brain teasers and conundrums that Sol loved so much. The puzzles were given in four (approximately half-hour) sessions containing three problems each. The twelve problems that we chose are presented in the first section below, edited slightly for the sake of clarity. The second section contains the solutions, many greatly improved from (and more complete than) those presented on the fly that evening.

II. PROBLEMS

A. 2-Power Tilings

Remove one square from a 2^n by 2^n checkerboard. Show that the resulting punctured checkerboard can be tiled by right

trominoes. (A right tromino is the figure formed by three 1 by 1 cells of a 2 by 2 grid.)

Remark. This was the first of a series of 6 problems on tilings by right trominoes in the very first "Golomb's Puzzle Column" in the IEEE newsletter [2].

B. Card Pairs

A warden gives two prisoners a pair of distinct cards from a deck that has (many copies of) three cards. The pairs given to the two prisoners are distinct (as sets). Each prisoner returns a card to the warden; if the cards are identical the prisoners both go free. The two prisoners know the contents of the deck and the rules, and can have a strategy session beforehand. What should they do to maximize their probability of success?

C. Nice Integers

Say that an integer n is nice if it can be written as a sum of a sequence of positive integers $a_1, \ldots, a_k$ such that

$$\frac{1}{a_1} + \frac{1}{a_2} + \cdots + \frac{1}{a_k} = 1.$$

Is 2017 nice?

Remark: For example, $9 = 3 + 3 + 3$ and $\frac{1}{3} + \frac{1}{3} + \frac{1}{3} = 1$, and $11 = 2 + 3 + 6$ and $\frac{1}{2} + \frac{1}{3} + \frac{1}{6} = 1$, so 9 and 11 are nice.

D. Batteries

There are 2017 flashlight batteries, of which 1009 are good and 1008 are bad. Your flashlight works only with two good batteries. What is the least (worst-case) number of trials needed to get the flashlight to work?

E. Three Random Variables

Find three random variables X, Y, and Z that are uniformly distributed on $[0, 1]$, and have constant sum.

Remark: Note that since the expectation of each random variable is $1/2$, the sum is $3/2$.

F. Cycle

A random walk on n equally spaced points on a circle goes one direction or the other according to the flip of a fair coin. Sooner or later every point has been visited. What is the distribution of the random variable "last point visited"?

Remark: We take the convention that the starting point has not been visited, so the first point visited is the end of the first step.

G. 545

The number 545 has the interesting property that if you erase any of its digits and write a new digit in that place (the new digit may be a leading zero or the same as the old digit), then you never get a multiple of 11. Find two more such integers.

Manuscript received June 28, 2017; accepted September 21, 2017. Date of publication December 29, 2017; date of current version March 15, 2018.

J. Buhler, A. Hales, and R. Stong are with the IDA Center for Communications Research, San Diego, CA 92121 USA (e-mail: buhler@ccrwest.org; hales@ccr-lajolla.org. stong@ccrwest.org).

P. Cuff is with the Electrical Engineering Department, Princeton University, Princeton, NJ 08544 USA (e-mail: cuff@princeton.edu).

Communicated by G. Gong, Guest Editor.

Digital Object Identifier 10.1109/TIT.2017.2769698

0018-9448 © 2017 IEEE. Personal use is permitted, but republication/redistribution requires IEEE permission.
See http://www.ieee.org/publications_standards/publications/rights/index.html for more information.

Authorized licensed use limited to: Institute for Defense Analyses. Downloaded on June 08,2022 at 16:07:25 UTC from IEEE Xplore. Restrictions apply.

*This chapter was originally published in *IEEE Transactions in Information Theory*, April 2018, Vol. 64, No. 4, pp. 2839–2843.

H. Contagion

A contagion is spreading on the squares of a checkerboard. If a cell has two or more immediate neighbors infected, then it becomes infected. What is the minimum number of starting cells that must be infected to spread the contagion over the whole board?

Remark: "Immediate neighbors" share an entire edge.

I. Penny Matching

Two players attempt to guess a random bit-string. One player is given the entire bit string beforehand. The bits are presented one by one. Before a bit is shown they must each predict the bit. If they are both right, they get a point. Find a strategy that gives an expected long-term fraction p of successes, for some $p > 1/2$.

Remark: The players can have a strategy session beforehand. Each player learns both guesses and the actual value of the bit.

J. Winning Paths

How many winning paths are there on the 4-dimensional $4 \times 4 \times 4 \times 4$ tic-tac-toe board?

Remark: A winning path consists of 4 consecutive cells in a line through the board.

K. Bit String

A warden gives two prisoners A and B each a bit string. Prisoner A returns a positive integer a to the warden and prisoner B similarly returns b. If the b-th bit of As string and the a-th bit of Bs string are both a 1, then the prisoners are freed. Prove that the optimal probability of success is at least one-third and at most three-eighths.

Remark: Each prisoner can only see his bit string. The players have a strategy session beforehand.

L. Painting

A group of $n \geq 3$ people decides to paint the outside of a large fence that encloses a circular field by the following curious method. Each person takes a bucket of paint to a random point on the circle and then, on a signal, begins to paint towards their farthest neighbor, stopping when they reach fresh paint. What is the expected fraction of the circle that will be left unpainted?

III. SOLUTIONS

A. 2-Power Tilings

The base case $n = 0$ is immediate. If $n > 0$ then put a tromino in the central 2×2 square, with its unused square in the quadrant of the removed square. The tiling problem on the $2^n \times 2^n$ board reduces to 4 problems on a punctured $2^{n-1} \times 2^{n-1}$ board, which we already know how to solve.

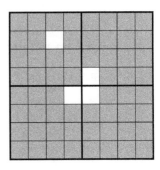

B. Card Pairs

Clearly the prisoners A and B cannot hope for a better than 50:50 chance of going free, since one of As cards does not match with either of Bs cards and hence is a sure loser. Either card is equally likely to be this loser and hence A has a 50:50 chance of picking it.

To achieve a 50:50 chance of going free, think of the cards as "rock", "paper", and "scissors". Prisoner A will return the winner out of his pair of cards and prisoner B will return the loser out of his pair. Then the prisoners will win in exactly half the cases. Indeed if they lose in one case, then swapping the pairs between them will give a case where they win and vice versa.

Remark: One of us heard this via Ron Rivest, and we do not know its original source.

C. Nice Integers

If the sum $n = a_1 + \cdots + a_k$ shows that n is nice then
$$\frac{1}{2a_1} + \frac{1}{2a_2} + \cdots + \frac{1}{2a_k} + \frac{1}{2} = 1$$
and
$$\frac{1}{2a_1} + \frac{1}{2a_2} + \cdots + \frac{1}{2a_k} + \frac{1}{3} + \frac{1}{6} = 1$$
show that $2n + 2$ and $2n + 9$ are nice.

To show that $2017 = 2 \cdot 1004 + 9$ is nice it suffices to show that 1004 is nice. Continuing in this way with $1004 = 2 \cdot 501 + 2$, $501 = 2 \cdot 246 + 9$, $246 = 2 \cdot 122 + 2$, $122 = 2 \cdot 60 + 2$, $60 = 2 \cdot 29 + 2$, $29 = 2 \cdot 10 + 9$, and $10 = 2 \cdot 4 + 2$, we reduce to needing to show that 4 is nice. But $4 = 2 + 2$ shows this.

In fact every integer larger than 23 is nice, though this takes checking more cases.

D. Batteries

Split the batteries into 1008 pairs and one singleton. First test the 1008 pairs. If every pair fails, then you know each pair had at least one bad battery in it. But since there are only 1008 bad batteries, you know each pair had exactly one bad battery. Thus the singleton must be a good battery. Trying it with the two batteries in the first pair in succession must make the flashlight work in at most 1010 trials.

Puzzles in Memory of Solomon Golomb

To see that you cannot guarantee to succeed in 1009 trials, we will show more generally that with $2n + 1$ batteries of which $n + 1$ are good and n bad you cannot guarantee success in $n + 1$ trials. This is easy to check for $n = 1$ (3 batteries with 1 bad). In general, suppose you could guarantee success for $2n + 1$ batteries in $n + 1$ trials and suppose we have the smallest such $n > 1$. Then counted with multiplicity you tested $2n + 2$ batteries, so there is some battery tested only once. If it was tested once, then we could imagine that it was good and the battery tested with it was bad. Then deleting these two batteries and any test involving one of them would give a strategy guaranteed to succeed with $2n - 1$ batteries and at most n trials, contradicting the minimality of n. Similarly if there is some untested battery we assume it is good and any other battery tested at least once is bad and we get a similar reduction.

E. Three Random Variables

In probabilist's terms, the goal is to describe a probability distribution on the intersection of the unit cube $[0, 1]^3$ in 3-space and the plane $x + y + z = 3/2$, such that the marginal projections onto the axes are uniform. There are many ways to do this, and we give three here.

Trinary

Write each of $X = 0.x_1x_2\ldots$, $Y = 0.y_1y_2\ldots$, and $Z = 0.z_1z_2\ldots$ in trinary and arrange that the three trits (x_i, y_i, z_i) are an independent random permutation of $\{0, 1, 2\}$. Alternatively, choose x uniformly at random, and then increment its trits to get y and z: let $y_i = 1 + x_i \bmod 3$ and $z_i = 1 + y_i \bmod 3$.

Unit Sphere

Let S be a uniformly distributed point on the unit sphere. It is well-known that for any unit vector v the orthogonal projection $v \cdot S$ onto the diameter parallel to v is uniformly distributed on $[-1, 1]$. Let u, v, w be three coplanar unit vectors that differ by 120 degree rotations. Then $u + v + w = 0$, so the projections $X' = u \cdot S$, $Y' = v \cdot S$, and $Z' = w \cdot S$ are uniform on $[-1, 1]$ and sum to zero. Letting $X = (1 + X')/2$, etc., gives the required example.

Direct

Let X be uniform on $[0, 1)$, let Y be $X + 1/2$ reduced modulo 1, and let $Z = 1 - 2X$ reduced modulo 1. Then $X + Y + Z = 3/2$ and it is easy to check that Y and Z are uniform on $[0, 1)$. (Technically, one should have Z take values in $(0, 1]$ here, otherwise $X + Y + Z = 3/2$ a.e., but not if $X = 0$ or $1/2$.)

Remark: One of us heard this from Yuval Peres in 2008, and the rest of us heard it at, or shortly after, the 2012 ITA workshop, where it was a popular topic of conversation.

F. Cycle

Suppose we took the opposite convention that the starting point is counted as visited. Look at any of the $n - 1$ other points X. There will be some moment when we first reach a neighbor of X, call it Y. Label X by 0, Y by 1, and keep going around the circle to label the other points $2, 3, \ldots, n-1$. At this point neither $X = 0$ nor $Z = n - 1$ has been visited (since Y is the first neighbor of X visited and Z is also a neighbor). Thus X will be the last point visited if and only if we reach $Z = n - 1$ before $X = 0$. This is a standard "gambler's ruin". We have a stake of 1 (we are at Y) and we gamble 1 on every flip of a fair coin. Since the game is fair, the chance that we make it to $Z = n - 1$ before losing all our money (hitting $X = 0$) is $1/(n-1)$. Thus the last point visited is uniform on the $n - 1$ non-starting points.

With our convention that the starting point is not visited, thing change only slightly. After the first step we are at one of the neighbors of the starting point and we are in the previous case. Thus we get a probability of $1/(n - 1)$ of ending at any point other than the neighbors of the starting point and $1/(2(n - 1))$ for either of the neighbors of the start.

Remark: We first heard this from Persi Diaconis.

G. 545

Write N in decimal as $N = \ldots x_2x_1x_0$. Since $10^k \equiv (-1)^k = \pm 1 \pmod{11}$, varying x_k over $0, \ldots, 9$ gives 10 consecutive remainders modulo 11. If none of these is a multiple of 11, then the next term in the sequence must be. Thus replacing x_k by 10 must give a multiple of 11. This says

$$0 \equiv N + (10 - x_0) \equiv N - (10 - x_1)$$
$$\equiv N + (10 - x_2) \equiv \ldots \pmod{11}.$$

After rearranging this becomes

$$x_0 \equiv x_2 \equiv x_4 \equiv \cdots \equiv 9 - x_1 \equiv 9 - x_3$$
$$\equiv \cdots \equiv N - 1 \pmod{11}.$$

The first part of this says that the digits of N alternate between two values, say a and b, that add to 9. We just need to get the final congruence.

If $N = baba$ has 4 digits, then we get $a \equiv 9 - b \equiv 2a - 2b - 1 \pmod{11}$ or $3a \equiv 19 \pmod{11}$ or $a \equiv 10 \pmod{11}$. Thus we get no solution with 4 digits. (One similarly finds no solutions with an even number of digits unless the number of digits is 1 modulo 11.)

If $N = ababa$ has 5 digits, then we get $a \equiv 9 - b \equiv 3a - 2b - 1 \pmod{11}$ or $4a \equiv 19 \pmod{11}$ or $a \equiv 2 \pmod{11}$. Thus $N = 27272$ is a second example. If $N = abababa$ has 7 digits we similarly get $6a \equiv 28 \pmod{11}$ or $a \equiv 1 \pmod{11}$. Thus $N = 1818181$ is a third example.

Remark: This is due to Mike Boshernitzan.

H. Contagion

More generally on an n by n checkerboard the answer is that n initially infected cells is enough. One way to infect every cell with n infected cells is to start with the entire diagonal infected. Then it is easy to see that the infection spreads along neighboring diagonals until it covers the entire board.

To see that you cannot infect the whole board with fewer than n initial cells, notice that when two (or more) cells

566   *The Wisdom of Solomon*

infect a neighbor, the perimeter of the infected region either stays the same (if 2 neighbors were infected) or decreases (if 3 or more were infected). Thus the perimeter of the infected region cannot increase. If the whole board is infected then the perimeter of the infected region is $4n$, and hence we must have started with at least n infected cells.

Remark: We saw this in Peter Winkler's first book [4] on (difficult) recreational math puzzles, and it apparently first appeared in print in *KVANT*.

I. Penny Matching

Split the string into blocks of 3. The second player predicts 0 for all bits in the block, *except* when the first player signals by making an incorrect prediction on the first bit, in which case the second player predicts "1" for the next two bits. A straightforward calculation shows that the probability of success is $5/8$. A different idea with variable size blocks gets $p = 2/3$. The first player, who knows the sequence, parses the string into a sequence of blocks that consist either of a single bit "0" or a pair of two bits "1X". For a "0" he correctly guesses zero. For a pair "1X" he guesses X twice. The second player, who does not know the sequence, guesses zero by default, but if his default guess is wrong at any stage, then at the next step he deviates from his default by guessing the same as the first players' last guess. With this strategy it is easy to see that the players get exactly one success for every block "0" or "1X". Since a random string of length N will have on average $N/3$ blocks of each kind, they expect about $2N/3$ successes.

More elaborate blocks and signaling, using error-correcting codes, can realize $p = 0.8107\ldots$, which is optimal.

Remark: Online searches for "matching pennies" show that this question has received serious attention in the last decade or so. Apparently its first appearance in print was in a 2003 paper by Gossner, Hernandez, and Neyman. The optimality result can be found in a 2011 paper *Coordination using Implicit Communication* by Cuff and Zhao [1].

J. Winning Paths

Imagine the $4 \times 4 \times 4 \times 4$ as sitting at the center of a $6 \times 6 \times 6 \times 6$ board, which we write as $\{0, 1, 2, 3, 4, 5\}^4$. Any winning path will be along a line that meets two outer cells of the $6 \times 6 \times 6 \times 6$ board. Given one of these two outer cells, we can find the other by switching every 0 coordinate to a 5 and every 5 coordinate to a 0. Thus the $6^4 - 4^4$ outer cells form pairs and each pair determines a winning path, thus there are $(6^4 - 4^4)/2 = 520$ winning paths.

For more on the mathematics of tic-tac-toe see, e.g., [3].

K. Bit Strings

For the lower bound A returns the index a of the first 1 in his string and B returns the index b of the first one in his string. Then the two prisoners will be freed if and only if $a = b$. Since the probability that $a = b = k$ is 2^{-2k}, this gives a probability of success of

$$\frac{1}{4} + \frac{1}{16} + \cdots = \frac{1}{3}.$$

For the upper bound, consider a pair of complementary strings from prisoner A, say a and $\bar{a}$, and suppose that in these cases A returns a and a', respectively. Condition on the assumption that prisoner A has either a or $\bar{a}$.

Suppose first that $a = a'$. If the a-th bit of Bs string is a zero, then the prisoners lose. If the a-th bit of Bs string is a one, then the prisoners win with probability $1/2$. Thus overall the prisoners win with probability only $1/4$ in this case.

Next suppose $a \neq a'$. Consider the pair of the a-th and a'-th bits of Bs string. If this pair is 00, then they lose. If it is 01 or 10, then they win for only one of the two possibilities for As string, hence with probability at most $1/2$. If it is 11, then whatever value b is returned by B will win with only one of a and $\bar{a}$. Thus again they win with probability at most $1/2$. Thus overall the prisoners win with probability at most $3/8$ in this case.

Remark: This is a disguised form of the 2-person version of a hats game due to Lionel Levine in 2011, where the players wear countably many hats that have can have two possible colors.

L. Painting

Measure distances on the circle as a fraction of the total circumference. Consider two painters i and j, where $1 \leq i$, $j \leq n$, and let U_{ij} be the random variable that is equal to u if the interval between the initial positions of the painters has the following properties: its length is u, it contains no other painters, and it is unpainted. If the interval is unpainted then all other painters were at least a distance u from the endpoints of the interval, i.e., the other painters started outside an interval of length $3u$; the probability of this is $(1 - 3u)^{n-2}$; note that this forces u to be at most $1/3$.

The total unpainted surface is $U = \sum_{i,j} U_{ij}$, and the expected value of U is therefore

$$n(n - 1) \int_0^{1/3} u \, (1 - 3u)^{n-2} \, du.$$

The integral is easy to compute by changing variables and its value is $1/9$. Thus one-ninth of the fence is left unpainted.

Remark: One of us heard this in one of Tom Cover's research group meetings in 2008.

ACKNOWLEDGMENT

The authors would like to thank Alon Orlitsky and the other organizers of the ITA meeting for suggesting the idea of celebrating Golomb's love of puzzles in a special session, and thank the audience for a scintillating evening.

REFERENCES

[1] P. Cuff and L. Zhao, "Coordination using implicit communication," in *Proc. IEEE Inf. Theory Workshop*, Oct. 2011, pp. 467–471.
[2] S. W. Golomb, "Tiling with right trominoes," *IEEE Inf. Theory Soc. Newslett.*, vol. 51, no. 3, p. 12, Sep. 2001.
[3] S. W. Golomb and A. W. Hales, "Hypercube tic-tac-toe," in *More Games of No Chance* (Mathematical Sciences Research Institute Publications), vol. 42. Cambridge, U.K.: Cambridge Univ. Press, 2002, pp. 167–182.
[4] P. Winkler, *Mathematical Puzzles: A Connoisseur's Collection*. Natick, MA, USA: A K Peters, Ltd., 2004, p. 79.

Puzzles in Memory of Solomon Golomb

Joe Buhler received his Ph.D. in mathematics from Harvard University. He was on the faculty at Reed College before serving as Deputy Director of the Mathematical Sciences Research Institute in Berkeley, starting in 1999. He became the Director the Director of the Center for Communications Research in San Diego in 2004.

Paul Cuff received the B.S. degree in electrical engineering from Brigham Young University in 2004 and the M.S. and Ph. D. degrees in electrical engineering from Stanford University in 2006 and 2009. He subsequently joined the Electrical Engineering Department at Princeton University as an Assistant Professor. In 2017, he joined Renaissance Technologies. As a graduate student, Dr. Cuff was awarded the ISIT 2008 Student Paper Award for his work titled Communication Requirements for Generating Correlated Random Variablesž and was a recipient of the National Defense Science and Engineering Graduate Fellowship and the Numerical Technologies Fellowship. As faculty, he received the NSF Career Award in 2014 and the AFOSR Young Investigator Program Award in 2015.

Al Hales received his B.S. and Ph.D. degrees from Caltech in 1960 and 1962. After postdoctoral positions at Cambridge and Harvard, he joined the UCLA mathematics department where he served as Chair from 1988–91. He was Director of IDA's Center for Communications Research in San Diego from 1992–2003, where he is now on the research staff. He was a junior coauthor of Sol Golomb's book on *Shift Register Sequences*.

Richard Stong is currently a researcher at the Center for Communications Research in La Jolla, California. His mathematical interests include combinatorics and four- dimensional topology.

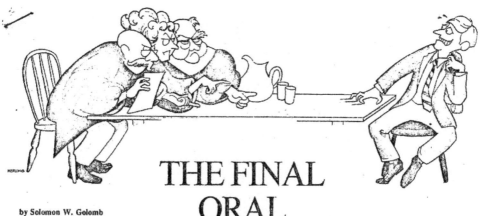

THE FINAL ORAL EXAM

by Solomon W. Golomb

"None of these questions is intended to be universally obscure," claims the author. "A Harvard senior who has taken the right courses could answer any one of them." Without recourse to reference works, can you?

1. Explain the origin and meaning of each of the following terms. (Which one, as a noun, could *not* refer to a person?)
   a. Jacobean
   b. Jacobian
   c. Jacobin
   d. Jacobite

2. What is T4, and how does it differ from T1, T2, or T3 (select as appropriate)? Answer from the point of view of each of the following:
   a. a virologist
   b. a communications engineer
   c. a topologist
   d. a clinical pathologist

3. Tokyo was formerly known as Edo, and Trondheim as Nidaros. Turku is also known as Åbo, and Aachen as Aix-la-Chapelle. Breslau is now called Wrocław, Königsberg is Kaliningrad, and Saigon is Ho Chi Minh City. Give *two* other names (typically one older and one newer) by which each of the following cities has been known:
   a. Constantinople
   b. Stalingrad
   c. Reval
   d. Hebron
   e. Pressburg

4. Identify the following terms from Norse mythology:
   a. Aesir
   b. Fafnir
   c. Fenrir
   d. Ragnarok
   e. Jotunheim
   f. Yggdrasil

5. In the Biblical Book of Genesis, which personage numbered the following ladies among his wives and/or concubines?
   a. Hagar and Keturah
   b. Bilhah and Zilpah
   c. Milcah and Reumah
   d. Adah and Zillah
   e. Aholibamah and Bashemath

6. Much of Greek mythology revolves around the doings of the House of Atreus. Draw a family tree that indicates the relationships of the following individuals to one another, including who murdered whom.
   a. Aegisthus
   b. Agamemnon
   c. Atreus
   d. Clytemnestra
   e. Electra
   f. Helen of Troy
   g. Iphigenia
   h. Menelaus
   i. Orestes
   j. Thyestes

7. Loan words reaching English via different routes may turn out to be cognates. In each of the following pairs of words there is a common root in a common donor or ancestor language. Trace the history of each word from that common root to its present English meaning. (The language from which each word entered English is indicated.)
   a. dragoon (French)—Dracula (Rumanian).
   b. tempo (Italian)—tempura (Japanese)
   c. minaret (Turkish)—menorah (Hebrew)
   d. mahatma (Hindi)—atmosphere (from Greek roots)
   e. azure (French)—lapis lazuli (Late Latin)

8. The poem that begins "Thirty days hath September" is probably the best-known mnemonic in the English language. Identify the application and interpretation of each of the following widely used mnemonics.
   a. Every good boy does fine.
   b. Oh, be a fine girl—kiss me.
   c. Bad boys rape our young girls behind victory garden walls.
   d. Kings play chess on fine-grained sand.
   e. On old Olympus' towering top/ A Finn and German viewed a hawk.
   f. How I want a drink, alcoholic of course, after the heavy lectures involving quantum mechanics.

9. Almost everyone knows what monarch was called "the Lionhearted," or "the Terrible." What European rulers had the following epithets appended to their names?
   a. the Unready
   b. the Magnificent
   c. the Bald
   d. Redbeard
   e. Forkbeard
   f. Fairhair
   g. Harefoot
   h. Barefoot

10. The nineteenth century was the golden age of opera. Match the titles of the operas (translated into English) in column 1 with the first names of the composers in column 2, and supply the last names of the composers.
    a. *The African Woman*   i. Alexander
    b. *Boris Godunov*   ii. Bedřich
    c. *Hansel and Gretel*   iii. Charles
    d. *Prince Igor*   iv. Engelbert
    e. *Samson and Delilah*   v. Giacomo
    f. *The Bartered Bride*   vi. Gioacchino
    g. *The Tales of Hoffmann* vii. Jacques
    h. *William Tell*   viii. Modest

Professor Golomb's answers may be found on pages 55 and 56.

---

*This chapter was originally published in the May–June 1980 issue of *Harvard Magazine*, pp. 19, 55–56.

570 — The Wisdom of Solomon

THE ANSWERS TO

THE FINAL ORAL EXAM

ADMINISTERED ON PAGE 19

1.a. *Jacobean*: pertaining to James I of England, the period when he was king, or a personage of this period (*Jacobus* being the Latin version of James).

b. *Jacobian*: in mathematics and mathematical physics, the differential element of area or volume associated with a coordinate transformation, named after the German mathematician Karl Gustav Jacobi, 1804-1851. (This one couldn't be a person.)

c. *Jacobin*: a political radical, after the Jacobin society of the French Revolution of 1789, so named because they met at the (Jacobin) Dominican friars' cloister. The French Dominicans were called Jacobins because their cloister was at the church of St. James (St. Jacques) of Compostela.

d. *Jacobite*: a follower of King James II of England after his abdication, or advocate of the restoration of his (Stuart) descendants to the throne.

2.a. The bacterial viruses, or bacteriophages, of *E. coli*, called coliphages, have been among the chief experimental "animals" in the development of molecular genetics. The best known of these are the T-series of phage (T1 to T7), of which T4, T2, and T3 have been the most extensively studied.

b. In telephony, T1 denotes the basic digital carrier system now in widespread intracity use (up to fifty miles), utilizing a single, twisted pair of copper wires, with repeaters at one-mile intervals, and achieving a data rate of 1.544 megabits per second. The T2 system, with four times the capacity of the T1, was developed for long-haul digital traffic. The more ambitious and higher-capacity T4 system was configured to provide the digital link needed for the Picturephone ®, but was never implemented because pictorial transmission by phone line proved a commercial flop.

c. In point-set topology, there is a hierarchy of increasingly restrictive "separation axioms" called T0, T1, T2, T3, and T4. In a T1 space, all sets consisting of single points are "closed" sets. In a T2 space, distinct points have disjoint open neighborhoods. T2 spaces are also called "Hausdorff spaces." T3 spaces are T1 spaces that are also "regular spaces," and T4 spaces, the most highly structured, are T1 spaces that are also "normal spaces." (See, for example, *General Topology* by John L. Kelley, Van Nostrand Reinhold, 1955.)

d. In clinical pathology, the T4 (tetraiodothyronine) test and the T3 (triiodothyronine) test are both used to measure thyroid activity. Specifically, they measure the concentration of the correspondingly named thyroid hormones.

3.a. Byzantium — Constantinople — Istanbul (Turkey).

b. Tsaritsyn — Stalingrad — Volgograd (Russia).

c. Kalevan — Reval — Tallinn (Estonia).

d. Qiryat 'Arba (Kirjath-arba) — Hebron — El Khalil (Israel, Occupied West Bank).

e. Pozsony — Pressburg — Bratislava (Czechoslovakia).

Changes in the name of a city usually result either from political upheavals (Tsaritsyn to Stalingrad to Volgograd), or transfer of nationality (Pozsony = Hungarian to Pressburg = German to Bratislava = Slovak). The name of Byzantium was changed to Constantinople ("City of Constantine") in honor of the emperor who converted to Christianity and made it the capital of the Roman Empire in 330 A.D. The Turks conquered the city in 1453, and in 1930 officially changed its name to Istanbul, possibly a Turkish corruption of the Greek εἰς τὴν πόλιν ("into the city").

The capital of Estonia is now known as Tallinn (from *Taan-linna*, Estonian for "Danish fortress," portions of the thirteenth-century city wall, towers, and other Danish fortifications surviving to this day). The ancient Estonian name was Kalevan, after the folk hero Kalev, whose name appears in the title of the Finnish folk epic the *Kalevala*, and whose exploits are described in the Estonian epic the *Kalevipoeg*. In between, the city was known as Reval (the Teutonic version) or Revel (the Slavic version).

Before he conquered Jerusalem, King David had his capital at Hebron (from a Hebrew root meaning "friendship") for seven years. The Old Testament frequently mentions that Hebron was earlier known as Qiryat 'Arba (Kirjath-arba in the King James rendition), which means "the City of Four," and there are many fanciful explanations as to what there may have been four of. Today the city is called Hebron by the Israelis and El Khalil by the Arabs. It is the largest city in the southern portion of the Occupied West Bank and contains the Tomb of the Patriarchs.

4.a. *Aesir*: the major gods collectively (Odin, Thor, Freya, Balder, etc.).

b. *Fafnir*: the dragon/giant who guarded the Nibelung treasure; slain by Sigurd.

c. *Fenrir*: one of Loki's monster offspring, a giant wolf, tied with a magic rope by the gods until Ragnarok.

d. *Ragnarok*: Götterdämmerung; Armageddon; Twilight of the Gods.

e. *Jotunheim*: the forbidding land of the giants.

f. *Yggdrasil*: the "world tree," a great ash with roots in the underworld, that serves as the axis on which the heavens pivot.

5.a. Abraham (whose principal wife was Sarah). Gen. 16:3 and 25:1.

b. Jacob (whose principal wives were Leah and Rachel). Gen. 30:4 and 9.

c. Nahor (the elder brother of Abraham). Gen. 22:20-24.

d. Lamech (who killed someone, possibly Cain, who killed Abel). Gen. 4:23.

e. Esau (a.k.a. Edom, the twin brother of Jacob). Gen. 36:2-3.

6. To avenge the treachery of his brother Thyestes, Atreus slew Thyestes' sons and served them to him at a banquet. Thyestes' illegitimate son, Aegisthus, then avenged these murders by killing his uncle, Atreus. The sons of Atreus (who were knows as the Atrides) were Agamemnon and Menelaus. When Menelaus' wife, Helen, was abducted by Paris to Troy, the Trojan War ensued, early in the course of which Agamemnon offered his daughter Iphigenia as a sacrifice to Artemis. Returning home to Mycenae, Agamemnon was killed by his wife, Clytemnestra, who was Helen's sister, abetted by Aegisthus, now her lover. Orestes, son of Agamemnon and Clytemnestra, egged on by his sister Electra, avenged his father's death by killing his mother and her consort. (Draw your own diagram, if you can!)

7.a. From Latin *draco*, a dragon, the diminutive Dracula ("little dragon") was the epithet applied to the bloody Walladian count, Vlad Țepeș. Bram Stoker's novel *Dracula* used little more than the nickname of the historical count.

In French, *draco* became *dragon* and was the name given to a type of musket. Soldiers armed with this musket were called *dragons*, or *dragoons*, a term later applied to various forms of armed, mounted infantry.

b. Latin *tempus* ("time"), pl. *tempora*, is the direct ancestor of Italian *tempo* ("time"), which entered English as a musical term.

When the Portuguese, the first Europeans to visit Japan, had to substitute seafood for meat during lent (in Latin, *tempora quadragesima*, the "times of the forty [days]"), the Japanese transferred the *tempura* to refer to the fried seafood.

c. From the Semitic root *nur* ("light"), a holder of light is *m'-nur-ah*, whence the Hebrew *menorah* ("candelabrum") and the Arabic *mena-*

MAY-JUNE 1980   55

FINAL EXAM ANSWERS
(continued)

rah ("lighthouse"). Substituting the Turkish noun ending *-et*, *menarah* became *minaret*, and the term was applied to the slender towers, resembling lighthouses, that adjoin mosques.

d. The Indo-European root *atmen* ("breath") is seen in German *atmen* ("breathe"), Sanskrit *atman* ("soul"), and Greek ἀτμός ("smoke" or "vapor"). As Hindi *maha-rajah* means a "great king," so *maha-atma*, contracted to *mahatma*, means a "great soul." *Atmosphere* is the "air sphere."

e. French *azur*, from Spanish *azul* (both meaning "blue"), comes from Late Latin *lazul(us)* (as if *l'azul*, and the initial *l* was dropped), in turn an adaptation of the Arabic *lāzaward*, from the Persian *laj-vard* (meaning "azure"). Thus, *lapis lazuli* means "blue stone" or "azure stone," where English *azure* means "sky blue." Curiously, Japanese *aozora* means "blue sky," but is not cognate, being compounded of *ao(i)* ("blue") and *sora* ("sky").

8. The first five mnemonics are based on using the first letters of the words, and the sixth, on the number of letters in each word.

a. E, G, B, D, F are the letters of the notes associated with the five lines in the treble clef. Similarly, "All cows eat grass" is used to remember the spaces in the bass clef. (The spaces in the treble clef spell *face*.)

b. O, B, A, F, G, K, M are the astronomical designations given to the spectral types of the main sequence stars, where O stars are at the blue-white end of the frequency spectrum, our own yellow sun is type G, and the wavelengths increase (become redder) as one progresses through the sequence.

c. Black, brown, red, orange, yellow, green, blue, violet, gray, white are the colors used to represent the digits 0, 1, 2, 3, 4, 5, 6, 7, 8, 9 on electrical resistors. A common variant of this mnemonic ends, "But Violet goes willingly."

d. Kingdom, phylum, class, order, family, genus, species compose the hierarchy of classification levels in biological taxonomy. Several competing mnemonics exist, e.g., "King Philip called out the family guard suddenly."

e. The twelve pairs of cranial nerves, the bane of the beginning-anatomy student's existence, from the top down, are: olfactory, optic, oculomotor, trochlear, trigeminal, abducens, facial, acoustic, glossopharyngeal, vagus, accessory, hypoglossal. The commonest variants end with "A fat-assed German viewed a hop" or "vaccinated a horse." A companion mnemonic, which indicates whether the corresponding nerves are sensory or motor or both, goes: "Some say marry money, but my brother say, 'Bad business marry money.'" (It seems the young lady has an oriental accent!)

f. 3.14159265358979, the first fifteen digits in the decimal expansion of π. Mnemonics in many languages, and for more than fifteen digits, abound. Thus:

Now I know a spell unfailing—
An artful charm for tasks availing,
Intricate results entailing,
Not in too exacting mood
(Poetry is pretty good)—
Try the talisman. Let be
Adverse ingenuity.

This verse dates back to volume 4 (1907-1908) of the British journal The Mathematical Gazette (page 103), and is attributed to "a mysterious F.R.S."

9.a. Æthelred (Ethelred) the Unready. King Æthelred II of the English, lived c. 968-1016, reign began 978.

b. Lorenzo de' Medici, Il Magnifico, lived 1449-1492, ruled Florence from 1469. OR, Suleiman the Magnificent, the Ottoman Sultan (Suleiman I), lived 1494-1566, reign began in 1520. (He was a European ruler in that his empire included the Balkans, and extended as far north as Budapest.)

c. Charles the Bald was the Holy Roman Emperor Charles II (Charles I was Charlemagne), king of the West Franks, lived 823-877, ruled from 843; not to be confused with Charles the Bold, also a Charles II, duke of Lorraine, who died in 1431.

d. Frederick Barbarossa (Redbeard) was the Holy Roman Emperor Frederick I, king of Germany and Italy, lived c. 1123-1190, ruled from 1152; drowned during the Third Crusade.

e. Sweyn Tveskæg (Forkbeard), King Sweyn I of Denmark, ruled from c. 985, ravaged the England of Æthelred the Unready. Sweyn was the father of Canute the Great, and died at Gainsborough in 1014.

f. Harald Haarfager (Fairhair), was Harald I, first king of Norway, lived c. 850-933, completed the unification of Norway by conquest in 872.

g. Harold Harefoot, King Harold I of England, illegitimate son of Canute the Great; claimed the throne on his father's death in 1035, was elected king in 1037, and died at Oxford in 1040. (It was Harold II who was defeated at Hastings in 1066.)

h. Magnus Barfot (Barefoot) was King Magnus III of Norway, who ruled from 1093 to 1103, conquered the Orkneys and the Hebrides for Norway, and died in Ireland.

10.a. v. *The African Woman (L'Africaine)*, by Giacomo Meyerbeer (born Jakob Beer), 1791-1864, his last major work, first performed after his death.

b. viii. *Boris Godunov*, by Modest Petrovich Mussorgsky, 1839-1881; several versions by the composer between 1869 and 1874, and a posthumous revision by Rimsky-Korsakov.

c. iv. *Hansel and Gretel (Hänsel und Gretel)*, by Engelbert Humperdinck, 1854-1921, first performed in 1893.

d. i. *Prince Igor (Kniaz' Igor)*, by Alexander Porfirevich Borodin, 1833-1887, unfinished when he died, was completed in 1889 by Rimsky-Korsakov and Glazunov.

e. iii. *Samson and Delilah (Samson et Dalila)*, by Charles Camille Saint-Saëns, 1835-1921, was first performed in 1877.

f. ii. *The Bartered Bride (Prodaná Nevěsta)*, by Bedřich Smetana, 1824-1884, first performed in 1866.

g. vii. *The Tales of Hoffmann (Les Contes d'Hoffmann)* by Jacques Offenbach, 1819-1880, was left unfinished at his death, and was completed by Ernest Guiraud.

h. vi. *William Tell (Guillaume Tell)*, by Gioacchino Antonio Rossini, 1792-1868, was the composer's last opera, produced in 1829.

Solomon W. Golomb, who received the Ph.D. degree from Harvard in 1957, is professor of electrical engineering and mathematics at the University of Southern California.

CPSIA information can be obtained
at www.ICGtesting.com
Printed in the USA
BVHW020247130723
667148BV00003B/12